CLINICAL, RESEARCH AND TREATMENT APPROACHES TO AFFECTIVE DISORDERS

Edited by **Mario Francisco Juruena**

Clinical, Research and Treatment Approaches to Affective Disorders
Edited by Mario Francisco Juruena

Published by InTech

Janeza Trdine 9, 51000 Rijeka, Croatia

Publishing Process Manager Anja Filipovic

Technical Editor Teodora Smiljanic

Cover Designer InTech Design Team

Additional hard copies can be obtained from orders@intechopen.com

Clinical, Research and Treatment Approaches to Affective Disorders,
Edited by Mario Francisco Juruena
p. cm.
ISBN 978-953-51-0177-2

For EU product safety concerns:
IN TECH d.o.o., Prolaz Marije Krucifikse Kozulić 3, 51000 Rijeka, Croatia,
info@intechopen.com or visit our website at intechopen.com.

Meet the editor

Mario Francisco Juruena graduated in medicine from the Pontifical Catholic University - RS, Brazil. He did his training in psychiatry at the Sao Pedro Hospital, and specialization in mental health at School of Public Health - RS. He received his master degree at the Department of Psychobiology, Federal University of São Paulo, and in Affective Neuroscience, Universiteit Maastricht, the Netherlands. He got his PhD degree in Psychiatry from the University of London, and training in cognitive psychotherapy at Beck Institute for Cognitive Behavior Therapy - USA. He is a specialist in adult psychiatry in the UK according to the Medical Royal Colleges and has published over 80 original research articles, reviews and book chapters, editing four books. British Association for Psychopharmacology awarded him the "Senior Clinical Psychopharmacology Award 2007". He also received the "Robert W. Kerwin Prize in 2010" from the Royal College of Psychiatrists for the best article published in the British Journal of Psychiatry. Juruena is currently a professor at the Department of Neurosciences and Behavior, University of São Paulo (USP-FMRP) and Honorary Senior Lecturer at the Institute of Psychiatry, Kings College London.

Contents

Preface

A fundamental problem in diagnosis is the fact that elaborate classification systems that exist today are solely based on subjective descriptions of symptoms. Such detailed phenomenology includes the description of multiple clinical subtypes; however, there is no biological feature that distinguishes one subtype from another. Moreover, it is recognized that a variety of disorders can exhibit similar clinical symptoms and that one disorder can manifest with distinct patterns in different people.

The Diagnostic and Statistical Manual of Mental Disorders (DSM) and the International Classification of Disease (ICD), the manuals that specify these diagnoses and the criteria for making them, are currently undergoing revision. These processes are involving a huge numbers of researchers from around the world; it is thus an appropriate time to question if neuroscience is prepared for the DSM-V and the ICD-11, and if they in turn are set for neuroscience. The presence of merely a few number of well-validated biomarkers and the early stage in which our understanding of neurobiology and genetics finds itself have obstructed the integration of neuroscience into psychiatric diagnosis to date. If we integrate a neurobiological approach that describes reliable neurobiological findings based on psychopathological syndrome, it will be more solid contrasted to a non-etiological system of classification. A future diagnostic criteria system in which aetiology and pathophysiology are essential in diagnostic decision-making would bring psychiatry closer to other specialties of medicine.

The relationship between stress and affective disorders is a strong example of a field of study that can be more fully understood from an integrative perspective. The potential of an integrative approach to contribute to improvements in human health and well being are more important than historical biases that have been associated with an integrative science approach. Approximately 60% of cases of depressive episodes are preceded by exposure to stressors, especially psychosocial stressors. Among the factors associated with depression in adulthood are exposure to childhood stressors such as the death of a parent or substitute, maternal deprivation, paternal abandonment, parental separation, and divorce. Psychological stress may change the internal homeostatic state of an individual. During acute stress, adaptive physiological responses occur, including increased adrenocortical hormone secretion, primarily cortisol. Whenever an acute interruption of this balance occurs, illness may result.

Particularly interesting are psychological stress (i.e., stress in the mind) and the interactions with the nervous, endocrine, and immune systems. For example childhood maltreatment is a major social problem. It is a complex global phenomenon that does not respect boundaries of class, race, religion, age, or educational level and can occur both publicly and privately, resulting in serious physical injury or even death. Moreover, its psychological consequences can acutely affect a child's mental health well into adulthood.

This approach says very clearly and without a doubt that the causes, development and outcomes of affective disorders are determined by the relationship of psychological, social and cultural factors with biochemistry and physiology. Biochemistry and physiology are not disconnected and different from the rest of our experiences and life events. This system is based on current studies that reported that the brain and its cognitive processes show a fantastic synchronization. Consequently, accepting the brain–body–mind complex is possible only when the three systems – nervous, endocrine and immune – have receptors on critical cells that can receive information (through messenger molecules) from each of the other systems. The fourth system, the mind (our thoughts, our feelings, our beliefs and our hopes), is part of the functioning of the brain integrating the paradigm. The interaction of the mind, an explicit functioning of the brain, with other body systems is critical for the maintenance of homeostasis and well being.

It is now broadly accepted that psychological stress may change the internal homeostatical state of an individual. During acute stress, adaptive physiological responses occur, which include hyperactivity of the hypothalamic–pituitary–adrenal (HPA) axis. Whenever there is an acute interruption of this balance, illness may result. The social and physical environments have an enormous impact on our physiology and behaviour, and they influence the process of adaptation or 'allostasis'. It is correct to state that at the same time that our experiences change our brain and thoughts, namely, changing our mind, we are changing our neurobiology. Of special interest are the psychological stress (stress in the mind) and the interactions of the nervous, endocrine and immune systems. Increased adrenocortical secretion of hormones, primarily cortisol in major depression, is one of the most consistent findings in psychiatry. A significant percentage of patients with major depression have been shown to exhibit increased concentrations of cortisol (the endogenous glucocorticoid in humans) in the plasma, urine, saliva and cerebrospinal fluid (CSF); an exaggerated cortisol response to adrenocorticotropic hormone (ACTH); and an enlargement of both the pituitary and adrenal glands. The maintenance of the internal homeostatic state of an individual is proposed to be based on the ability of circulating glucocorticoids to exert negative feedback on the secretion of hypothalamic-pituitary-adrenal (HPA) hormones through binding to mineralocorticoid receptors (MRs) and glucocorticoid receptors (GRs), limiting the vulnerability to diseases related to psychological stress in genetically predisposed individuals. The HPA axis response to stress can be thought of as a mirror of the organism's response to stress: acute responses are generally adaptive, but excessive or prolonged responses can lead to deleterious effects.

Generally, HPA axis changes appear in chronic depressive and more severe episodes. Moreover, HPA axis changes appear to be state-dependent, tending to improve upon resolution of the depressive syndrome. Interestingly, persistent HPA hyperactivity has been associated with higher rates of relapse.

There is an increasing data supporting that depressive disorders include a group of conditions which may be different with regard to the activity of the HPA axis, immune functions and treatment response. Melancholia, for instance, a syndrome with a long history and distinctive psychopathological features, is differentiated from major depression by the DSM-IV specifiers and partly described in the ICD-10th edition. Nevertheless, it has a distinctive psychopathology and biological homogeneity in clinical experience and laboratory test markers, and it is differentially responsive to specific treatment interventions according to international studies. In the last few years an important movement proposes to reinstitute the definition of melancholia, set a duration criterion and add as secondary criteria the associated laboratory findings of dexamethasone non-suppression of cortisol, high night-time cortisol levels, or decreased REM latency or other characteristic sleep abnormalities.

The lack of correlations between clinical and biological data continues to be, according to several authors, one of the great unsolved problems of psychiatry today and could be solved by recovering the value of traditional psychopathological analysis based on fundamental and thorough clinical assessment, which should support aetiological research and treatment decisions.

Therefore, I am greatly pleased to edit this book where the authors achieve a balance among diagnostic, research, clinical and new treatment approaches to Affective Disorders.

Mario Francisco Juruena, MD, MSc, MPhil, PhD
Stress and Affective Disorders Programme (SAD Programme)
Department of Neurosciences and Behaviour
Faculty of Medicine Ribeirao Preto, University of Sao Paulo
Brazil

Part 1

Clinical

1

Biological Prediction of Suicidal Behavior in Patients with Major Depressive Disorder

Yong-Ku Kim
Department of Psychiatry, College of Medicine, Korea University
Republic of Korea

1. Introduction

Suicide is a major public health issue and a significant cause of death worldwide. Most suicides (about 90%) occur in the context of psychiatric disorders, most commonly major depressive disorder, which is associated with approximately 60% of all suicides (Carlson et al. 1991). Prediction of suicidal risk in major depressive disorder is very important for preventing suicide, but current approaches to predicting suicidal behavior are based on clinical history and have low specificity. Accordingly, biological markers may provide a more specific means of identifying individuals at high risk of suicide with major depressive disorder (Lee and Kim 2011). Despite the high lifetime rate of suicide in patients with major depressive disorder (estimated to be 10-15%; Wulsin et al. 1999), most never attempt suicide. This raises the question of why some people with major depression are at risk of suicide and others are not, and suggests that the predisposition toward suicidal behavior is independent of psychiatric disorders. Other factors that increase the risk of suicidal behavior include psychosocial stressors, aggressive and impulsive traits, hopelessness, pessimism, substance abuse and dependence, physical or sexual abuse during childhood, and a history of head injury or neurological disorders. In considerations of these risk factors, suicidal behavior has been conceptualized into stress-diathesis and state-trait interaction models (Mann et al. 1999; Van Heeringen and Marusic 2003). Figure 1 illustrates the stress-diathesis model of suicidal behavior.

These models suggest that acute psychological stressors act on the diathesis, or traits of suicidal behavior, and that the complicated interactions between stress and diathesis gradually evolve into suicidal behavior over time. Previous research has explored potential biological markers and predictors of suicide and suicidal behavior, especially in the context of major depression. Although work in this area has been inconclusive, many animal, post-mortem, clinical, and genetic studies have produced results implicating at least 3 neurobiological systems in the pathogenesis of suicidal behavior in major depression: deficiency in the serotonergic system, hyperactivity of the hypothalamic-pituitary-adrenal axis, and decreased brain derived neurotrophic factor (BDNF) metabolism. Additionally, other neurotransmitters, cholesterol, nitric oxide (NO) and cytokines may be associated with suicide and suicidal behavior in major depression. Specifically, diathesis or trait-dependent risk factors are associated with dysfunctions in the serotonin system; however, the stress response (i.e., state-dependent factors) is related to hypothalamic-pituitary adrenal(HPA)

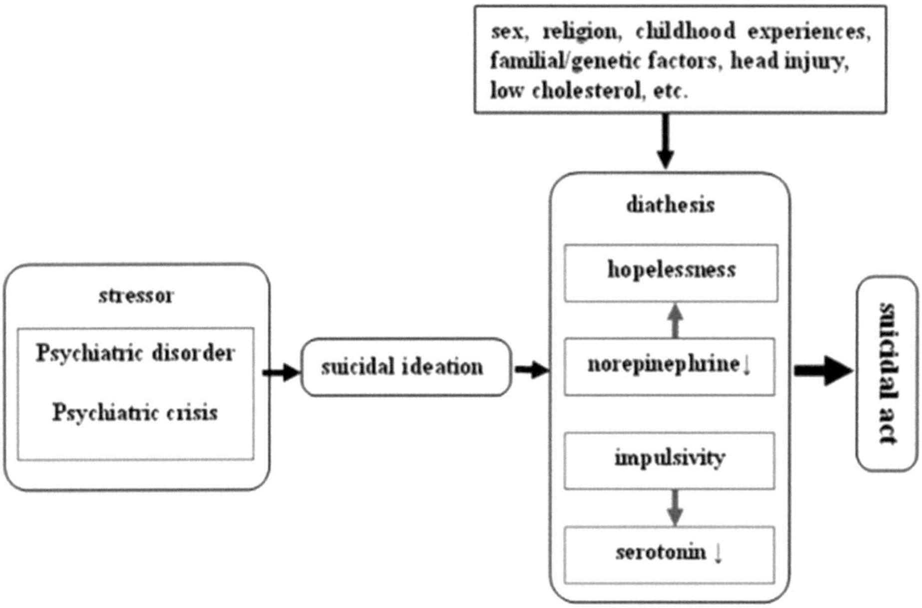

Fig. 1. Stress-diathesis model of suicidal behavior

axis hyperactivity. Decreases in cholesterol and BDNF levels are associated with impaired brain plasticity among individuals with suicidal behavior in major depressive disorder. In this chapter, I discuss peripheral biological markers involved in the pathogenesis of suicidal behavior in major depressive disorder and propose a model to predict the risk of suicidal behavior in these patients.

2. The neurotransmitter system

2.1 Serotonin

The serotonin system has been widely investigated in studies of major depression and suicide. The innervations of the serotonin system project from the dorsal raphe nucleus to all of the regions of the brain, including the cerebral cortex and hippocampus. Decreased function and activity of the serotonergic system in suicide victims and patients with major depression who attempt suicide have been confirmed in postmortem, serotonin transporter, serotonin receptor and cerebrospinal fluid (CSF) studies and neuroendocrine challenge tests. Post-mortem studies of the brains of suicide victims provide evidence of reduced serotonin transporter sites in the prefrontal cortex, hypothalamus, occipital cortex and brainstem (Purselle and Nemeroff 2003). In an autoradiographic study, this abnormality was found to be localized to the ventromedial prefrontal cortex (Arango et al. 1995). Abnormalities were also observed at the receptor level, as postsynaptic 5-HT1A and 5-HT2A receptors were found to be upregulated in the prefrontal cortex. It has been hypothesized that this increase may be a compensatory mechanism to counter the low activity of serotonergic neurons (Mann 2003). It is interesting to note that this serotonin dysfunction appears to be localized to the ventral prefrontal cortex, a region that is involved in behavioral and cognitive inhibition. Thus, low serotonergic input may contribute to impaired inhibition, creating a greater propensity to act on suicidal or aggressive feelings (Mann 2003).

Tryptophan hydroxylase (TPH), which has two isoforms (TPH1 and TPH2), is one of the rate limiting factors in serotonin synthesis, Postmortem studies have reported significantly higher numbers and higher densities of TPH immunoreactive neurons in the dorsal raphe nuclei of depressed suicide victims (Underwood et al. 1999) and in the same regions of alcohol dependent, depressed suicide victims (Bonkale et al. 2006) when compared to controls. We have found that the TPH2 -703G/T SNP may have an important effect on susceptibility to suicidal behavior in those with major depressive disorder. Additionally, an increased frequency of the G allele of the TPH2 SNP is associated with elevated risk of suicidal behavior itself rather than with the diagnosis of major depression, and may increase the risk of suicidality, independent of diagnosis (Yoon and Kim 2009). Collectively, TPH, serotonin transporter, and serotonin receptor studies suggest that deficient or impaired serotonin activity is involved in suicidal behavior. Increased activity in TPH and postsynaptic 5-HT2A receptors may be compensatory results of decreased central levels of serotonin. Notably, serotonin dysfunction appears to be localized in the ventral prefrontal cortex among suicide victims (Mann et al. 2000), as well as in individuals who make suicide attempts (Leyton et al. 2006).

The prefrontal cortex has been implicated in both behavioral and cognitive inhibition, as well as in willed action and decision-making. A meta-analysis examining 27 prospective and retrospective reports found that individuals who attempt suicide, and particularly those who use violent methods, had lower cerebrospinal fluid 5-hydroxyindoleacetic acid (CSF 5-HIAA) levels when compared to psychiatric controls (Lester 1995). Additionally, a meta-analysis of prospective biological studies estimated the odds ratio for the prediction of suicide completion to be 4.5-fold greater for individuals with low levels of CSF 5-HIAA than individuals with high levels of CSF 5-HIAA among patients with mood disorders (Mann et al. 2006). CSF 5-HIAA may serve as a predictor of future suicide attempts and completions, as findings associating CSF 5-HIAA levels with suicidal behavior have been relatively consistent. Additionally, levels of CSF 5-HIAA are relatively stable and therefore believed to be under substantial genetic control (Rogers et al. 2004). Blunted prolactin response to the fenfluramine challenge test has been observed among young (<30 years) inpatients with major depression and histories of suicide attempts (Mann et al. 1995). Other work has shown significantly lower prolactin responses to fenfluramine challenge tests among depressed patients with histories of suicide attempts than among patients without such histories or healthy controls (Correa et al. 2000; Mann et al. 1995). Further, decreased prolactin response has been reported among patients with histories of high-lethality suicide attempts (Malone et al. 1996). These results suggest that blunted prolactin response to fenfluramine, which indicates reduced serotonin function, may serve as a marker for suicidality among individuals with major depressive disorder.

2.2 The noradrenergic and dopaminergic systems

Few post-mortem studies have examined alterations in the noradrenergic or dopaminergic systems in suicide victims. Studies have found decreased noradrenalin (NA) levels in the brainstem and increased α2-adrenergic receptor densities in suicide victims (Ordway et al. 1994a). One study found that tyrosine hydroxylase (TH), the rate-limiting enzyme for NA and dopamine (DA) synthesis, is higher in suicide victims (Ordway et al. 1994b), however another study found the opposite (Biegon and Fieldust 1992). Increased TH and α2-

adrenergic receptor densities could be indicative of noradrenergic depletion compensatory to increased NA release. Increased NA release may be explained by the relationship between the noradrenergic system and stress response, as severe anxiety and agitation are associated with noradrenergic overactivity, higher suicide risk, and overactivity of the hypothalamic-pituitary-adrenal (HPA) axis (Mann 2003).

Few studies have examined the dopaminergic system. Overall, no alterations were found in mRNA levels of the D1, D2 and D4 receptors that bind in the caudate nuclei of suicide victims (Hurd et al. 1997; Sumiyoshi et al. 1995). A recent investigation exploring homovanillinic acid (HVA) in the CSF of depressed suicide attempters found reduced HVA levels in attempters, but not in depressed non-attempters (Sher et al. 2006). Thus, the dopamine system seems to be hypofunctional in major depression (Kapur and Mann 1992).

3. Neurotrophic factors

3.1 Brain derived neurotrophic factor (BDNF)

Neurotrophic factors including BDNF, nerve growth factor (NGF) and neurotrophin (NT)- 3, 4/5, play an important physiological role in the maintenance and growth of neurons and synaptic plasticity in the adult brain (Lewin and Barde 1996) and are known to be involved in the pathogenesis of depression and suicide (Duman et al. 1997; Nestler et al. 2002). In particular, BDNF mRNA expression levels are significantly decreased in animals subjected to forced swimming and chronic immobilization stress (Russo-Neustadt et al. 2001; Xu et al. 2002). Moreover, chronic antidepressant treatment increases the expression of BDNF and neurogenesis in adult rat hippocampi (Duman et al. 1997; Malberg et al. 2000). Several clinical studies have found differing BDNF levels in the blood sera or plasma of patients with major depression and patients who have attempted suicide. Deveci and colleagues (2007) investigated serum BDNF levels among suicide attempters without major psychiatric disorders, patients with major depression, and healthy subjects. They found that serum BDNF levels were lower among both suicide attempters and depressed patients than among healthy controls. Our research group has also examined plasma BDNF levels among patients with major depression who both have and have not attempted suicide. One study found that plasma BDNF levels were significantly lower among depressed patients than among normal controls (Lee et al. 2007). Plasma BDNF levels were also significantly lower among suicidal patients than non-suicidal patients with major depression, and that suicidal patients had the lowest levels of BDNF among all of the groups assessed (Lee et al. 2007). Further, Kim and colleagues (2007b) measured plasma BDNF levels in patients with depression who had recently attempted suicide, non-suicidal patients with depression, and healthy controls. BDNF levels were significantly lower among suicidal patients with depression than non-suicidal patients with depression and healthy controls. However, BDNF levels did not differ between individuals who made fatal and nonfatal suicide attempts (Kim et al. 2007b). One study examining BDNF mRNA expression in peripheral blood mononuclear cells revealed that patients with major depression and recent suicide attempts had decreased BDNF mRNA expression, compared to patients who had not attempted suicide (Lee and Kim 2010). Measurements of BDNF levels in sera or plasma in previous studies have been challenged, as it is questionable whether BDNF in the blood is released from the brain or from other sources. To address this issue, Dawood and colleagues (2007) used direct blood sampling from the internal jugular vein and the brachial artery and found that veno-arterial BDNF plasma concentration gradient acts as an index of brain BDNF production. Based on this determination, the veno-arterial BDNF concentration gradient was shown to be significantly reduced among patients at medium to high suicide risk compared to those at low risk. Additionally, this gradient was negatively correlated with suicide risk among untreated patients with depression. As such, BDNF level in sera or plasma appears to be decreased among suicidal individuals soon after attempted suicide, which is consistent with the changes observed in brain BDNF levels that have been reported in postmortem studies. These results suggest that BDNF may play an important role in the neurobiology of suicide and suicidal behavior in major depression.

3.2 Other neurotrophic factors

One study has found that BDNF and neurotrophin-3 (NT-3) levels are decreased in postmortem brains of suicide victims (Karege et al. 2005). Additionally, mRNA levels of nerve growth factor (NGF), NT-3, NT-4/5, cyclophilin, and neuron-specific enolase are decreased in the hippocampi of suicide victims (Dwivedi et al. 2005). Few studies have investigated other neurotropic factors, and further studies in suicidal depression are necessary.

4. The hypothalamic–pituitary–adrenal (HPA) axis and cortisol

The HPA axis is the major biological system involved in the acute stress response. The stress-related theory of depression states that chronic stress may lead to long-term activation of the HPA axis, which may then result in reductions in the volume or impairments to the function of the hippocampus (Holsboer 1988). Corticotropin-releasing hormone (CRH) levels in the CSF tend to be increased among suicide victims, suggesting an increase in HPA axis activity among individuals with suicidal behaviors (Arato et al. 1989). However, this association remains controversial and other research has shown that patients who make repeated suicide attempts may have even lower CSF CRH levels than patients who do not (Traskman-Bendz et al. 1992).

The dexamethasone suppression test (DST) is one of the most useful assessments of HPA axis activity. During normal HPA axis activity, administration of dexamethasone, an exogenous synthetic glucocorticoid hormone, leads to negative feedback to the HPA axis. This negative feedback results in suppression of the release of adrenocorticotropic hormone (ACTH) from the hypothalamus, which results in suppression of the release of cortisol from the adrenal gland. The reduction in cortisol levels as measured in plasma results in a positive result on the DST test. Many studies have shown that cortisol non-suppression in response to the DST is a strong predictor of suicidal behavior (Coryell and Schlesser 2001; Kunugi et al. 2004; Yerevanian et al. 2004). Specifically, some reports have demonstrated that patients with non-suppression engage in more serious suicide attempts (Coryell 1990; Norman et al. 1990) or use more violent methods (Roy 1992) than those who do not exhibit non-suppression. Jokinen and Nordström (2008) found that DST non-suppression is associated with suicide attempts among young adult and elderly inpatients with mood disorders. However, Black and colleagues (2002) found no significant differences in the frequency of suicidal ideation or completed suicides between patients demonstrating DST suppression and those demonstrating non-suppression (Black et al. 2002). A long-term follow-up study spanning 15 years has shown that patients with depression and DST non-suppression have a roughly 14-fold higher risk of suicide than do patients with DST suppression (Coryell and Schlesser 2001). A meta-analysis estimated the odds ratio of suicide completion to be 4.5-fold greater among non-suppressors than suppressors in patients with mood disorders (Mann et al. 2006). Moreover, other long-term follow-up studies have suggested that the DST is a useful predictor of suicidal behaviors and attempts among individuals with mood disorders, depressed inpatients, and patients with manifest suicidality, but not among the general population (Jokinen et al. 2007) or in patients displaying DST suppression (Coryell et al. 2006). Further, Jokinen and colleagues (2008b) suggested that a different threshold for cortisol levels following dexamethasone may require defining DST non-suppression for the prediction of suicide among individuals experiencing melancholic depression. Yerevanian and colleagues (2004) also reported that DST non-suppression identifies unipolar depressed patients at higher risk of future suicide completion or hospitalization for suicidality. Overall, evidence suggests that HPA axis hyperactivity may influence the overactivity of the adrenergic system and alternations of the serotonergic system (Mann 2003; Meijer and de Kloet 1998).

5. Cholesterol

Trials of cholesterol-lowering drugs revealed increased mortality due to accidents, violence, and suicide among subjects who received the drugs (Kaplan et al. 1997; Muldoon et al. 1990). Kaplan and colleagues (1997) suggest that serum cholesterol reduction achieved by changing the serum composition or concentration of lipoproteins, could affect brain levels of fat-soluble micronutrient supply, structural lipids, cellular communication, or neurotransmitters, including serotonin. However, a second meta-analysis revealed only a modest, non-significant increase in deaths due to suicide and violence among patients receiving trials of dietary interventions and non-statin drugs (Muldoon et al. 2001).

Clinical studies of psychiatric subjects indicate a relationship between lower total cholesterol levels and suicidal behavior. Specifically, it has been reported that suicide attempters tend to have significantly lower cholesterol levels than non-suicidal psychiatric inpatients and individuals experiencing accidental injuries (Kunugi et al. 1997). Plasma cholesterol levels among acutely suicidal patients with mood disorders were found to be lower than among non-suicidal inpatients with mood disorders and healthy subjects (Papassotiropoulos et al. 1999). Additionally, a study of serum cholesterol levels showed that serum cholesterol is 30% lower among violent suicide attempters, in comparison to non-violent suicide attempters and healthy subjects (Alvarez et al. 2000). Of note, studies of Korean subjects found that serum total cholesterol levels and densities of lipoproteins tend to be lower among parasuicidal individuals, and that serum triglyceride levels tend to be lower among suicide attempters than non-suicidal patients with major depressive disorder (Kim et al. 2002a; Lee and Kim 2003). Moreover, our data suggest two cut-off points for serum cholesterol levels in patients with depression: 180 mg/dl, which may serve as a point for high sensitivity of possible risk of suicide, and 150 mg/dl, a point with a high specificity of probable risk of suicide (Kim and Myint 2004). However, studies in the general Korean population have failed to report consistent findings linking low cholesterol levels and suicidal behavior (Ellison and Morrison 2001; Iribarren et al. 1995). If suicidal behavior is associated with reductions in serum or plasma cholesterol levels, this may be explained because low cholesterol levels are related to decreased serotonin activity, which may increase tendencies toward impulsive, aggressive, and suicidal behavior (Heron et al. 1980;

Kaplan et al. 1997; Ringo et al. 1994). Another possible explanation is that decreased cholesterol in peripheral blood may reduce cholesterol levels in the brain, which may lead to reduced synaptic plasticity and brain dysfunction associated with impaired neurobehavioral consequences (Mauch et al. 2001; Pfrieger 2003).

6. Nitric oxide and cytokines

Nitric oxide (NO) is an endogenous gas that is known to influence cerebral monoaminergic activity, including serotonin activity (Montague et al. 1994; Yamada et al. 1995). In patients with major depression, the total amount and density of neurons with immunoreactivity to nitric acid synthase (NOS) were reduced in paraventricular neurons (Bernstein et al. 1998), and NOS activity was decreased in the prefrontal cortex (Xing et al. 2002). A previous study revealed that plasma NO levels were dramatically lower in patients with major depressive disorder compared to healthy controls (Chrapko et al. 2004). However, another study detected elevated NO levels in patients with major depression compared to patients with anxiety disorder and normal control subjects (Jozuka et al. 2003). We found that increased NO production in plasma is associated with suicide attempts in depressed patients (Kim et al. 2006).

It has been postulated that major depression is accompanied by significant changes in cell-mediated and humoral immunity and that these changes are related to the pathophysiology or pathogenesis of the illness (Miller and O'Callaghan 2005; Myint and Kim 2003; Schiepers et al. 2005). Pro-inflammatory cytokines including IL-1β, IL-6, IL-12, and TNF-α are increased in the blood in major depression (Kim et al. 2002b; Thomas et al. 2005; Tuglu et al. 2003; Viljoen and Panzer 2005). These findings suggest that innate immunity is activated by secretion from monocytes and macrophages during major depression. A previous study measured cytokine secretion of T-cells of suicidal and non-suicidal depressed patients and healthy controls and found that the T-cells of suicidal depressed patients have Th1 characteristics, while the T-cells of non-suicidal depressed patients have Th2 characteristics (Mendlovic et al. 1999). A new hypothesis concerning the relationships between serum lipids, depression, suicide and atherosclerosis suggests that IL-2 plays important roles in lipid metabolism, depression, suicide and atherosclerosis (Colin et al. 2003; Penttinen 1995). Our group found that Th1 and Th2 cytokine imbalances are observed in a subpopulation of depressed patients (Myint et al. 2005). We also found that Th1 cytokine (IL-2 and IL-6) levels were significantly lower in suicidal depressed patients than in non-suicidal depressive patients and normal controls (Kim et al. 2007a). Collectively, NO and cytokines may be candidates for biological markers of suicidal behavior in major depression, but they have not yet been investigated extensively.

7. Can we predict suicidal behavior in major depression?

Many studies have tried to identify biological etiologies and predictors of suicidal behavior in major depression, but this task has been difficult because most suicide risk factors have low specificity and the rate of suicide completion is relatively low the in the general population (Cohen 1986). These difficulties can be addressed when combinations of risk factors for suicide are used to estimate the suicide risk of individuals. For instance, several researchers have examined combinations of two biological risk factors for suicide simultaneously. Specifically, researchers have studied the coupling of CSF 5-HIAA and DST non-suppression (Jokinen et al. 2008a; Jokinen et al. 2009; Mann et al. 2006) and the coupling of serum cholesterol and DST non-suppression (Coryell and Schlesser 2007). These combined factors may be useful because they reflect diverse aspects of suicidal phenomena. Specifically, Jokinen and colleagues (2008a) suggest that CSF 5-HIAA and DST non-suppression are independent biomarkers and that CSF 5-HIAA may reflect short-term suicide risk, while dysregulation of the HPA axis may be a more long-term predictor of suicidal behavior. These findings appear to be even better predictors among individuals with major depression or with previous histories of attempted suicide. Mann and colleagues (2006) also suggested that low CSF 5-HIAA and serotonin dysfunction are markers of the diathesis and that DST non-suppression and HPA axis hyperactivity are markers of the acute stress response.

Additionally, reduced cholesterol and BDNF levels in blood serum or plasma may be associated with impaired brain plasticity among individuals with suicidal behavior and ideation. In the future, it will be useful to examine multiple tests and risk factors, including CSF 5-HIAA, DST, cholesterol, and BDNF levels, as well as patient history of attempted suicide, in the prediction of suicide risk, especially among patients with depression.

We propose a model that predicts suicide risk that also considers several factors. We based this model on the Child-Pugh classification system of severity of chronic liver disease (Pugh et al. 1973) and the model is presented in Table 1. Abnormal findings associated with serotonin or HPA activity are more significant among individuals with major depression or with previous histories of attempted suicide (Coryell and Schlesser 2007; Coryell et al. 2006). Additionally, an interaction effect of childhood abuse and gene polymorphisms of serotonin transporters and BDNF has also been reported to influence the risk for suicidal behavior (Currier and Mann 2008). Suicide is associated with dysfunction in the prefrontal cortex, which is related to poor executive function. Such dysfunction can be measured with the

Parameter	Points assigned			
	0	1	2	3
Childhood abuse history	negative		positive	
Current depression	negative	positive		
Previous attempt	0	1		≥ 2
Genetic factors	negative		positive	
CSF 5-HIAA	> cut-off		≤ cut-off	
DST	suppression		Non-suppression	
BDNF levels	normal		decreased	
Cholesterol levels (mg/dl)	≥ 180	180-150	< 150	
Wisconsin Card Sort Test	Normal		Abnormal	

BDNF, brain-derived neurotrophic factor, DST, dexamethasone suppression test, CSF, cerebrospinal fluid, 5-HIAA, 5-hydroxyindoleacetic acid. Our hypothesis is that the total score of these parameters is correlated with current risk of suicide in major depression.

Table 1. Proposed classification of multiple factors to explain risk of suicide in major depression.

Wisconsin Card Sorting Test, and reported deficits in executive functioning are associated with high-lethality suicidal attempts among individuals with major depression (Keilp et al. 2001). Table 1 outlines nine risk factors for suicidal behavior and assigns one point to each factor. It is hypothesized that the total score of these risk factors is correlated with current risk of suicide.

8. Conclusions

Suicide is a complicated phenomenon that results from the interaction of several factors, including neurobiological changes, genetic predisposition, and psychological factors. Postmortem and clinical studies suggest that serotonin dysfunction is a form of diathesis or trait style-risk factor while HPA dysfunction is associated with stress response or state-dependency. Decreased cholesterol and BDNF levels are also related to brain dysfunction among suicidal individuals. Decreased serotonin functioning among suicidal individuals has been measured with CSF 5-HIAA, fenfluramine challenge studies, and levels of platelet 5-HT2A receptors. HPA axis dysfunction has been evaluated using the DST. Cholesterol and BDNF levels can be measured in blood serum or plasma. Additionally, serotonin dysfunction and lower BDNF activity has been found in the prefrontal cortex of the brain in suicidal individuals. Impairment in this region may be associated with behavioral disinhibition and executive dysfunctions, which is often examined with neurocognitive tests. We have proposed a model that incorporates present research on biological factors that may contribute to suicide risk. Clinical studies are needed to evaluate the validity of our risk scale for suicide, but we believe that based on current evidence, this provides a comprehensive screen.

It remains challenging to identify neurobiological predictors of suicidal behavior that are promising and easily assessable. Since suicidal behavior is a complex phenomenon, a multi-dimensional approach, including the above assessments, may be required to predict suicide risk, especially among individuals with major depression. A better understanding of the neurobiology of suicide in major depression will help detect at risk individuals or populations, and help develop better treatment interventions.

9. References

Alvarez, J.C., Cremniter, D., Gluck, N., Quintin, P., Leboyer, M., Berlin, I., Therond, P., Spreux-Varoquaux, O. (2000). Low serum cholesterol in violent but not in non-violent suicide attempters. *Psychiatry Research*, Vol.95, No.2, pp. 103-108, ISSN 0165-1781

Arango, V., Underwood, M.D., Gubbi, A.V., Mann, J.J. (1995). Localized alterations in pre- and postsynaptic serotonin binding sites in the ventrolateral prefrontal cortex of suicide victims. *Brain Research*, Vol.688, No.1-2, pp. 121-133, ISSN 0006-8993

Arato, M., Banki, C.M., Bissette, G., Nemeroff, C.B. (1989). Elevated CSF CRF in suicide victims. *Biological Psychiatry*, Vol.25, No.3, pp. 355-359, ISSN 0006-3223

Bernstein, H.G., Stanarius, A., Baumann, B., Henning, H., Krell, D., Danos, P., Falkai, P., Bogerts, B. (1998). Nitric oxide synthase-containing neurons in the human hypothalamus: reduced number of immunoreactive cells in the paraventricular nucleus of depressive patients and schizophrenics. *Neuroscience*, Vol.83, No.3, pp. 867-875, ISSN 0306-4522

Biegon, A., Fieldust, S. (1992). Reduced tyrosine hydroxylase immunoreactivity in locus coeruleus of suicide victims. *Synapse*, Vol.10, No.1, pp. 79-82, ISSN 0887-4476

Black, D.W., Monahan, P.O., Winokur, G. (2002). The relationship between DST results and suicidal behavior. *Annals of Clinical Psychiatry* Vol.14, No.2, pp. 83-88, ISSN 1040-1237

Bonkale, W.L., Turecki, G., Austin, M.C. (2006). Increased tryptophan hydroxylase immunoreactivity in the dorsal raphe nucleus of alcohol-dependent, depressed suicide subjects is restricted to the dorsal subnucleus. *Synapse*, Vol.60, No.1, pp. 81-85, ISSN 0887-4476

Carlson, G.A., Rich, C.L., Grayson, P., Fowler, R.C. (1991). Secular trends in psychiatric diagnoses of suicide victims. *Journal of Affective Disorders*, Vol.21, No.2, pp. 127-132, ISSN 0165-0327

Chrapko, W.E., Jurasz, P., Radomski, M.W., Lara, N., Archer, S.L., Le Melledo, J.M. (2004). Decreased platelet nitric oxide synthase activity and plasma nitric oxide metabolites in major depressive disorder. *Biological Psychiatry*, Vol.56, No.2, pp. 129-134, ISSN 0006-3223

Cohen, J. (1986). Statistical approaches to suicidal risk factor analysis. *Annals of the New York Academy of Sciences*, Vol.487, pp. 34-41, ISSN 0077-8923

Colin, A., Reggers, J., Castronovo, V., Ansseau, M. (2003). [Lipids, depression and suicide]. *L'Encephale*, Vol.29, No.1, pp. 49-58, ISSN 0013-7006

Correa, H., Duval, F., Mokrani, M., Bailey, P., Tremeau, F., Staner, L., Diep, T.S., Hode, Y., Crocq, M.A., Macher, J.P. (2000). Prolactin response to D-fenfluramine and suicidal behavior in depressed patients. *Psychiatry Research*, Vol.93, No.3, pp. 189-199, ISSN 0165-1781

Coryell, W. (1990). DST abnormality as a predictor of course in major depression. *Journal of Affective Disorders*, Vol.19, No.3, pp. 163-169, ISSN 0165-0327

Coryell, W., Schlesser, M. (2001). The dexamethasone suppression test and suicide prediction. *The American Journal of Psychiatry*, Vol.158, No.5, pp. 748-753, ISSN 0002-953X

Coryell, W., Schlesser, M. (2007). Combined biological tests for suicide prediction. *Psychiatry Research*, Vol.150, No.2, pp. 187-191, ISSN 0165-1781

Coryell, W., Young, E., Carroll, B. (2006).Hyperactivity of the hypothalamic-pituitary-adrenal axis and mortality in major depressive disorder. *Psychiatry Research*, Vol.142, No.1, pp. 99-104, ISSN 0165-1781

Currier, D., Mann, J.J. (2008). Stress, genes and the biology of suicidal behavior. *The Psychiatric Clinics of North America*, Vol.31, No.2, pp. 247-269, ISSN 1558-3147

Dawood, T., Anderson, J., Barton, D., Lambert, E., Esler, M., Hotchkin, E., Haikerwal, D., Kaye, D., Lambert, G. (2007). Reduced overflow of BDNF from the brain is linked with suicide risk in depressive illness. *Molecular Psychiatry*, Vol.12, No.11, pp. 981-983, ISSN 1359-4184

Deveci, A., Aydemir, O., Taskin, O., Taneli, F., Esen-Danaci, A. (2007). Serum BDNF levels in suicide attempters related to psychosocial stressors: a comparative study with depression. *Neuropsychobiology*, Vol.56, No.2-3, pp. 93-97, ISSN 1423-0224

Duman, R.S., Heninger, G.R., Nestler, E.J. (1997). A molecular and cellular theory of depression. *Archives of General Psychiatry*, Vol.54, No.7, pp. 597-606, ISSN 0003-990X

Dwivedi, Y., Mondal, A.C., Rizavi, H.S., Conley, R.R. (2005). Suicide brain is associated with decreased expression of neurotrophins. *Biological Psychiatry*, Vol.58, No.4, pp. 315-324, ISSN 0006-3223

Ellison, L.F., Morrison, H.I. (2001). Low serum cholesterol concentration and risk of suicide. *Epidemiology*, Vol.12, No.2, pp. 168-172, ISSN 1044-3983

Heron, D.S., Shinitzky, M., Hershkowitz, M., Samuel, D. (1980). Lipid fluidity markedly modulates the binding of serotonin to mouse brain membranes. *Proceedings of the National Academy of Sciences of the United States of America*, Vol.77, No.12, pp. 7463-7467, ISSN 0027-8424

Holsboer, F. (1988). Implications of altered limbic-hypothalamic-pituitary-adrenocortical (LHPA)-function for neurobiology of depression. *Acta Psychiatrica Scandinavica Supplementum*, Vol.341, pp. 72-111, ISSN 0065-1591

Hurd, Y.L., Herman, M.M., Hyde, T.M., Bigelow, L.B., Weinberger, D.R., Kleinman, J.E. (1997). Prodynorphin mRNA expression is increased in the patch vs matrix compartment of the caudate nucleus in suicide subjects. *Molecular Psychiatry*, Vol.2, No.6, pp. 495-500, ISSN 1359-4184

Iribarren, C., Reed, D.M., Wergowske, G., Burchfiel, C.M., Dwyer, J.H. (1995). Serum cholesterol level and mortality due to suicide and trauma in the Honolulu Heart Program. *Archives of Internal Medicine*, Vol.155, No.7, pp. 695-700, ISSN 0003-9926

Jokinen, J., Carlborg, A., Martensson, B., Forslund, K., Nordstrom, A.L., Nordstrom, P. (2007). DST non-suppression predicts suicide after attempted suicide. *Psychiatry Research*, Vol.150, No.3, pp. 297-303, ISSN 0165-1781

Jokinen, J., Martensson, B., Nordstrom, A.L., Nordstrom, P. (2008a). CSF 5-HIAA and DST non-suppression -independent biomarkers in suicide attempters? *Journal of Affective Disorders*, Vol.105, No.1-3, pp. 241-245, ISSN 0165-0327

Jokinen, J., Nordstrom, A.L., Nordstrom, P. (2008b). ROC analysis of dexamethasone suppression test threshold in suicide prediction after attempted suicide. *Journal of Affective Disorders*, Vol.106, No.1-2, pp. 145-152, ISSN 0165-0327

Jokinen, J., Nordstrom, A.L., Nordstrom, P. (2009). CSF 5-HIAA and DST non-suppression--orthogonal biologic risk factors for suicide in male mood disorder inpatients. *Psychiatry Research*, Vol.165, No.1-2, pp. 96-102, ISSN 0165-1781

Jokinen, J., Nordstrom, P. (2008). HPA axis hyperactivity as suicide predictor in elderly mood disorder inpatients. *Psychoneuroendocrinology*, Vol.33, No.10, pp. 1387-1393, ISSN 0306-4530

Jozuka, H., Jozuka, E., Suzuki, M., Takeuchi, S., Takatsu, Y. (2003). Psycho-neuro-immunological treatment of hepatocellular carcinoma with major depression--a single case report. *Current Medical Research and Opinion*, Vol.19, No.1, pp. 59-63, ISSN 0300-7995

Kaplan, J.R., Muldoon, M.F., Manuck, S.B., Mann, J.J. (1997). Assessing the observed relationship between low cholesterol and violence-related mortality. Implications for suicide risk. *Annals of the New York Academy of Sciences*, Vol.836, pp. 57-80, ISSN 0077-8923

Kapur, S., Mann, J.J. (1992). Role of the dopaminergic system in depression. *Biological Psychiatry*, Vol.32, No.1, pp. 1-17, ISSN 0006-3223

Karege, F., Vaudan, G., Schwald, M., Perroud, N., La Harpe, R. (2005). Neurotrophin levels in postmortem brains of suicide victims and the effects of antemortem diagnosis and psychotropic drugs. *Brain Research Molecular Brain Research*, Vol.136, No.1-2, pp. 29-37, ISSN 0169-328X

Keilp, J.G., Sackeim, H.A., Brodsky, B.S., Oquendo, M.A., Malone, K.M., Mann, J.J. (2001). Neuropsychological dysfunction in depressed suicide attempters. *The American Journal of Psychiatry*, Vol.158, No.5, pp. 735-741, ISSN 0002-953X

Kim, Y.K., Jung, H.G., Myint, A.M., Kim, H., Park, S.H. (2007a). Imbalance between pro-inflammatory and anti-inflammatory cytokines in bipolar disorder. *Journal of Affective Disorders*, Vol.104, No.1-3, pp. 91-95, ISSN 0165-0327

Kim, Y.K., Lee, H.J., Kim, J.Y., Yoon, D.K., Choi, S.H., Lee, M.S. (2002a). Low serum cholesterol is correlated to suicidality in a Korean sample. *Acta Psychiatrica Scandinavica*, Vol.105, No.2, pp. 141-148, ISSN 0001-690X

Kim, Y.K., Lee, H.P., Won, S.D., Park, E.Y., Lee, H.Y., Lee, B.H., Lee, S.W., Yoon, D., Han, C., Kim, D.J., Choi, S.H. (2007b). Low plasma BDNF is associated with suicidal behavior in major depression. *Progress in Neuro-Psychopharmacology & Biological Psychiatry*, Vol.31, No.1, pp. 78-85, ISSN 0278-5846

Kim, Y.K., Myint, A.M. (2004). Clinical application of low serum cholesterol as an indicator for suicide risk in major depression. *Journal of Affective Disorders*, Vol.81, No.2, pp. 161-166, ISSN 0165-0327

Kim, Y.K., Paik, J.W., Lee, S.W., Yoon, D., Han, C., Lee, B.H. (2006). Increased plasma nitric oxide level associated with suicide attempt in depressive patients. *Progress in Neuro-Psychopharmacology & Biological Psychiatry*, Vol.30, No.6, pp. 1091-1096, ISSN 0278-5846

Kim, Y.K., Suh, I.B., Kim, H., Han, C.S., Lim, C.S., Choi, S.H., Licinio, J. (2002b). The plasma levels of interleukin-12 in schizophrenia, major depression, and bipolar mania: effects of psychotropic drugs. *Molecular Psychiatry*, Vol.7, No.10, pp. 1107-1114, ISSN 1359-4184

Kunugi, H., Takei, N., Aoki, H., Nanko, S. (1997). Low serum cholesterol in suicide attempters. *Biological Psychiatry*, Vol.41, No.2, pp. 196-200, ISSN 0006-3223

Kunugi, H., Urushibara, T., Nanko, S. (2004). Combined DEX/CRH test among Japanese patients with major depression. *Journal of Psychiatric Research*, Vol.38, No.2, pp. 123-128, ISSN 0022-3956

Lee, B.H., Kim, H., Park, S.H., Kim, Y.K. (2007). Decreased plasma BDNF level in depressive patients. *Journal of Affective Disorders*, Vol.101, No.1-3, pp. 239-244, ISSN 0165-0327

Lee, B.H., Kim, Y.K. (2010). BDNF mRNA expression of peripheral blood mononuclear cells was decreased in depressive patients who had or had not recently attempted suicide. *Journal of Affective Disorders*, Vol.125, No.1-3, pp. 369-373, ISSN 1573-2517

Lee, B.H., Kim, Y.K. (2011). Potential peripheral biological predictors of suicidal behavior in major depressive disorder. *Progress in Neuro-Psychopharmacology & Biological Psychiatry*, Vol.35, No.4, pp. 842-847, ISSN 1878-4216

Lee, H.J., Kim, Y.K. (2003). Serum lipid levels and suicide attempts. *Acta Psychiatrica Scandinavica*, Vol.108, No.3, pp. 215-221, ISSN 0001-690X

Lester, D. (1995). The concentration of neurotransmitter metabolites in the cerebrospinal fluid of suicidal individuals: a meta-analysis. *Pharmacopsychiatry*, Vol.28, No.2, pp. 45-50, ISSN 0176-3679

Lewin, G.R., Barde, Y.A. (1996). Physiology of the neurotrophins. *Annual Review of Neuroscience*, Vol.19, pp. 289-317, ISSN 0147-006X

Leyton, M., Paquette, V., Gravel, P., Rosa-Neto, P., Weston, F., Diksic, M., Benkelfat, C. (2006). alpha-[11C]Methyl-L-tryptophan trapping in the orbital and ventral medial prefrontal cortex of suicide attempters. *European Neuropsychopharmacology* Vol.16, No.3, pp. 220-223, ISSN 0924-977X

Malberg, J.E., Eisch, A.J., Nestler, E.J., Duman, R.S. (2000). Chronic antidepressant treatment increases neurogenesis in adult rat hippocampus. *The Journal of Neuroscience* Vol.20, No.24, pp. 9104-9110, ISSN 1529-2401

Malone, K.M., Corbitt, E.M., Li, S., Mann, J.J. (1996). Prolactin response to fenfluramine and suicide attempt lethality in major depression. *The British Journal of Psychiatry* Vol.168, No.3, pp. 324-329, ISSN 0007-1250

Mann, J.J. (2003). Neurobiology of suicidal behaviour. *Nature Reviews Neuroscience*, Vol.4, No.10, pp. 819-828, ISSN 1471-003X

Mann, J.J., Currier, D., Stanley, B., Oquendo, M.A., Amsel, L.V., Ellis, S.P. (2006). Can biological tests assist prediction of suicide in mood disorders? *The International Journal of Neuropsychopharmacology* Vol.9, No.4, pp. 465-474, ISSN 1461-1457

Mann, J.J., Huang, Y.Y., Underwood, M.D., Kassir, S.A., Oppenheim, S., Kelly, T.M., Dwork, A.J., Arango, V. (2000). A serotonin transporter gene promoter polymorphism (5-HTTLPR) and prefrontal cortical binding in major depression and suicide. *Archives of General Psychiatry*, Vol.57, No.8, pp. 729-738, ISSN 0003-990X

Mann, J.J., McBride, P.A., Malone, K.M., DeMeo, M., Keilp, J. (1995). Blunted serotonergic responsivity in depressed inpatients. *Neuropsychopharmacology* Vol.13, No.1, pp. 53-64, ISSN 0893-133X

Mann, J.J., Waternaux, C., Haas, G.L., Malone, K.M. (1999). Toward a clinical model of suicidal behavior in psychiatric patients. *The American Journal of Psychiatry*, Vol.156, No.2, pp. 181-189, ISSN 0002-953X

Mauch, D.H., Nagler, K., Schumacher, S., Goritz, C., Muller, E.C., Otto, A., Pfrieger, F.W. (2001). CNS synaptogenesis promoted by glia-derived cholesterol. *Science*, Vol.294, No.5545, pp. 1354-1357, ISSN 0036-8075

Meijer, O.C., de Kloet, E.R. (1998). Corticosterone and serotonergic neurotransmission in the hippocampus: functional implications of central corticosteroid receptor diversity. *Critical Reviews in Neurobiology*, Vol.12, No.1-2, pp. 1-20, ISSN 0892-0915

Mendlovic, S., Mozes, E., Eilat, E., Doron, A., Lereya, J., Zakuth, V., Spirer, Z. (1999). Immune activation in non-treated suicidal major depression. *Immunology Letters*, Vol.67, No.2, pp. 105-108, ISSN 0165-2478

Miller, D.B., O'Callaghan, J.P. (2005). Depression, cytokines, and glial function. *Metabolism: Clinical and Experimental*, Vol.54, No.5 Suppl 1, pp. 33-38, ISSN 0026-0495

Montague, P.R., Gancayco, C.D., Winn, M.J., Marchase, R.B., Friedlander, M.J. (1994). Role of NO production in NMDA receptor-mediated neurotransmitter release in cerebral cortex. *Science*, Vol.263, No.5149, pp. 973-977, ISSN 0036-8075

Muldoon, M.F., Manuck, S.B., Matthews, K.A. (1990). Lowering cholesterol concentrations and mortality: a quantitative review of primary prevention trials. *BMJ*, Vol.301, No.6747, pp. 309-314, ISSN 0959-8138

Muldoon, M.F., Manuck, S.B., Mendelsohn, A.B., Kaplan, J.R., Belle, S.H. (2001). Cholesterol reduction and non-illness mortality: meta-analysis of randomised clinical trials. *BMJ*, Vol.322, No.7277, pp. 11-15, ISSN 0959-8138

Myint, A.M., Kim, Y.K. (2003). Cytokine-serotonin interaction through IDO: a neurodegeneration hypothesis of depression. *Medical hypotheses*, Vol.61, No.5-6, pp. 519-525, ISSN 0306-9877

Myint, A.M., Leonard, B.E., Steinbusch, H.W., Kim, Y.K. (2005). Th1, Th2, and Th3 cytokine alterations in major depression. *Journal of Affective Disorders*, Vol.88, No.2, pp. 167-173, ISSN 0165-0327

Nestler, E.J., Barrot, M., DiLeone, R.J., Eisch, A.J., Gold, S.J., Monteggia, L.M. (2002). Neurobiology of depression. *Neuron*, Vol.34, No.1, pp. 13-25, ISSN 0896-6273

Norman, W.H., Brown, W.A., Miller, I.W., Keitner, G.I., Overholser, J.C. (1990). The dexamethasone suppression test and completed suicide. *Acta Psychiatrica Scandinavica*, Vol.81, No.2, pp. 120-125, ISSN 0001-690X

Ordway, G.A., Smith, K.S., Haycock, J.W. (1994a). Elevated tyrosine hydroxylase in the locus coeruleus of suicide victims. *Journal of Neurochemistry*, Vol.62, No.2, pp. 680-685, ISSN 0022-3042

Ordway, G.A., Widdowson, P.S., Smith, K.S., Halaris, A. (1994b). Agonist binding to alpha 2-adrenoceptors is elevated in the locus coeruleus from victims of suicide. *Journal of Neurochemistry*, Vol.63, No.2, pp. 617-624, ISSN 0022-3042

Papassotiropoulos, A., Hawellek, B., Frahnert, C., Rao, G.S., Rao, M.L. (1999). The risk of acute suicidality in psychiatric inpatients increases with low plasma cholesterol. *Pharmacopsychiatry*, Vol.32, No.1, pp. 1-4, ISSN 0176-3679

Penttinen, J. (1995). Hypothesis: low serum cholesterol, suicide, and interleukin-2. *American Journal of Epidemiology*, Vol.141, No.8, pp. 716-718, ISSN 0002-9262

Pfrieger, F.W. (2003). Cholesterol homeostasis and function in neurons of the central nervous system. *Cellular and Molecular Life Sciences* Vol.60, No.6, pp. 1158-1171, ISSN 1420-682X

Pugh, R.N., Murray-Lyon, I.M., Dawson, J.L., Pietroni, M.C., Williams, R. (1973). Transection of the oesophagus for bleeding oesophageal varices. *The British Journal of Surgery*, Vol.60, No.8, pp. 646-649, ISSN 0007-1323

Purselle, D.C., Nemeroff, C.B. (2003). Serotonin transporter: a potential substrate in the biology of suicide. *Neuropsychopharmacology* Vol.28, No.4, pp. 613-619, ISSN 0893-133X

Ringo, D.L., Lindley, S.E., Faull, K.F., Faustman, W.O. (1994). Cholesterol and serotonin: seeking a possible link between blood cholesterol and CSF 5-HIAA. *Biological Psychiatry*, Vol.35, No.12, pp. 957-959, ISSN 0006-3223

Rogers, J., Martin, L.J., Comuzzie, A.G., Mann, J.J., Manuck, S.B., Leland, M., Kaplan, J.R. (2004). Genetics of monoamine metabolites in baboons: overlapping sets of genes influence levels of 5-hydroxyindolacetic acid, 3-hydroxy-4-methoxyphenylglycol, and homovanillic acid. *Biological Psychiatry*, Vol.55, No.7, pp. 739-744, ISSN 0006-3223

Roy, A. (1992). Hypothalamic-pituitary-adrenal axis function and suicidal behavior in depression. *Biological Psychiatry*, Vol.32, No.9, pp. 812-816, ISSN 0006-3223

Russo-Neustadt, A., Ha, T., Ramirez, R., Kesslak, J.P. (2001). Physical activity-antidepressant treatment combination: impact on brain-derived neurotrophic factor and behavior in an animal model. *Behavioural Brain Research*, Vol.120, No.1, pp. 87-95, ISSN 0166-4328

Schiepers, O.J., Wichers, M.C., Maes, M. (2005). Cytokines and major depression. *Progress in Neuro-Psychopharmacology & Biological Psychiatry*, Vol.29, No.2, pp. 201-217, ISSN 0278-5846

Sher, L., Mann, J.J., Traskman-Bendz, L., Winchel, R., Huang, Y.Y., Fertuck, E., Stanley, B.H. (2006). Lower cerebrospinal fluid homovanillic acid levels in depressed suicide attempters. *Journal of Affective Disorders*, Vol.90, No.1, pp. 83-89, ISSN 0165-0327

Sumiyoshi, T., Stockmeier, C.A., Overholser, J.C., Thompson, P.A., Meltzer, H.Y. (1995). Dopamine D4 receptors and effects of guanine nucleotides on [3H]raclopride binding in postmortem caudate nucleus of subjects with schizophrenia or major depression. *Brain Research*, Vol.681, No.1-2, pp. 109-116, ISSN 0006-8993

Thomas, A.J., Davis, S., Morris, C., Jackson, E., Harrison, R., O'Brien, J.T. (2005). Increase in interleukin-1beta in late-life depression. *The American Journal of Psychiatry*, Vol.162, No.1, pp. 175-177, ISSN 0002-953X

Traskman-Bendz, L., Ekman, R., Regnell, G., Ohman, R. (1992). HPA-related CSF neuropeptides in suicide attempters. *European Neuropsychopharmacology*, Vol.2, No.2, pp. 99-106, ISSN 0924-977X

Tuglu, C., Kara, S.H., Caliyurt, O., Vardar, E., Abay, E. (2003). Increased serum tumor necrosis factor-alpha levels and treatment response in major depressive disorder. *Psychopharmacology*, Vol.170, No.4, pp. 429-433, ISSN 0033-3158

Underwood, M.D., Khaibulina, A.A., Ellis, S.P., Moran, A., Rice, P.M., Mann, J.J., Arango, V. (1999). Morphometry of the dorsal raphe nucleus serotonergic neurons in suicide victims. *Biological Psychiatry*, Vol.46, No.4, pp. 473-483, ISSN 0006-3223

Van Heeringen, C., Marusic, A. (2003). Understanding the suicidal brain. *The British Journal of Psychiatry*, Vol.183, pp. 282-284, ISSN 0007-1250

Viljoen, M., Panzer, A. (2005). Proinflammatory cytokines: a common denominator in depression and somatic symptoms? *Canadian Journal of Psychiatry*, Vol.50, No.2, pp. 128, ISSN 0706-7437

Wulsin, L.R., Vaillant, G.E., Wells, V.E. (1999). A systematic review of the mortality of depression. *Psychosomatic Medicine*, Vol.61, No.1, pp. 6-17, ISSN 0033-3174

Xing, G., Chavko, M., Zhang, L.X., Yang, S., Post, R.M. (2002). Decreased calcium-dependent constitutive nitric oxide synthase (cNOS) activity in prefrontal cortex in schizophrenia and depression. *Schizophrenia Research*, Vol.58, No.1, pp. 21-30, ISSN 0920-9964

Xu, H., Qing, H., Lu, W., Keegan, D., Richardson, J.S., Chlan-Fourney, J., Li, X.M. (2002). Quetiapine attenuates the immobilization stress-induced decrease of brain-derived neurotrophic factor expression in rat hippocampus. *Neuroscience Letters*, Vol.321, No.1-2, pp. 65-68, ISSN 0304-3940

Yamada, K., Noda, Y., Nakayama, S., Komori, Y., Sugihara, H., Hasegawa, T., Nabeshima, T. (1995). Role of nitric oxide in learning and memory and in monoamine metabolism in the rat brain. *British Journal of Pharmacology*, Vol.115, No.5, pp. 852-858, ISSN 0007-1188

Yerevanian, B.I., Feusner, J.D., Koek, R.J., Mintz, J. (2004). The dexamethasone suppression test as a predictor of suicidal behavior in unipolar depression. *Journal of Affective Disorders*, Vol.83, No.2-3, pp. 103-108, ISSN 0165-0327

Yoon, H.K., Kim, Y.K. (2009). TPH2 -703G/T SNP may have important effect on susceptibility to suicidal behavior in major depression. *Progress in Neuro-Psychopharmacology & Biological Psychiatry*, Vol.33, No.3, pp. 403-409, ISSN 0278-5846

2

Self-Reported Symptoms Related to Depression and Suicidal Risk

Kouichi Yoshimasu, Shigeki Takemura,
Jin Fukumoto and Kazuhisa Miyashita
*Department of Hygiene, School of Medicine,
Wakayama Medical University
Japan*

1. Introduction

Depression is often accompanied by a wide variety of somatic symptoms even when there is no evidence of any organic disorder that can cause such symptoms. Though the underlying mechanism still remains unclear, there are two assumptions explaining the associations between somatic symptoms and depression. One is that depression itself causes several somatic symptoms. That is, people with depression may express their mental conflicts in various somatic symptoms. This hypothesis could apply especially to those who have a vague feeling of stigma or prejudice against mental disorders, and who are reluctant to frankly express their mental symptoms. From another point of view, this might mean that some chronic somatic symptoms are magnified due to the person's depressive state. Another possible hypothesis is that those who have had chronic somatic symptoms of unknown origin, in other words, functional somatic symptoms, are likely to be depressed since they cannot always receive effective medical treatment for such symptoms.

Likewise, people at risk of suicide sometimes express somatic symptoms instead of obvious psychiatric symptoms, such as depressive moods, loss of interest, anxiety, or irritation. Since depression is strongly associated with suicidal risk, it may often be that those who do not show any signs of a depressive state suddenly attempt suicide, especially when they harbor strong feelings against expressing emotional conflicts of the kind described above. Therefore, general physicians should pay close attention to such suicide-related somatic symptoms as possible signs of suicide, and if necessary, take appropriate action, including referring such patients to a psychiatrist.

Previous epidemiological studies suggest that Japanese are generally more likely than Westerners to share a strong prejudice against mental disorders or suicidal ideation and to suppress their emotions and mental symptoms accordingly (Griffiths et al., 2006; Kawakami et al., 2008). Police officers are considered to have this tendency more strongly than the general population (Royle, Keenan, & Farrell, 2009). In such populations, some particular self-reported somatic symptoms may serve as an alternative for detecting depressive disorders or suicidal signs. Furthermore, if such people have a prejudice against mental disorders, rather than visit a psychiatric clinic, they are more likely to visit a psychosomatic clinic of the kind that has become popular in Japan.

Although many studies have evaluated the associations between somatic symptoms and depression or suicidal risk (Smolderen et al., 2009; Spiegel, Schoenfeld & Naliboff, 2007; Wang et al., 2007, 2009; Yoon et al., 2011), few have evaluated gender differences of such symptoms (Silverstein, 1999, 2002). Likewise, few investigations have evaluated the associations between a wide range of somatic symptoms and depression in both general and clinical populations (Haug, Mykletun & Dahl, 2004; Simon et al., 1999). In one of these reports (Simon et al., 1999), Japanese patients were reported to show the lowest number of depression-related psychological and somatic symptoms among those of 14 countries at the primary care setting. Thus, Japanese cultural characteristics may exert some influence on the relationship between somatic symptoms and depression.

We have conducted a series of epidemiological studies regarding this issue using separate samples of new outpatients visiting a psychosomatic clinic, community residents, and a working population (male police officers). The purpose of the present study is to evaluate the associations among various subjective somatic symptoms and depression as well as the suicidal risk among Japanese clinical, community and working populations. That involves establishing the key contributing elements that might aid in discovering depressive/suicidal signs in the pre-clinical or primary care stage by extracting key somatic symptoms associated with such risk in those populations. Furthermore, focusing on the outpatients who have a major depressive disorder, we evaluated the gender differences in psychiatric symptoms related to suicidal ideation.

2. Methods

Three separate samples were included in the present study; i.e., new outpatients of a psychosomatic clinic, community dwellers aged 40 or older, and male police officers in one prefectural police organization. In previously published studies of outpatients (Sugahara et al., 2004; Yoshimasu et al., 2006, 2009), we used several psychological tests, such as State-Trait Anxiety Inventory (STAI) (Spielberger, 1972) or Zung's self-rating depression scale (SDS) (Zung, Richards & Short, 1965), for evaluating their mental and physical status. Since outpatients had too many mental and somatic symptoms, those related to depression or suicidal ideation were narrowed down by an appropriate statistical method (i.e., stepwise selection).

Annual health check-up data were available for the latter two samples (community dwellers and male police officers). In those samples, the relation between each symptom (both mental and somatic) and depression or suicidal risk was assessed. Because depressive symptoms were regarded as confounding factors between somatic symptoms and suicidal risk, we adjusted for depression in multivariate analyses when suicidal risk/ideation was used as an outcome variable.

We also examined the gender differences of somatic symptoms related to suicidal ideation in outpatients visiting the psychosomatic clinic. Furthermore, focusing on the patients who have a major depressive disorder, the gender differences in psychiatric symptoms related to suicidal ideation were evaluated. Those studies were generally approved by the institutional review boards of each corresponding institution.

2.1 Outpatients visiting a psychosomatic clinic

A total of 914 consecutive new patients had check-ups at the Department of Psychosomatic Medicine in a university hospital in the Kyushu area of Japan during the period from June 2000 to March 2001. The Department usually treats primary cases with psychosomatic disorders or mild psychiatric disorders. Patients with psychotic diseases such as schizophrenia or severe depression are not treated in the Department. Those with such disorders are rather treated in the Department of Neuropsychiatry, which is distinct from the Department of Psychosomatic Medicine. In the first stage, an admitting physician interviewed the outpatients. After the interview, the patients were assigned a separate physician (physician in charge). Both the admitting physician and the physician in charge diagnosed each patient independently based on the Diagnostic and Statistical Manual of Mental Disorders Fourth Edition (DSM-IV) (American Psychiatric Association, 1994; Japanese edition, 1996).

2.1.1 Diagnosis of depression

Patients were diagnosed with depression if both the admitting physician and the physician in charge confirmed that their symptoms met the diagnostic criteria of major depressive disorder based on the DSM-IV. A total of 335 patients were diagnosed with depression according to those criteria. If both physicians did not diagnose the patients with a major depressive disorder, they were classified in a non-depression group (n = 423). The remaining patients (n = 156) were excluded from the analysis since a definitive diagnosis of a major depressive disorder could not be obtained.

2.1.2 Assessment of suicidal ideation

The patients were requested to answer questions based on the Kyudai Medical Index (KMI), Kyushu University's original medical index (Matsuoka, 1990), which was developed as a modification of the Cornell Medical Index Health Questionnaire for rapid screening. Each patient was requested to choose between dichotomous answers (yes/no), and to reveal their true intentions with a guarantee of confidentiality. They were also requested to give intuitive responses since prolonged thinking might confuse them. The question regarding suicidal ideation was included in the KMI that asked: "Do you often think you want to die?" If the patients answered "yes," they were regarded as having suicidal ideation. McNemar's test did not reveal any significant differences between this question and the corresponding suicide-related question included in the SDS (data not shown). After patients with missing data of suicidal ideation were excluded, 820 (304 men and 516 women) remained in the analyses.

2.1.3 Assessment of somatic symptoms

KMI includes questions for 45 subjective somatic symptoms (two concerning menstruation were added for women). The patients confirmed the presence of each symptom by yes/no answers to the corresponding questions. At the same time, they were asked to note the three symptoms that were causing them the most distress, and how long they had been suffering from these symptoms, since information regarding chief complaints is important in clinical setting. The three most distressing symptoms were checked with the original health questionnaire, and were also later identified by the admitting physician at the first interview.

2.1.4 Statistical analysis

In our analyses of the associations between somatic symptoms and depression, those somatic symptoms identified as the three most distressing ones were used. Since some of the three most distressing symptoms were psychiatric, we also evaluated the associations between those symptoms and depression. Somatic symptoms included in the KMI were used in the analyses for associations between those somatic symptoms and suicidal ideation. In the multivariate logistic regression models, depression and suicidal ideation were used as outcome variables, while somatic symptoms were used as the explanatory variables. A stepwise method was applied to narrow down the somatic symptoms that were significantly associated with depression or suicidal ideation. The number of subjects included in the regression models varied according to each calculating algorithm due to the missing values of the relevant factors to be adjusted for. In addition, the gender differences in psychiatric symptoms related to suicidal ideation were assessed in patients with major depressive disorders by logistic regression analysis using the stepwise method. In this analysis, the candidates' psychiatric symptoms considered to be related to suicidal ideation were chosen from the KMI.

2.2 Community dwellers

Town A (total population, approximately 8000) and Town B (total population, approximately 7000) are both located in the middle of Wakayama Prefecture, the Kinki area in Japan. To detect lifestyle-related diseases such as metabolic syndrome, those towns provided an annual health check-up program for self-employed community dwellers and their family members aged 40-74 years. In 2008, a total of 3656 people aged 40-74 years were eligible for the annual health check-up program, and 686 underwent a health check-up from May through August. Among those, 452 agreed to participate in the study.

2.2.1 Assessment of depression and suicidal risk

The Mini-International Neuropsychiatric Interview (M.I.N.I.), Japanese version 5.0.0 (2003) (Sheehan et al., 1998; Sheehan & Lecrubier, 2003), a conveniently structured tool designed to identify cases of mental disorder, was used for the present interview survey. The reliability and validity of the Japanese version of the M.I.N.I. were reported to be satisfactory (Otsubo et al., 2005). A total of nine interviewers, all of whom were licensed doctors or nurses, were enrolled as competent to conduct the interviews. The first author (KY), a psychiatrist, trained them in essential interview skills, including didactic sessions of a general interview, or reviews of the instrument sections. The first author also checked the interviewers and corrected them as the need arose during the sessions so that the interview could be appropriately conducted. Thus, the diagnosis of major depressive disorder was conducted according to the diagnostic criteria of DSM-IV.

The suicidal risk of each person was measured by six relevant items included in the M.I.N.I., five of which were concerned with suicidal thoughts or behaviors within the previous one-month, while one item dealt with lifetime experiences of suicidal attempts according to the weighted value of each question; points 1, 2, 6, and 10 (comprising two questions) were allotted for each response to the former five questions, and point 4 was allotted to the last response regarding lifetime experiences of suicidal attempts. More concretely, they were: a wish to die (point 1), a desire to harm oneself (point 2), suicidal thoughts (point 6), having a suicide plan (point 10), suicide attempts (point 10) (all five of which were events occurring in the past month), and life-time suicide attempts (point 4). This scoring system is in accordance with the M.I.N.I. 5.0.0 (January 1, 2003). Thus, a total score of 33 showed the maximum points for suicidal risk. All questions included in the M.I.N.I. were coded as two categories according to the respondents' yes/no answers. Subjects who scored more than 0 were regarded as possibly having a suicidal risk.

2.2.2 Assessment of somatic symptoms

An annual health examination for self-employed community dwellers in those towns was conducted during the period from May to July 2008. The health examination was comprised of several basic examinations and a doctor's check-up based on one's self-reported medical history and symptoms as confirmed by a self-administered questionnaire. The questionnaire included items regarding lifestyle factors, past and current illnesses as well as their current treatment status and self-reported symptoms. Because this health examination mainly focused on the secondary prevention of lifestyle-related diseases, a checklist included in the questionnaire for such symptoms contained a variety of 18 current physical symptoms, including those of the respiratory, cardiovascular, or digestive organs. Based on the information from this checklist, any associations between self-reported physical symptoms and depression as well as suicidal risk were assessed.

2.2.3 Statistical analysis

Logistic regression analyses were conducted using depression and suicidal risk as outcome variables and somatic symptoms as explanatory variables. Age, sex, and two basic depressive symptoms that were used for the screening of major depressive disorders were controlled when suicidal risk was used as an outcome variable. Using logistic regression analysis, we also evaluated the associations between psychiatric symptoms of major depressive disorder as well as dysthymia and suicidal risk.

2.3 Male police officers

A total of 2399 employees at 18 stations of one prefectural police organization in the Kinki area of Japan underwent annual health checkups from May to July 2008. The number of police officers amounted to approximately 1% of all Japanese police officers. However, the characteristics of each prefectural police organization in Japan are standardized and strictly controlled by the National Police Agency. Thus, our sample would be representative of the entire Japanese police organization. Of these, 2100 (87.5%) agreed to participate in the study. After excluding female police officers whose suicidal rate is negligibly few, and clerical workers that might impede the homogenous characteristic of the study sample, 1718 male police officers remained in the analyses. Further 33 officers with missing information regarding suicidal risk were excluded in the analyses for evaluating the associations between somatic symptoms and suicidal risk.

2.3.1 Assessments of depression and suicidal risk

Assessments of depression and suicidal risk were conducted in the same manner as for community dwellers using M.I.N.I. as described above. Furthermore, assessments of post-traumatic stress disorders (PTSD) by M.I.N.I. were included for the police officers.

2.3.2 Assessment of somatic symptoms

As with community dwellers, annual health check-up data confirmed by self-administered questionnaires were used for the assessment of somatic symptoms. Since the secondary prevention of the lifestyle-related diseases was the main purpose of the health examination in both community dwellers and the workplace, this checklist included in the questionnaire also contained a variety of 22 current physical symptoms, including those of the respiratory, cardiovascular, or digestive organs.

2.3.3 Statistical analysis

As in the previous two samples described above, logistic regression analyses were performed using depression and suicidal risk as the outcome variables and somatic symptoms as the explanatory variables. Age, two basic depressive and three PTSD symptoms that were used for the screening of major depressive disorder and PTSD, respectively, were controlled when suicidal risk was used as an outcome variable. We also evaluated the associations between psychiatric symptoms of major depressive disorder, PTSD as well as dysthymia, and suicidal risk by logistic regression analyses.

3. Results

Several somatic symptoms were shown to be significantly associated with suicidal ideation or suicidal risk in the three populations, even after adjusting for depression and/or PTSD. Among outpatients, women showed a wider variety of somatic symptoms related to depression than men. Some differences among the outpatients were observed regarding somatic symptoms associated with depression between men and women.

In community dwellers and male police officers, the somatic symptoms significantly associated with the suicidal risk were limited. On the other hand, a variety of symptoms were significantly associated with depression in those populations.

3.1 Somatic symptoms associated with suicidal ideation in outpatients visiting a psychosomatic clinic

Table 1 shows somatic symptoms significantly associated with suicidal ideation among the outpatients separately for men and women. Women showed a somewhat wider variety of somatic symptoms compared to men. There were no common symptoms significantly associated with suicidal ideation in either men or women.

3.2 Somatic symptoms associated with suicidal risk in community dwellers and male police officers

Those somatic symptoms associated with suicidal risk in community dwellers and male police officers were shown in Table 2.

Men (n=177)	Women (n=216)
Somatic symptoms (OR, 95% CI)[a]	
General fatigue (4.1, 1.0-16.3)	Lack of persistence (3.5, 1.6-8.1)
Difficulty breathing (9.3, 2.6-33.4)	Chest pain (2.6, 1.0-6.7)
	Edema (4.5, 1.8-11.1)
	Tinnitus (3.8, 1.5-9.8)
	Difficulty falling asleep (2.8, 1.3-6.1)
	Frequent dreams (3.3, 1.2-8.9)

[a] Variables are selected using a stepwise method in simultaneously adjusted multivariate analyses. Adjusted for age and total score of a self-rating depression scale.

Table 1. Somatic symptoms significantly associated with suicidal ideation in outpatients visiting a university psychosomatic clinic in Japan (Yoshimasu et al., 2009).

Community dwellers (n=452)[a]	Male police officers (n=1685)
Somatic symptoms (OR, 95% CI)[b]	
Dysesthesia, arthralgia, and swelling (2.7, 1.1-6.5)	Headaches (2.7, 1.2-6.4)
	Feelings of constriction in the throat (2.7, 1.0-6.9)
	Abdominal pain (3.2, 1.3-7.8)

[a] n=183 for men, 269 for women.
[b] Adjusted for sex, age, and basic symptoms of major depressive disorder for community dwellers; age and basic symptoms of major depressive disorder as well as post-traumatic stress disorder for male police officers.

Table 2. Somatic symptoms significantly associated with suicidal risk among community dwellers (Takemura et al., 2011) and male police officers in Japan (Yoshimasu et al., 2011).

Once mental symptoms of depression and/or PTSD were adjusted for, the somatic symptoms significantly associated with suicidal risk in those populations were diminished. Interestingly, pain-related symptoms were significantly associated with an increased suicidal risk in both community dwellers and police officers.

3.3 Somatic symptoms associated with depression in outpatients visiting a psychosomatic clinic

Table 3 shows somatic symptoms significantly associated with depression in outpatients separately for men and women. Sleep disturbance, loss of appetite, and general fatigue were common symptoms associated with depression in both men and women. Diarrhea, excessive sweating, and weight loss were significantly associated with an increased risk of depression only in men, while headaches and dysesthesia were associated with such a risk only in women.

Men (*n*=259)	Women (*n*=471)
Somatic symptoms (OR, 95% CI)[a]	
Sleep disturbance (2.2, 1.2-4.2)	Sleep disturbance (3.9, 2.4-6.2)
Loss of appetite (5.8, 1.8-18.3)	Loss of appetite (5.4, 2.3-12.5)
General fatigue (5.0, 1.4-18.5)	General fatigue (2.9, 1.3-6.7)
Diarrhea (5.0, 1.7-14.5)	Headaches (3.4, 1.7-6.6)
Excessive sweating (10.5, 1.9-57.6)	Dysthesia (3.7, 1.1-12.7)
Weight loss (5.1, 1.0-25.0)	

[a] Variables are selected using a stepwise method in simultaneously adjusted multivariate analyses. Age was also adjusted for.

Table 3. Somatic symptoms significantly associated with depression in outpatients visiting a university psychosomatic clinic in Japan (Sugahara et al., 2004).

3.4 Somatic symptoms associated with depression in community dwellers and male police officers

Somatic symptoms significantly associated with depression in community dwellers and male police officers are shown in Table 4. A variety of somatic symptoms were significantly associated with depression in those samples. Fatigue and pain-related symptoms, such as headaches, abdominal pain, and chest pain, were commonly associated with depression in those populations. In contrast to the symptoms related to suicidal risk, police officers showed somewhat fewer somatic symptoms associated with depression.

Community dwellers (*n*=452)[a]	Male police officers (*n*=1718)
Somatic symptoms (OR, 95% CI)[b]	
General fatigue (5.5, 1.4-21.3)	Easily fatigued (5.6, 2.1-15.3)
Insomnia (4.9, 1.1-21.7)	Headaches (9.0, 3.2-25.2)
Abdominal pain (20.3, 3.2-129)	Chest pain during exercise (6.4, 1.4-29.4)
Heavy stomach (6.3, 1.2-34.3)	Dizziness (8.4, 2.6-27.4)
Nausea/heartburn (5.8, 1.1-31.4)	Feeling of constriction in throat (4.4, 1.2-15.8)
Headache/heavy headedness/ eye strain/shoulder stiffness (5.2, 1.2-21.7)	Abdominal pain (4.1, 1.2-15.0)
Vertigo/dizziness (45.6, 9.6-217)	
Palpitation/shortness of breath (12.5, 2.8-56.5)	
Pain or constriction in the chest (11.0, 1.1-112)	
Dysesthesia, arthralgia, and swelling (6.8, 1.6-28.1)	

[a] *n*=183 for men, 269 for women.
[b] Adjusted for sex and age for community dwellers and age for male police officers.

Table 4. Somatic symptoms significantly associated with depression in community dwellers (Takemura et al., 2011) and male police officers in Japan (unpublished data).

3.5 Gender differences in psychiatric symptoms related to suicidal ideation in outpatients with major depressive disorder

Psychiatric symptoms significantly associated with suicidal ideation in patients with major depressive disorder were shown separately for men and women in Table 5. Low perceived support from workplace or family members and depersonalization were significantly associated with suicidal ideation in men, whereas derealization, depressive mood, and state anxiety assessed by STAI were significantly associated with such ideation in women.

Men (*n*=66)	Women (*n*=133)
Psychiatric symptoms (OR, 95% CI)[a]	
Low perceived social/family support (13.2, 2.1-84.6)	Derealization (3.2, 1.4-7.4)
Depersonalization (5.1, 1.5-17.6)	Depressive mood (4.9, 1.8-13.1)
	State anxiety (4.2, 1.2-14.4)

[a] Variables are selected using a stepwise method in simultaneously adjusted multivariate analyses. Age was also adjusted for.

Table 5. Psychiatric symptoms significantly associated with suicidal ideation in outpatients with a major depressive disorder (Yoshimasu et al., 2006).

3.6 Associations between depressive symptoms and suicidal risk

Associations between depression-related psychiatric symptoms and suicidal risk were assessed among the community dwellers and male police officers using diagnostic criteria of DSM-IV described in the M.I.N.I. "Depressive mood" and "loss of interest" which are DSM-IV screening symptoms for major depressive disorder, were strongly associated with suicidal ideation in both community dwellers and male police officers (data not shown). Those two symptoms identified as the three most distressing ones were also found to be significantly or marginally significantly associated with depression in both male and female outpatients (data not shown). In addition, "chronic depressive mood," which is a DSM-IV screening symptom for dysthymia, was significantly associated with suicidal risk in male police officers (OR 49.2, 95% CI 9.5-254.9).

4. Discussion

4.1 Somatic symptoms associated with suicidal ideation or suicidal risk

In patients visiting a psychosomatic clinic, women showed a broader range of somatic symptoms related to suicidal ideation compared to men. Two of those were symptoms related to sleep disturbance. Together with other symptoms, such as feelings of edema and chest pain, which seemed to be related to autonomic ataxia, women are prone to express symptoms related to somatoform autonomic dysfunction defined by the International Statistical Classification of Diseases and Related Health Problems, 10th Revision (ICD-10), as possible signs of suicide.

Pain-related symptoms in clinical, community, and occupational samples, such as chest pain, headaches, abdominal pain, and arthralgia, were significantly associated with suicidal ideation/risk even after adjusting for depression. These results suggest that pain-related symptoms should be regarded as critical signs of suicide even when the subjects show no obvious psychiatric symptoms of depression.

In a comparison between community dwellers and male police officers, it should be noted that community dwellers showed only one symptom associated with suicidal risk, while police officers had three somatic symptoms even after adjusting for the effect of depression. In contrast, more somatic symptoms associated with depression were observed in community dwellers compared to male police officers. These findings suggest a hypothesis that the associations between somatic symptoms and suicidal risk in community dwellers can be explained by the effects of depression. In other words, the somatic symptoms associated with suicidal risk in those people might be due to depression intervening between those somatic symptoms and suicidal risk.

On the other hand, three somatic symptoms remained significant after adjusting for depression in male police officers. In general, police officers are more likely to harbor a prejudice or stigma against mental disorders compared to the general population (Royle, Keenan, & Farrell, 2009) since they hate to be regarded as mentally as well as physically weak. Consequently, even if they had some suicide-related mental symptoms (i.e., depressive symptoms), such symptoms might be replaced by somatic ones that the police officers did not want to conceal. In another case, those somatic symptoms might be connected with "masked depression," which has somatic rather than mental symptoms, and which is also considered to be associated with suicidal risk (Sarai., 1994). As mentioned above, since police officers tend to suppress their mental problems, it is very likely that they may express somatic symptoms related to masked depression rather than mental symptoms, even when they are actually suffering from depression. In any case, it is clear that the relationship between somatic symptoms and suicidal risk in police officers could not be adequately explained by mental symptoms such as depression or post-traumatic stress disorder when compared to community dwellers.

4.2 Somatic symptoms associated with depression

Since outpatients had exhibited very miscellaneous somatic symptoms associated with depression, those symptoms were narrowed down by a stepwise method. Problems related to sleep, appetite, and general conditions were associated with depression in both men and women. Diarrhea and weight loss were significantly associated with depression in male patients, which suggests that they are vulnerable to such stressors related to the digestive organs. On the other hand, headaches remained significantly associated with depression in women. Thus, such symptoms might be useful for detecting depression at the primary care stage.

Community dwellers and male police officers often exhibited similar somatic symptoms associated with depression, many of which were pain-related, such as headaches, chest pain, and abdominal pain. Although we did not always confirm organic disorders among such people, pain-causing conditions such as stomach or duodenal ulcers were reported to be significantly associated with depression, and functional disability could, to a great extent, explain those associations (Stegmann et al., 2010). Such functional disability related to physical pain might pose a burden to mental conditions among those people.

In addition, it should be noted that vertigo or dizziness were significantly associated with depression in both community dwellers and police officers, while 95% confidence intervals were extremely wide due to the small number of subjects showing such symptoms. Mild vertigo or dizziness itself is generally not troublesome enough to lead people to consult a physician as long as those symptoms are not accompanied by other painful symptoms. Thus, the strong associations between those symptoms and depression observed in non-clinical populations have a very significant meaning. Namely, people with dizziness suffering from depression may not consult even a general practitioner, to say nothing of a psychiatrist, resulting in delaying the detection of their depression.

From another point of view, we could integrate all of those depression-related somatic symptoms into a somatoform autonomic dysfunction on the assumption that such people did not suffer from any organic diseases. In Japan, a diagnosis of "autonomic ataxia," which is a similar concept with somatoform autonomic dysfunction, is often used instead of depression even when the patients actually suffer from depression, since patients dislike such a diagnosis to be recorded on their medical certificates (Ohkuma, 2008). The present findings, however, suggest that using a diagnosis of somatoform autonomic dysfunction instead of depression is not necessarily wrong. Community dwellers and male police officers with a major depressive disorder were found to have somatic symptoms related to the somatoform autonomic dysfunction described in ICD-10.

As mentioned in the Introduction, Japanese patients at the primary care stage were shown to report fewer depression-related mental and somatic symptoms (e.g., sleep or appetite problems) compared to those in other countries (Simon et al., 1999). Therefore, significant associations between some somatic symptoms and a diagnosis of depression observed in the current samples are considered to have a significant meaning for detecting mood disorders at an early stage.

4.3 Gender differences in psychiatric symptoms related to suicidal ideation in outpatients with major depressive disorder

By using KMI that includes a wide variety of somatic and psychiatric symptoms, we can evaluate the gender differences in psychiatric symptoms associated with suicidal ideation in patients with major depression. Symptoms related to emotions and one's identity were significantly associated with suicidal ideation in women, while depersonalization was associated with suicidal ideation in men. It should be noted that perceived low support from family or colleagues was significantly associated with suicidal ideation in men but not in women. This means that men with depression are more likely than women to seek help or support from those close to them. It is important to provide appropriate support for men with depression to prevent suicidal inclinations.

5. Limitations

Three major limitations could be considered in the current sequence of studies. One is representative of the present samples. The clinical sample of outpatients could represent patients of Japanese psychosomatic clinics in other university hospitals, since the university hospital enrolled in the current study was one of only several such hospitals that had a psychosomatic clinic in Japan at the time of the survey. However, it is uncertain whether those outpatients were representative of the psychosomatic clinics of general practitioners. We confirmed, however, that approximately half of them were walk-in patients, although new patients were to be referrals at the university hospital.

Any representatives of the other two samples were also somewhat problematic. Community dwellers were limited to those aged 40 or older and self-employed in a rural area, resulting in a dearth of information regarding young salaried workers. Furthermore, in such a small community, people are generally reluctant to speak about their mental problems since they worry about what their neighbors will think of them. This tendency is considered to be conspicuous among elderly people who have scant knowledge about mental disorders. Thus, our sample of the community dwellers is not always considered to represent the community population as a whole.

Police officers are also unique in the kind of jobs described above. Police organization is characterized by its hierarchical ethical system, and the police officers are constrained by strict rules and regulations. Topics regarding suicide are taboo within the organization. It is thought outrageous that police officers should commit suicide using their handgun, since such an action would lead to a loss of prestige for the police organization. In such an atmosphere, it is difficult for them to speak frankly about their mental conflict to one another. Given these constraints, it is unlikely that police officers are typical of other male salaried workers.

Another major limitation is the clinical usefulness of somatic symptoms that were found to be associated with depression or suicidal risk. That is, such symptoms are not always considered to show sufficient sensitivity and specificity for detecting depression or suicidal risk. In other words, those somatic symptoms are widespread, and could be encountered in ordinary lives and a variety of physical disorders. However, clinicians do not make diagnoses of depression nor predict suicide based only on somatic symptoms. They always integrate various kinds of information such as clinical history, life history, lifestyle, and family history. During the process of diagnosing patients, some somatic symptoms may be seen to play an important role for clinicians in predicting depression or suicidal risk, especially when patients exhibit no organic disorders or obvious mood problems. In our series of studies, several symptoms were shown to be strongly associated with depression or suicidal risk (odds ratio of five or more). Such symptoms could provide clues for detecting depression or suicidal risk.

Last, although we assumed that our samples did not include severe organic disorders that might accompany painful somatic symptoms, we did not confirm this, especially in community dwellers and male police officers. Therefore, some of those somatic symptoms associated with mental adversity may directly stem from the organic disorders that produce those symptoms. This might be plausible in the community dwellers given relatively old age of them (mean age = 62.5). Indeed, somatic illness was reported to be one of the main predictors of depressive disorders in elderly people (Djernes, 2006). However, we think that our sample of community dwellers were generally healthy as the purpose of the annual health check-up was secondary prevention of the diseases, and those who had severe somatic disorders are considered not to have such a health check-up since they are usually followed by physicians in hospitals. Even if some of the community dwellers had severe organic disorders, we consider that the association of those symptoms with depression or suicidal risk have important clinical implications for the early detection of a mental crisis.

6. Conclusions

It is quite natural to conclude that depressive psychiatric symptoms are associated with suicidal risk, a finding also observed in our samples. However, as shown in the findings from male police officers, some somatic symptoms were associated with suicidal risk independently of depressive symptoms. This indicates that people without any depressive signs might possibly commit suicide. In that sense, such somatic symptoms associated with suicidal risk may provide very useful markers for suicide prevention, which is why we described issues regarding suicidal risk ahead of depression.

In summary, there are two models regarding the relationships among somatic symptoms, depression, and suicidal risk (Fig. 1). Model A can be applied to community dwellers in which the associations between suicidal risk and somatic symptoms are explained by interactive effects between those symptoms. On the other hand, model B shows a typical pattern of police organization in which regulations rigidly constrain the police officers. In this model, at first glance, some somatic symptoms are directly connected with suicide independently of depression. However, there may also be latent effects of depression in this pattern (though they can not be necessarily confirmed).

Model A

(Somatic Symptoms ⇔ Depression) ⇒ Suicide

Model B

Somatic Symptoms ⇔ (Latent Depression) ⇒ Suicide

Fig. 1. Relationships among somatic symptoms, depression, and suicide: two models are assumed.

In model A, it would be appropriate to treat them together with somatic symptoms and depression for suicide prevention. In model B, it is difficult to readily understand the relation between somatic symptoms and suicidal risk. However, some of those somatic symptoms may provide critical signs of a potential suicidal risk, especially among those who are reluctant to frankly express their mental problems due to some kind of stigma or prejudice. Clinicians should pay close attention to such symptoms; for example, those shown in our police officers' sample, to prevent suicide among people at the primary care or pre-clinical stages.

7. Acknowledgments

The authors wish to acknowledge the valuable contribution of the medical staff of the Department of Psychosomatic Medicine, Graduate School of Medical Sciences, Kyushu

University, for providing data and offering appropriate advice. A part of this work was supported by JSPS KAKENHI (Grant No. 19590645).

8. References

American Psychiatric Association. (1994). *Diagnostic and Statistical Manual of Mental Disorders Fourth Edition (Japanese edition, 1996),* Translated from English by S. Takahashi; Y. Ono. & T. Someya. Igaku Shoin, ISBN 4-260-11804-8, Tokyo, Japan

Djernes, J.K. (2006). Prevalence and predictors of depression in populations of elderly: a review. *Acta Psychiatrica Scandinavica,* Vol. 113, pp. 372-387, ISSN 1600-0447

Griffiths, K.M.; Nakane, Y. ; Christensen, H. ; Yoshioka, K. ; Jorm, A.F. & Nakane, H. (2006). Stigma in response to mental disorders : a comparison of Australia and Japan. *BMC Psychiatry,* Vol.6, pp. 21, ISSN 1471-244X

Haug, T.T.; Mykletun, A. & Dahl, A.A. (2004). The association between anxiety, depression, and somatic symptoms in a large population : The HUNT-II Study. *Psychosomatic Medicine,* Vol.66, pp. 845-851, ISSN 0033-3174

Kawakami, N.; Takeshima, T.; Ono, Y.; Uda, H.; Nakane, Y.; Nakamura, Y.; Tachimori, H.; Iwata, N.; Nakane, H.; Watanabe, M.; Naganuma, Y.; Furukawa, T.A.; Hata, Y.; Kobayashi, M.; Miyake, Y. & Kikkawa, T. (2008). Twelve-month prevalence, severity, and treatment of common mental disorders in communities in Japan: The World Mental Health Japan 2002-2004 Survey. In: *The WHO World Mental Health Surveys,* R.C. Kessler & T.B. Üstün (Ed.), pp. 474-485, Cambridge University Press, ISBN 978-0-521-88419-8, New York, USA

Matsuoka, Y. (1990). KMI, In: *Psychological Test for Psychosomatic Medicine,* T. Kawano, H. Suematsu, & R. Shinzato, (Ed.), pp. 45-49, Asakura Shoten, ISBN 4-254-32137-6, Tokyo, Japan (in Japanese)

Royle, L.; Keenan, P. & Farrell, D. (2009). Issues of stigma for first responders accessing support for post traumatic stress. *International Journal of Emergency Mental Health,* Vol. 11, pp. 79-85, ISSN 1522-4821

Sarai, K. (1994). Masked depression. *Nippon Rinsho,* Vol. 52, pp. 1273-1278, ISSN 0047-1852 (in Japanese)

Sheehan, D.V. & Lecrubier, Y. (2003). *Mini-International Neuropsychiatric Interview, Japanese edition,* Translated from English by T. Otsubo; H. Miyaoka. & K. Kamijima. Seiwa Shoten, ISBN 4-7911-0410-2, Tokyo, Japan

Silverstein, B. (1999). Gender differences in the prevalence of clinical depression: the role played by depression associated with somatic symptoms. *American Journal of Psychiatry,* Vol.156, pp. 480-482, ISSN 0002-953X

Silverstein, B. (2002). Gender differences in the prevalence of somatic vs. pure depression: a replication. *American Journal of Psychiatry,* Vol.159, pp. 1051-1052, ISSN 0002-953X

Simon, G.E.; VonKorff, M.; Piccinelli, M.; Fullerton, C. & Ormel, J. (1999). An international study of the relation between somatic symptoms and depression. *The New England Journal of Medicine,* Vol.341, pp. 1329-1335, ISSN 0028-4793

Smolderen, K.G.; Spertus, J.A.; Reid, K.J.; Buchanan, D.M.; Krumholz, H. M.; Denollet, J.; Vaccarino, V. & Chan, P.S. (2009). The association of cognitive and somatic depressive symptoms with depression recognition and outcomes after myocardial infarction. *Circulation: Cardiovascular Quality and Outcomes,* Vol.2, pp. 328-337, ISSN 1941-7705

Spiegel, B.; Schoenfeld, P. & Naliboff, B. (2007). Systematic review : The prevalence of suicidal behavior in patients with chronic abdominal pain and irritable bowel syndrome. *Alimentary Pharmacology & Therapeutics,* Vol.26, pp. 183-193, ISSN 1365-2036

Spielberger, C.D. (1972). Anxiety as an emotional state, In: *Anxiety-Current trends and theory,* C.D. Spielberger, (Ed.), pp. 3-20, Academic Press, New York, USA

Stegmann, M.E.; Ormel, J.; de Graaf, R.; Haro, J.M.; de Girilamo, G.; Demyttenaere, K.; Kovess, V.; Matschinger, H.; Vilagut, G.; Alonso, J.; Burger, H. & ESEMED/MHEDEA Investigators. (2010). Functional disability as an explanation of the associations between chronic conditions and 12-month major depressive episode. *Journal of Affective Disorders,* Vol.124, pp. 38-44, ISSN 0165-0327

Sugahara, H.; Akamine, M.; Kondo, T.; Fujisawa, K.; Yoshimasu, K.; Tokunaga, S. & Kubo, C. (2004). Somatic symptoms most often associated with depression in an urban hospital medical setting in Japan. *Psychiatry Research,* Vol.128, pp. 305-311, ISSN 0165-1781

Takemura, S.; Yoshimasu, K.; Fukumoto, J.; Yamamoto, H. & Miyashita, K. (2011). Subjective symptoms related to depression and suicidal risk in a Japanese community : a cross-sectional study. *Open Journal of Psychiatry,* Vol.1, pp. 40-48, ISSN 2161-7325

Wang, S.J.; Fuh, J.L.; Juang, K.D. & Lu, S.R. (2009). Migraine and suicidal ideation in adolescents aged 13-15. *Neurology,* Vol. 72, pp. 1146-1152, ISSN 0028-3878

Wang, S.J.; Juang, K.D.; Fuh, J.L. & Lu, S.R. (2007). Psychiatric comorbidity and suicide risk in adolescents with chronic daily headache. *Neurology,* Vol. 68, pp. 1468-1473, ISSN 0028-3878

Yoon, J.C.; Crane, P.K.; Ciechanowski, P.S.; Harrington, R.D.; Kitahata, M.M. & Crane, H.M. (2011). Somatic symptoms and the association between hepatitis C infection and depression in HIV-infected patients. *AIDS Care,* (Epub ahead of print), ISSN 1360-0451

Yoshimasu, K.; Fukumoto, J.; Takemura, S.; Shiozaki, M.; Yamamoto, H. & Miyashita, K. (2011). Subjective symptoms related to suicidal risk in Japanese male police officers. *Suicidology Online,* Vol.2, pp. 38-47, ISSN 2078-5488

Yoshimasu, K.; Kondo, T.; Tokunaga, S.; Kanemitsu, Y.; Sugahara, H.; Akamine, M.; Fujisawa, K.; Miyashita, K. & Kubo, C. (2009). Mental and somatic symptoms related to suicidal ideation in patients visiting a psychosomatic clinic in Japan. *International Journal of General Medicine,* Vol.2, pp. 163-170, ISSN 1178-7074

Yoshimasu, K.; Sugahara, H.; Tokunaga, S.; Akamine, M.; Kondo, T.; Fujisawa, K.; Miyashita, K. & Kubo, C. (2006). Gender differences in psychiatric symptoms related to suicidal ideation in Japanese patients with depression. *Psychiatry and Clinical Neurosciences,* Vol.60, pp. 563-569, ISSN 1323-1316

Zung, W.W.; Richards, C.B. & Short, M.J. (1965). Self-rating depression scale in an outpatient clinic. Further validation of the SDS. *Archives of General Psychiatry,* Vol.13, pp. 508-515, ISSN 0003-990x

3

Chronobiological Aspects of Mood Disorders

Rosa Levandovski[1], Ana Harb[1,2],
Fabiana Bernardi[1] and Maria Paz Loayza Hidalgo[1,*]
[1]*Universidade Federal do Rio Grande do Sul (UFRGS),*
Laboratory of Chronobiology of HCPA
[2]*Universidade do Vale do Rio dos Sinos (UNISINOS),*
Health Department, São Leopoldo
Brazil

1. Introduction

Mood disorders, especially those related to depression, are among the most prevalent mental impairments. Therefore, understanding their etiology is important for prevention and improving therapy. Sub-clinical depressive symptoms, which predict the onset of mood and anxiety disorders (Gentil et al., 2007), are correlated with slow information processing, poor memory functioning (Simons et al., 2009) and cardiovascular dysfunction (Taillard et al., 1993; Wassertheil-Smoller et al., 2004).

Depression is a common disorder that affects more than 121 million people worldwide. A recent epidemiological study reported a worldwide prevalence of approximately 10% to 15% (Lepine & Briley, 2011) The prevalence of depression appears to be independent of culture. The prevalence of depression is 8.2% in Canada, 8.7% in the United States, 8.6% in Europe and 10% in Southern Brazil (Mari & Williams, 1986; Patten, 2007), and varies from 3% to 9% in Japan (Lepine & Briley, 2011). According to the World Health Organization, major depressive disorder is among the leading cause of disability among Americans aged 15 to 44 years old (Kessler et al., 2005). This condition is the second leading cause of disabling illness in the world, second only to ischemic heart disease (Sullivan et al., 2000).

One of the primary findings of depression epidemiology is the high female to male sex ratio, especially through the reproductive years. French data from the European Study of the Epidemiology of Mental Disorders (ESEMeD) support these results: Approximately twice as many women suffer from depression as men, and this trend is found all over the world. However, some exceptions were found for specific ages; for example, there was approximate parity between the sexes among 18- to 24-year-olds. The authors of that study suggested that the reduced depression rate in women was the result of increased opportunities for women in education, employment, birth control and other factors associated with rising gender equality (Lepine & Briley, 2011).

* Corresponding Author

Depressive disorders are more common in individuals with chronic diseases such as obesity, cardiovascular disease, diabetes, asthma, arthritis, and cancer. Unhealthy behaviors (e.g., smoking, physical inactivity, and binge drinking) are also included in these risk factors. All depressive disorder studies emphasize the significance of the condition's associated mortality and morbidity. The risk of suicide in individuals with depression is more than 20 times greater than for the general population (Lepine & Briley, 2011).

Some researchers have shown that the consequences of depression are a risk factor for cardiovascular death. In addition, depressive symptoms are connected with an appreciably higher risk of mortality, including due to cardiovascular death and stroke (Lepine & Briley, 2011). Beyond mortality, the functional impairment and disability associated with depression is also important to note. Depression increases absenteeism and decreases productivity, which results in lowered profits or greater unemployment. Both absenteeism and presenteeism (present but working at a suboptimal level) cause losses of $36.6 billion per year in the US. The World Health Organization projects that unipolar major depression will be the leading cause of disease burden worldwide by the year 2030 (Lepine & Briley, 2011).

The genetic effects on depression were previously evaluated in a study of 4,639 adult twins from Australia. In this survey, genetic effects accounted for 29% of the variance in seasonality (Madden et al., 1996). The daily modulations of pathological symptoms as well as the relationship between the circadian clock and health have been extensively documented (Foster & Wuff, 2005; Scheer et al., 2009; Ramsey & Bass, 2009; Wulff et al., 2010; Selvi et al., 2010). For example, patients with depressive episodes show variations in daily mood, abnormal patterns in sleep-wake behavior, cortisol secretion, and adrenocorticotropic hormone (ACTH) levels as well as a daily modulation of other endocrine-metabolic parameters (Soria & Urretavizcaya, 2009). Additionally, clock-gene variants have been associated with abnormal sleep time and mood disorders (Wulff et al., 2010), which indicate that they share pathways. Based on the importance of depressive disorder and the recently contribution of the studies of rhythms, this chapter review the interphase between chronobiology and depressive disorder and point to some clinical evidence.

2. Concept of chronobiology

The survival of a species depends upon its ability to adapt to its environment. One of most important adaptations is the ability to anticipate events. This adaptation allows a living being to develop physiological and behavioral responses to predict the changes that occur in nature. The light and dark cycle due to the rotation of the earth is the most reliable and salient of such changes. The ratio of light to dark is dependent upon season and geographic location (i.e., latitude and longitude). Therefore, an internal circadian system able to measure the time or environmental cues is an important response to these natural processes. The endogenous rhythm in the body that is in line with environmental signals (i.e., the circadian system) is such an adaptation (Aschoff *et al.*, 1971).

In normal environmental conditions, the various rhythm functions in the body maintain a stable relationship among these phases. This relationship demonstrates a temporal synchronization that features an internal temporal order (Aschoff, 1960). The term "biological clock" is used in chronobiology to designate a pacemaker and refers to the entire structure which operation oscillates cyclically over time. Through nervous and humoral pathways, it is able to transmit this oscillation throughout the body as well as impose its rhythm on other structures. The biological clock cannot directly trace most biological rhythms; however, it can trace the first rhythm of a successive chain or cascade of events. The biological clock exerts control at several physiological levels, from gene expression to complex behavior. In constant conditions, the biological clock acts freely with an endogenous period of approximately 24 hours (Aschoff *et al.*, 1971).

Ruppert and Weaver (Reppert *et al.*, 2001) describe the circadian system in mammals as a hierarchy of multiple circadian oscillators dispersed by the organism. The "command center" of this timing system is restrained in the neurons of the suprachiasmatic nuclei (SCN). Most of these neurons may work as "cell clocks" or "circadian oscillators". The SCN receive photic information through the retinohypothalamic tract, synchronizing for 24 hours; in turn, this information coordinates the actions of oscillators present in other brain areas and "peripheral oscillators" such as the lung, kidney, and liver. These synchronized oscillators regulate the local physiological rhythms that constitute multi-oscillator entities (28). In humans, the circadian rhythms can be synchronized by photic and non-photic cues (Aschoff, 1960).

Aschoff designated the external events that adjust circadian variables *"Zeitgebers"* (a German word meaning (*Give time*), Halberg and collaborators called them "synchronizers", and Pittendrigh dubbed them "entrainment agents" (Reinberg *et al.*, 2001). Pittendrigh conducted the first systematic experiment on entrainment in the 1960s. In constant darkness with brief exposures to light pulses at different times throughout the day, Pittendrigh observed that exposure to light promoted phase changes dependent on the time of the stimulus. Later, other experiments found the same effect in humans (Beersma *& Dann, 1993*).

The SCN are synchronized to outside rhythms by light as well as social and behavioral factors. In humans, the *zeitgebers* associated with social cycles have a special relevance (Aschoff *et al.*, 1971). Daily variations in the intensity of ambient noise; the timing of meals, home and business routines; and social interactions are examples of the rhythms that synchronize zeitgebers in humans (Aschoff *et al., 1971*).

When these biological rhythms are no longer in phase with external cycles, external desynchronizations occur, such as those observed during the performance of night work or rapid time zone crossing. If the situation persists, external desynchronizations may occur in some people, resulting in a dissociation or desynchronization of the internal phase relationships between biological rhythms.

The ability to predict seasonal changes and other circadian rhythms is based on autonomous regulation of cell transcription and translation via feedback loops of cell clocks, proteins and their bases. The identification of circadian rhythm-based clock genes in *Drosophila melanogaster* and *Neurospora crassa* was a fundamental step in the search for homologues in other animal species. Three clock genes have been found in mammals: 1) CLOCK (*Circadian Locomotor Output Cycle Kaput*) and BMAL1 (Brain and Muscle Arnt-Like; also known as MOP3) are involved in the timing system; 2) Per1 (*Period*), Per2 and Per3 are rhythmically expressed in the SCN (Per1 and Per2 induce the response to pulses of light in the SCN, whereas Per3 has an oscillatory pattern of expression independent of light) (Ohdo, 2007); and 3) Cry1 (*cryptochrome*) and Cry2 are members of the blue-light receptor family in plants. These genes were first classified as potential candidates for circadian photoreceptors; however, additional analyses showed that these genes have an independent light effect in the circadian system of mammals (Ohdo, 2007).

These clock genes found in other transcription factors such as BMAL1, CLOCK, CRY1 and Cry2. All of these proteins, including Per1, Per2, and Per3 establish oscillation periods of approximately 24 hours. The CLOCK-BMAL1 heterodimer is a linked DNA sequence that stimulates the transcription of genes that regulate behavior and metabolism and maintain physiological states. Moreover, the CLOCK-BMAL1 heterodimer activates the transcription of the PER and CRY genes. The PER-CRY complex is formed in the cytoplasm and translocated to the nucleus where it regulates the transcriptional activity of the CLOCK-BMAL1 heterodimer. A second feedback loop has also been identified that involves Rev-erbα and the transcription repression of BMAL1. The expression, transcription suppression, and formation of this gene, as well as the heterodimerization of CLOCK and BMAL1, results in a 24-hour rhythmicity (Mignot & Takahashi, 2007). Moreover, BMAL1 is a transcription factor that controls circadian rhythms and contributes to the control of adipogenesis and the lipid metabolism activity of mature adipocytes. The level of BMAL1 mRNA increases during adipogenesis. BMAL1 is related to the induction and promotion of the activation of many factors involved in 3T3-L1 adipocyte lipogenesis. Maximizing the expression of BMAL1 increases the synthesis activity of lipids involved in the regulation, differentiation and maturation of adipocytes (Ohdo, 2007).

Animal studies have demonstrated that variations in clock genes can cause abnormalities in the regulation of circadian rhythms. Investigations in humans have observed that this system regulates many physiological processes such as temperature, feeding behavior, sleep disorders, hormone secretion, metabolism and drug use, glucose homeostasis, tumor formation and cell cycle progression (Takahashi et al., 2008; Wassertheil-Smoller et al., 2004). One hypothesis that may help to understand the abnormalities in the regulation of circadian rhythms suggests that changes in the length of endogenous period can lead to changes in the phase angle of light synchronization. Circadian phenotypes similar to those found in animals are also seen in humans, which allows scientific research to examine the relationship between changes in clock genes and circadian phenotypes (Harmer et al., 2001).

In 1810, Sprengel observed that depression is typically related to daily circadian rhythmicity such that greater fatigue occurs in the early morning hours versus at night (Lemmer, 2009). This disturbance in the temporal order of the circadian system was associated with genetic vulnerability, gender and age. Mood variations during the day, sleep disorders, and changes in the circadian rhythm of cortisol, ACTH, melatonin and other endocrine and metabolic parameters were observed in patients with depressive episodes (Soria & Urretavizcaya, 2009). Individual chronobiology characteristics and preferences related to allocating activities during the day were also related to mood disorders. Using the Horne and Ostberg Morningness-Eveningness Questionnaire, studies in our laboratory discovered an association between evening-type and depressive symptoms in healthy people (Bernardi et al., 2009; Hidalgo et al., 2009). Bipolar depression was also associated with evening preferences (Giglio *et al., 2010*).

Bipolar disorder (classically known as manic depression) presents with fluctuations between manic and depressive episodes. The oscillations observed in many patients occur in regular cycles and can last from several days to several years. This discovery has led some researchers to attribute the origin of the depressive phase of this syndrome to variation in the photoperiod, thus contributing to the explanation that this disease has a natural rhythm. Another example of rhythmic manifestation can be seen in seasonal depression, which Rosenthal (Rosenthal *et al., 1984*) describes as an infradian circannual rhythmicity disease. Specifically, the symptoms begin in the fall, reach their maximum expression in winter and progressively decrease with the arrival of spring.

Seasonal depression is a circadian rhythm disorder caused by the desynchronization between the light/dark cycle and the human biological clock during seasons with shorter photoperiods (Rosenthal et al., 1984). During these periods, the production of melatonin increases. This increase is related to seasonal depression because photic treatment, which occurs with exposure to light in the morning, inhibits the production of melatonin, thereby reducing the symptoms of depression (Lewy et al., 1987; Wirz-Justice et al., 2005).

From a biological point of view, the temporal character of depression can be observed by changes in the profiles of several circadian rhythms during depression, such as changes in body temperature rhythm and sleep architecture. Some studies have observed changes in the circadian secretion of some hormonal such as cortisol, melatonin, prolactin and growth hormones (Soria & Urretavizcaya, 2009). However, there is no unanimity among these studies. The variety of physiological and clinical phenomena present in mood disorders justifies this lack of agreement. The diversity of experimental findings could be a result of the various types of depression studied and the inclusion of patients with undifferentiated depression who have diagnoses that vary according to the different classification criteria used by the researcher. Other aspects that could influence the results are related to different treatment scheme, disease duration, the duration of changes in the photoperiod and the climate of the study sites, or other differences on the methodology such as design.

3. Circadian rhythms and mood disorders

Biological rhythms can affect mood states; conversely, they may be affected by them and thus contribute to the manifestation of depressive symptoms. Additionally, symptoms may occur at different frequencies: daily, monthly, seasonally or annually. These data imply that there is an association between psychopathology and the pre-existing rhythmic processes in humans. In contrast, early studies in individuals with depression did not find abnormal circadian rhythms (Soria & Urretavizcaya, 2009). As alteration on appetite, motility and memory are important functions that are manifest on depression we describe the relation between this brain capacity and circadian rhythm.

3.1 Feeding and circadian rhythms

Mood disorders are associated with changes in appetite including decreases or increases as well as cravings for carbohydrates. Symptoms of seasonal affective disorder (Sacharczuk et al., 2009) include increased sleep and the need for sleep, increased appetite with carbohydrate craving and weight gain, and decreased energy and increased fatigue (Peiser, 2009). An association between vitamin B12 deficiency and depression has been suggested for older adults. (Kohyama, 2009).

It has been long known that there is a higher frequency of people who are overweight or obese among those with depression compared with the general population. Obesity is associated with several symptoms of major depressive disorder (Zimmerman et al., 2011). In addition to changes in appetite and weight, depression is often associated with endocrine problems such as hypothyroidism and hyperthyroidism, Cushing syndrome, and night eating syndrome (NES), among others. NES is not only associated with depression but also with, anxiety and poor sleep quality (Bernardi, 2009).

The plasma concentrations of metabolic products and the hormones related to energy balance vary according to dynamic conditions. Fasting is the most frequently used approach to study whether a rhythm in glucose metabolism is independent of the (disturbing/masking) effect of feeding because the daily rhythms that depend on feeding should disappear in food-deprived animals (Polakof et al., 2006).

One of the mechanisms by which the processes of some syndromes, such as NES, might be explained includes the desynchronization of temporal order and internal difficulties in synchronizing with environmental cycles, such as delayed eating times. Preliminary studies in humans suggest that polymorphisms or mutations in the CLOCK and BMAL1 genes are associated with obesity and metabolic syndrome (Green et al., 2007; Takahashi et al., 2008). Numerous genes are also involved in the metabolism of fatty acids, cholesterol, and glucose in the liver. These substances are regulated by circadian patterns, which indicate that this clock as a large role in regulating metabolism (Green et al., 2007). Many behaviors, including energy intake, vary in intensity throughout the day. Studies that record daily food intake have shown significant and substantial changes in eating behavior in the natural environment; specifically, there is an increase in food intake and energy consumption (greater than 150%) at night compared with morning. This increase occurs in conjunction with an increase in meal size as the day progresses, producing a modest positive correlation between the time of day and meal size (de Castro, 2004). The volume of the meal as it increases throughout the day can affect the length of the intervals between meals, decreasing the time between them and interfering with satiety. These intervals become shorter as the night progresses for both humans and rats. Clearly, increasing an individual's vulnerability to excessive intake of food during the night reduces satiety (de Castro, 2004). The link between mood disorder and food behavior appears to be connected via hormones that regulate not only feeding but also the neuronal activity in the mesolimbic dopaminergic pathway. In particular, leptin reduces the firing rates of dopaminergic neurons, which indicates that leptin exerts a direct influence on dopamine neurons via leptin receptors (Dibner et al., 2010).

Leptin and ghrelin are hormones that act directly on the timing of food ingestion and regulate appetite. A study that investigated the relationship between leptin levels and ghrelin as well as symptoms of depression, anxiety and stress in women demonstrated a significantly negative relationship between leptin and the severity of depressive symptoms and anxiety but not between leptin and ghrelin (Lawson et al., 2011).

3.2 Activity/rest and circadian rhythm

Studies have shown that amount of physical activity is inversely related to depressive symptoms and weight gain (Kohyama, 2009).

Physical exercise influenced the internal phase relationship of the circadian system: as the minimum temperature rhythm advances phase, mood improved in people after weeks of training in the early morning (Peiser, 2009). Other researchers have found similar results and suggested that treating SAD (Seasonal Affective Disorder) using physical exercise, with or without bright light treatment, can alleviate its symptoms. Physical exercise may work for people with SAD because it can facilitate a circadian phase advance. Nevertheless, neither exercise alone, nor exercise in combination with light therapy, has been reported to positively affect seasonal mood deterioration or its physiological, hormonal, social and psychological symptoms (Peiser, 2009). Numerous recent studies suggest that physical exercise provides an effective and easily accessible treatment for patients who suffer from SAD. Concluding, there is no consensus as to which components of exercise produce beneficial effects on mental health.

Moreover, the activity and rests has been assed by actigraph as a way to improve the diagnose and differentiate among types of depression (See table 1). There were controversial findings on the result, probably related to difference on the samples. Some studies describe depressive patient with higher amplitude than control group and difference on acrophase, other authors describe the opposite (Table 1).

Study	objective	Nº pacientes	Using	Groups	Conclusion
Circadian activity rhythm abnormalities in ill and recovered bipolar I disorder patients Bipolar Disord. 2008 Mar;10(2):256-65	compared circadian activity rhythms in **BPD I** patients during ill and **recovered** states to those of normal **controls**	36 adults with bipolar disorder during acute mania or mixed states and to 32 healthy controls	wrist-worn, piezoelectric actigraphic monitoring for 72 h	BPD Recovered Controls	abnormalities of circadian motility patterns found in states of acute manic or mixed BPD illness persisted during full and sustained clinical recovery with euthymia, and remained highly significantly different from normal controls
Patterns of circadian activity rhythms and their relationships with fatigue and anxiety/depression in women treated with breast cancer adjuvant chemotherapy. Support Care Cancer. 2009 Apr 19	examine patterns of circadian activity rhythms and their relationship with fatigue, anxiety/depression, and demographic/medical variables in women receiving breast cancer adjuvant therapy treatments (Tx) at three times within a randomized clinical trial (RCT) designed to improve sleep and modify fatigue.	219 women with stage I-IIIA breast cancer who were randomized 2 days prior to starting chemotherapy to a behavioral therapy sleep intervention or healthy eating control group	wrist actigraphy for 7 days at three times: the start (Tx 1), continuation (Tx 3), and recovery (30 days after last Tx) of chemotherapy	Tx 1 Tx 3 30Days	Circadian activity rhythm parameters at three times in both groups were disrupted compared to healthy adults, but similar to values of cancer patients. Significant changes in mesor, amplitude, peak activity, and 24 h autocorrelation values were found over time in both groups
Phase advance is an actimetric correlate of antidepressant response to sleep deprivation and light therapy in bipolar depression. Chronobiol Int. 2007;24(5):921-37	To known if the clinical effects are paralleled by changes in biological rhythms in sleep deprivation (TSD) and light therapy (LT) in bipolar depression and rapid antidepressant effects The action has been hypothesized to involve the enhancement of all of the monoaminergic systems targeted by antidepressant drugs	39 in patients with Type I Bipolar Disorder	Wrist actigraphy	Responders (day 0) No Responders (day 0) Responders (day 7) No Responders (day 7)	Responders showed an increase in daytime activity, phase-advance of the activity-rest rhythm of 57 min compared to the pre-treatment baseline, and reduced nighttime sleep. Non-responders did not show significant changes in the parameters. Phase advance of the activity-rest rhythm is an actimetric correlate of the antidepressant response to TSD and LT in bipolar

Table 1. Results of actigraphic assessment in psychiatric disorders

3.3 Memory and circadian rhythm

Memory is one of the most important brain functions related not only to survival but also to the notion of time. Borella et al. (2011) examined the effect of testing time on adult differences in resistance to interference, working memory, processing speed, and vocabulary. Their results showed specifically, older adults tested in the afternoon were more susceptible to interference compared with younger adults. In contrast, processing speed, and vocabulary were not modulate by time of the day (Borella et al., 2011). Using the fractional desynchronization technique to induce internal desynchronization in humans, was observed a period of 21 hours in some performance even when the participant's physiological rhythms are apparently synchronized to the 24-hour day (Folkard et al., 1985)

Memory is link with mood disorders. The normal function of the circadian clock is disrupted in jetlag. This syndrome is characterized by signs including sleep disorders, concentration impairment, dysphoria, asthenia, irritability and memory alterations (Mahé & Chevalier, 1995)

Memory performances are not constant; rather, they undergo fluctuations with a periodicity that can be analyzed. Infradian, circadian and ultradian rhythms are involved in memory disorders. In addition, the hypothesis that memory impairment is associated with sleep disturbance has been proposed. These rhythms include the light-dark and social cycles, a circadian hypothalamic oscillator that alternates virtually independently of behavior, and a homeostatic oscillator driven primarily by sleep-wake behavior. Both types of internal oscillators contribute to the variation in many aspects of sleep and wakefulness as well as performance parameters including attention and memory. Neither oscillator can independently predict sleep or performance; however, both constructs critically depend on their phase relationship and amplitude. Thus, the amplitude of the observed circadian variation in sleep and performance depends upon the length of time we have been asleep or awake. Paradoxical sleep represents a preferential period in which certain processes (e.g., the activation of the central nervous system) may facilitate the learning process. The relationship between sleep and memory leads to a discussion of the daily course of cerebral activation and its effects on the variation in our capacity to remember (Leconte, 1989)

Also, inter-individual differences in sleep timing, duration, and morning or evening preferences are associated to changes in circadian processes, sleep homeostatic processes, or both. Some of these inter-individual differences in molecular genetic correlates, including polymorphisms in clock genes, are emerging (Dijk & von Schantz, 2005).

Certain structures are involved in the process of retaining information. Izquierdo has demonstrated that memory and mood are correlated with the amygdala and hippocampus as one such structure (Izquierdo et al., 1997). Ma et al. (2007) reported that constant light exposure for 3 weeks disrupted the 24-h cycle of locomotion activity in an open-field test. Compared with control group, experiment group showed shorter escape latencies but impaired hippocampus-dependent spatial memory that did not affect the visual platform learning task in a Morris water maze (MWM) during the initial phase of spatial learning (Ma et al., 2007).

Fujioka et al. examined whether an environment with constant light affected hippocampal neurogenesis in mice. The number of BrdU-labeled cells (proliferating cells) as well as BrdU and Class-III β-tubulin double-labeled cells (newborn neurons) significantly decreased in the granule cell layer of the L/L group compared with that of the L/D group. Also Fujioka found that the exposure to an L/L treatment for 3 weeks impaired spatial learning task performance; These findings demonstrate that constant light conditions impaired hippocampal neurogenesis and cognitive performance (Fujioka et al., 2011).

4. Experimental evidence of desynchronizations in circadian rhythms and mood disorders

Animal studies have demonstrated that variations in clock genes can cause abnormalities in the regulation of circadian rhythms (Bunney & Bunney, 2000). Studies in humans have shown that this system regulates many physiological processes such as temperature control, feeding behavior, sleep disorders, hormone secretion, metabolism, glucose homeostasis, tumor formation, and cell cycle progression (Takahashi et al., 2008).

Recent findings in the fields of molecular biology and genetics regarding the complex machinery that regulates biological clocks support the role of disrupted circadian rhythms in the pathophysiology of depression. Transgenic mice that overexpress the glycogen synthase-kinase-3β (GSK-3β) gene, a recently recognized central regulator of circadian rhythms and target of the mood stabilizer lithium (Prickaerts et al., 2006), show some depression-like behavior and an increased startle response reminiscent of mania-like behavior (Nestler & Hyman, 2010). Moreover, mice that carry a CLOCK gene mutation exhibit mania-like behaviors such as a decreased need for sleep and motor hyperactivity, which revert with the chronic administration of lithium. In addition, these behaviors are resolved when a functional CLOCK protein is specifically expressed in the ventral tegmental area of these mutant mice (Monteleone et al., 2011).

5. Clinical evidence for desynchronizations in circadian rhythms and mood disorders

The concept of internal circadian phase disturbances in individuals with depression can be explained in the previous signs observed during flights across multiple time zones, known as jet lag, and night-shift workers. These transitory symptoms are similar to those observed during depressive episodes. The problem results from the inability of the body's clock to adapt promptly to abrupt changes in zeitgebers during periods when there is a desynchronization between the internal and external rhythms. A temporal loosening in the coordination of these rhythms, which is generated by the environmental and social systems, results in mood variability after the desynchronization period.

One of the first circadian hypotheses of major depressive disorder suggested that patients with depression have a significantly later circadian sleep phase compared with healthy people (Kripke, 1984; Wehr *et al., 1979*). Several follow-up studies showed that circadian timing is drastically delayed in patients with bipolar disorder (Giglio *et al.*, 2010; Soreca *et al., 2009*), SAD (Elmore *et al., 1993*), and major depressive disorder (Drennan *et al.*, 1991; Soria & Urretavizcaya, 2009). SAD is also associated with internal circadian misalignment (Lewy *et al., 2006*). Although the underlying mechanisms are poorly understood, they are clinically and therapeutically relevant as evidenced by the antidepressant effects of light therapy, sleep deprivation, and rigid sleep schedules (Benedetti *et al.*, 2007; Boivin *et al.*, 1996; Haynes *et al.*, 2005; Lewy *et al.*, 1987; Terman & Terman, 2005; Zeitzer *et al.*, 2000).

In addition, circadian misalignment and sleep disruptions in patients with mood disorders have been linked to abnormal daily patterns of gene expression, hormonal secretion, body temperature, cognitive and behavioral functions (Wulff *et al., 2010*). An additional indication that circadian misalignment contributes to depressive symptoms is the finding that people who live in urban settings are prone to depressive symptoms (Chelminski *et al.*, 1999;

Giannotti *et al.*, 2002; Hidalgo *et al.,2009*). The circadian clock in this population is delayed compared to that in rural populations (Roenneberg *et al., 2007*), which accentuates the discrepancies between social and biological time (Wittmann *et al., 2006*). This internal desynchronization may be a major factor with regard to mood.

New findings on the desynchronization found in night-shift workers and transmeridian travelers are now evident in an epidemiology study of depression (Levandovski et al., 2011). In this study, late chronotypes exhibited higher levels of depressive symptoms compared with intermediate and early chronotypes. Moreover, jetlag was positively correlated with depressive symptoms: participants with more than 2 h of jetlag showed significantly high levels of depression (Levandovski et al, 2011).

6. Conclusion

Until now, the literature stated that the major environmental signals able to synchronize circadian rhythms in animals and humans are the light/dark cycle and the phototransduction performed by retinal ganglion cells (Hirota & Fukada, 2004). Photic signals can also synchronize humans, but social signals are the primary determinant (Mistlberger & Skene, 2004). The relationship between social rhythm and circadian rhythm has been known since Aschoff (Aschoff, 1960). Based on his theory, several researchers have developed non-invasive measurement tools that include social rhythms (Monk et al., 1990). Currently, social rhythms are known to be able to interfere with the synchronization of biological rhythms. These rhythms are characterized by regular social zeitgebers whose exposure is defined as an external synchronic agent inserted into a known social context. The study of social rhythms leads to reflect upon the temporal organization of society, thereby revealing the noxious effects of the incompatibility between social demands and biological rhythms in the workplace and how this relationship can affect quality of life. Moreover, it may allow people to think about their career and social organization.

In summary, this study indicated that the misalignment of the circadian and social rhythms is a risk factor for developing depression and there are an amount of epidemiological, clinical and experimental evidence to support a new approach on diagnose and treatment of depressive disorder.

7. References

Aschoff J. (1960). Exogenous and endogenous components in circadian rhythms. Cold Spring Harb Symp Quant Biol. 25:11-28

Aschoff J, Fatranská M, Giedke H, Doerr P, Stamm D & Wisser H. (1971). Human circadian rhythms in continuous darkness: entrainment by social cues. Science. Jan 15;171(967):213-5.

Beersma DG & Daan S. (1993). Strong or weak phase resetting by light pulses in humans? J Biol Rhythms. Winter;8(4):340-7.

Benedetti F, Barbini B, Colombo C & Smeraldi E. (2007). Chronotherapeutics in a psychiatric ward. Sleep Med Rev. Dec;11(6):509-22.

Bernardi, F., Harb, A.B.C., Levandovski, R.M. & Hidalgo, M.P. (2009). Transtornos alimentares e padrão circadiano alimentar: uma revisão. Rev. psiquiatr. Rio Gd. Sul Sept./Dec 31, no.3, 170-176.

Boivin DB, Duffy JF, Kronauer RE & Czeisler CA. (1996). Dose-response relationships for resetting of human circadian clock by light. Nature. Feb 8;379(6565):540-2.

Borella E., Ludwig C., Dirk J. & de Ribaupierre A.(2011). The influence of time of testing on interference, working memory, processing speed, and vocabulary: age differences in adulthood. Exp Aging Res. Jan;37(1):76-107.

Bunney WE & Bunney BG. (2000). Molecular clock genes in man and lower animals: possible implications for circadian abnormalities in depression. Neuropsychopharmacology. Apr;22(4):335-45.

Chelminski I, Ferraro FR, Petros TV & Plaud JJ. (1999). An analysis of the "eveningness-morningness" dimension in "depressive" college students. J Affect Disord. Jan-Mar;52(1-3):19-29.

de Castro JM. (2004). The time of day of food intake influences overall intake in humans. J Nutr. Jan;134(1):104-11.

Dibner C, Schibler U, & Albrecht U (2010). The mammalian circadian timing system: organization and coordination of central and peripheral clocks. Annu Rev Physiol. Mar 17;72:517-49.

Dijk DJ & von Schantz M. (2005).Timing and consolidation of human sleep, wakefulness, and performance by a symphony of oscillators. J Biol Rhythms. Aug;20(4):279-90.

Drennan MD, Klauber MR, Kripke DF & Goyette LM. (1991). The effects of depression and age on the Horne-Ostberg morningness-eveningness score. J Affect Disord. Oct;23(2):93-8.

Elmore SK, Dahl K, Avery DH, Savage MV & Brengelmann GL. (1993). Body temperature and diurnal type in women with seasonal affective disorder. Health Care Women Int. Jan-Feb;14(1):17-26.

Folkard S, Marks M, Minors DS & Waterhouse JM. (1985). Circadian rhythms in human performance and affective state. Acta Psychiatr Belg. Sep-Oct;85(5):568-81.

Foster RG & Wulff K. (2005). The rhythm of rest and excess. Nat Rev Neurosci. May;6(5):407-14.

Fujioka A, Fujioka T, Tsuruta R, Izumi T, Kasaoka S & Maekawa T. (2011). Effects of a constant light environment on hippocampal neurogenesis and memory in mice. Neurosci Lett. Jan 13;488(1):41-4.

Gentil V, Zilberman ML, Lobo D, Henna E, Moreno RA & Gorenstein C (2007). Clomipramine-induced mood and perceived performance changes in selected healthy individuals. J Clin Psychopharmacol 27:314-5.

Giannotti F, Cortesi F, Sebastiani T & Ottaviano S. (2002). Circadian preference, sleep and daytime behaviour in adolescence. J Sleep Res. Sep;11(3):191-9.

Giglio LM, Magalhães PV, Andersen ML, Walz JC, Jakobson L & Kapczinski F. (2010). Circadian preference in bipolar disorder. Sleep Breath. Jun;14(2):153-5.

Green CB, Douris N, Kojima S, Strayer CA, Fogerty J, Lourim D, Keller SR & Besharse JC. (2007). Loss of Nocturnin, a circadian deadenylase, confers resistance to hepatic steatosis and diet-induced obesity. Proc Natl Acad Sci U S A. Jun 5;104(23):9888-93.

Haynes PL, Ancoli-Israel S & McQuaid J. (2005). Illuminating the impact of habitual behaviors in depression. Chronobiol Int. 22(2):279-97.

Harmer SL, Panda S & Kay SA. (2001).Molecular bases of circadian rhythms. Annu Rev Cell Dev Biol. 2001;17:215-53.

Hidalgo MP, Caumo W, Posser M, Coccaro SB, Camozzato AL & Chaves ML. (2009). Relationship between depressive mood and chronotype in healthy subjects.Psychiatry Clin Neurosci. Jun;63(3):283-90.

Hirota T & Fukada Y. (2004). Resetting mechanism of central and peripheral circadian clocks in mammals. Zoolog Sci. Apr;21(4):359-68.

Izquierdo I, Quillfeldt JA, Zanatta MS, Quevedo J, Schaeffer E, Schmitz PK & Medina JH. (1997). Sequential role of hippocampus and amygdala, entorhinal cortex and parietal cortex in formation and retrieval of memory for inhibitory avoidance in rats. Eur J Neurosci. Apr;9(4):786-93.

Kessler RC, Chiu WT, Demler O, Merikangas KR & Walters EE. (2005). Prevalence, severity, and comorbidity of 12-month DSM-IV disorders in the National Comorbidity Survey Replication. Arch Gen Psychiatry. Jun;62(6):617-27.

Kohyama J. (2009). A newly proposed disease condition produced by light exposure during night: asynchronization. Brain Dev. Apr;31(4):255-73. Epub 2008 Aug 30.

Kripke DF. (1984). Critical interval hypotheses for depression. Chronobiol Int. 1(1):73-80.

Lawson EA, Miller KK, Blum JI, Meenaghan E, Misra M, Eddy KT, Herzog DB & Klibanski A. (2011). Leptin Levels Are Associated With Decreased Depressive Symptoms in Women Across the Weight Spectrum, Independent of Body Fat. Clin Endocrinol (Oxf). Jul 22.

Leconte P. (1989). Chronobiological rhythm constraints of memory processes. Arch Gerontol Geriatr Suppl.;1:21-35.

Lemmer B. (2009). Discoveries of rhythms in human biological functions: a historical review. Chronobiol Int. Aug;26(6):1019-68.

Lépine JP & Briley M. (2011). The increasing burden of depression. Neuropsychiatr Dis Treat. 7(Suppl 1):3-7. Epub 2011 May 31.

Levandovski R, Dantas G, Fernandes LC, Caumo W, Torres I, Roenneberg T, Hidalgo MP & Allebrandt KV. (2011). Depressive scores associate with chronotype and social jetlag in a rural population. Chronobiol Int, 46, in press.

Lewy AJ, Lefler BJ, Emens JS & Bauer VK. (2006). The circadian basis of winter depression. Proc Natl Acad Sci U S A. May 9;103(19):7414-9.

Lewy AJ, Sack RL, Miller LS & Hoban TM. (1987). Antidepressant and circadian phase-shifting effects of light. Science. Jan 16;235(4786):352-4.

Ma WP, Cao J, Tian M, Cui MH, Han HL, Yang YX & Xu L. (2007). Exposure to chronic constant light impairs spatial memory and influences long-term depression in rats. Neurosci Res. Oct;59(2):224-30.

Mahé V & Chevalier JF. (1995). Role of biological clock in human pathology. Presse Med. Jun 17;24(22):1041-6.

Madden PA, Heath AC, Rosenthal NE & Martin NG. (1996). Seasonal changes in mood and behavior. The role of genetic factors. Arch Gen Psychiatry. Jan;53(1):47-55.

Mari JJ & Williams P. (1986). A validity study of a psychiatric screening questionnaire (SRQ-20) in primary care in the city of Sao Paulo. Br J Psychiatry. Jan;148:23-6.

Mignot E &Takahashi JS. (2007). A circadian sleep disorder reveals a complex clock. Cell. Jan 12;128(1):22-3.

Mistlberger RE & Skene DJ. (2004). Social influences on mammalian circadian rhythms: animal and human studies. Biol Rev Camb Philos Soc. Aug;79(3):533-56.

Monk T.H., Flaherty J.F., Frank E., Hoskinson K. & Kupfer D.J. (1990). The Social Rhythm Metric. An instrument to quantify the daily rhythms of life. J Nerv Ment Dis.Feb;178(2):120-6.

Monteleone P, Martiadis V & Maj M. (2011). Circadian rhythms and treatment implications in depression. Prog Neuropsychopharmacol Biol Psychiatry. Aug 15;35(7):1569-74.

Nestler EJ & Hyman SE. (2010). Animal models of neuropsychiatric disorders. Nat Neurosci. Oct;13(10):1161-9. Epub 2010 Sep 27.

Ohdo S. (2007). Chronopharmacology focused on biological clock. Drug Metab Pharmacokinet. Feb 25;22(1):3-14.

Patten SB. (2007). An animated depiction of major depression epidemiology. BMC Psychiatry. Jun 8;7:23.

Peiser B. (2009). Seasonal affective disorder and exercise treatment: a review. Biological Rhythm Research, 40, 85–97.

Polakof S, Ceinos RM, Fernández-Durán B, Míguez JM & Soengas JL. (2007). Daily changes in parameters of energy metabolism in brain of rainbow trout: dependence on feeding. Comp Biochem Physiol A Mol Integr Physiol. Feb;146(2):265-73.

Prickaerts J, Moechars D, Cryns K, Lenaerts I, van Craenendonck H, Goris I, Daneels G, Bouwknecht JA & Steckler T. (2006). Transgenic mice overexpressing glycogen synthase kinase 3beta: a putative model of hyperactivity and mania. J Neurosci. Aug 30;26(35):9022-9.

Ramsey, K.M. & Bass, J. (2009). Obeying the clock yields benefits for metabolism. Proc Natl Acad Sci U S A, 106, 4069-70.

Ramsey KM & Bass J. (2009). Obeying the clock yields benefits for metabolism. Proc Natl Acad Sci U S A. Mar 17;106(11):4069-70. Epub 2009 Mar 10.

Reinberg AE, Lewy H & Smolensky M. (2001). The birth of chronobiology: Julien Joseph Virey 1814. Chronobiol Int. Mar;18(2):173-86.

Reppert SM & Weaver DR. (2001). Molecular analysis of mammalian circadian rhythms. Annu Rev Physiol. 63:647-76.

Roenneberg T, Kuehnle T, Juda M, Kantermann T, Allebrandt K, Gordijn M & Merrow M. (2007). Epidemiology of the human circadian clock. Sleep Med Rev. Dec;11(6):429-38.

Rosenthal NE, Sack DA, Gillin JC, Lewy AJ, Goodwin FK, Davenport Y, Mueller PS, Newsome DA & Wehr TA. (1984). Seasonal affective disorder. A description of the syndrome and preliminary findings with light therapy. Arch Gen Psychiatry. Jan;41(1):72-80.

Sacharczuk M, Juszczak G, Swiergiel AH, Jaszczak K, Lipkowski AW & Sadowski B. (2009). Alcohol reverses depressive and pronociceptive effects of chronic stress in mice with enhanced activity of the opioid system. Acta Neurobiol Exp (Wars). 69(4):459-68.

Scheer FA, Hilton MF, Mantzoros CS & Shea SA. (2009). Adverse metabolic and cardiovascular consequences of circadian misalignment. Proc Natl Acad Sci U S A. Mar 17;106(11):4453-8.

Selvi Y, Aydin A, Boysan M, Atli A, Agargun MY & Besiroglu L. (2010). Associations between chronotype, sleep quality, suicidality, and depressive symptoms in patients with major depression and healthy controls. Chronobiol Int. 2010 Oct;27(9-10):1813-28.

Simons CJ, Jacobs N, Derom C, Thiery E, Jolles J, van Os J & Krabbendam L. (2009). Cognition as predictor of current and follow-up depressive symptoms in the general population. Acta Psychiatr Scand. 2009 Jul;120(1).

Soreca I, Fagiolini A, Frank E, Goodpaster BH & Kupfer DJ. (2009). Chronotype and body composition in bipolar disorder. Chronobiol Int. 2009 May;26(4):780-8.

Soria V & Urretavizcaya M. (2009). [Circadian rhythms and depression]. Actas Esp Psiquiatr. 2009 Jul-Aug;37(4):222-32.

Sullivan PF, Neale MC & Kendler KS. (2000). Genetic epidemiology of major depression: review and metanalysis. Am J Psychiatry. 2000 Oct;157(10):1552-62.

Taillard J, Lemoine P, Boule P, Drogue M & Mouret J. (1993). Sleep and heart rate circadian rhythm in depression: the necessity to separate. Chronobiol Int. 1993 Feb;10(1):63-72.

Taillard J, Philip P & Bioulac B. (1999). Morningness/eveningness and the need for sleep. J Sleep Res. 1999 Dec;8(4):291-5.

Takahashi JS, Hong HK, Ko CH, & McDearmon EL (2008). The genetics of mammalian circadian order and disorder: implications for physiology and disease. Nat Rev Genet. Oct;9(10):764-75.

Terman M & Terman JS. (2005). Light therapy for seasonal and nonseasonal depression: efficacy, protocol, safety, and side effects. CNS Spectr. 2005 Aug;10(8):647-63; quiz 672.

Wassertheil-Smoller S, Shumaker S, Ockene J, Talavera GA, Greenland P, Cochrane B, Robbins J, Aragaki A & Dunbar-Jacob J. (2004). Depression and cardiovascular sequelae in postmenopausal women. The Women's Health Initiative (WHI). Arch Intern Med. 2004 Feb 9;164(3):289-98.

Wehr TA, Wirz-Justice A, Goodwin FK, Duncan W & Gillin JC. (1979). Phase advance of the circadian sleep-wake cycle as an antidepressant. Science. Nov 9;206(4419):710-3.

Wirz-Justice A, Benedetti F, Berger M, Lam RW, Martiny K, Terman M & Wu JC. (2005). Chronotherapeutics (light and wake therapy) in affective disorders. Psychol Med. Jul;35(7):939-44.

Wittmann M, Dinich J, Merrow M & Roenneberg T. (2006). Social jetlag: misalignment of biological and social time. Chronobiol Int. 23(1-2):497-509.

Wulff K, Gatti S, Wettstein JG & Foster RG. (2010). Sleep and circadian rhythm disruption in psychiatric and neurodegenerative disease. Nat Rev Neurosci. Aug;11(8):589-99.

Zeitzer JM, Dijk DJ, Kronauer R, Brown E & Czeisler C. (2000). Sensitivity of the human circadian pacemaker to nocturnal light: melatonin phase resetting and suppression. J Physiol. Aug 1;526 Pt 3:695-702.

Zimmerman M, Hrabosky JI, Francione C, Young D, Chelminski I, Dalrymple K & Galione JN. (2011). Impact of obesity on the psychometric properties of the Diagnostic and Statistical Manual of Mental Disorders, Fourth Edition criteria for major depressive disorder. Compr Psychiatry. Mar-Apr;52(2):146-50

Mood Disorders in Individuals with Genetic Syndromes and Intellectual Disability

Maria Cristina Triguero Veloz Teixeira,
Maria Luiza Guedes de Mesquita, Marcos Vinícius de Araújo,
Laís Pereira Khoury and Luiz Renato Rodrigues Carreiro
Developmental Disorders Program, Center for Health and Biological Sciences,
Mackenzie Presbyterian University, São Paulo
Brazil

1. Introduction

1.1 Mood disorders

Mental disorder is a behavioral pattern that causes psychological suffering or incapacity (American Psychiatric Association [APA], 2003). This concept also underlines that whatever the cause of the problem, a mental disorder is the manifestation of a behavioral, psychological or biological dysfunction. The manual offers an organizational plan that groups mental disorders in 16 diagnostic categories. These categories are described according to common traits that allow the specialist to identify the disorder or make differential diagnosis when necessary (APA, 2003). DSM follows the International Statistical Classification of Diseases and Related Health Problems (ICD-10) developed by the World Health Organization (World Health Organization [WHO], 2000).

There are no doubts about the advantages of DSM IV over its prior versions. Criteria have become clearer and more precise and, at the same time, they allow the consideration of much additional information that goes further than the diagnosis *per se*. However, many challenges to the improvement of these criteria have still been discussed in work groups in order to create a new edition in 2013 (Joyce, 2008; APA, 2010).

Mood disorders constitute one of the 16 diagnostic categories. They are a kind of health problem whose main trait is a mood alteration that can follow two distinct directions: a) depressive mood, b) expansive mood. The disorders charactized by depressive mood are: Major Depressive Disorder, Dysthymic Disorder and Depressive Disorder Not Otherwise Specified. Expansive mood comprises Bipolar Disorders Type I and Type II, Ciclothymic Disorder and Bipolar Disorder Not Otherwise Specified (APA, 2003).

The main clinical characteristics of Depressive Disorders are sadness, tearfulness, hopelessness, irritability, lack of interests, diminished appetite, sleep disorders, psychomotor agitation or retardation, diminished energy, tiredness, guilty feelings, attention and concentration problems, memory problems, thoughts of death, suicidal ideation, and/or suicide attempts. Bipolar Disorders also present a persistent, elevated and abnormal mood alteration, but in two opposite directions, i.e. expansive or irritable and depression. Criteria for a Manic Episode include report inflated self-esteem, euphoria and elevated enthusiasm, grandiosity feelings, diminished need for sleep, flight of ideas, distractibility, psychomotor agitation, excessive involvement in pleasurable activities with high risk potential, pressured speech, among others (APA, 2003). The specificities of typically depressive, manic or mixed frames require an attentive and careful clinical interview to determine characteristics of the disorder, its course and, when necessary, a differential diagnosis.

Mood Disorders due to a General Medical Condition are more likely to present a physical condition of acute beginning. In this case, clinical traits also involve depressive mood, prominently diminished interest or pleasure, and expansive, irritable or elevated mood. These characteristics allow the specialist to classify the disorder according to the main symptomatic frame: a) depressive traits; b) depressive type episode; c) maniac traits, c) mixed traits. The manual specifies that in every mood disorder related to a medical condition it must be verified the presence of different clinical manifestations of depressive, maniac or mixed disorders. Nevertheless, these manifestations do not fulfill all criteria for the disorders and there must be a confirmation of an etiologically related general medical condition.

1.2 Intellectual disability

In this chapter, we chose to use the term intellectual disability (ID) as a synonym for mental retardation that has been classically described by DSM IV as a Disorder Usually First Diagnosed in Infancy, Childhood, or Adolescence. The manual describes MR as a disorder marked by significantly subaverage intellectual functioning and an Intellectual Quotient (IQ) of aproximately 70 or below, with onset before the age of 18. It can be codified as Mild, Moderate, Severe, Profound Mental Retardation, and Mental Retardation Severity Unspecified (APA, 2003). MR's main characteristic is subaverage intellectual functioning (usually standard deviation 2 below average) and significant impairment in adaptive functioning. Such impairments can be observed through significant limitations in three domains of daily life: conceptual, social and practical abilities. The 11th edition of American Association Intellectual and Developmental Disability (2011) define that intellectual disability is characterized by significant limitations both in intellectual functioning and adaptive behavior as expressed in conceptual, social, and practical skills. This disability originates before age 18 (American Association on Intellectual and Developmental Disabilities, 2011).

Genetic syndromes are considered biological risk factors for ID that occurs in pre-natal phase of development. Mood disorders and ID are both complex conditions and involve an interdisciplinary professional work, especially when such conditions are associated. Thus, when such conditions are to be assessed, two diagnostic approaches are needed: an etiological, to identify possible genetic cause for ID, and a functional, to determine the amount of adaptive resource related to environmental needs (Moescheler, 2008).

2. Behavioral phenotype and psychiatric phenotype

The development of molecular biology techniques has allowed progress in the study of etiological factors related to genetic evidences in psychiatric disorders. In this context, two mental disorders stand out: schizophrenia and bipolar disorder (Basset et al., 2003; Craddock et al., 2010; Pun et al., 2011). Nevertheless, to establish a possible causal relation it is necessary to deepen the study of several issues. For instance, why is it that not all individuals with psychiatric disorders present genetic alterations? Why is it that not everyone who has genetic alterations develops psychiatric disorders? (Basset et al., 2003; Leyfer et al., 2006; Mota et al., 2010; Walker et al., 2011). The advances of molecular biology are enormous, however many questions are difficult to be answered. For example, in 1987 mood disorders were studied in several genetic researches that aimed finding the linkage between different genes and bipolar disorder (Egeland et al., 1987). This study was classic in the search for markers in chromosome 11 in a genetically isolated community in Old Order Amish, Pennsylvania. Notwithstanding, this finding could not be confirmed by further research (Kelsoe et al., 1989).

In individuals with genetic syndromes researchers have found the presence of indicators of psychiatric disorders, especially those associated with ID (Antonell et al., 2006; Bilder et al., 2009; Jarvinen-Pasley et al., 2008; Ridha, 2011). Some of the most studied psychiatric symptoms belong to the clinical frames of psychosis, mood disorders and anxiety disorders (Siegel & Smith, 2010). This has determined an operational definition that helps the recognition of psychiatric alterations according to the type of genetic syndrome. This definition is known as Psychiatric Phenotype (PP). It must be emphasized that PP is different from the concept of Behavioral Phenotype (BP). This difference will be treated in the following paragraphs.

Historically, the first description of behavioral characteristics in genetic syndromes was made by Langdon Down in 1866 in his observations of a group of people that have typical mongoloids that shared physical, mental and behavioral characteristics (Down, 1995). After Langdon Down's original description (1866) the syndrome's typical personality and behavioral traits have become known (Collacott et al., 1998). Almost 100 years have passed until a new study appeared and cast suspicion on the existence of genetically based behavior patterns when, in 1963, Money described the visuospatial difficulties in Turner Syndrome (Artigas-Pallarés, 2002). Despite the introduction of the genotype-phenotype relation, the concept of behavioral phenotype associated with genetic disorders was firstly proposed only in 1976 (Nyhan, 1976). The researcher proposed a relation between the presence of self-injury behavior and the syndrome later known as Lesch-Nyhan (Holland et al., 2003). The children affected by this syndrome present a typical behavioral pattern characterized by stereotypies and self-injury behaviors (Nyhan, 1976). The concept of BP proposed by Nyhan was kept and is still used (Artigas-Pallarés et al., 2006). However, during the 1970's and 1980's few researches showed interest in the study this concept. As from the 1990's new studies were developed in the field of genetics, contributing to the understanding of BP concept and the genotype-phenotype relation. In this relation behavioral problems have been underlined and many of them configure indicators of psychiatric disorders (Artigas-Pallarés, 2002; Artigas-Pallláres et al., 2006; Bachega et al., 2004; Bilder et al., 2009; Collacot et al., 1998).

The explanation for the origin of different types of BP has two sides. The first appeared with Nyhan's proposal in 1976, which states that traits of behavioral phenotypes are biologically determined, not learned and has genetic origin (Artigas-Pallláres, 2002; Holland et al., 2003; Moldavsk et al., 2001; Nyhan, 1972). The other variant suggests that BP patterns in genetic syndromes are determined by biological factors and influenced by environmental factors, as described in the following definition: "Behavioral, cognitive, linguistic, social and also motor patterns that have a biological basis, without disregarding the influence of the environment in their development (Ruggieri & Arberas, 2003, p. 240).

When a syndrome is described with a certain behavioral, linguistic and cognitive pattern one should not presupose that all individuals affected by the syndrome will present equivalent typical behaviors, once the expression of BP is determined by biological and environmental factors (Ruggiere & Arberas, 2003; O'Brien, 2000). All of these factors operate in a complex of interactions that contribute to the manifestation at times heterogeneous of some behavioral problems (Gallardo-Pujol, et al., 2009; Jarvinen- Pasley et al., 2008; O'Brien, 2000; Ruggieri & Arberas, 2003). Following this discussion, BP was defined as the higher probability for people with genetic syndromes to present certain behavioral and developmental patterns, when compared to people without the syndrome (Dykens & Hodapp, 2007; Flint, 1998; Tartaglia et al., 2010).

The definition of Psychiatric Phenotype, in its turn, involves that set of psychiatric problems which is more frequently associated with a genetic syndrome and in a larger proportion when compared to people of the same age, gender and level of intellectual disability that do not present other biologically based disease (Clarke, 2003). This definition aims at explaining why some individuals that present genetic alterations develop certain psychiatric disorders. One example is the case of the association between Autistic Disorder (AD) and Fragile X Syndrome. People with this syndrome present in a more consistent way autistic behavioral patterns and, likewise, fragile X mutations rates are raised in children diagnosed with AD (Fombonne et al., 1997). This is one example on how genetic factors can increase the vulnerability to develop certain types of psychiatric disorders.

3. Using the classification system of mood disorders of the diagnostic and statistical manual of mental disorders (DSM IV) in populations with intellectual disability

If we observe in detail the clinical criteria of depressive manifestations on DSM IV, it is possible to verify that the assessment of many of them require a minimum of preserved intellectual abilities. Such abilities include organization, planning, competence in problem solving, abstract thought, working memory, attentional abilities, among others (American Association on Mental Retardation, 2006). Take the example of an adult with intellectual functioning significantly below population average, diagnosed before 18 years old and who presents varied deficits in adaptive functioning with losses in the areas of communication, personal care, social abilities, academic performance, safety and leisure, as well as behavioral indicators of sadness and depression. Suppose this individual has a genetic syndrome. In this case, how is it possible to use depressive disorder clinical criteria that have been developed for normative population? How can a person with moderate or severe ID talk about feelings of emptiness or sadness, lack of interest in activities, diminished appetite, fatigue, difficulties in concentration, feelings of uselessness, death thoughts, suicidal ideation and others (APA, 2003). The raising of these questions has lead to different discussions workgroups. One question is related to the procedures to assess the presence of psychiatric disorder in ID condition (Cooper, 2003;

Hermans & Evenhuis, 2010). The other, seeks new proposals for diagnostic classification of psychiatric disorders in people with learning disabilities and ID (Cooper et al., 2003). Both questionings will be treated next.

ID, encoded on DSM IV as MR, is a clinical condition that belongs to the category of Disorders Usually First Diagnosed in Infancy, Childhood, or Adolescence. In the multi-axial system of DSM IV, MR is classified in Axis II together with Personality Disorders (APA, 2003). The manual specifies that, although MR belongs to Axis II, depending on the condition it can be the main diagnosis. However, this classification does not seem appropriate, for it compromises the understanding of what is the ID condition itself. DSM V work group dedicated to childhood and adolescence has already proposed a new category for MR, i.e. Intellectual Developmental Disorder within the condition of Neurodevelopmental Disorders (APA, 2010). According to such discussions, Intellectual Developmental Disorder involves both an intellectual deficit - marked by deficits in mental abilities such as reasoning, problem-solving, planning, abstract thinking, judgment, academic learning and learning from experience - and a deficit in adaptive functioning, which limits participation or restricts performance in one or more aspects of daily activities, such as communication, social participation, functioning at school or work. The onset of these traits must occur during developmental period.

Mood disorders in the ID condition are usually underestimated. There is a tendency to attribute every mood change to the ID condition and this can lead to an obscuration of the diagnosis (American Association on Mental Retardation, 2006). When indicators of mood disorders are attributed to the ID condition, the need for clinical assessment of the patient's mental health and the possibility of intervention are denied. Furthermore, when mental health problems affect people with ID, damages in the adaptive functioning tend to increase (American Association on Mental Retardation, 2006).

Behavioral Phenotype of an individual with a genetic syndrome indicates aspects of the etiological diagnosis which are valuable for professionals that assist him. Unfortunately, not all syndromes have been studied from the psychiatric point of view as they should. Probably, this is associated with the low rates of prevalence of these disorders in general population. In many of them there is consistent data on indicators of mood disorders (Arron et al., 2010; Leyfer et al., 2006; Sinnema et al., 2011; Walker et al., 2011). One of the essential clinical criteria for a health problem to be considered a mental disorder is the persistence of abnormalities that impair several areas of the individual's life (APA, 2003). It is noteworthy that people with genetic syndromes and ID present significant limitations in social, academic and work areas. Hence the use of the manual's diagnostic classifications requires from the specialist expertise in this type of clinical judgement.

Despite the broad acceptance DSM IV and ICD 10, what are their advantages for mood disorder assessment in people with genetic syndromes associated with mental retardation? The multi-axial classification of DSM-IV TR involves an assessment in five axes: Axis I - mental disorders; Axis II - personality disorders and mental retardation; Axis III - medical conditions; Axis IV - psychosocial and environmental problems; and Axis V - global assessment of functioning scale (APA, 2003). As it has been said previously, many of the clinical criteria for depression or bipolar disorder are not easy to be identified in populations with moderate to severe ID. This limitation has lead many groups of specialists to work with them in a different way, considering typical phenotype traits and limitations in intellectual functioning. The assessment of a depressive or maniac symptom is very important, for it indicates the presence or absence of these symptoms, as well as their severity degree. And, precisely in this aspect, DSM IV clinical criteria can not always be applied in individuals with atypical development (Cooper, 2003).

To respond to the limitations of DSM IV, specialists from London Royal College of Psychiatrists (Cooper, 2003) have stand out for developing a classification system of Diagnostic Criteria for Psychiatric Disorders for Use with Adults with Learning Disabilities/Mental Retardation (DC-LD). The group has developed this system based on clinical and research criteria. About the research it is pointed out the need for a diagnostic system different from DSM's for the following reasons: (1) lack of clearness in the diagnostic criteria to help the development of epidemiological studies of incidence and prevalence; (2) need of clinical essay to assess the efficiency of interventions; (3) need of data about the effectiveness of interventions among groups and (4) to identify risk and protection factors for psychiatric disorders in populations with ID. About clinical criteria, the system proposes that the diagnostic process is made in three basic stages: (1) measuring psychopathology; (2) classifying the psychopathology using operationalized diagnostic criteria (descriptive diagnosis) and (3) determining the etiology of the descriptive diagnosis (often multifactorial), using a biological psychological-social-developmental framework (Cooper, et al., 2003, p. 4).

DC-LD has a critical view of DSM IV. The group considers that the classification of disorders in Axis I is ambiguous in case of ID condition. It suggests that, in people with ID, instead of assessing symptoms by language abilities, it should be used their behavioral equivalents. DC-LD recommends avoiding the use of diagnostic subcategories for each disorder. A careful analysis is also made on the behavioral phenotype described for the genetic syndrome. It is not recommended consider in the psychopathology of a disorder some behavioral pattern of the genetic disorder. To illustrate this, it is possible to observe the hyperphagia pattern that marks Prader-Willi Syndrom, the bizarre/aberrant behavior presented by people with Smith-Magenis Syndrome or the impulsiveness described in Cri du Chat Syndrome (Clarke, 2003).

According to DC-LD, for adults with mild intellectual disability it is suggested the use of DSM IV classification (Cooper et al., 2003). Finally, the DC-LD system presents a multi-axial structure that includes a classification guideline for psychiatric disorders associated with genetic syndromes for moderate to severe ID cases. Axis I: severity of intellectual disabilities, Axis II: cause of intellectual disabilities and Axis III: presence of additional psychiatric disorders. It is noteworthy that Axis I and II can adopt diagnostic categories from ICD-10 and DSM IV.

4. Assessment of mood disorders in people with mental retardation

Both DSM IV and DC-LD's new system are supported by clinical judgments to raise diagnostic hypothesis for mood disorders. This judgment is a special type of assessment that is based on empirical evidence. However, there are other procedures that help this clinical judgment. The observational method and psychological instruments (scales, inventories or semi-structured interviews) are some examples. In any case, they must not be used as the only source of information for the diagnosis.

The observational method helps professional practices for modifying and management behavioral problems (Martin & Pear, 2009; Silvares, 2000). This type of assessment procedure requires that the observer has been previously trained in three areas: clinical traits of the mood disorder he wishes to observe the ID condition and the Behavioral Phenotype described for the genetic syndrome. To observe mood indicators of an individual with ID requires the specialist to focus on the way the person behaves, the frequency of the behavior's occurrence, intensity and also environmental factors possibly related to mood change. The main advantage of the observational method is that it allows the control of variables directly related to the observed behavior, whether it happens in a natural environment or in laboratories. Otherwise, behavioral observation is not always sensitive for the assessment of mood states and feelings (Martin & Pear, 2009).

The indirect assessment procedures use standardized tools that rely on information provided by informers/caregivers. For people with ID it is necessary to consider the vast number of standardized tools available to assess general behavioral alterations in children and adults with atypical development. These instruments track only behavioral problems associated with the behavioral phenotype of genetic syndromes, i. e. aggression, self-injury, behavioral stereotypy, repetitive responses, tantrums, irritability, agitation, lethargy, among others (Aman et al.,1985a; Brinkley et al., 2007; Brown et al., 2002; Hill et al., 2008; Kaland et al., 2008; Karabekiroglu & Aman, 2009; Keil et al., 2009; Narita et al., 2004; Polk et al., 2008; Rojahn et al., 1997). Some tools for this assessment are: Behavior Problems Inventory (BPI-01) (Rojahn, J. et al., 2001), Aberrant Behavior Checklist (ABC) (Aman et al., 1985b, Communication and Symbolic Behavior Scales Developmental Profile (CSBS) (Wetherby et al., 2002), The Repetitive Behavior Scale Revised (RBS-R) (Bodfish, et al., 2000), Questions about Behavior Function Scale (QABF) (Matson et al., 1999). These instruments offer discerning indicators of behaviors and are of relatively fast application and easy correction.

However, other tools are specifically designed to assess behavioral indicators of psychiatric disorders, such as mood disorders. In childhood and adolescence some of the most used instruments for this end are the behavioral assessment inventories of the Achenbach System of Empirically Based Assessment - ASEBA (Achenbach & Rescorla, 2001). Its application involves asking parents or caregivers about the child's competence in different areas, as well as the assessment of the existence of diseases and disabilities, parental worries about the child and positive aspects they recognize in them. Another set of questions assesses emotional and behavioral problems encoded in the scales of internalization, externalization and total problems. These inventories have been translated, adapted and internationally validated (Achenbach & Rescorla, 2001; Dykens et al., 2002, 2007; Graham et al., 2005). Some of the most commonly used instruments in childhood and adolescence are: Child Behavior Checklist (CBCL), Youth Self-Report (YSR) and Teacher's Report Form (TRF). All three inventories provide specific DSM-oriented scales. In the affective problems scale, for example, statements about the behavior of the assessed child are consistent with DSM's clinical criteria described for Major Depressive Disorder and Dysthymic Disorder. The inventory requires the informant to say if the statement is: not true of the child, somewhat or sometimes true, very true or often true of the child. Some of the statements found in the inventory are: there is very little he/she enjoys, he/she cries a lot, he/she deliberately harms him/herself or attempts suicide, he/she doesn't eat well, he/she feels worthless or inferior, he/she feels too guilty or overtired without good reason (Achenbach System of Empirically Based Assessment [ASEBA], 2011).

In the form of a semi-structured psychiatric interview, the Schedule for Affective Disorders and Schizophrenia for School-Age Children - Present and Lifetime Version (K-SADS-PL) is another tool designed to assess the symptomatology of psychiatric disorders in the present and throughout the life of children between 6 and 18 years old. The questions from the first version of this interview were based on DSM III and DSM III R's clinical criteria for psychiatric disorders in childhood (APA, 1980; APA, 1987; Chambers et al., 1985). After the publication of the manual's fourth edition, Kaufman and colleagues developed a new version of the interview that comprised the assessment of the child's symptoms in the present and throughout his/her lifetime (Present and Lifetime Version), according to clinical criteria of Disorders Usually First Diagnosed in Infancy, Childhood or Adolescence (Kaufman et al., 1997). K-SADS-PL provides objective questions and criteria to assess each symptom. Some of the assessed diagnoses are: Major Depression, Dysthymia, Mania, Hypomania, Cyclothymia, Bipolar Disorders, Schizoaffective Disorder, Attention Deficit Hyperactivity Disorder, Conduct Disorder, among others. This interview collects data from parents/informants and also from the child. A conclusive score about the symptomatology is based on the summary of all this information. For example, on the field of depressive disorders the interview assesses if the child has ever felt very sad and depressed, and how often it has happened; subjective feelings of irritability; bad mood; anger; lack of interest; little motivation; thoughts of death; suicidal ideation, among others. Concerning maniac affective states, the following problems are assessed: exalted mood, excessively optimistic attitude, excessive happiness, diminished need for sleep, among others.

For adulthood there are also several instruments to track mood disorders indicators in people with ID. In 2010, Research in Developmental Disabilities published a review article that aimed at obtaining information on psychometric properties of screening instruments for depression in adults with ID (Hermans & Evenhuis, 2010). The period of time covered by the review was of 28 years, and the databases revised were PUBMED, PSYCHINFO and EMBASE. Only articles in English were included. A quality grade was granted to instruments according to their psychometric properties. Among the instruments that reached the best grades were Glasgow Depression Scale for people with a Learning Disability (GDS-LD), Reiss Screen for Maladaptive Behavior (RSMB), Diagnostic Assessment for the Severely Handicapped-II (DASH-II), Mini Psychiatric Assessment Schedule for Adults with Intellectual Disabilities (Mini PAS-ADD), Assessment of Dual Diagnosis (ADD), Psychiatric Assessment Schedule for Adults with a Developmental Disability (PAS-ADD), Glasgow Depression Scale - Carer Supplement (GDS-CS), Self-Report Depression Questionnaire (SRDQ) and Psychopathology Instrument for Mentally Retarded Adults (PIMRA). The authors draw attention to the need for many instruments to be assessed from the point of view of their sensibility and specificity to identify affective problems in people with ID, since not all of the studies fulfilled properly these psychometric properties.

The previous theoretical considerations allow us to present next a brief synthesis of the main mood disorders identified in people with ID and genetic syndromes, such as Prader Willi syndrome, Down syndrome and Williams syndrome.

5. Mood disorders indicators in genetic syndromes

The proportion of mental health problems in individuals with intellectual disabilities associated with genetic syndromes is twice to three times greater than general in the general population. Despite that, the identification of emotional problems is complex due to the expression of behavior phenotype variability (Sturmey et al., 2010; Clarke, 2003). However, in individuals with genetic syndromes and ID the acknowledgement of psychopathological conditions such as depression and bipolar disorder is relatively recent compared to typically developing people (Adams & Oliver, 2011; Hayes et al., 2011; Matson & Laud, 2007; Mutkins et al., 2011; Tsiouris et al., 2011).

In ID condition, mood disorders have received special attention due to the impairments that this kind of problem causes in the accomplishment of daily life activities. People with intellectual disabilities may exhibit significant limitations in adaptive behavior which can impair their social life (Denis et al., 2011; Koskentausta et al., 2007). These limitations can be caused by difficult management of psychiatric and emotional problems, such as shyness, phobias, mood changes, nervousness, somatic complaints, sadness, withdrawal, worries (Bolsoni-Silva et al., 2010; Walker et al., 2011).

Within the category of mood disorders, depressive disorders have received special attention in genetic syndromes accompanied by ID. Due to the interference in language expressive abilities that can be caused by the intellectual impairment, it is recommended, whenever it is possible, the assessment of these individuals according to behavioral equivalents for depression. Previous studies have described some of these behavioral equivalents which usually have acute beginning and do not have medical causes. Some examples are: agressive behavior, tearfulness, self-injury, diminished daily life abilities, social isolation, somatic complaints, hyponchondria, appetite alterations, weight changes, psychomotor agitation or retardation (Benson & Brooks, 2008; Cooper, et al., 2007; Smiley & Cooper, 2003). It will be presented next a brief characterization of genetic syndromes and their main associated mood disorders.

5.1 Prader Willi syndrome

Prader Willi Syndrome (PWS) is a genetic disorder caused by the lack of expression of paternally inherited imprinted genes on chromosomal region 15q11-q13. Genes form this region have diferential expression according to the paternal origin so that the maternal and paternal copies must be present for normal genetic expression. Criteria of PWS are well established and laboratory diagnosis is made in 95% of the cases: in 70% of the patients paternal deletion is found in chromossome region 15q11-q13 and 25% have uniparental disomy. The incidence ranges from 1:10.000 to 15.000 births (Cassidy & Driscoll, 2009; Jin, 2011). The syndrome is characterized by two clinical phases. In the first cardinal symptoms are neonatal hypotonia, diffculties to be fed, lethargy, weak cry and hyporeflexia. In the second phase, from six months on, there is an improvement of hypotonia, weight gain, progressive development of hyperfagia and obesity. Genital alterations such as cryptorchidism, micropenis and scrotal hypoplasia are found in male children and hypoplasia of female external genitals in female children (Holm et al., 1993).

The Behavioral Phenotype described in people with this disease indicates many behavioral problems in childhood, adolescence and adulthood such as: obsessive-compulsive spectrum behaviors, anger and agressive bursts, tendence to rigidness, oppositional behavior, lying and stealing, mild to moderate ID, neuropsychological alterations associated with prefrontal lobe activity in attentional abilities, executive functions (for instance, inhibioty control and selective attention) and visuospatial organization. The most frequent psychological complaints are of the emotional-affective type, psychotic disorders spectrum complaints, withdrawal, mood changes and depression (Benarroch et al., 2007; Dykens & Roof, 2008; Goldstone et al., 2004; Goos & Ragsdale, 2008; Gunay-Aygun et al. , 2001; Hinton et al., 2006; Jauregi et al., 2007; Koenig et al., 2004; Semenza et al., 2008.)

Altough the interest for the sudy of mood disorders in people with PWS dates back to the 1990's (Dykens & Cassidy, 1995), recent research have been trying to assess the association between certain types of mood disorders and two of the main genetic subtypes of the disease: deletion and maternal uniparental disomy (Hartley et al., 2005). For instance, a recent study has put forward the hypothesis that deletion subtype is more associated with depressive disorders without psychotic symptoms, as opposed to maternal uniparental disomy that presented more association with bipolar disorder and psychotic symptoms (Soni, et al., 2008). Notwithstanding such findings are still contradictory.

A recently published cohort study with an expressive sample of 68 individuals with clinical and genetic diagnosis of PWS has allowed researchers to share the sample in four age groups: < 25 years old, 25-34 years old, 35-44 years old, ≥ 45 years old. They were also able to make comparative analysis for age and genetic subtype according to the prevalence of psychiatric disorders, among them mood disorders and psychotic disorders. The instrument used to verify psychopatological indicators was the Developmental Behavior Checklist for Adults, which was answered by the main caregiver of each participant (Sinnema et al., 2011). This study compared mood disorders distribution of the diagnoses according to the genetic subtype of deletion and maternal uniparental disomy. The findings indicate that, whatever the genetic subtype is, it was observed a relatively equivalent distribution of 16 to 20% of the sample in the fulfilling of typical bipolar disorder symptoms with psychotic symptoms and psychotic disorder isolatedly. The depressive symptoms observed were isolation, diminished interest or pleasure in making activities, impairment of self-care abilities, depressive mood or sadness. In the scale of the inventory that asseses clinical indicators of depression it was observed that the number of affective problems increased with the age of the participants. Among the main symptoms verified in the frame of bipolar disorder with psychotic symptoms were: mood fluctuations that involve hypomania and depression, paranoid ideas and hallucination (Sinnema et al., 2011).

So far, one Brazilian study was published about behavioral phenotype of PWS (Mesquita et al., 2010). The sample of this reserach was composed of 11 children and adolescents with clinical and molecular cytogenetic diagnosis of PWS and it explored indicators of mood disorders through the use of Achenbach & Rescorla's Child Behavior Checklist 6-18 - CBCL/6-18 (Achenbach & Rescorla, 2001). It was verified that out of CBCL/6-18's four DSM-oriented scales, the affective problems scale indicated the presence isolation, depression, sadness, excessive dependence of adults and affective problems. It was also observed positive and statistically significant correlation beteween social problems and the scales of isolation and depression (Mesquita et al., 2010). Despite the restricted sample and convenience criteria for sample selection, this was the first Brazilian research to describe the behavioral phenotype of children and adolescents with the syndrome, and some of the findings were consistent with previous studies (Lindgren et al., 2000; Semenza et al., 2008).

5.2 Williams syndrome

Another target genetic disorder in the study of depressive disorders is Williams syndrome (WS). WS is a genetic disorder caused the deletion of multiple genes of the long arm of chromosome 7 (region 7q11-23) (Sugayama et al., 2007). Previous studies registered incidence rates of 1:20.000 to 1:50.000 live births and prevalence between 1:7.500 and 1:20.000 live births (Meyer-Lindenberg et al., 2006; Sugayama et al., 2007). Initially, the diagnosis of the syndrome is based on clinical criteria such as typical facial dysmorphism during childhood (flat mid-face, periorbital fullness or swelling, upturned nose, prominent cheeks); cardiovascular diseases (supravalvar aortic stenosis); connective tissue and calcium level alterations (Cardoso-Martins & Silva, 2008; Herreros et al., 2007; Jarvinen-Pasley et al., 2008; Martens et al., 2008; Rossi, et al., 2006, 2009). However, diagnostic confirmation can only happen by means of cytogenetic examination (Fluorescence *in situ* Hybridization Test - FISH) (Sugayama et al., 2007).

The syndrome is characterized by different cognitive and behavioral alterations (Porter & Dodd, 2011). Typical cognitive patterns of the syndrome's phenotype are: intellectual disability in various degrees, neuropsychomotor development delay in the first years of life, good performance in tasks that require the use of expressive language, impairment in receptive language abilities, syntactic-pragmatic, structural and functional alterations of language, use of clichés, sound effects, intonation resources, echolalia, deficits in visuospatial abilities, attention deficit, alterations in executive functions that involve work memory, inhibitory control and planning of task accomplishment (Mervis & John, 2010; Teixeira et al. ,2010).

About the Behavioral Phenotype, studies with children from 5 years old, report significant contrast between deficits and excess of abilities that interfere considerably in their adaptation in familiar, social and school environment (Artigas-Pallarés, 2002; Bellugi et al, 2000; Cardoso-Martins & Silva, 2008; Davies et al.,1998; Jarvinen-Pasley et al., 2008). Among the excesses it is found empathy, hipersociability, agitation, various behavioral stereotypies and anxiety. Among behavioral deficits some traits are difficulties in establishing and maintaining relationships, establishing limits for physical proximity, maintaining cooperative games, sharing interests and waiting for his/her turn in conversations.

Although the most prevalent mental disability in the syndrome is Attention Deficit Hyperactivity Disorder, the behavioral and emotional phenotype of individuals with Williams Syndrome is characterized by other affective problems, such as Generalized Anxiety Disorder, phobias, fears and social withdrawal (Martens et al., 2008; Mervis & Becerra, 2007; Leyfer et al. 2006; Martens et al., 2008). Previous studies that used behavior problems assessment scales verified the prevalence of behavioral indicators of depression, almost always associated with anxiety and somatic complaints (Pérez-García, et al., 2011). Also about the contrasts between excesses and deficits in abilities, in a systematic review study on WBS (Mervis & John, 2010), underscore the high rate of social interactions that they establish and social abilities deficits observed during these interactions. Some of these contrasts are likely to contribute to the development of affective problems, especially concerning hypersociability and, at the same time, lack of social abilities that can end up with the individual's withdrawal.

A previous study with an expressive number of participants (n=190) has allowed the author to share the sample in three age groups: 6-9, 10-13 and 14-18. The instruments used were

Child Behavior Checklist (CBCL/6-18), Behavior Assessment System for Children (BASC), Revised Children's Manifest Anxiety Scale (RCMAS) and Yale-Brown Obsessive-Compulsive Symptom Checklist (Y-BOCS). Concerning depressive disorders indicators it was found that the group of adolescents between 14-18 years old reached the highest scores in the anxiety/depression scale of CBCL when compared to the remaining groups (Switaj, 2000).

Other studies have compared behavioral indicators of mood disorders among people with different genetic syndromes. One example is a research that involved three groups of syndromes: Fragile X syndrome, Down syndrome and Williams syndrome. This study shows CBCL clinical profile for the total WBS cohort. The scales with highest percentage of individuals within subclinical range (T-scores between 65 and 69) and clinical range (T-scores higher than 70) were anxiety/depression and attention problems. Global internalizing and total scales also included a significant percentage of patients in either subthreshold or clinical range. Although this research aimed at contributing to the description of WBS behavioral phenotype, as in previous studies, anxiety and depression appeared with the highest mean scores in WBS with many individuals within clinical range (Pérez-García, 2011).

In Brazil there are only two published studies on the identification of behavior problems and mood disorders indicators in children and adolescents with WS. The first aimed at describing behavioral, cognitive and language profiles and identifying autistic behavior in a group of 10 children and adolescents with WBS, age range 5-16 years old, and 10 typically developing children and adolescents. The instruments used were: Test of Nonverbal Intelligence (Leiter-R); Child Behavior Checklist (CBCL/1½-5; CBCL/6–18); Language Exam (TIPITI); and Autism Screening Questionnaire (ASQ). One-way ANOVAs were applied to compare the mean of compatible scales of CBCL/1½ -5 and CBCL/6-18 inventories between groups with WBS and CG. There were statistically significant differences between groups on the scales of internalizing problems, total emotional/behavioral problems and affective problems (Teixeira et al., 2010). The second study was made with 22 children and adolescents with WS, age range 7 years 5 months old to 18 years 3 months old, with clinical and molecular cytogenetic diagnosis for WS confirmed by Fluorescence *in situ* Hybridization (FISH) techniques. All participantes were members of the Brazilian Association of Williams Syndrome. The sample's intellectual performance was assessed through block design and vocabulary subtests of Wechsler Intelligence Scale for Children (WISC-III). The result was an estimate intelligence quotient compatible with inferior classifications that characterize mild to moderate intellectual disability. The results concerning behavior problems and mood disorders were assessed through the use of Child Behavior Checklist (CBCL/6-18) and indicated a bordering classification in the scales of anxiety and depression, affective problems and in the internalizing scale of the inventory, which is formed by the subscales of anxiety and depression, isolation and depression, and somatic complaints. The scale of total emotional/behavioral problems indicated clinical range compatible scores (Segin, 2010).

5.3 Down syndrome

Down syndrome (DS) is another genetic disorder that is associated with psychiatric disorders with difficult social and familiar handling. The disease is caused by chromosome 21 trisomy and is the most common genetic cause for ID. It is also one of the few aneuploidy conditions compatible with post-natal surviving chances. The syndrome has a prevalence of aproximately 1:700 births and is associated with more than 80 clinical conditions that include congenital cardiac malformations, duodenal stenosis, hypotonia, immune system deficiency, raised risk for dementia of the Alzheimer type in adulthood, among other conditions of higher or lower severity according to the variability of each case (Carr, 1994; Harris et al., 1996; Sherman et al., 2007; Sommer & Henrique-Silva, 2008). Some typical patterns of DS's behavioral phenotype in children and adolescents are increased sociability, mood oscillations and stubbornness. Among cognitive alterations it is described ID, joint attention problems, deficits in verbal processing of information, deficits in working memory and several executive functions, inattention and expressive language deficits (Adams et al., 2008; Balci & Ahmet, 2007; Fidler, 2005; Ruggieri & Arberas, 2003).

In 2007 it was published a sistematic review study that identified from other researches the main behavioral alterations and psychiatric disorders in individuals with DS in three development phases: childhood, adolescence and adulthood (Dykens, 2007). It is noteworthy the fact that children with DS do not present high rates of behavioral problems, emotional and psychiatric alterations when compared to other groups of children with ID. However, when compared to typically developing children, a larger incidence of externalizing behaviors is reported for the DS group (manipulation, oppositionism, concentration problems, impulsiveness, attention deficit hyperactivity disorder, oppositional defiant disorder, low rates of co-occurrence of autistic spectrum disorders). In adolescence, the author warns about improvements that can happen in the course of some cognitive symptoms which are identified in childhood. For instance: hyperactivity, concentration problems and inattention. Nevertheless, the beginning or aggravation of affective problems that indicate mood disorders (such as internalizing symptoms, especially withdrawal, being more secretive and quiet, preference to be alone) happen in this phase of development. In adulthood, although there is less literature material, studies point to the increase of neurological diseases such as Dementia of the Alzheimer Type with aggressiveness symptoms, high rates of depression, isolation, negative mood, passiveness, insomnia, diminished appetite, schizophrenia and obsessive-compulsive disorder.

Another recent literature review study analized 30 articles on depression in DS (Walker et al., 2011). The work draws attention to precautions for the diagnosis of depressive disorders in people with DS, once there are two frequently associated medical conditions: dementia and hypothyroidism. However, people with DS present several compatible symptoms with mood disorders of Major Depression Disorder type, such as reduced interest and pleasure, depressed affect, psychomotor retardation, tearfulness, reduced energy, loss of appetite, sleep disorders, hypochondria, aggression or tantrums and reduced speech. As in PWS, it can be identified an association of depressive symptoms with psychotic ones, such as hallucinations, obsessive ideas and compulsion. This makes a proper diagnosis even more difficult. The study concluded that there are many "blind spots" in our current clinical knowledge on depression in DS. Further studies should include systematic assessment of treatment strategies for depression in DS and incorporate the investigation of the role of social support, psychotherapy and biological treatments (Walker et al., 2011).

6. Conclusion

There are several specificities in the assessment of mood disorders in genetic syndromes accompanied by ID, especially due to the difficulty in having access to reliable verbal reports on the degree of psychological suffering of these individuals. Seeing that, it should be taken into account the use of standardized instruments in the assessment process to allow access of this information through informant's reports. Thus, it is suggested that the diagnosis of mood disorders in people with ID to be done mainly with the use of behavioral and/or psychiatric assessment instruments that allow adaptations for using DSM-IV's clinical criteria. Besides, in ID populations it is necessary to assess adaptive behaviors and risk and protection factors involved in the etiology and in the course of a mood disorder.

The chapter has pointed the need of interdisciplinary aproaches in the assessment and intervention of mood disorders in populations with genetic syndromes associated with ID. Such approach allows the identification of impaired or preserved areas of cognitive, affective, behavioral and social functioning and, therefore, planning interventions that incorporate more appropriate repertoire to their functioning.

Although prevalence rates of genetic syndromes are low, especially WS and PWS, it is imperative to develop epidemiologic studies that can contribute to more accurate knowledge on the incidence and prevalence of mood disorders in this population.

The table 1 summarizes the contents of this chapter. There are reported the main problems and recommendations in the assessment of mood disorders in genetic syndromes associated with intellectual disability.

Problems	Recomendations
Mood disorders and ID are both complex conditions.	The assessment process ought to involve an interdisciplinary professional team.
Mood disorders in the ID condition are usually underestimated. There is a tendency to attribute every mood change to the ID condition and this can lead to an obscuration of the diagnosis.	The clinical judgments may be established by special type of assessments that have to be based on empirical evidence.
How to access mood disorders indicators in individuals with ID?	There are some procedures that can help clinical judgment. The observational method and use of psychological instruments (scales, inventories or semi-structured interviews) are some examples. In any case, they must not be used as the only source of information for the diagnosis.
How to assess depression disorders indicators in individuals with expressive language impairment?	It is recommended to assess these individuals according to behavioral equivalents for depression.

Problems	Recomendations
Some times, it is common to confound typical behavioral patterns of a syndrome with signs of mood disorders.	Improve knowledge about behavioral phenotype of the syndrome to stablish differences between typical behavioral patterns and mood disorders indicators.
High incidence of mental health problems in Prader Willi syndrome.	It is necessary to consider that the most frequent mental health and psychological complaints in PWS are the emotional-affective problems as withdrawal, mood changes and depression, in addition to psychotic disorders spectrum.
High incidence of language alterations and depression in the Williams syndrome.	During assessment of the Williams syndrome, it is necessary to consider that the language domain is characterized by strength in concrete vocabulary and grammatical abilities but weakness in relational/conceptual and pragmatic language. Thus, it is necessary to be careful about assessing verbal report of these individuals.
High incidence of attention deficit hyperactivity disorder in the Williams syndrome.	During assessment of the Williams syndrome, it is necessary to consider that performance on cognitive tests may be influenced by the signs of this disorder.
High incidenceof mental health problems in Down syndrome.	It is necessary to consider that people with DS may present several compatible symptoms with mood disorders, such as reduced interest and pleasure, depressed affect, psychomotor retardation, tearfulness, reduced energy, loss of appetite, sleep disorders, hypochondria, aggression or tantrums and reduced speech.

Table 1. Main problems and recommendations in the assessment of mood disorders in genetic syndromes associated with intelectual disability.

7. References

Achenbach, T (2011). Achenbach System of Empirically Based Assessment. In: *ASEBA,* august, 2011, Available from: < http://aseba.org/ >.

Achenbach, T. & Rescorla, L. (2001). *Manual for the ASEBA School-Age Forms & Profiles.* ISBN: 0938565737, University of Vermont, Burlington, VT: Research Center for Children, Youth & Families.

Adams, D.; Oliver, C.; Kalsy, S.; Peters, S.; Broquard, M.; Basra, T.; Konstandinidi, E. & Mcquillan, S. (2008). Behavioural characteristics associated with dementia assessment referrals in adults with Down syndrome. *Journal of Intellectual Disability Research,* Vol.52, No. 4, (april, 2008), pp. 358–368, ISSN: 0964-2663.

Adams, D. & Oliver, C. (2011). The Expression and Assessment of Emotions and Internal States in Individuals with Severe or Profund Intellectual Disabilities.*Clinical Psychology Review,* Vol. 31, No. 3, (april, 2011), pp. 293-306, ISSN: 02727358.

Aman, M.; Singh, N.; Stewart, A. & Field, C. (1985). Psychometric Characteristics of the Aberrant Behavior Checklist. *American Journal of Mental Deficiency*,Vol. 89, No.5, (march, 1985), pp. 492-502, ISSN: 0002-9351.

Aman, M.; Singh, N.; Stewart, A. & Field, C. (1985). The aberrant behavior checklist: a behavior rating scale for the assessment of treatment effects. *American Journal on Mental Deficiency*, Vol. 89, No. 5, (march, 1985), pp. 491-495, ISSN: 0002-9351.

American Association on Intellectual and Developmental Disabilities, (2011). In: *American Association on Intellectual and Developmental Disabilities*, august, 2011, Available from: < http://aaidd.org/ >.

American Association on Mental Retardation.(2006). *Retardo Mental Definição, Classificação e Sistemas de Apoio* (10 edição), Artmed, ISBN 85-363-0711-0, Porto Alegre, Brasil.

American Psychiatric Association. (2003). *Manual Diagnóstico e Estatístico de Transtornos Mentais-DSM IV-TR*. Texto Revisado. 4ª edição, Artmed, ISBN: 8573079851, Porto Alegre, Brasil.

American Psychiatric Association, (2010). In: American Psychiatric Association, DSM V. Proposed Revision, august, 2011 . Available from: <http://www.dsm5.org/ProposedRevision/Pages/proposedrevision>.

American Psychiatric Association. (1980). *Diagnostic and Statistical Manual of Mental Disorders - DSM-III*, (3 ed.) American Psychiatric Association, *ISBN:* 0-89042-051-3, Washington, DC.

American Psychiatric Association. (1987). *Diagnostic and Statistic Manual of Mental Disorders-III-revised*. (3ed.). American Psychiatric Association, ISBN: 0890420181, Washington, DC.

Antonell, A.; Campos, M.; Flores, R.; Campuzano, V. & Pérez-Jurado, L. (2006). A Síndrome de Williams: Aspectos Clínicos y Bases Molecurares. *Revista de Neurologia*, Vol. 42, No. 1, (january, 2006), pp. 69-75. ISSN: 0210-0010.

Arron, K.; Oliver, C.; Moss, J.; Berg, K. & Burbidge, C. (2011). The Prevalence and Phenomenology of Self-Injurious and Aggressive Behaviour in Genetic Syndromes. *Journal of Intellectual Disability Research*, Vol. 55, No. 2, (February, 2011), pp. 109-120, ISSN: 0964-2633.

Artigas-Pallarés, J. (2002). Fenotipos Conductales. *Revista de Neurología*, Vol 34, No 1, (february, 2002), pp. 38-48. ISSN: 1576-6578.

Artigas-Pallarés, J.; Gabau-Vila, E. & Guitart-Feliubadaló, M. (2006). Fenotipos Conductuales em el Retraso Mental de Origen Genético. *Revista de Neurologia*, Vol. 42, No. 1, (january, 2006), pp. 15-19, ISSN 1576-6578.

Bachega, T.; Biillerbeck A.; Prente, E.; Marini, S.; Baptista, T.; Mello, M.; Guerra, G.; Kuperman, H.; Setian, N.; Damiani, D.; Torres, N.; Castro, M. & Mendonça, B. (2004). Estudo Multicêntrico de Pacientes Brasileiros com Deficiência da 21-Hidroxilase: Correlação do Genótipo com o Fenótipo. *Arquivo Brasileiro de Endocrinologia e Metabogia*, Vol. 48, No. 5, (october, 2004), pp. 697-704. ISSN: 0004-2730.

Balci, T. & Ahmet, A. (2007). Correlative Study of the Cognitive Impairment Regional Cerebral Blood Flow and Electroencephalogram Abnormalities in Children with Down´s Syndrome. *The International Journal of Neuroscience*. Vol. 117, No.3, (march, 2007), pp. 327-335, ISSN: 0020-7454.

Basset, A.; Chow, E.; Malik, P.; Gheorghiu, M.; Husted, J. & Weksberg, R. (2003). The Schizophrenia Phenotype in 22q11 Deletion Syndrome. *The American Journal of Psychiatry,* Vol 160, No 9, (September, 2003), pp. 1580-1586, ISSN: 0002-953X

Bellugi, U.; Lichtenberger, L.; Jones, W.; Lai, Z. & George, M. (2000). The neurocognitive profile of Williams syndrome: a complex pattern of strengths and weaknesses. *Journal of Cognitive Neuroscience,* Vol.12, No 1, pp. 7-29, ISSN: 0898-929X.

Benarroch, F.; Hirsch, H.; Genstil, L.; Landay, Y. & Gross-Tsur, V. (2007). Prader-Willi syndrome: medical prevention and behavioral challenges. *Child and Adolescent Psychiatric clinics of North America,* Vol. 16, No. 3, (july, 2007), pp.695-708, ISSN: 1056-4993.

Benson, B. & Brooks, W. (2008). Aggressive Challeging Behaviour and Intellectual Disability. *Current Opinion in Psychiatry,* Vol.21, No.5, (September, 2008), pp. 454-458, ISSN: 0951-7367.

Bilder, R.; Sabb, F.; Cannon, E.; London, E.; Jentsch, J.; Parker, D.; Poldrack, R.; Evans, C. & Freimer, N. (2009) Phenomics: The Systematic Study of Phenotypes on a Genome-Wide Scale. *Neuroscience,* Vol. 164, No. 1(November, 2009) pp. 30-42. ISSN: 0306-4522.

Bodfish, J.; Symons, F.; Parker, D. & Lewis, M. (2000). Varieties of repetitive behavior in autism: Comparisons to mental retardation. *Journal of Autism and Developmental Disorders,* Vol. 30, No.3, (june, 2000),pp. 237–243. ISSN: 0162-3257.

Bolsoni-Silva, A.; Rodrigues, O.; Abramides, D.; Souza, L. & Loureiro, S. (2010) Práticas educativas parentais de crianças com deficiência auditiva e de linguagem. *Revista Brasileira de Educação Especial,* v. 16, n. 2, (may/August, 2010) p. 265-282, ISSN: 1413-6538

Brinkley, J.; Nations, L.; Abramson, R.; Hall, A.; Wright, H.; Gabriels, R.; Gilbert, J.; Pericak-Vance, M. & Cuccaro, M. (2007). Factor analysis of the Aberrant Behavior Checklist in Individuals with Austim Sperctrum Disorders. *Journal of Autism and Developmental Disorders,* Vol.37, No. 10, (November, 2007) pp.1949–1959, ISSN: 0162-3257.

Brown, C.; Aman, M. & Havercamp, S. (2002). Factor analysis and norms for parent ratings on the Aberrant Behavior Checklist-Community for young people in special education. *Research in Developmental Disabilities,* Vol. 23, No. 1, (January/February, 2002) pp. 45-60, ISSN: 0891-4222.

Cardoso-Martins, C.; & Silva, J. (2008). A relação entre o processamento fonológico e a habilidade de leitura: evidência da síndrome de Down e da síndrome de Williams. *Psicoogia Reflexão e Crítica,* Vol. 21, No. 1, pp.151-159, ISSN: 0102-7972.

Carr, J. (1994). Long term outcome for people with Down syndrome. *J Child Psychol Psychiatry,* Vol.35, No. 3, (march, 1994), pp.425- 439 , ISSN: 0021-9630.

Cassidy, S. & Driscoll, D. (2009). Prader-Willi Syndrome. *European journal of human genetics,* Vol 17, No 1, (2009), pp. 3-13. ISSN: 1018-4813.

Chambers, W.; Puig-Antich, J.; Hirsch, M.; Paez, P.; Tabrizi, M. & Davies, M. (1985). The Assessment of Affective Disorders in Children and Adolescents by Semistructured Interview: Test-Retest Reliability of the Schedule for Affective Disorders and Schizophrenia for School-Age Children, Present Episode Version. *Archives of General Psychiatry.* Vol.42, No 7, (july, 1985), pp: 696-702, ISSN: 0003-990X.

Clarke, D. (2003). Diagnostic Criteria for Psychiatric Disorders for Use with Adults with Learning Disabilities/ Mental Retardation (DC-LD) and Psychiatric Phenotypes.

Journal of Intellectual Disability Research, Vol. 47, No. 1 , (september/October, 2003), pp. 43-49. ISSN: 0964-2633.

Collacott, R.; Cooper, S.; Branford, D. & McGrother, C. (1998). Behaviour Phenotype for Down´s Syndrome. *The British Journal of Psychiatry,* Vol. 172, No. 1, (January, 1998), pp. 85-89. ISSN 0007-125.

Cooper, S. (2003). The Diagnostic Criteria for Psychiatric Disorders for Use with Adults with Learning Disabilities/Mental Retardation (DC-LD) papers. *Journal of Intellectual Disability Research,* Vol. 47, No.1, (September/October, 2003), pp. 1-2. ISSN: 0964-2633.

Cooper, S.; Melville, C. & Einfeld, S. (2003). Psychiatric Disorders, Intelectual disabilities and Diagnostic Criteria for Psychiatric Disorder for use with Adults with Learning Disabilities/ Mental Retardation (DC-LD). *Journal of Intellectual Disability Research,* Vol, 41, No. 1, (September, 2003), pp. 3-15, ISSN: 0964-2633.

Cooper, S.; Smiley, E.; Morrison, J.; Williamson, A. & Allan, L. (2007). An Epidemiological Investigation of Affective Disorders with a Population-Based Cohort of 1023 Adults with Intellectual Disabilities. *Psychological Medicine,* Vol 37, No 6, (June, 2007), pp. 873-882, ISSN: 0033-2917.

Craddock, N.; Jones, L.; Jones, IR.; Kirov, G.; Green, EK.; Grozeva, D.; Moskvina, V.; Nikolov, I.; Hamshere, ML; Vukcevic, D.; Caesar, S.; Gordon-Smith, K.; Fraser, C.; Russel, E.; Norton, N.; Breen, C.; St Clair, D.; Collier, DA.; Young, AH.; Ferrier, IN.; Farmer, A.; McGuffin, P.; Holmans, PA.; Donnelly, P.; Owen, MJ. & O´Donovan, MC. (2010). Strong Genetic Evidence for a Selective Influence of GABA Receptors on a Component of the Bipolar Disorder Phenotype. *Molecular Psychiatry,* Vol 15, No 2, (july, 2010), pp. 146-153, ISSN: 1359-4184.

Davies, M.; Udwin, O. & Howlin, P. (1998). Adults with Williams syndrome: preliminary study of social, emotional and behavioural difficulties. *The British Journal of Psychiatry*, Vol.172, No.1, (march, 1998), pp. 273-276, ISSN: 0007-1250.

Denis, J.; Noortgate, V. & Maes, B. (2011). Self-Injurious Behavior in People with Profund Intellectual Disabilities: A Meta-Analysis of Single-Case Studies. Research in Developmentl Disabilities, Vol. 32, No. 3 (may/june, 2011), pp. 911-923, ISSN: 0891-4222.

Down, J. (1995). Observations on an Ethnic Classification of Idiots. *Mental Retardation,* Vol. 33, No. 1, (February, 1995), pp.54-56 ISSN: 0047-6765.

Dykens, E. (2007). Psychiatric and behavioral disorders in persons with Down Syndrome. *Mental Retardation and Developmental Disabilities Research Reviews,* Vol.13, No.3, (2007), p.272-278, 2007.

Dykens, E. & Roof, E. (2008). Behavior in Prader-Willi syndrome: relationship to genetic subtypes and age. *Journal of Child Psychology and Psychiatry, and Allied Disciplines,* Vol. 49, No.9, (September, 2008), pp.1001-1008, ISSN: 0021-9630.

Dykens, E.; & Cassidy, S. (1995). Correlates of Maladaptive Behavior in Children and Adults with Prader-Willi Syndrome. *American journal of medical genetics,* Vol. 60, No. 6, (December, 1995), pp. 546-549, ISSN: 0148-7299.

Dykens, E. & Hodapp, R. (2007). Three steps toward improving the measurement of behavior in behavioral phenotype research. *Child and Adolescent psychiatric Clinics of North America,* Vol. 16, No. 3, (jully, 2007), pp.617–630, ISSN: 1056-4993.

Dykens, E.; Schwenk, K.; Maxwell, M. & Myatt, B. (2007). Vanderbilt Kennedy Center for Research on Human Development, Vanderbilt University, Nashville,Tennessee, USA The Sentence Completion and Three Wishes tasks: windows into the inner lives of people with intellectual disabilities. *Journal of Intellectual Disability Research,* Vol. 51 No. 8, (august, 2007) , pp. 588–597, ISSN: 0964-2633.

Dykens, E.; Shah, B.; Sagun, J.; Beck, T. & King, B. (2002). Maladaptive behaviour in children and adolescents with Down's syndrome. *Journal of intellectual disability research,* Vol. 46, No. 6,(september, 2002) pp. 484-492, ISSN: 0964-2633.

Egeland, J.; Gerhard, D.; Pauls D.; Sussex, J.; Kid, K.; Allen, C.; Hostetter, AM. & Housman, DE. (1987). Bipolar Affective Disorders linked to DNA Markers on Chromosome 11, *Nature,* Vol. 325, No:6107 (February, 1987) , pp. 783-787, ISSN: 0028-0836.

Fidler, D. (2005). The Emerging Down Syndrome Behavioral Phenotype in Early Childhood: Implications for Practice. *Infants & Young Children,* Vol. 18, No. 2, pp. 86–103, ISSN: 0896-3746.

Flint, J. (1998). Behavioral Phenotypes: Conceptual and Methodological Issues. *American Journal of Medical Genetics,* Vol. 81, No. 3, (may, 1998), pp. 235-240, ISSN: 0148-7299.

Fombonne, E.; Mazaubrun, C.; Cans, C. & Grandjean, H. (1997). Autism and associated medical disorders in a French epidemiological survey. *Journal of the American Academy of Child and Adolescent Psychiatry,* Vol. 36, No 11, (November, 1997), pp. 1561-1569. ISSN: 0890-8567.

Gallardo-Pujol, D.; Forero, C.; Maydeu-Olivares, A. & Andrés-Pueyo, A. Desa. (2009). Desarrollo del comportamiento antisocial: factores psicobiológicos, ambientales e interacciones genotipo-ambiente. *Revista de Neurologia,* Vol.48, No. 4, (February, 2009), pp.191-198, ISSN 0210-0010.

Goldstone, A.; Thomas, E.; Brynes, A.; Castroman, G.; Edwards, R.; Ghatei, M.; Frost, G.; Holland, A.; Grossman, A.; Korbonits, M.; Bloom, S. & Bell, J. (2004). Elevated Fasting Plasma in Prader-Willi Syndrome Adults is Not Solely Explained by Their Reduced Visceral Adiposity and Insulin Resistance. *The Journal of Clinical Endocrinology and Metabolism,* Vol. 89, No. 4, (april, 2004), pp. 1718-1726, ISSN: 0021-972X.

Goos, L. & Ragsdale, G. (2008). Genomic imprinting and human psychology: cognition, behavior and pathology. *Advances in Experimental Medicine and Biology,* Vol. 626, (2008), pp.71-88, ISSN: 0065-2598.

Graham, J.; Rosner, B.; Dykens, E. & Visootsak. J. (2005). Behavioral features of Charge syndrome (Hall-Hittner syndrome) comparison with Down syndrome, Prader-Willi syndrome, and Williams syndrome. *American Journal of Medical Genetics,* Vol.15, No. 133A3, (march, 2005) pp. 240-247, ISSN: 1552-4825.

Gunay-Aygun, M.; Schwartz, S.; Heeger, S.; O'Riordan, M.; & Cassidy, S. (2001). The Changing Purpose of Prader-Willi Syndrome Clinical Diagnostic Criteria and Proposed Revised Criteria. *Pediatrics,* Vol. 108, No.5, (November, 2001), ISSN: 0031-4005.

Harris, S.; Kasari, C. & Sigman, M. (1996). Join attention and language gains in children with Down syndrome. *American Journal of Mental Retardation,* Vol.100, No.6, (may, 1996), pp. 608- 619, ISSN: 0895-8017.

Hartley, S.; Maclean, W.; Butler, M.; Zarcone, J. & Thompson, T. (2005). Maladaptive Behaviors and Risk Factors among the Genetic Subtypes of Prader-Willi Syndrome.

American Journal of Medical Genetics Part A, Vol. 136, No. 2, (july, 2005), pp. 140-145, ISSN: 15524825.

Hayes, S.; McGuire, B.; O´Neill, M.; Oliver, C. & Morrison, T. (2011). Low Mood and Challenging Behaviour in People with Severe and Profund Intellectual Disabilities. *Journal of Intellectual Disability Research*, Vol. 55, No. 2, (February, 2011), pp. 182-189, ISSN: 0964-2633.

Hermans, H. & Evenhuis, H. (2010). Characteristics of Instruments Screening for depression in Adults with Intellectual Disabilities: Sistematic Review. *Research in Developmental Disabilities,* Vol. 31, No. 6.(November-December, 2010). pp.1109-1120. ISSN: 0891-4222.

Herreros, M.; Ascurra, M.; & Franco, R. (2007). El Síndrome de Williams-Reporte de tres casos. *Memorias del Instituto del Investigaciones en Ciencias de la Salud,* Vol.5, No.1, (june, 2007), pp.45-49, ISSN: 1812-9528 .

Hill, J.; Powlitch, S. & Furniss, F. (2008). Convergent Validity of the Aberrant Behavior Checklist and Behavior Problems Inventory with People with Complex Needs. *Research in Developmental Disabilities,* Vol. 29, No. 1,(January/February, 2008) pp. 45–60, ISSN: 0891-4222.

Hinton, E.; Holland, A.; Gellatly, M.; Sonil, S. & Owen, A. (2006). An investigation into food preferences and the neural basis of food-related incentive motivation in Prader-Willi syndrome. *Journal of Intellectual Disability Research.* Vol. 50, No. 9, (September, 2006), pp. 633-642, ISSN: 0964-2633.

Holland, A.; Whittington, J.; Butler, J.; Webb, T.; Boer, H. & Clarke, D. (2003). Behavior Phenotypes Associated with Specific Genetic Disorders: Evidence from a Population-Based Study of People with Prader-Willi Syndrome. *Psychological Medicine* , Vol. 33, No 1, (January, 2003), pp. 141-153, ISSN 0033-2917.

Holm, V.; Cassidy, S.; Butler, M.; Hanchett, J.; Greenswag, L.; Whitman, B. & Greenberg, F. (1993). Prader-Willi syndrome: concensus diagnostic criteria. *Pediatrics.* Vol. 91:, No.2 (february, 1993), pp.398-402, ISSN: 0031-4005.

Jarvinen-Pasley, A.; Bellugi, U.; Reilly, J.; Mills, D.; Galaburda, A.; Reiss, A. & Korenberg, J. (2008). Defining the Social Phenotype in Williams Syndrome: A Model for Linking Gene, the Brain, and Behavior. *Development and Psychopathology,* Vol .20, No. 1, pp. 1-35. ISSN: 0954-5794.

Jauregi, J.; Arias, C.; Vegas, O.; Martinez, S.; Copet, P. & Thuilleaux, D. (2007). A neuropsychological assessment of frontal cognitive functions in Prader-Willi syndrome. . *Journal of Intellectual Disability Research,* Vol. 51, No.5, (may, 2007), pp. 350-65. ISSN: 0964-2633.

Jin, D. (2011). Systematic Review of the Clinical and Genetic Aspects of Prader-Willi Syndrome. *Korean Journal of Pediatrics,* Vol. 42, No. 2, (February, 2011), pp. 55-63. ISSN: 1738-1061.

Joyce, P. (2008). Classification of Mood Disorders in DSM-V and DSM-VI. *Australian and New Zealand Journal of Psychiatry.* Vol. 42, No.10, (October, 2008), pp. 851-862, ISSN: 0004-8674.

Kaland, N.; Smith, L. & Mortensen, E. (2008). Brief report: cognitive flexibility and focused attention in children and adolescents with Asperger syndrome or high-functioning autism as measured on the computerized version of the Wisconsin Card Sorting

Test. *Journal of Autism and Developmental Disorders*, Vol.38, No.6, (july, 2008) pp. 1161-1165, ISSN: 0162-3257.

Karabekiroglu, K. & Aman, M. (2009).Validity of the Aberrant Behavior Checklist in a clinical Sample of Toddlers. *Child Psychiatry and Human Development*, Vol 40, No. 5, (march, 2009), pp. 99-110, ISSN: 0002-9351.

Kaufman, J.; Birmaher, B.; Brent, D.; Rao, U.; Flynn, C.; Moreci, P.; Williamson, D.; & Ryan, N. (1997). Schedule for Affective Disorders and Schizophrenia for School-Age Children-Present and Lifetime Version (K-SADS-PL): Initial Reliability and Validity Data. *Journal of the American Academy of Child & Adolescent Psychiatry*. Vol. 36, No. 7, (july, 1997), pp. 980-988, ISSN: 0890-8567.

Keil, M.; Merke, D.; Wiggs, E.; Obunse, K.; & Stratakis, C. (2009). Quality of life in children and adolescents 1-year after cure of Cushing syndrome: a prospective study. *Clinical Endocrinology*, Vol. 71, No. 3, (September, 2009), pp.326-333, ISSN: 0300-0664.

Kelsoe, J.; Ginns, E.; Egeland, J.; Gerhard, D.; Goldstein, A.; Bale, S.; Pauls, D.; Long, R.; Kidd, K. & Conte, G. (1989). Re-evaluation of the Linkage Relationship Between Chromosome 11p loci and the Gene for Bipolar Affective Disorder in the Old Order Amish. *Nature*, Vol. 342, No. 6247, (November, 1989), pp. 238-243, ISSN: 0028-0836.

Koenig, K.; Klein, A. & Schultz, R. (2004). Deficits in social attribution ability in Prader-Willi syndrome. *Journal of Autism and Developmental Disorders*, Vol.34, No.5, (October, 2004), pp. 573-582, ISSN: 0162-3257.

Koskentausta, T.; Livanainen, M. & Almqvist. (2007). Risk Factors for Psychiatric Disturbance in Children with Intellectual Disability. *Journal of Intellectual Disability Research*, Vol. 51, No. Pt1, (January, 2007), pp. 43-53, ISSN: 0964-2633

Leyfer, O.; Woodruff-Borden, J.; Klein-Tasman, B.; Fricke, J. & Mervis, C. (2006). Prevalence of psychiatric disorders in 4-16 years- olds with Williams syndrome. *American Journal of Medical Genetics Part B: Neuropsychiatric Genetics*, Vol.141B, No.6, (September, 2006) pp. 615-622, ISSN: 1552-4841.

Lindgren, A.; Barkeling, B.; Hãgg, A.; Ritzén, E.; Marcus, C. & Rõssner, S. (2000). Eating Behavior in Prader-Willi Syndrome, Normal Weight, and Obese Control Groups., *The Journals of Pediatrics*, Vol. 137, No.1, (july, 2000), pp. 50-55, ISSN: 0022-3476.

Martens, M.; Wilson, S. & Reutens, D. (2008). Research Review: Williams syndrome: a critical review of the cognitive, behavioral, and neuroanatomical phenotype. *Journal of Child Psychology and Psychiatry, and Allied Disciplines*, Vol. 49, No.6, (june, 2008), pp. 576-608, 2008. ISSN:0021-9630.

Martin, G.; Pear, J. (2009). *Modificação do Comportamento: O que é e como Fazer*. 8ª.ed., Roca, ISBN: 8572418253,São Paulo, SP, Brasil.

Matson, J. & Laud. R. (2007).Assessment and Treatment Psychopatology among people with Developmental Delays. In: *Handbook of Intellectual and Developmental Disabilities*. Jacobson, J., Mulick, J., Rojahn, J (Eds), pp. 507- 539,Springer Science, ISBN: 0387329307 , New York, USA.

Matson, J.; Bamburg, J.; Cherry, K. & Paclawskyj, T. (1999). A Validity Study on the Questions about Behavioral Function (QABF) Scale: Predicting Treatment Success for Self-Injury, Agression, and Stereotypies. *Research in Developmental Disabilities*, Vol. 20, No.2, (march/april, 1999), pp. 163-175, ISSN: 0891-4222.

Mervis, C. & Becerra, A. (2007). Language and communicative Development in Williamns Syndrome. *Mental Retardatino and Developmental Disabilites Research Reviews*, Vol.13, No. 1,(2007), pp.3-15, ISSN: 1080-4013.

Mervis, C. & John, A. (2010). Cognitive and Behavioral Characteristics of Children With Williams Syndrome: Implications for Intervention Approaches. *American Journal of Medical Genetics Part C* (Seminars in Medical Genetics), Vol.154C, No.2, (may, 2010),pp. 229–248, ISSN: 1552-4868.

Mesquita M. , Brunoni D., Neto J. M. P., Kim C. A., Melo M. H .S., Teixeira M .C. T. V. (2010), Behavioral Phenotype Of Children And Adolescentes With Prader-Willi Syndrome. *Revista Paulista de Pediatria* 28, Vol. 28 , No.1 , (march, 2010), pp. 63-69, ISSN: 0103-0582.

Meyer-Lindenberg, G.; Mervis, C. & Berman, K. (2006). Neural Mechanisms in Williams Syndrome: a Unique Window to Genetic Influences on Cognition and Behaviour. *Nature Reviews Neuroscience,* Vol. 7, No 5, (may, 2006) pp. 380-393. ISSN: 1471-003X.

Moldavsky, M.; Lev, D. & Lerman-Sagie, T. (2001). Behavioral Phenotypes of Genetic Syndromes: A Reference Guide for Psychiatrists. *Journal of the American Academy of Child and Adolescent Psychiatry,* Vol. 40, No. 7, (july, 2001), pp. 749-761. ISSN: 0890-8567.

Moeschler, J. (2008). Genetic Evaluation of Intellectual Disabilities. *Seminars in Pediatric Neurology,* Vol. 15, No.1 (march, 2008), pp. 2-9, ISSN: 1071-9091.

Mutkins, E.; Brown, R. & Thorsteinsson, E. (2011). Stress, Depression, Workplace and Social Supports and Burnout in Intellectual Disability Support Staff. *Journal of Intellectual Disability Research,* Vol. 55, No. 5, (may, 2011), pp. 500-510, ISSN: 09642633.

Mota, E.; Bertola, D.; Kim, C. & Teixeira, M. (2010). Behavioral Disturbances in Noonan Syndrome: Brazilian Preliminary Data. *Rev Latino Americana Psicologia,* Vol. 42, No. 1, (april, 2010), pp. 87-95. ISSN: 0120-0534.

Narita, H.; Odawara, T.; Iseki, E.; Kosaka, K. & Hirayasu, Y. (2004). Psychomotor retardation correlates with frontal hypoperfusion and the Modified Stroop Test in patients under 60-years-old with major depression. *Psychiatry and Clinical Neurosciences,* Vol. 58, No. 4, (august, 2004), pp. 389-395, ISSN: 1323-1316.

Nyhan, W. (1972). Behavioral Phenotypes in Organic Genetic Disease. Presidential address to the Society for Pediatric Research 1971.*Pediatric Research,* Vol.6, No.1, (January, 1972), pp. 1-9, ISSN: 0031-3998.

Nyhan, W. (1976). Behavior in the LeschNyhan syndrome. *Journal of Autism and Childhood Schizophrenia,* Vol.6, No.3, (September, 1976), p.235-252, ISSN: 0021-9185.

O´Brien, G. (2000). Behavioural Phenotypes. *A Journal of the Royal Society of Medicine,* Vol. 93, No.12, (December, 2000), pp. 618-620. ISSN: 0141-0768.

Organização Mundial da Saúde. Cid-10 (2000): *classificação estatística internacional de doenças e problemas relacionados à saúde.* 8. ed. EDUSP, ISBN: 8531401933,São Paulo, Brasil.

Pérez-García, D.; Granero, R.; Gallastegui, F.; Pérez-Jurado, L. & Brun-Gasca. (2011) Behavioral features of Williams Beuren syndrome compared to Fragile X syndrome and subjects with intellectual disability without defined etiology. *Research in Developmental Disabilities,* Vol. 32, No 2, (march/april, 2011), pp.643-652, ISSN: 0891-4222.

Polk, T.; Drake, R.; Jonildes, J; Smith, M. & Smith, E. (2008). Attention enhances the neural processing of relevant features and suppresses the processing of irrelevant features in humans: a functional magnetic resonance imaging study of the Stroop task. *Neuroscience,* Vol.17; N. 28,(December, 2008), pp.13786-13792, ISSN: 0270-6474.

Porter, M. & Dodd, H. (2011). A Longitudinal Study of Cognitive Abilities in Williams Syndrome. *Developmental Neuropsychology*. Vol. 36, No.2, (February, 2011), pp. 255-272, ISSN: 8756-5641.

Pun, F.; Zhao, C.; Lo, W.; Ng, S.; Tsang, S.; Nimgaonkar, V.; Chung, W., Ungvari, G.; & Xue, H. (2011). Imprinting in the Schizophrenia Candidate Gene GABRB2 Encoding GABA Receptor B2 Subunit. *Molecular Psychiatry,* Vol. 16, No 5, (may, 2011), pp. 557-568, ISSN: 1359-4184.

Ridha, J. (2011). The 1000 Genomes Project: Deep Genomic Sequencing Waiting for Deep Psychiatric Phenotyping. *Journal of Psychiatry and Neuroscience,* Vol 36, No 3, (may, 2011), pp. 147-149. ISSN: 1180-4882.

Rojahn, J.; Matson, J.; Lott, D.; Esbensen, A. & Smalls, Y. (2001). The behavior problems inventory: An instrument for the assessment of self-injury, stereotyped behavior and aggression / destruction in individuals with developmental disabilities. *Journal of Autism and Developmental Disorders,* Vol. 31, No. 6, (December, 2001), pp. 577–588, ISSN: 0162-3257.

Rojahn, J.; Tassé, M. & Sturmey, P. (1997).The Stereotyped Behavior Scale for Adolescents and Adults with Mental Retardation. *American Journal on Mental retardation,* Vol.102, No. 2, (September, 1997), pp. 137-156, ISSN: 0895-8017.

Rossi, N.; Moretti-Ferreira, D.; & Giacheti, C. (2009). Perfil da fluência da fala na síndrome de Williams-Beuren: estudo preliminar. *Pró-Fono Revista de Atualização Científica,* Vol.21, No.2, (june, 2009), pp.107-112, ISSN: 0104-5687.

Rossi, N.; Moretti-Ferreira, D. & Giacheti, C. (2006). Genética e linguagem na Síndrome de Williams-Beuren: uma condição neuro-cognitiva peculiar. *Pró-Fono Revista de Atualização Científica,* Vol.18, No.3, (october, 20006) pp. 331-38, ISSN: 0104-5687.

Ruggieri, V. & Arberas, C. (2003). Fenotipos conductales. Patrones neuropsicológicos biológicamente determinados. *Revista de Neurologia,* Vol 37, No. 3, (august, 2003), pp. 239-253, ISSN: 0210-0010.

Segin, M. (2010). *Caracterização de habilidades lingüísticas de crianças e adolescentes com Síndrome de Williams-Beuren.* Dissertação de mestrado, Universidade Presbiteriana Mackenzie, São Paulo, SP, Brasil.

Semenza, C.; Pignatti, R.; Bertella, L.; Ceriani, F.; Mori, I.; Molinari, E.; Giardino, D.; Malvestiti, F. & Grugni, G.(2008). Genetics and mathematics: evidence from Prader-Willi syndrome. *Neuropsychologia,* Vol 46, No.1, (january, 2008), pp. 206-12, ISSN: 0028-3932.

Sherman, S.; Allen, E.; Bean, L. & Freeman, S. (2007). Epidemiology of Down Syndrome. *Mental Retardation and Developmental Disabilities Research Reviews.* Vol. 13, No 3, pp. 221-227, ISSN: 1080-4013.

Siegel, M. & Smith, W. (2010). Psychiatric Features in Children with Genetic Syndromes: Toward Functional Phenotypes. *Child Adolesc Psychiatric Clin N Am.* Vol. 19, No.2 , (april, 2010), pp. 229-261, ISSN: 1056-4993.

Silvares, E. (2000). *Estudos de Caso em psicologia Clínica Comportamental Infantil.* 5ª. Ed., Papirus, ISBN: 9788530805999, Campinas, São Paulo, Brasil.

Sinnema, M.; Einfeld, S.; Schrander-Stumpel, C.; Maaskant, M.; Boer, H. & Curfs, L. (2011). Behavioral Phenotypes in Adults with Prader-Willi Syndrome. *Research in Developmental Disabilities,* Vol. 32, No.2, (January, 2011), pp. 604-612,ISSN: 08914222.

Smilley, E. & Cooper, S. (2003). Intellectual disabilities, depressive episode, diagnostic criteria and Diagnostic Criteria for Psychiatric Disorders for Use with Adults with Learning Disabilities/Mental Retardation (DC-LD). *Journal of Intellect Disability Research,* Vol. 47, No. 1, (September, 2003), pp. 62-71, ISSN: 0964-2633.

Sommer, C. & Henrique-Silva. (2008). Trisomy 21 and Down syndrome: a short review. *Brazilian Journal of Biology,* Vol. 68, No.2, (may, 2008), p.447-452, ISSN: 1519-6984.

Soni, S.; Whittington, J.; Holland, A.; Webb, T.; Maina, E.; Boer, H. & Clarke, D. (2008). The Phenomenology and Diagnosis Of Psychiatric Illness in People with Prader-Willi Syndrome. *Psychological Meicine,* Vol. 38, No. 10, (october, 2008), pp. 1505-1514, ISSN: 0033-2917.

Sturmey, P.; Laud, R.; Cooper, C.; Matson, J. & Fodstad, J. (2010).Challenging behaviors should not be considered depressive equivalents in individuals with intellectual disabilities. II. A replication study. *Research in Developmental Disabilities,* Vol. 31, No.5, (September/October, 2010), pp. 1002-1007, ISSN: 0891-4222.

Sugayama, S.; Leone, C.; Chauffaille, M.; Okay, T. & Kim, C. (2007). A Síndrome de Williams: proposta de sistema de pontuação para diagnóstico clínico. *Clinics,* Vol.62, No.2, pp.159-166, ISSN: 1807-5932.

Switaj, (2000). DM. *Identification and measurement of anxiety and obsessive-compulsive tendencies in the Williams syndrome behavioral phenotype.* Philadelphia: Temple University., ISBN: 0599706341, Retrieved from <http://sunzi1.lib.hku.hk/ER/detail/hkul/2986472>.

Tartaglia, N.; Cordeiro, L.; Howell, S.; Wilson, R. & Janusz, J. (2010). The Spectrum of the Behavioral Phenotype in Boys and Adolescents 47, XXY (Klinefelter Syndrome). *Pediatric Endocrinology Reviews,* Vol. 8, No.1, (December, 2010), pp. 151-159, ISSN: 1565-4753.

Teixeira, M.; Monteiro, C.; Velloso, R.; kim, C. & Carrero, L. (2010). Fenótipo comportamental e cognitivo de crianças e adolescentes com Síndrome de Williams-Beuren. *Pró-Fono Revista de Atualização Científica,* Vol. 22, No. 3, (July/september, 2010), pp. 215-220, ISSN: 0104-5687.

Tsiouris, J.; Kim, S.; Brown, W. & Cohen, I. (2011). Association of Aggressive Behaviours with Psychiatric Disorders, Age, Sex and Degree of Intellectual Disability: a Large-Scale Survey. *Journal of Intellectual Disability Research,* Vol. 55, No. 7, (july, 2011), pp. 636- 649, ISSN: 0964-2633.

Walker, J.; Dosen, A.; Buitelaar, J. & Janzing, J. (2011). Depression in Down Syndrome: A review of the literature. *Research in Developmental Disabilities,* Vol.32, No.3, (September/October, 2011), pp. 1432–1440, ISSN: 0891-4222.

Wetherby, A.; Allen, L.; Cleary, J.; Kublin, K. & Goldstein, H. (2002). Validity and Reliability of the Communication and Symbolic Behavior Scales Developmental Profile with Very Young Children. *Journal of speech, language, and hearing research,* Vol. 45, No.6, (December, 2002), pp. 1202-1218, ISSN: 1092-4388.

5

Mood Disorders and Cardiovascular Disease

Jennifer L. Gordon, Kim L. Lavoie,
André Arsenault, Blaine Ditto and Simon L. Bacon
*McGill University, Université du Québec à Montréal,
Montreal Heart Institute, Concordia University,
Montreal Behavioural Medicine Centre
Canada*

1. Introduction

1.1 Cardiovascular disease (CVD)

Cardiovascular disease (CVD) accounts for 29% of global deaths, making it the leading cause of mortality worldwide. In 2004, 17.1 million people died from CVD and this number is expected to increase to 23.6 million by 2030 (World Health Organization, 2007). The most important mechanism contributing to the development and progression of CVD is atherosclerosis, the process by which fatty plaques accumulate on the inner walls of arteries, leading to their narrowing and loss of elasticity. This process is thought to generally begin with damage to the vascular endothelium, which is a layer of cells lining the vascular wall. The endothelium plays a critical role in many important functions, such as the dilation and constriction of blood vessels and arteries, thrombus (blood clot) formation, and inflammation (Quyyumi, 2003). When damage to the endothelium occurs, low-density-lipoproteins (LDL's), commonly known as "bad" cholesterol, permeate the endothelial lining, allowing them to enter the inner layer of the arteries. An inflammatory response, the immune system's attempt to promote self-repair, is triggered. As part of this inflammatory response, macrophages, whose function is to ingest and decompose pathogens found in the body, engulf these LDL particles and form what are called "foam cells". The accumulation of foam cells in the arterial wall form a "fatty streak", a yellowish slightly raised area that is the precursor to atherosclerotic plaques. If the endothelium continues to be damaged, triggering an escalation of the inflammatory response, a fibrous cap eventually covers the lesion, forming a hard plaque. The arterial wall calcifies and hardens. The formation of the plaque and the hardening of the artery cause obstruction of blood flow. If a plaque cap is unstable and becomes damaged, a thrombus can form, increasing the risk of a myocardial infarction (heart attack) or stroke (Stanner, 2005), which occur when blood flow to the heart or brain are completely obstructed. Given that endothelial damage and inflammation are so critical to the atherosclerotic process, factors that damage the endothelium or promote the inflammatory response indirectly contribute to atherosclerosis and therefore CVD.

1.2 Risk factors for CVD

A variety of factors have been shown to contribute to the development of CVD. Hypertension, for example, is thought to contribute to CVD by impairing endothelial function. It is believed to do so by decreasing the endothelium's production of an important substance called nitric oxide (Panza et al., 1993). Nitric oxide plays an important role in promoting vasodilation and inhibiting inflammation and platelet aggregation and is key in maintaining good endothelial function and therefore preventing CVD.

Insulin resistance, a condition where normal levels of insulin are inadequate to trigger glucose absorption by liver, muscle and fat cells, has also been shown to contribute to endothelial dysfunction (Laine et al., 1998; Steinberg et al., 1996) and is also thought to do so by impairing nitric oxide production (Petrie et al., 1996). Insulin resistance, and the hyperglycemia (elevated blood sugar) which often accompanies it, is also believed to promote inflammation and fatty streak formation in a variety of complex mechanisms (Nigro et al., 2006; Reusch & Draznin, 2007).

Dyslipidemia, characterized by high LDL and triglyceride levels, also contributes to CVD by damaging the endothelium and by being directly involved in the formation of foam cells (Maggi et al., 1994). Having low levels of HDL (high density lipoprotein or "good" cholesterol) is also detrimental since HDL protects against atherosclerosis. In fact, research suggests that increasing HDL levels may be a therapeutic method to reversing plaque accumulation (Lee & Choudhury, 2007), partly through its beneficial effect on endothelial function (O'Connell & Genest, 2001).

Several psychological variables have also been found to contribute to CVD development. For example, the famous INTERHEART study found that psychosocial factors represented the third most important risk factor for myocardial infarction, with only smoking and high cholesterol being associated with more risk (Yusuf et al., 2004). Numerous studies have since confirmed that several psychosocial factors, including depression, anxiety, hostility, lack of social support, and chronic life stress (Rozanski et al., 1999), play an important role in contributing to CVD. However, depression is arguably the most important of these factors.

1.3 Depression and CVD

Depressed individuals exhibit many of the risk factors that contribute to atherosclerosis, including hypertension (Davidson et al., 2000; Jonas et al., 1997; Rabkin et al., 1983) and insulin resistance (Everson-Rose et al., 2004; Timonen et al., 2005). Unsurprising then, both depressive symptoms (Sherwood et al., 2005) and diagnosed major depression (Rajagopalan et al., 2001) are associated with endothelial dysfunction as well as vascular inflammation (Joynt et al., 2003). As mentioned above, this increased incidence of CVD risk factors in depressed individuals translates to increased risk of CVD. One meta-analysis of prospective studies examining the development of CVD in initially healthy individuals found that major depression was associated with a 2.5-fold increased risk of developing CVD compared to non-depressed individuals (Van der Kooy et al., 2007). Another meta-analysis found that clinical depression in CVD patients was also associated with a 2- to 2.5-fold increased risk of all-cause mortality compared to those without depression (van Melle et al., 2004).

Although it is clear that depression is associated with many CVD risk factors and that it predicts the development of CVD, the mechanisms behind this association are unclear. Two main pathways have been proposed (Joynt et al., 2003): the first assumes that depression is linked to CVD and its risk factors through an imbalance or dysregulation in the physiological systems involved in the stress response: the autonomic nervous system (ANS) and the hypothalamic-pituitary-adrenal (HPA) axis. The second proposes that depressed patients' tendency to engage in poor health behaviours, such as smoking and over-eating, is responsible for their unhealthy cardiovascular profiles. In this chapter, we will examine the evidence supporting each of these potential pathways (see Fig. I).

2. Physiological pathways

2.1 The autonomic nervous system (ANS) and sympathoadrenal (SA) system

2.1.1 The ANS, SA system, and CVD

One of the most important physiological systems involved in the stress response is the ANS. There are two branches that work together to form the ANS. On the one hand, the parasympathetic nervous system (PNS) releases the neurotransmitter acetylcholine to promote a resting state, decreasing heart rate and blood pressure. On the other hand, the sympathetic nervous system (SNS) releases the neurotransmitter adrenaline to prepare the body for action during times of stress or alertness by increasing heart rate, blood pressure, and the heart's contractile force (D. S. Goldstein, 2006). The SNS branch of the ANS also promotes the release of adrenaline and noradrenaline in the form of hormones through the sympathoadrenal (SA) system. Though both the neurotransmitter and hormonal forms of adrenaline have similar functions, the hormonal form has a more systemic effect on the body.

While the SNS plays an important role in preparing the body to react to danger, chronic activation of the SNS can have negative consequences for the cardiovascular system. Many indicators of dominance of the SNS over the PNS have been shown to predict future cardiac mortality. For example, reduced heart rate variability (variation in the amount of time between heart beats), an indicator of increased SNS relative to PNS activity, predicts myocardial infarction and cardiac mortality (La Rovere et al., 1998; Nolan et al., 1998; Tsuji et al., 1994). Impaired heart rate recovery (i.e. drop in heart rate) after exercise (Cole et al., 2000; Farrell et al., 1992; Mora et al., 2003; Morshedi-Meibodi et al., 2002; Nishime et al., 2000) as well as reduced baroreflex sensitivity, the ANS' ability to alter heart rate in response to changes in blood pressure, two known indicators of sympathetic dominance, have similarly been found to predict cardiac outcomes (Farrell et al., 1992; La Rovere et al., 1998). SNS hyperactivity is likely related to cardiac mortality by causing instability in the heart's electrical system, causing the heart to beat abnormally (arrhythmias), which can cause sudden cardiac death (Coker et al., 1984). In addition, sympathetic dominance likely increases one's risk of CVD through disruption of the SA system.

The SA system, as mentioned above, is also greatly involved in the body's stress response. When the hormonal forms of adrenaline and noradrenaline are released into the bloodstream, some of their main effects on the cardiovascular system include altering heart rate and the heart's contractile force, as well as promoting the release of stored sugar into the blood stream. Adrenaline and noradrenaline also impact the cardiovascular system by stimulating the HPA axis, whose effects are described below.

As is the case with the neurotransmitter form of adrenaline, chronically high levels of plasma adrenaline and noradrenaline can contribute to the development of many of the

CVD risk factors observed in depressed patients. For example, chronically high levels of adrenaline and noradrenaline can induce hypertension by increasing blood flow (Esler, 2000) and promote the development of insulin resistance by preventing muscle cells from absorbing the insulin that is in the blood stream, (Lembo et al., 1993; Mancia et al., 2007). Plasma adrenaline and noradrenaline can also cause blood platelets to aggregate, an essential step in the formation of a thrombus (Anfossi & Trovati, 1996).

2.1.2 ANS and depression

Much research supports the hypothesis that an imbalance in the ANS, and subsequently the SA system, is a mechanism linking depression to CVD. For example, numerous studies have found both clinical depression (Carney et al., 2001) and self-reported depressive symptoms (Guinjoan et al., 2007) to be associated with poorer heart rate variability, indicating SNS dominance. Depressed mood has also been shown to be associated with markers of decreased parasympathetic cardiac control in reaction to laboratory stressors (Hughes & Stoney, 2000). Depression has also been associated with elevated resting heart rate (Dawson et al., 1977; Lahmeyer & Bellur, 1987), poor heart rate recovery after exercise (Gordon et al., 2011), reduced baroreflex cardiac control (Watkins & Grossman, 1999) and increased systolic blood pressure during exercise (Pelletier et al., 2009), once again, indicating sympathetic dominance. Finally, depression is associated with electrical instability of the heart (Nahshoni et al., 2000; Yeragani et al., 2000), likely a consequence of ANS dysfunction, increasing the risk of arrhythmias (Carney et al., 1993).

Depressed individuals have also been found to exhibit signs of a hyperactive SA system. For example, we have long known that depressed patients exhibit higher plasma noradrenaline levels, thought to parallel SA activity, compared to non-depressed controls (Esler et al., 1982; Lake et al., 1982; Roy et al., 1988; Veith et al., 1984; Wyatt et al., 1971). One study had 60 healthy women complete the Beck Depression Inventory (BDI) and compared the catecholamine levels of the 15 women with highest scores to the 15 with the lowest scores in response to a public speaking task. The "depressed" women were found to have higher plasma noradrenaline levels (in addition to greater blood pressure and several measures of sympathetic activation) during the task compared to the "non-depressed" group (Light et al., 1998). Another study examined BDI scores and physiological recovery to a stress-inducing public speaking task in non-depressed women. It found that although all participants exhibited high plasma adrenaline levels immediately after the stressor, only the women with scores in the high-normal range had high adrenaline levels 45 minutes post-stressor (Gold et al., 2004). In contrast, participants with low BDI scores had similar adrenaline levels to non-stressed controls within 30 minutes of the stressor. These analyses adjusting for perceived stress during the task, reducing the possibility that the more "depressed" participants simply found the task more psychologically stressful. Collectively, these studies suggest that depressive symptoms, even when in the sub-clinical range, are associated with enhanced SA system activation, both in general and in response to stressful tasks. Furthermore, the latter study suggests that this physiological response does not result simply from depressed people feeling more psychologically distressed in response to stressful tasks than non-depressed people. Instead, depressive symptoms appear to be associated with an imbalance in ANS cardiac control.

2.2 The HPA axis

2.2.1 The HPA axis and CVD

The HPA axis, through its release of hormones called glucocorticoids, the most important of which is cortisol in humans, also plays a pivotal role in the body's stress response, though its effects are delayed but longer-lasting than the ANS'. As with adrenaline, increased cortisol release is helpful during stress but is thought to contribute to the development of CVD when chronically activated (G. E. Miller et al., 2002). This is most apparent in patients with Cushing's Syndrome, a disorder characterized by the hypersecretion of cortisol, who are at four times the risk of developing CVD compared to the general population (Arnaldi et al., 2004; Mancini et al., 2004; Whitworth et al., 2000). Even people with sub-clinical Cushing's Syndrome (Tauchmanova et al., 2002) and people who have been cured from this disorder for five years (Colao et al., 1999) are at an increased risk of developing CVD. Cortisol exposure while awake (Dekker et al., 2008) and decline in cortisol levels throughout the day (Matthews et al., 2006) are also independently associated with atherosclerosis. A recent study has also shown that urinary cortisol levels predict cardiac mortality - in fact those in the highest tertile of urinary cortisol exhibit a 5-fold increased risk in CV death (Vogelzangs et al., 2010).

Elevated cortisol levels are thought to increase one's risk of developing CVD through several mechanisms. For example, chronically elevated levels of cortisol can lead to the development of hypertension, which has long been known to be a consequence of Cushing's Syndrome (Whitworth et al., 2000). Many placebo-controlled studies have confirmed that a dose-response relationship exists between cortisol and blood pressure in healthy individuals as well (Whitworth et al., 1989; Whitworth et al., 1984; Williamson et al., 1996). Cortisol's hypertension-inducing effect can be partially explained by its tendency to increase both salt and water retention (Panarelli et al., 1998) and to constrict the blood vessels by inhibiting its production of nitric oxide, a substance that dilates the blood vessels (Kelly et al., 1998). High cortisol levels can also contribute to insulin resistance by impairing insulin-dependent glucose uptake and enhancing glucose production in the pancreas (Andrews & Walker, 1999; Reynolds & Walker, 2003).

Cortisol levels are associated with the development of endothelial dysfunction in both clinical and sub-clinical cases of Cushing's Syndrome (Baykan et al., 2007; Colao et al., 1999). Studies blocking the production of cortisol have discovered that it is also responsible for the endothelial dysfunction observed in healthy participants in response to acute stress, for example, a public speaking task (Broadley et al., 2005). Though this effect may be partially due to cortisol's effect on blood pressure and insulin resistance, it is also thought to induce endothelial dysfunction by inducing cell apoptosis, a series of biochemical events that cause endothelium cell death (Vogt & Schmid-Schonbein, 2001) and decreasing the endothelium's production of nitric oxide (Johns et al., 2001).

Elevated cortisol levels also causes the accumulation of abdominal obesity, a well-known risk factor for CVD (Hubert et al., 1983). The effect of cortisol on fat deposition is not only clear in Cushing's Syndrome patients, studies also show cortisol levels to be associated with abdominal obesity in healthy people (Fraser et al., 1999). Research suggests this may help explain the association between excess cortisol and high cholesterol and triglyceride levels (Walker et al., 2000).

2.2.2 The HPA axis and depression

There is a great deal of research suggesting that depressed individuals have a hyperactive HPA axis. Numerous studies have found that depressed patients exhibit abnormally high cortisol and corticotropin releasing factor levels, another byproduct of the HPA axis (Plotsky et al., 1998). Depressed coronary artery disease patients have also been found to exhibit lower cortisol levels in the morning and higher levels in the evening (i.e. "flatter" cortisol rhythms) compared to both non-depressed coronary artery disease patients and depressed patients without coronary artery disease (Bhattacharyya et al., 2008). That depressed individuals fail to show glucocorticoid suppression in response to the dexamethasone suppression test demonstrates that the abnormally high cortisol levels seen in depressed patients could be due to a dysfunctional feedback system (Carroll et al., 1968; Schatzberg et al., 1984).

Depressed people have also been found to exhibit abnormal cortisol responses to psychological stressors. A meta-analysis of seven laboratory studies (Burke et al., 2005) found that depressed people exhibit much higher post-stress recovery cortisol levels compared to non-depressed controls. Another study examining cortisol responses to daily stressors (Peeters et al., 2003) found that depressed individuals exhibited no change in cortisol levels in response to the stressors. Similarly, a study examining cortisol responses to orthopedic surgery found that while chronically depressed patients had significantly higher cortisol levels than non-depressed controls before the surgery, their cortisol levels failed to increase in response to the stress as did the non-depressed patients' (Kudoh et al., 2000).

Abnormalities in the HPA axis are also believed to be involved in the vascular inflammation observed in depressed patients. Although depressed patients are believed to have abnormally high levels of circulating glucocorticoids, which have anti-inflammatory properties, their glucocorticoid immune receptors are thought to be desensitized, perhaps because of their chronic over-exposure to cortisol, and therefore less responsive to the anti-inflammatory actions of circulating glucocorticoids (Cooney & Dinan, 1996).

In summary, there are several physiological pathways by which depression may confer risk for CVD. However, evidence suggests that health behaviours may also be important in explaining the relationship between depression and CVD. These health behaviours will be discussed in the following section.

3. Behavioural pathways

Depressed individuals tend to engage in several poor health behaviours known to contribute to CVD, which may explain their increased risk of developing CVD. In this chapter, we will discuss four behaviours in relation to depression: excess calorie consumption (i.e., obesity), physical inactivity, smoking, and excess alcohol consumption, all of which are important risk factors for CVD. In fact, in the INTERHEART study, including nearly 30 000 participants, these four health behaviours were found to explain 75% of the population attributable risk (PAR) for myocardial infarction (Yusuf et al., 2004).

3.1 Smoking

3.1.1 Smoking and CVD

Smoking is arguably the most important and well-established risk factor for CVD (Yusuf et al., 2004), with the number of pack years (# of packs smoked/ day X # years of smoking) a person has smoked being consistently associated with the level of severity of atherosclerotic development (Herbert, 1975; Ramsdale et al., 1985; Wang et al., 1994). Smoking is thought to contribute to atherosclerotic development via several pathways. One pathway is by damaging the endothelium. Numerous studies have confirmed that smokers have significantly impaired endothelial function compared to non-smokers (Barua et al., 2001; Esen et al., 2004; Kiowski et al., 1994; Zeiher et al., 1995) and that this dysfunction can last up to three months after a smoker quits (Celermajer et al., 1993). Smoking is thought to have this impact on the endothelium through two main mechanisms: by increasing oxidative stress, which deactivates nitric oxide (Puranik & Celermajer, 2003) and by causing direct endothelial cell damage (Bernhard et al., 2003; Hoshino et al., 2005).

Smoking also contributes to the inflammatory response involved in CVD development. Cigarette smoking has been associated with elevated levels of several markers of inflammation (Bermudez et al., 2002; Mendall et al., 1997; Tappia et al., 1995; Tracy et al., 1997). Cigarette smoke extract has also been found to double the rate at which macrophages are created and engulf LDL's to form foam cells (Shen et al., 1996).

Smoking may also contribute to atherosclerosis by altering one's lipid profile. Studies have found smokers to have higher levels of LDL and triglycerides and lower levels of HDL compared to non-smokers (Craig et al., 1989). Several studies have confirmed that cigarette smoke extract increases the oxidation of LDL cholesterol, a step that allows the LDL to cross the endothelial layer (Frei et al., 1991; Heitzer et al., 1999; Pech-Amsellem et al., 1996; Yokode et al., 1988).

Smoking also increases thrombosis (blood clotting) through several mechanisms, including increasing platelet aggregation (Blache, 1995; Fusegawa et al., 1999; Rival et al., 1987), altering the release of antithrombotic and prothrombotic factors by endothelial cells (Kannel et al., 1987; Sambola et al., 2003; Smith et al., 1997) as well as altering fibrinolysis, the process by which blood clots are broken down (Barua et al., 2002; Newby et al., 1999; Pretorius et al., 2002). In having such an effect on thrombosis, smoking greatly increases the chances that atherosclerosis will lead to a myocardial infarction (Ambrose & Barua, 2004).

3.1.2 Smoking and depression

There is a strong association between depression and smoking. Several studies have found that depression in adolescence is associated with a greater risk of smoking initiation (Lam et al., 2005; Weiss et al., 2005), and that this risk increases with increasing depression severity (Escobedo et al., 1998). Several studies find that depressed patients are also less successful at quitting smoking compared to non-depressed patients (Burgess et al., 2002; Ginsberg et al., 1995), but that their ability to quit is improved with cognitive behaviour therapy (S. M. Hall et al., 1994). It is therefore unsurprising that cross-sectional studies find a strong link between depression and smoking (Almeida & Pfaff, 2005; Escobedo et al., 1996). Given that smoking is such an important contributor to CVD and that depressed patients are more likely to smoke than non-depressed individuals, smoking may explain why depressed patients are at an increased risk for CVD.

3.2 Obesity

3.2.1 Obesity and CVD

Obesity, defined as a body mass index (BMI) of 30 kg/m^2 or larger (Soodini & Hamdy, 2004), or a waist circumference greater or equal to 88 cm for women and 102 cm for men (Grundy, 2006), is a well-known risk factor for CVD. Obese individuals have been found to have increased mortality rates (Ajani et al., 2004; Hu et al., 2004; Widlansky et al., 2004) and are 2.4 times more likely to develop CVD compared to normal weight individuals (Lawlor & Leon, 2005). This is unsurprising given the long list of CVD risk factors associated with obesity, including hypertension, insulin resistance, dyslipidemia, and endothelial dysfunction (Smoak et al., 1987). Although obesity is influenced by several factors, studies have found caloric consumption to be the most important predictor of weight loss or gain and caloric restriction to be the most successful means of inducing weight loss (Blumenthal et al., 2000; Brownell, 1999; Elfhag & Rossner, 2005; Jakicic et al., 2003; Petersen & Harper, 2004). In this chapter, "obesity" will therefore be considered a proxy of long-term excess caloric consumption although it is not a behaviour per say.

The increase in CVD risk factors in obese individuals is believed to be a consequence of adipose tissue's release of signaling proteins called adipokines (Gordon et al., 2008). For example, most adipokines (e.g., leptin, resistin) are known to induce endothelial dysfunction by decreasing nitric oxide availability, leading to endothelial cell death (Beltowski et al., 2004; Bouloumie et al., 1999; Clapp et al., 2004; Meldrum, 1998; Mercurio & Manning, 1999). In addition, several adipokines, promote the inflammatory response that contributes to atherosclerosis (Lee & Pratley, 2007). Exceptionally, one beneficial adipokine, adiponectin, which promotes the production of nitric oxide, is downregulated in obese individuals (Avogaro & de Kreutzenberg, 2005; Chen et al., 2003).

Adipokines may also contribute to CVD through their tendency to induce hypertension. One adipokine called leptin, for example, is thought to induce hypertension by promoting sympathetic activation (Hall et al., 2001). Another adipokine called interleukin (IL)-6, may also contribute to the development of hypertension through its stimulation of the HPA axis (Yudkin et al., 2000). Finally, Angiotensin-II, another adipokine, contributes to hypertension by constricting the blood vessels (Brasier et al., 2002).

Several adipokines also contribute to insulin resistance. For example, one study found that the acute administration of the adipokine resistin induces glucose intolerance. The same study also found an improvement in blood glucose levels and insulin sensitivity when obese mice were given an anti-resistin antibody (Steppan et al., 2001). Adipokines also contribute to obesity-induced insulin resistance by inhibiting insulin's signal to fat and muscle cells to absorb glucose (Senn et al., 2002; Uysal et al., 1997). Finally, adiponectin, down-regulated in obese individuals, is known to reduce the risk of insulin resistance (Yamauchi et al., 2001). Obesity may also contribute to CVD through its association with high LDL cholesterol levels (Denke et al., 1993; Morrison et al., 1999), although the mechanism behind this association is not completely understood.

3.2.2 Obesity and depression

Studies have consistently found depression to be associated with obesity. For example, one found that in a nationally representative sample, overweight and obese women and overweight (but not obese) men were more likely to report depressive mood compared to participants with a healthy weight (Heo et al., 2006). Another study comparing clinically depressed individuals to controls with no history of psychiatric illness also found that the depressed individuals had a significantly greater average BMI compared to controls (V. M. Miller & Vanhoutte, 1988). Several longitudinal studies aimed at determining whether the onset of depression precedes or follows the onset of obesity also found that depressive symptoms predict obesity later in life. For example, one such study measured depressed mood ratings and BMI in adolescents and found that, after controlling for baseline BMI and other confounding variables (e.g. self-esteem and physical activity levels), 'depressed' adolescents (measured using the Center for Epidemiologic Studies Depression Scale (CES-D scale)) were twice as likely to be obese at one-year follow-up compared to non-depressed adolescents (Goodman & Whitaker, 2002). Another found similar results, with depressed younger adolescents being more than twice as likely to become obese towards the end of adolescence (Richardson et al., 2003). Childhood depression has also been found to be predictive of adult obesity after adjusting for childhood BMI (Pine et al., 2001). However, a recent meta-analysis of longitudinal studies suggests that obesity also predicts the onset of depression, suggesting a bi-directional relationship (Luppino et al., 2010).

Given that obesity is closely associated with depression and is such a well-known contributor to CVD, it represents a plausible mechanism explaining the relationship between depression and CVD. For these same reasons, physical inactivity may also potentially explain the link between depression and CVD.

3.3 Physical inactivity

3.3.1 Physical inactivity and CVD

It has long been known that maintaining an active lifestyle is beneficial for CV health. However, perhaps less commonly known is that although regular exercise is beneficial partly through its calorie-burning effects, its benefits on the cardiovascular system extend beyond this. In fact, increased physical activity has been associated with lower CVD risk independent of body weight (Haapanen-Niemi et al., 2000).

Much of this benefit is believed to be due to exercise's effect on the endothelium. Many studies have found increased physical activity to be associated with better endothelial function (Kingwell et al., 1996; Rinder et al., 2000; Rywik et al., 1999). Intervention studies have supported the idea that increased physical activity predicts improvements in endothelial function, even without weight loss (Clarkson et al., 1999). It is thought to do so mainly through exercise's tendency to increase nitric oxide availability. Through an increase in blood flow, exercise stimulates the increased production of nitric oxide by activating sensors located on the endothelium that are meant to detect blood flow and cause the endothelium to release nitric oxide to dilate the arteries when blood flow is high (Garcia-Cardena et al., 1998; Schwartz & Lechene, 1992). Furthermore, exercise produces antioxidant factors (Fukai et al., 2000), which prevent the oxidation and destruction of nitric oxide.

Physical activity may also protect against atherosclerosis through its anti-inflammatory effect. Numerous cross-sectional studies have found increased physical activity to be associated with lower levels of inflammatory markers (Church et al., 2002; Ford, 2002; Reuben et al., 2003; Wannamethee et al., 2002). Furthermore, while only a few studies have examined the effect of an exercise intervention on inflammation, these have found a significant reduction in inflammatory markers after exercise training (Mattusch et al., 2000; J. K. Smith et al., 1999; Tisi et al., 1997). Most recently, one study found inflammation to be the most important mechanism linking physical activity and CVD, accounting for 59% of the inverse association (Mora et al., 2007). However, it is important to note that while this study considered many potential mechanisms linking physical activity and inflammation, it did not measure endothelial function. The extent to which the relationship between exercise and inflammation is explained by exercise's beneficial effect on the endothelium is therefore unknown.

In addition to directly benefiting the endothelium, physical activity also likely reduces the risk of CVD through its effects on other CVD risk factors. For example, randomized controlled trials suggest that three to five 30-60 minute bouts of physical activity per week result in a clinically significant reduction in blood pressure (Arroll & Beaglehole, 1992; Fagard, 2001). In the short-term, physical activity is thought to reduce blood pressure by dilating the arteries and in the long-term, it reduces SNS activity and results in a fitter, stronger heart that is able to more easily pump blood into the body. Physical activity is also known to reduce the risk of developing insulin resistance as it promotes glucose uptake by muscle cells (for energy) and makes them more sensitive to insulin: even a single exercise session is sufficient to improve insulin sensitivity for 1-2 days (Thompson et al., 2001). One study found that an exercise intervention could prevent the development of type II diabetes in a group of insulin resistant middle-aged men and women (Tuomilehto et al., 2001). Increased physical activity also increases HDL cholesterol as well as lowers LDL cholesterol and triglyceride levels (Thompson et al., 2001).

3.3.2 Physical inactivity and depression

Research suggests that depressed individuals are less active than the general population. Cross-sectional studies have found self-reported physical activity levels to be inversely related to levels of depressive symptoms in adolescents (Piko & Keresztes, 2006), fibromyalgia patients (Oliver & Cronan, 2002), as well as in the general adult population (De Moor et al., 2006; Galper et al., 2006). One study of 4493 CVD-free elderly (≥65 years) men and women found a negative association between depressive symptoms and number of blocks walked in the last week (Ariyo et al., 2000). Another study of 6247 elderly men and women initially free of disability corroborates this finding: participants high in depressive symptoms reported more physical disability and less mobility every year during the six-year follow-up compared to those with low depressive symptoms (Penninx et al., 1999). A recent study by the same research group found that individuals suffering from a mood disorder spend significantly more time in front of the computer and television (de Wit et al., 2011), both of which are negatively correlated to physical activity.

Although there is much evidence suggesting that exercise can decrease depressive symptoms (Brosse et al., 2002), it is likely that a bi-directional relationship exists between depression and physical activity. This is supported by studies such as the one conducted by

Penninx et al. (1999), finding that depression often precedes physical inactivity. Given the benefits of physical activity on CV health and given that depressed patients tend to be relatively inactive, physical inactivity represents a plausible mechanism linking depression and CVD.

3.4 Alcohol consumption

3.4.1 Alcohol consumption and CVD

The contribution of alcohol consumption to CVD is not as straightforward as the other three health behaviours discussed above. This is because researchers have found that while larger quantities of alcohol can damage the cardiovascular system, smaller quantities may actually be beneficial (Fuchs et al., 1995; Gaziano et al., 2000; Keil et al., 1997; Murray et al., 2002). These beneficial effects of small quantities of alcohol are likely due to its tendency to improve endothelial function by increasing the production of nitric oxide (Davda et al., 1993; Hendrickson et al., 1999) as well as increasing HDL levels (Watts et al., 1996) though this evidence is still considered somewhat controversial.

In larger quantities, though, the negative cardiovascular effects of alcohol outweigh the beneficial effects. Consuming the equivalent of four alcoholic beverages or more in one sitting is enough to induce endothelial dysfunction (Bau et al., 2005). Many studies have confirmed that chronic alcohol abuse is associated with important impairments in endothelial function (Di Gennaro et al., 2007; Maiorano et al., 1999; Zilkens et al., 2003). This effect is thought to be due to the increased oxidative stress produced by excessive amounts of alcohol, causing endothelial cell apoptosis (Croft et al., 1996; Soardo et al., 2005; Spyridopoulos et al., 2001; Sun & Mayhan, 2001). Large quantities of alcohol may also impair the production of nitric oxide (Persson & Gustafsson, 1992).

Excess alcohol consumption may also contribute to atherosclerosis through inflammation. Several studies have found that while moderate alcohol consumption is associated with lower levels of several inflammatory markers, consumption of large amounts of alcohol is associated with increased levels of these markers (Imhof et al., 2001; Imhof et al., 2004).

It is well-established that alcohol consumption contributes to the development of hypertension (Beilin, 1995; Beilin & Puddey, 2006; McFadden et al., 2005), regardless of beverage type (Zilkens et al., 2005). A meta-analysis of randomized controlled trials examining the effect of alcohol reduction on blood pressure found that a reduction in alcohol intake is consistently associated with a significant drop in blood pressure, and that a dose-response relationship can be seen such that the greater the drop in alcohol consumption, the greater the drop in blood pressure observed (Xin et al., 2001).

3.4.2 Alcohol consumption and depression

Depression is associated with heavier alcohol consumption. Epidemiological studies have consistently found major depression and alcohol dependence to be highly co-morbid (B. I. Goldstein & Levitt, 2006; Grant, 1995; Kessler et al., 1997). One study found that 32.3% of Canadian adults who had had a depressive episode in the last year were alcohol-dependent while only 9.5% of non-depressed adults were alcohol-dependent (Lukassen & Beaudet, 2005). Similarly, one prospective study found that among 1383 women at risk for heavy alcohol use, those with a history of a depressive disorder were 2.6 times more likely to be heavy alcohol users one year later compared to those with no history of a depressive disorder (Dixit & Crum, 2000). Another large study of 22 954 adults from four different countries also found an odds ratio between 2 and 5 depending on the country when examining the comorbidity of alcoholism within the past 12 months with lifetime depressive disorders (Swendsen et al., 1998). Increased alcohol consumption may therefore contribute to depressed patients' risk for developing CVD.

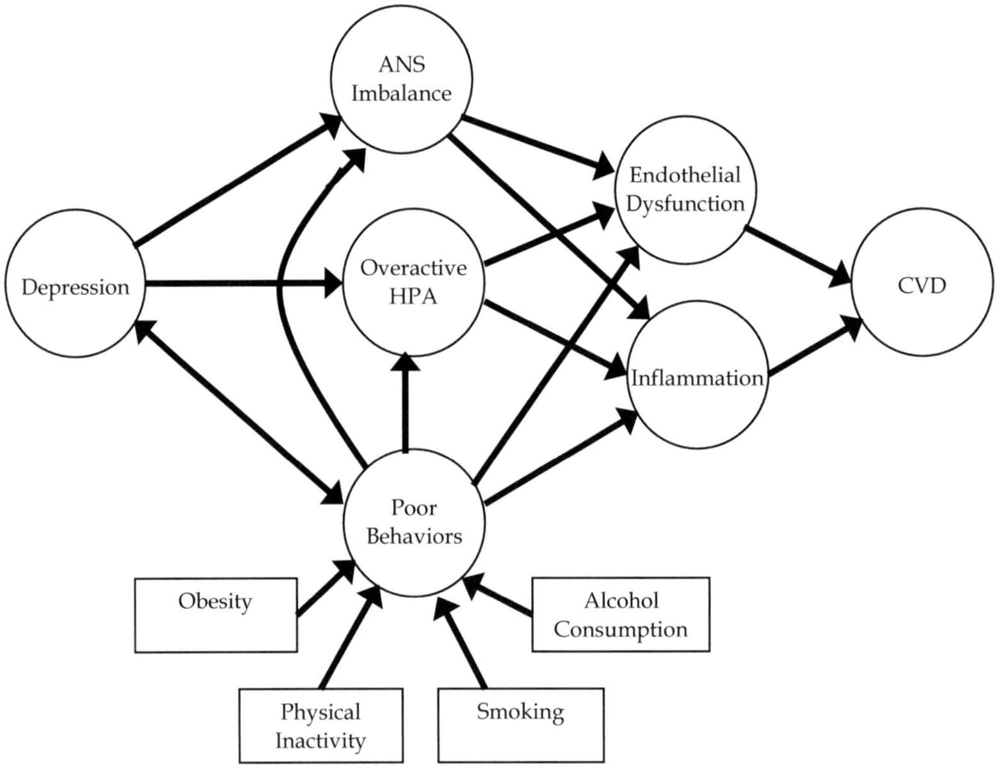

Fig. 1. Proposed model explaining how depression contributes to CVD

4. Conclusion

While our understanding of the mechanisms involved in linking depression to CVD has expanded greatly in the last decade, our knowledge of the relative importance of these mechanisms remains limited. To date there have been only two studies that have aimed to determine their relative importance. The first study of 1017 coronary heart disease patients found that smoking, medication non-adherence and physical inactivity accounted for 48% of the effect of depressive symptoms on the risk of CV events in the following 6 years. Physical inactivity accounted for the largest proportion of the variance (31.7%) (Whooley et al., 2008). A second similar study in 6576 healthy participants also found that smoking, physical inactivity and alcohol consumption accounted for 65% of the variance in the association between psychological distress and CVD (Hamer et al., 2008). However, since this study measured psychological distress and not depressive symptoms, one should be cautious in generalising its findings to depression. Nonetheless, these studies seem to suggest that health behaviours account for at least half of the relationship between psychological variables and CVD. However, further research is required to replicate these findings.

To date, there have been a small number of studies attempting to improve the CV prognosis of depressed post-myocardial infarction patients through depression treatment. The largest study to date (Berkman et al., 2003), called the ENRICHD trial, included 2481 patients who had recently suffered a myocardial infarction and were either depressed or reported low levels of social support. Half of the participants received usual care by their physician while the other half received individual cognitive behavioural therapy lasting a maximum of 6 months. Following individual therapy, if necessary, participants could also receive 12 weeks of group therapy and up to 12 months of antidepressant treatment. The intervention was successful in reducing depressive symptoms, though results were somewhat modest: while depressed patients having received therapy experienced an overall decrease of 8.6 points on the Beck Depression Inventory, their usual care counterparts experienced a 5.8-point decrease. However, despite the improvement in depressive symptoms, this did not translate to an improvement in their CV prognosis: patients in the intervention group did not differ from the usual care group in subsequent CV events of mortality.

A second study, the CREATE trial (Lesperance et al., 2007), examined the efficacy of citalopram and interpersonal psychotherapy (IPT) among 284 coronary artery disease patients diagnosed with depression. Participants were randomised to four interventions: 1) IPT plus clinical management and citalopram; 2) IPT plus clinical management and placebo pill; 3) clinical management only and citalopram; 4) clinical management and placebo pill. It was found that while IPT was not more successful than clinical management alone in treating depression, citalopram was successful in reducing depressive symptoms with the citalopram group exhibiting a 14.7-point decrease on the BDI-II and the placebo group exhibiting a 11.1-point decrease. Yet, despite the improvement in depressive symptoms, the four groups did not differ in CV outcomes following treatment.

These and other studies (Frasure-Smith et al., 1997; Glassman et al., 2002) suggest that even when interventions successfully decrease depressive symptoms in cardiac patients, this does not translate to improved prognosis. This, despite research finding that antidepressants have been found to result in modest improvements in ANS balance (Khaykin et al., 1998; McFarlane et al., 2001). Though there has been much speculation as to why this is (Joynt & O'Connor, 2005; Sheps et al., 2003), further research is needed to uncover the answer. One potential explanation may relate to poor health behaviours persevering despite improvements in depressive symptoms. Based on the studies by Whooley et al. (2008) and Hamer et al. (2008), health behaviours appear to account for a significant proportion of the relationship between depression and CVD outcomes. However, treating depression may not automatically translate to immediate improvements in health behaviours. Obesity, smoking, excessive alcohol consumption, and physical inactivity, which depressed individuals are particularly prone to developing, are not particularly easy behaviours to alter. It is therefore possible that in addition to psychotherapy targeted at depressive symptoms, depressed patients may also require interventions specifically targeted at improving their health behaviours in order to exhibit noticeable improvements in their CV health. Considering the bi-directionality of the relationship between depression and obesity (Luppino et al., 2010) and depression and physical activity (Penninx et al., 1999), targeting health behaviours directly may also lead to improvements in depressive symptoms.

Future intervention studies may therefore be improved by assessing health behaviours and implementing interventions specifically aimed at altering health behaviours, such as motivational interviewing (W. R. Miller & Rollnick, 2002). Assessing participants' ANS and HPA axis functioning throughout interventions may also be useful in identifying the critical components of a successful intervention and the toxic components of depression that we should target. Until this research is conducted, clinicians treating depressed individuals at high risk for CVD might consider routinely integrating interventions targeting health behaviour change into their practice.

5. Glossary

Adrenaline and noradrenaline: substances called catecholamines released by the sympathetic nervous system (as neurotransmitters) and the sympathoadrenal system (as hormones) to induce the fight-or-flight response; both forms have similar cardiovascular effects but as hormones, their effects are more systemic and longer-lasting

Apoptosis: process by which cells die

Autonomic nervous system (ANS): biological system key in orchestrating the body's response to stress; is composed of two opposing branches - the sympathetic nervous system and the parasympathetic nervous system

Cortisol: steroid hormone and type of glucocorticoid released into the blood by the hypothalamic-pituitary-adrenal axis in response to stress

Endothelium: a single layer of cells lining the interior of the blood vessels; is involved in many important aspects of cardiovascular health, including vessel dilation and constriction, inflammation and blood clot formation

High-density lipoproteins (HDL): commonly known as "good" cholesterol, protects against atherosclerosis by removing LDL cholesterol from the bloodstream

Hypothalamic-pituitary-adrenal (HPA) axis: circuit involving the hypothalamus, pituitary gland and adrenal gland resulting in the release of hormones called glucocorticoids into the bloodstream, the most important of which is cortisol in humans; is involved in the stress response

Low-density lipoproteins (LDL): commonly known as "bad" cholesterol, contributing to atherosclerosis

Myocardial infarction: technical term for "heart attack"; occurs when a blood clot completely obstructs blood flow to the heart and heart cells die as a result

Nitric oxide: substance produced by the endothelium that is key in maintaining its health

Parasympathetic nervous system (PNS): branch of the autonomic nervous system that promotes a resting state through the release of acetylcholine; its activation reduces heart rate and blood pressure

Stroke: occurs when a blood clot completely obstructs blood flow to the brain and brain cells die as a result

Sympathetic nervous system (SNS): branch of the autonomic nervous system that induces the fight-or-flight response in response to stress through the release of adrenaline and noradrenaline; its activation raises heart rate and blood pressure

Thrombus: technical term for "blood clot"

6. References

Ajani, U. A., Lotufo, P. A., Gaziano, J. M., Lee, I.-M., Spelsberg, A., Buring, J. E., et al. (2004). Body mass index and mortality among US male physicians. *Annals of Epidemiology, 14*, 10, pp. 731-739, 1047-2797

Almeida, O. P., & Pfaff, J. J. (2005). Depression and smoking amongst older general practice patients. *Journal of Affective Disorders, 86*, 2-3, pp. 317-321, 0165-0327

Ambrose, J. A., & Barua, R. S. (2004). The pathophysiology of cigarette smoking and cardiovascular disease: an update. *Journal of the American College of Cardiology, 43*, 10, pp. 1731-1737, 0735-1097

Andrews, R. C., & Walker, B. R. (1999). Glucocorticoids and insulin resistance: old hormones, new targets. *Clinical Science, 96*, 5, pp. 513-523, 0143-5221

Anfossi, G., & Trovati, M. (1996). Role of catecholamines in platelet function: pathophysiological and clinical significance. *European Journal of Clinical Investigation, 26*, 5, pp. 353-370, 0014-2972

Ariyo, A. A., Haan, M., Tangen, C. M., Rutledge, J. C., Cushman, M., Dobs, A., et al. (2000). Depressive symptoms and risks of coronary heart disease and mortality in elderly Americans. Cardiovascular Health Study Collaborative Research Group. *Circulation, 102*, 15, pp. 1773-1779, 0009-7322

Arnaldi, G., Mancini, T., Polenta, B., & Boscaro, M. (2004). Cardiovascular risk in Cushing's syndrome. *Pituitary, 7*, 4, pp. 253-256, 1386-341X

Arroll, B., & Beaglehole, R. (1992). Does physical activity lower blood pressure: a critical review of the clinical trials. *Journal of Clinical Epidemiology, 45*, 5, pp. 439-447, 0895-4356

Avogaro, A., & de Kreutzenberg, S. V. (2005). Mechanisms of endothelial dysfunction in obesity. *Clinica Chimica Acta, 360*, 1-2, pp. 9-26, 0009-8981

Barua, R. S., Ambrose, J. A., Eales-Reynolds, L. J., DeVoe, M. C., Zervas, J. G., & Saha, D. C. (2001). Dysfunctional endothelial nitric oxide biosynthesis in healthy smokers with impaired endothelium-dependent vasodilatation. *Circulation, 104*, 16, pp. 1905-1910, 0009-7322

Barua, R. S., Ambrose, J. A., Saha, D. C., & Eales-Reynolds, L. J. (2002). Smoking is associated with altered endothelial-derived fibrinolytic and antithrombotic factors: an in vitro demonstration. *Circulation, 106*, 8, pp. 905-908, 0009-7322

Bau, P. F., Bau, C. H., Naujorks, A. A., & Rosito, G. A. (2005). Early and late effects of alcohol ingestion on blood pressure and endothelial function. *Alcohol, 37*, 1, pp. 53-58, 0741-8329

Baykan, M., Erem, C., Gedikli, O., Hacihasanoglu, A., Erdogan, T., Kocak, M., et al. (2007). Impairment of flow-mediated vasodilatation of brachial artery in patients with Cushing's Syndrome. *Endocrine, 31*, 3, pp. 300-304, 0969-711X

Beilin, L. J. (1995). Alcohol and hypertension. *Clinical & Experimental Pharmacology & Physiology, 22*, 3, pp. 185-188, 0305-1870

Beilin, L. J., & Puddey, I. B. (2006). Alcohol and hypertension: an update. *Hypertension, 47*, 6, pp. 1035-1038, 0362-4323

Beltowski, J., Jochem, J., Wojcicka, G., & Zwirska-Korczala, K. (2004). Influence of intravenously administered leptin on nitric oxide production, renal hemodynamics and renal function in the rat. *Regulatory Peptides, 120*, 1-3, pp. 59-67, 0167-0115

Berkman, L. F., Blumenthal, J., Burg, M., Carney, R. M., Catellier, D., Cowan, M. J., et al. (2003). Effects of treating depression and low perceived social support on clinical events after myocardial infarction: the Enhancing Recovery in Coronary Heart Disease Patients (ENRICHD) Randomized Trial. *Journal of the American Medical Association, 289*, 23, pp. 3106-3116, 0098-7484

Bermudez, E. A., Rifai, N., Buring, J. E., Manson, J. E., & Ridker, P. M. (2002). Relation between markers of systemic vascular inflammation and smoking in women. *American Journal of Cardiology, 89*, 9, pp. 1117-1119, 0002-9149

Bernhard, D., Pfister, G., Huck, C. W., Kind, M., Salvenmoser, W., Bonn, G. K., et al. (2003). Disruption of vascular endothelial homeostasis by tobacco smoke: impact on atherosclerosis. *FASEB Journal, 17*, 15, pp. 2302-2304, 0892-6638

Bhattacharyya, M. R., Molloy, G. J., & Steptoe, A. (2008). Depression is associated with flatter cortisol rhythms in patients with coronary artery disease. *Journal of Psychosomatic Research, 65*, 2, pp. 107-113, 0022-3999

Blache, D. (1995). Involvement of hydrogen and lipid peroxides in acute tobacco smoking-induced platelet hyperactivity. *American Journal of Physiology, 268*, 2 Pt 2, pp. H679-685, 0002-9513

Blumenthal, J. A., Sherwood, A., Gullette, E. C., Babyak, M., Waugh, R., Georgiades, A., et al. (2000). Exercise and weight loss reduce blood pressure in men and women with mild hypertension: effects on cardiovascular, metabolic, and hemodynamic functioning. *Archives of Internal Medicine, 160*, 13, pp. 1947-1958, 0003-9926

Bouloumie, A., Marumo, T., Lafontan, M., & Busse, R. (1999). Leptin induces oxidative stress in human endothelial cells. *FASEB Journal, 13*, 10, pp. 1231-1238, 0892-6638

Brasier, A. R., Recinos, A., 3rd, & Eledrisi, M. S. (2002). Vascular inflammation and the renin-angiotensin system. *Arteriosclerosis, Thrombosis & Vascular Biology, 22*, 8, pp. 1257-1266, 1079-5642

Broadley, A. J., Korszun, A., Abdelaal, E., Moskvina, V., Jones, C. J., Nash, G. B., et al. (2005). Inhibition of cortisol production with metyrapone prevents mental stress-induced endothelial dysfunction and baroreflex impairment. *Journal of the American College of Cardiology, 46*, 2, pp. 344-350, 0735-1097

Brosse, A. L., Sheets, E. S., Lett, H. S., & Blumenthal, J. A. (2002). Exercise and the treatment of clinical depression in adults: recent findings and future directions. *Sports Medicine, 32*, 12, pp. 741-760, 0112-1642

Brownell, K. D. (1999). The central role of lifestyle change in long-term weight management. *Clinical Cornerstone, 2*, 3, pp. 43-51, 1098-3597

Burgess, E. S., Brown, R. A., Kahler, C. W., Niaura, R., Abrams, D. B., Goldstein, M. G., et al. (2002). Patterns of change in depressive symptoms during smoking cessation: who's at risk for relapse? *Journal of Consulting & Clinical Psychology, 70*, 2, pp. 356-361, 0022-006X

Burke, H. M., Davis, M. C., Otte, C., & Mohr, D. C. (2005). Depression and cortisol responses to psychological stress: a meta-analysis. *Psychoneuroendocrinology, 30,* 9, pp. 846-856, 0306-4530

Carney, R. M., Blumenthal, J. A., Stein, P. K., Watkins, L., Catellier, D., Berkman, L. F., et al. (2001). Depression, heart rate variability, and acute myocardial infarction. *Circulation, 104,* 17, pp. 2024-2028, 0009-7322

Carney, R. M., Freedland, K. E., Rich, M. W., Smith, L. J., & Jaffe, A. S. (1993). Ventricular tachycardia and psychiatric depression in patients with coronary artery disease. *American Journal of Medicine, 95,* 1, pp. 23-28, 0002-9343

Carroll, B. J., Martin, F. I., & Davies, B. (1968). Pituitary-adrenal function in depression. *Lancet, 1,* 7556, pp. 1373-1374, 0140-6736

Celermajer, D. S., Sorensen, K. E., Georgakopoulos, D., Bull, C., Thomas, O., Robinson, J., et al. (1993). Cigarette smoking is associated with dose-related and potentially reversible impairment of endothelium-dependent dilation in healthy young adults. *Circulation, 88,* 5 Pt 1, pp. 2149-2155, 0009-7322

Chen, H., Montagnani, M., Funahashi, T., Shimomura, I., & Quon, M. J. (2003). Adiponectin stimulates production of nitric oxide in vascular endothelial cells. *Journal of Biological Chemistry, 278,* 45, pp. 45021-45026, 0021-9258

Church, T. S., Barlow, C. E., Earnest, C. P., Kampert, J. B., Priest, E. L., & Blair, S. N. (2002). Associations between cardiorespiratory fitness and C-reactive protein in men. *Arteriosclerosis, Thrombosis & Vascular Biology, 22,* 11, pp. 1869-1876, 1079-5642

Clapp, B. R., Hingorani, A. D., Kharbanda, R. K., Mohamed-Ali, V., Stephens, J. W., Vallance, P., et al. (2004). Inflammation-induced endothelial dysfunction involves reduced nitric oxide bioavailability and increased oxidant stress. *Cardiovascular Research, 64,* 1, pp. 172-178, 0008-6363

Clarkson, P., Montgomery, H. E., Mullen, M. J., Donald, A. E., Powe, A. J., Bull, T., et al. (1999). Exercise training enhances endothelial function in young men. *Journal of the American College of Cardiology, 33,* 5, pp. 1379-1385, 0735-1097

Coker, R., Koziell, A., Oliver, C., & Smith, S. E. (1984). Does the sympathetic nervous system influence sinus arrhythmia in man? Evidence from combined autonomic blockade. *Journal of Physiology, 356,* pp. 459-464, 0022-3751

Colao, A., Pivonello, R., Spiezia, S., Faggiano, A., Ferone, D., Filippella, M., et al. (1999). Persistence of increased cardiovascular risk in patients with Cushing's disease after five years of successful cure. *Journal of Clinical Endocrinology & Metabolism, 84,* 8, pp. 2664-2672, 0021-972X

Cole, C. R., Foody, J. M., Blackstone, E. H., & Lauer, M. S. (2000). Heart rate recovery after submaximal exercise testing as a predictor of mortality in a cardiovascularly healthy cohort. *Annals of Internal Medicine, 132,* 7, pp. 552-555, 0003-4819

Cooney, J. M., & Dinan, T. G. (1996). Type II (glucocorticoid) receptors mediate fast-feedback inhibition of the hypothalamic-pituitary-adrenal axis in man. *Life Sciences, 59,* 23, pp. 1981-1988, 0024-3205

Craig, W. Y., Palomaki, G. E., & Haddow, J. E. (1989). Cigarette smoking and serum lipid and lipoprotein concentrations: an analysis of published data. *BMJ, 298,* 6676, pp. 784-788, 0959-8138

Croft, K. D., Puddey, I. B., Rakic, V., Abu-Amsha, R., Dimmitt, S. B., & Beilin, L. J. (1996). Oxidative susceptibility of low-density lipoproteins--influence of regular alcohol use. *Alcoholism: Clinical & Experimental Research, 20*, 6, pp. 980-984, 0145-6008

Davda, R. K., Chandler, L. J., Crews, F. T., & Guzman, N. J. (1993). Ethanol enhances the endothelial nitric oxide synthase response to agonists. *Hypertension, 21*, 6 Pt 2, pp. 939-943, 0194-911X

Davidson, K., Jonas, B. S., Dixon, K. E., & Markovitz, J. H. (2000). Do depression symptoms predict early hypertension incidence in young adults in the CARDIA study? Coronary Artery Risk Development in Young Adults. *Archives of Internal Medicine, 160*, 10, pp. 1495-1500, 0003-9926

Dawson, M. E., Schell, A. M., & Catania, J. J. (1977). Autonomic correlates of depression and clinical improvement following electroconvulsive shock therapy. *Psychophysiology, 14*, 6, pp. 569-578, 0048-5772

De Moor, M. H., Beem, A. L., Stubbe, J. H., Boomsma, D. I., & De Geus, E. J. (2006). Regular exercise, anxiety, depression and personality: a population-based study. *Preventive Medicine, 42*, 4, pp. 273-279, 0091-7435

de Wit, L., van Straten, A., Lamers, F., Cuijpers, P., & Penninx, B. (2011). Are sedentary television watching and computer use behaviors associated with anxiety and depressive disorders? *Psychiatry research, 186*, 2-3, pp. 239-243, 0165-1781

Dekker, M. J. H. J., Koper, J. W., van Aken, M. O., Pols, H. A. P., Hofman, A., de Jong, F. H., et al. (2008). Salivary cortisol is related to atherosclerosis of carotid arteries. *Journal of Clinical Endocrinology & Metabolism, 93*, 10, pp. 3741-3747, 0021-972X

Denke, M. A., Sempos, C. T., & Grundy, S. M. (1993). Excess body weight. An underrecognized contributor to high blood cholesterol levels in white American men. *Archives of Internal Medicine, 153*, 9, pp. 1093-1103, 0003-9926

Di Gennaro, C., Biggi, A., Barilli, A. L., Fasoli, E., Carra, N., Novarini, A., et al. (2007). Endothelial dysfunction and cardiovascular risk profile in long-term withdrawing alcoholics. *Journal of Hypertension, 25*, 2, pp. 367-373, 0263-6352

Dixit, A. R., & Crum, R. M. (2000). Prospective study of depression and the risk of heavy alcohol use in women. *American Journal of Psychiatry, 157*, 5, pp. 751-758, 1090-1248

Elfhag, K., & Rossner, S. (2005). Who succeeds in maintaining weight loss? A conceptual review of factors associated with weight loss maintenance and weight regain. *Obesity Reviews, 6*, 1, pp. 67-85, 1467-7881

Escobedo, L. G., Kirch, D. G., & Anda, R. F. (1996). Depression and smoking initiation among US Latinos. *Addiction, 91*, 1, pp. 113-119, 0965-2140

Escobedo, L. G., Reddy, M., & Giovino, G. A. (1998). The relationship between depressive symptoms and cigarette smoking in US adolescents. *Addiction, 93*, 3, pp. 433-440, 0965-2140

Esen, A. M., Barutcu, I., Acar, M., Degirmenci, B., Kaya, D., Turkmen, M., et al. (2004). Effect of smoking on endothelial function and wall thickness of brachial artery. *Circulation Journal, 68*, 12, pp. 1123-1126, 1346-9843

Esler, M. (2000). The sympathetic system and hypertension. *American Journal of Hypertension, 13*, 6 Pt 2, pp. 99S-105S, 0895-7061

Esler, M., Turbott, J., Schwarz, R., Leonard, P., Bobik, A., Skews, H., et al. (1982). The peripheral kinetics of norepinephrine in depressive illness. *Archives of General Psychiatry, 39*, 3, pp. 295-300, 0003-990X

Everson-Rose, S. A., Meyer, P. M., Powell, L. H., Pandey, D., Torrens, J. I., Kravitz, H. M., et al. (2004). Depressive symptoms, insulin resistance, and risk of diabetes in women at midlife. *Diabetes Care, 27,* 12, pp. 2856-2862, 0149-5992

Fagard, R. H. (2001). Exercise characteristics and the blood pressure response to dynamic physical training. *Medicine & Science in Sports & Exercise, 33,* 6 Suppl, pp. S484-492; discussion S493-484, 0195-9131

Farrell, T. G., Odemuyiwa, O., Bashir, Y., Cripps, T. R., Malik, M., Ward, D. E., et al. (1992). Prognostic value of baroreflex sensitivity testing after acute myocardial infarction. *British Heart Journal, 67,* 2, pp. 129-137, 0007-0769

Ford, E. S. (2002). Does exercise reduce inflammation? Physical activity and C-reactive protein among U.S. adults. *Epidemiology, 13,* 5, pp. 561-568, 1044-3983

Fraser, R., Ingram, M. C., Anderson, N. H., Morrison, C., Davies, E., & Connell, J. M. (1999). Cortisol effects on body mass, blood pressure, and cholesterol in the general population. *Hypertension, 33,* 6, pp. 1364-1368, 0194-911X

Frasure-Smith, N., Lesperance, F., Prince, R. H., Verrier, P., Garber, R. A., Juneau, M., et al. (1997). Randomised trial of home-based psychosocial nursing intervention for patients recovering from myocardial infarction. *Lancet, 350,* 9076, pp. 473-479, 0140-6736

Frei, B., Forte, T. M., Ames, B. N., & Cross, C. E. (1991). Gas phase oxidants of cigarette smoke induce lipid peroxidation and changes in lipoprotein properties in human blood plasma. Protective effects of ascorbic acid. *Biochemical Journal, 277,* Pt 1, pp. 133-138, 0264-6021

Fuchs, C. S., Stampfer, M. J., Colditz, G. A., Giovannucci, E. L., Manson, J. E., Kawachi, I., et al. (1995). Alcohol consumption and mortality among women. *New England Journal of Medicine, 332,* 19, pp. 1245-1250, 0028-4793

Fukai, T., Siegfried, M. R., Ushio-Fukai, M., Cheng, Y., Kojda, G., & Harrison, D. G. (2000). Regulation of the vascular extracellular superoxide dismutase by nitric oxide and exercise training. *Journal of Clinical Investigation, 105,* 11, pp. 1631-1639, 0021-9738

Fusegawa, Y., Goto, S., Handa, S., Kawada, T., & Ando, Y. (1999). Platelet spontaneous aggregation in platelet-rich plasma is increased in habitual smokers. *Thrombosis Research, 93,* 6, pp. 271-278, 0049-3848

Galper, D. I., Trivedi, M. H., Barlow, C. E., Dunn, A. L., & Kampert, J. B. (2006). Inverse association between physical inactivity and mental health in men and women. *Medicine & Science in Sports & Exercise, 38,* 1, pp. 173-178, 0195-9131

Garcia-Cardena, G., Fan, R., Shah, V., Sorrentino, R., Cirino, G., Papapetropoulos, A., et al. (1998). Dynamic activation of endothelial nitric oxide synthase by Hsp90. *Nature, 392,* 6678, pp. 821-824, 0028-0836

Gaziano, J. M., Gaziano, T. A., Glynn, R. J., Sesso, H. D., Ajani, U. A., Stampfer, M. J., et al. (2000). Light-to-moderate alcohol consumption and mortality in the Physicians' Health Study enrollment cohort. *Journal of the American College of Cardiology, 35,* 1, pp. 96-105, 0735-1097

Ginsberg, D., Hall, S. M., Reus, V. I., & Munoz, R. F. (1995). Mood and depression diagnosis in smoking cessation. *Experimental and Clinical Psychopharmacology, 3,* 4, pp. 389-395, 1064-1297

Glassman, A. H., O'Connor, C. M., Califf, R. M., Swedberg, K., Schwartz, P., Bigger, J. T., Jr., et al. (2002). Sertraline treatment of major depression in patients with acute MI or unstable angina. *Journal of the American Medical Association, 288*, 6, pp. 701-709, 0098-7484

Gold, S. M., Zakowski, S. G., Valdimarsdottir, H. B., & Bovbjerg, D. H. (2004). Higher Beck depression scores predict delayed epinephrine recovery after acute psychological stress independent of baseline levels of stress and mood. *Biological Psychology, 67*, 3, pp. 261-273, 0301-0511

Goldstein, B. I., & Levitt, A. J. (2006). Is current alcohol consumption associated with increased lifetime prevalence of major depression and suicidality? Results from a pilot community survey. *Comprehensive Psychiatry, 47*, 5, pp. 330-333, 0010-440X

Goldstein, D. S. (2006). *Adrenaline and the inner world*. The John Hopkins University Press. 0801882885, Baltimore

Goodman, E., & Whitaker, R. C. (2002). A prospective study of the role of depression in the development and persistence of adolescent obesity. *Pediatrics, 110*, 3, pp. 497-504, 0079-0400

Gordon, J. L., Ditto, B., Lavoie, K. L., Arsenault, A., & Bacon, S. L. (2011). The effect of major depression on post-exercise cardiovascular recovery. *Psychophysiology, 48*, 11, pp. 1604-1609, 0048-5772

Gordon, J. L., Lavoie, K. L., Arsenault, A., Ditto, B., & Bacon, S. L. (2008). Health behaviors and endothelial function. *Journal of Behavioral Medicine, 31*, 1, pp. 5-21, 0160-7715

Grant, B. F. (1995). Comorbidity between DSM-IV drug use disorders and major depression: Results of a national survey of adults. *Journal of Substance Abuse, 7*, 4, pp. 481-497, 0899-3289

Grundy, S. M. (2006). Metabolic Syndrome: Connecting and Reconciling Cardiovascular and Diabetes Worlds. *Journal of the American College of Cardiology, 47*, 6, pp. 1093-1100, 0735-1097

Guinjoan, S. M., Castro, M. N., Vigo, D. E., Weidema, H., Berbara, C., Fahrer, R. D., et al. (2007). Depressive symptoms are related to decreased low-frequency heart rate variability in older adults with decompensated heart failure. *Neuropsychobiology, 55*, 3-4, pp. 219-224, 0302-282X

Haapanen-Niemi, N., Miilunpalo, S., Pasanen, M., Vuori, I., Oja, P., & Malmberg, J. (2000). Body mass index, physical inactivity and low level of physical fitness as determinants of all-cause and cardiovascular disease mortality. *International Journal of Obesity & Related Metabolic Disorders, 24*, 11, pp. 1465-1474, 0307-0565

Hall, J. E., Hildebrandt, D. A., & Kuo, J. (2001). Obesity hypertension: role of leptin and sympathetic nervous system. *American Journal of Hypertension, 14*, 6 Pt 2, pp. 103S-115S, 0895-7061

Hall, S. M., Munoz, R. F., & Reus, V. I. (1994). Cognitive-behavioral intervention increases abstinence rates for depressive-history smokers. *Journal of Consulting & Clinical Psychology, 62*, 1, pp. 141-146, 0022-006X

Hamer, M., Molloy, G. J., & Stamatakis, E. (2008). Psychological distress as a risk factor for cardiovascular events: pathophysiological and behavioral mechanisms. *Journal of the American College of Cardiology, 52*, 25, pp. 2156-2162, 0735-1097

Heitzer, T., Yla Herttuala, S., Wild, E., Luoma, J., & Drexler, H. (1999). Effect of vitamin E on endothelial vasodilator function in patients with hypercholesterolemia, chronic smoking or both. *Journal of the American College of Cardiology, 33*, 2, pp. 499-505, 0735-1097

Hendrickson, R. J., Cahill, P. A., Sitzmann, J. V., & Redmond, E. M. (1999). Ethanol enhances basal and flow-stimulated nitric oxide synthase activity in vitro by activating an inhibitory guanine nucleotide binding protein. *Journal of Pharmacology & Experimental Therapeutics, 289,* 3, pp. 1293-1300, 0022-3565

Heo, M., Pietrobelli, A., Fontaine, K. R., Sirey, J. A., & Faith, M. S. (2006). Depressive mood and obesity in US adults: comparison and moderation by sex, age, and race. *International Journal of Obesity, 30,* 3, pp. 513-519, 0307-0565

Herbert, W. H. (1975). Cigarette smoking and arteriographically demonstrable coronary artery disease. *Chest, 67,* 1, pp. 49-52, 0012-3692

Hoshino, S., Yoshida, M., Inoue, K., Yano, Y., Yanagita, M., Mawatari, H., et al. (2005). Cigarette smoke extract induces endothelial cell injury via JNK pathway. *Biochemical & Biophysical Research Communications, 329,* 1, pp. 58-63, 0006-291X

Hu, F. B., Willett, W. C., Li, T., Stampfer, M. J., Colditz, G. A., & Manson, J. E. (2004). Adiposity as Compared with Physical Activity in Predicting Mortality among Women. *New England Journal of Medicine, 351,* 26, pp. 2694-2703, 0028-4793

Hubert, H. B., Feinleib, M., McNamara, P. M., & Castelli, W. P. (1983). Obesity as an independent risk factor for cardiovascular disease: a 26-year follow-up of participants in the Framingham Heart Study. *Circulation, 67,* 5, pp. 968-977, 0009-7322

Hughes, J. W., & Stoney, C. M. (2000). Depressed mood is related to high-frequency heart rate variability during stressors. *Psychosomatic Medicine, 62,* 6, pp. 796-803, 0033-3174

Imhof, A., Froehlich, M., Brenner, H., Boeing, H., Pepys, M. B., & Koenig, W. (2001). Effect of alcohol consumption on systemic markers of inflammation. *Lancet, 357,* 9258, pp. 763-767, 0140-6736

Imhof, A., Woodward, M., Doering, A., Helbecque, N., Loewel, H., Amouyel, P., et al. (2004). Overall alcohol intake, beer, wine, and systemic markers of inflammation in western Europe: results from three MONICA samples (Augsburg, Glasgow, Lille). *European Heart Journal, 25,* 23, pp. 2092-2100, 0195-668X

Jakicic, J. M., Marcus, B. H., Gallagher, K. I., Napolitano, M., & Lang, W. (2003). Effect of exercise duration and intensity on weight loss in overweight, sedentary women: a randomized trial. *Journal of the American Medical Association, 290,* 10, pp. 1323-1330, 0098-7484

Johns, D. G., Dorrance, A. M., Tramontini, N. L., & Webb, R. C. (2001). Glucocorticoids inhibit tetrahydrobiopterin-dependent endothelial function. *Experimental Biology & Medicine, 226,* 1, pp. 27-31, 1535-3702

Jonas, B. S., Franks, P., & Ingram, D. D. (1997). Are symptoms of anxiety and depression risk factors for hypertension? Longitudinal evidence from the National Health and Nutrition Examination Survey I Epidemiologic Follow-up Study. *Archives of Family Medicine, 6,* 1, pp. 43-49, 1063-3987

Joynt, K. E., & O'Connor, C. M. (2005). Lessons from SADHART, ENRICHD, and other trials. *Psychosomatic medicine, 67 Suppl 1,* pp. S63-66, 0033-3174

Joynt, K. E., Whellan, D. J., & O'Connor, C. M. (2003). Depression and cardiovascular disease: mechanisms of interaction. *Biological Psychiatry, 54,* 3, pp. 248-261, 0006-3223

Kannel, W. B., D'Agostino, R. B., & Belanger, A. J. (1987). Fibrinogen, cigarette smoking, and risk of cardiovascular disease: insights from the Framingham Study. *American Heart Journal, 113,* 4, pp. 1006-1010, 0002-8703

Keil, U., Chambless, L. E., Doring, A., Filipiak, B., & Stieber, J. (1997). The relation of alcohol intake to coronary heart disease and all-cause mortality in a beer-drinking population. *Epidemiology, 8,* 2, pp. 150-156, 1044-3983

Kelly, J. J., Mangos, G., Williamson, P. M., & Whitworth, J. A. (1998). Cortisol and hypertension. *Clinical & Experimental Pharmacology & Physiology - Supplement, 25,* pp. S51-56, 0143-9294

Kessler, R. C., Crum, R. M., Warner, L. A., Nelson, C. B., & et al. (1997). Lifetime co-occurrence of DSM-III-R alcohol abuse and dependence with other psychiatric disorders in the National Comorbidity Survey. *Archives of General Psychiatry, 54,* 4, pp. 313-321, 0003-990X

Khaykin, Y., Dorian, P., Baker, B., Shapiro, C., Sandor, P., Mironov, D., et al. (1998). Autonomic correlates of antidepressant treatment using heart-rate variability analysis. *Canadian journal of psychiatry, 43,* 2, pp. 183-186, 0706-7437

Kingwell, B. A., Tran, B., Cameron, J. D., Jennings, G. L., & Dart, A. M. (1996). Enhanced vasodilation to acetylcholine in athletes is associated with lower plasma cholesterol. *American Journal of Physiology, 270,* 6 Pt 2, pp. H2008-2013, 0002-9513

Kiowski, W., Linder, L., Stoschitzky, K., Pfisterer, M., Burckhardt, D., Burkart, F., et al. (1994). Diminished vascular response to inhibition of endothelium-derived nitric oxide and enhanced vasoconstriction to exogenously administered endothelin-1 in clinically healthy smokers. *Circulation, 90,* 1, pp. 27-34, 0009-7322

Kudoh, A., Ishihara, H., & Matsuki, A. (2000). Inhibition of the cortisol response to surgical stress in chronically depressed patients. *Journal of Clinical Anesthesia, 12,* 5, pp. 383-387, 0952-8180

La Rovere, M. T., Bigger, J. T., Jr., Marcus, F. I., Mortara, A., & Schwartz, P. J. (1998). Baroreflex sensitivity and heart-rate variability in prediction of total cardiac mortality after myocardial infarction. ATRAMI (Autonomic Tone and Reflexes After Myocardial Infarction) Investigators. *Lancet, 351,* 9101, pp. 478-484, 0140-6736

Lahmeyer, H. W., & Bellur, S. N. (1987). Cardiac regulation and depression. *Journal of Psychiatric Research, 21,* 1, pp. 1-6, 0022-3956

Laine, H., Yki-Jarvinen, H., Kirvela, O., Tolvanen, T., Raitakari, M., Solin, O., et al. (1998). Insulin resistance of glucose uptake in skeletal muscle cannot be ameliorated by enhancing endothelium-dependent blood flow in obesity. *Journal of Clinical Investigation, 101,* 5, pp. 1156-1162, 0021-9738

Lake, C. R., Pickar, D., Ziegler, M. G., Lipper, S., Slater, S., & Murphy, D. L. (1982). High plasma norepinephrine levels in patients with major affective disorder. *American Journal of Psychiatry, 139,* 10, pp. 1315-1318, 0002-953X

Lam, T. H., Stewart, S. M., Ho, S. Y., Lai, M. K., Mak, K. H., Chau, K. V., et al. (2005). Depressive symptoms and smoking among Hong Kong Chinese adolescents. *Addiction, 100,* 7, pp. 1003-1011, 0965-2140

Lawlor, D. A., & Leon, D. A. (2005). Association of Body Mass Index and Obesity Measured in Early Childhood With Risk of Coronary Heart Disease and Stroke in Middle Age: Findings From the Aberdeen Children of the 1950s Prospective Cohort Study. *Circulation, 111,* 15, pp. 1891-1896, 0009-7322

Lee, & Choudhury. (2007). Prospects for atherosclerosis regression through increase in high-density lipoprotein and other emerging therapeutic targets. *Heart, 93*, 5, pp. 559-564, 0953-0495

Lee, Y. H., & Pratley, R. E. (2007). Abdominal obesity and cardiovascular disease risk: the emerging role of the adipocyte. *Journal of Cardiopulmonary Rehabilitation & Prevention, 27*, 1, pp. 2-10, 1932-7501

Lembo, G., Rendina, V., Iaccarino, G., Lamenza, F., Volpe, M., & Trimarco, B. (1993). Insulin reduces reflex forearm sympathetic vasoconstriction in healthy humans. *Hypertension, 21*, 6 Pt 2, pp. 1015-1019, 0194-911X

Lesperance, F., Frasure-Smith, N., Koszycki, D., Laliberte, M.-A., van Zyl, L. T., Baker, B., et al. (2007). Effects of citalopram and interpersonal psychotherapy on depression in patients with coronary artery disease: the Canadian Cardiac Randomized Evaluation of Antidepressant and Psychotherapy Efficacy (CREATE) trial. *Journal of the American Medical Association, 297*, 4, pp. 367-379, 0098-7484

Light, K. C., Kothandapani, R. V., & Allen, M. T. (1998). Enhanced cardiovascular and catecholamine responses in women with depressive symptoms. *International Journal of Psychophysiology, 28*, 2, pp. 157-166, 0167-8760

Lukassen, J., & Beaudet, M. P. (2005). Alcohol dependence and depression among heavy drinkers in Canada. *Social Science & Medicine, 61*, 8, pp. 1658-1667, 0037-7856

Luppino, F. S., de Wit, L. M., Bouvy, P. F., Stijnen, T., Cuijpers, P., Penninx, B. W., et al. (2010). Overweight, obesity, and depression: a systematic review and meta-analysis of longitudinal studies. *Archives of General Psychiatry, 67*, 3, pp. 220-229, 003X-990X

Maggi, E., Chiesa, R., Melissano, G., Castellano, R., Astore, D., Grossi, A., et al. (1994). LDL oxidation in patients with severe carotid atherosclerosis. A study of in vitro and in vivo oxidation markers. *Arteriosclerosis & Thrombosis, 14*, 12, pp. 1892-1899, 1049-8834

Maiorano, G., Bartolomucci, F., Contursi, V., Minenna, F. S., Di Mise, R., Palasciano, A., et al. (1999). Noninvasive detection of vascular dysfunction in alcoholic patients. *American Journal of Hypertension, 12*, 2 Pt 1, pp. 137-144, 0895-7061

Mancia, G., Bousquet, P., Elghozi, J. L., Esler, M., Grassi, G., Julius, S., et al. (2007). The sympathetic nervous system and the metabolic syndrome. *Journal of Hypertension, 25*, 5, pp. 909-920, 0263-6352

Mancini, T., Kola, B., Mantero, F., Boscaro, M., & Arnaldi, G. (2004). High cardiovascular risk in patients with Cushing's syndrome according to 1999 WHO/ISH guidelines. *Clinical Endocrinology, 61*, 6, pp. 768-777, 0300-0664

Matthews, K., Schwartz, J., Cohen, S., & Seeman, T. (2006). Diurnal cortisol decline is related to coronary calcification: CARDIA study. *Psychosomatic Medicine, 68*, 5, pp. 657-661, 0033-3174

Mattusch, F., Dufaux, B., Heine, O., Mertens, I., & Rost, R. (2000). Reduction of the plasma concentration of C-reactive protein following nine months of endurance training. *International Journal of Sports Medicine, 21*, 1, pp. 21-24, 0172-4622

McFadden, C. B., Brensinger, C. M., Berlin, J. A., & Townsend, R. R. (2005). Systematic review of the effect of daily alcohol intake on blood pressure. *American Journal of Hypertension, 18*, 2 Pt 1, pp. 276-286, 0895-7061

McFarlane, A., Kamath, M. V., Fallen, E. L., Malcolm, V., Cherian, F., & Norman, G. (2001). Effect of sertraline on the recovery rate of cardiac autonomic function in depressed patients after acute myocardial infarction. *American heart journal, 142*, 4, pp. 617-623, 0002-8703

Meldrum, D. R. (1998). Tumor necrosis factor in the heart. *American Journal of Physiology, 274*, 3 Pt 2, pp. R577-595, 0002-9513

Mendall, M. A., Patel, P., Asante, M., Ballam, L., Morris, J., Strachan, D. P., et al. (1997). Relation of serum cytokine concentrations to cardiovascular risk factors and coronary heart disease. *Heart, 78*, 3, pp. 273-277, 1355-6037

Mercurio, F., & Manning, A. M. (1999). Multiple signals converging on NF-kappaB. *Current Opinion in Cell Biology, 11*, 2, pp. 226-232, 0955-0674

Miller, G. E., Cohen, S., & Ritchey, A. K. (2002). Chronic psychological stress and the regulation of pro-inflammatory cytokines: a glucocorticoid-resistance model. *Health Psychology, 21*, 6, pp. 531-541, 0278-6133

Miller, V. M., & Vanhoutte, P. M. (1988). Enhanced release of endothelium-derived factor(s) by chronic increases in blood flow. *American Journal of Physiology, 255*, 3 Pt 2, pp. H446-451, 0002-9513

Miller, W. R., & Rollnick, S. (2002). *Motivational Interviewing: Preparing People for Change* (2 ed.), Guilford Press, 1572305630, New York

Mora, S., Cook, N., Buring, J. E., Ridker, P. M., & Lee, I. M. (2007). Physical activity and reduced risk of cardiovascular events: potential mediating mechanisms. *Circulation, 116*, 19, pp. 2110-2118, 0009-7322

Mora, S., Redberg, R. F., Cui, Y., Whiteman, M. K., Flaws, J. A., Sharrett, A. R., et al. (2003). Ability of exercise testing to predict cardiovascular and all-cause death in asymptomatic women: a 20-year follow-up of the lipid research clinics prevalence study. *Journal of the American Medical Association, 290*, 12, pp. 1600-1607, 0098-7484

Morrison, J. A., Sprecher, D. L., Barton, B. A., Waclawiw, M. A., & Daniels, S. R. (1999). Overweight, fat patterning, and cardiovascular disease risk factors in black and white girls: The National Heart, Lung, and Blood Institute Growth and Health Study. *Journal of Pediatrics, 135*, 4, pp. 458-464, 0022-3476

Morshedi-Meibodi, A., Larson, M. G., Levy, D., O'Donnell, C. J., & Vasan, R. S. (2002). Heart rate recovery after treadmill exercise testing and risk of cardiovascular disease events (The Framingham Heart Study). *American Journal of Cardiology, 90*, 8, pp. 848-852, 0002-9149

Murray, R. P., Connett, J. E., Tyas, S. L., Bond, R., Ekuma, O., Silversides, C. K., et al. (2002). Alcohol volume, drinking pattern, and cardiovascular disease morbidity and mortality: is there a U-shaped function? *American Journal of Epidemiology, 155*, 3, pp. 242-248, 0002-9262

Nahshoni, E., Aizenberg, D., Strasberg, B., Dorfman, P., Sigler, M., Imbar, S., et al. (2000). QT dispersion in the surface electrocardiogram in elderly patients with major depression. *Journal of Affective Disorders, 60*, 3, pp. 197-200, 0165-0327

Newby, D. E., Wright, R. A., Labinjoh, C., Ludlam, C. A., Fox, K. A., Boon, N. A., et al. (1999). Endothelial dysfunction, impaired endogenous fibrinolysis, and cigarette smoking: a mechanism for arterial thrombosis and myocardial infarction. *Circulation, 99*, 11, pp. 1411-1415, 0009-7322

Nigro, J., Osman, N., Dart, A. M., & Little, P. J. (2006). Insulin resistance and atherosclerosis. *Endocrine Reviews, 27*, 3, pp. 242-259, 0163-769X

Nishime, E. O., Cole, C. R., Blackstone, E. H., Pashkow, F. J., & Lauer, M. S. (2000). Heart rate recovery and treadmill exercise score as predictors of mortality in patients referred for exercise ECG. *Journal of the American Medical Association, 284,* 11, pp. 1392-1398, 0098-7484

Nolan, J., Batin, P. D., Andrews, R., Lindsay, S. J., Brooksby, P., Mullen, M., et al. (1998). Prospective study of heart rate variability and mortality in chronic heart failure: results of the United Kingdom heart failure evaluation and assessment of risk trial (UK-heart). *Circulation, 98,* 15, pp. 1510-1516, 0009-7322

O'Connell, B. J., & Genest, J., Jr. (2001). High-density lipoproteins and endothelial function. *Circulation, 104,* 16, pp. 1978-1983, 0009-7322

Oliver, K., & Cronan, T. (2002). Predictors of exercise behaviors among fibromyalgia patients. *Preventive Medicine, 35,* 4, pp. 383-389, 0091-7435

Panarelli, M., Holloway, C. D., Fraser, R., Connell, J. M., Ingram, M. C., Anderson, N. H., et al. (1998). Glucocorticoid receptor polymorphism, skin vasoconstriction, and other metabolic intermediate phenotypes in normal human subjects. *Journal of Clinical Endocrinology & Metabolism, 83,* 6, pp. 1846-1852, 0021-972X

Panza, J. A., Casino, P. R., Kilcoyne, C. M., & Quyyumi, A. A. (1993). Role of endothelium-derived nitric oxide in the abnormal endothelium-dependent vascular relaxation of patients with essential hypertension. *Circulation, 87,* 5, pp. 1468-1474, 0009-7322

Pech-Amsellem, M. A., Myara, I., Storogenko, M., Demuth, K., Proust, A., & Moatti, N. (1996). Enhanced modifications of low-density lipoproteins (LDL) by endothelial cells from smokers: a possible mechanism of smoking-related atherosclerosis. *Cardiovascular Research, 31,* 6, pp. 975-983, 0008-6363

Peeters, F., Nicholson, N. A., & Berkhof, J. (2003). Cortisol responses to daily events in major depressive disorder. *Psychosomatic Medicine, 65,* 5, pp. 836-841, 0033-3174

Pelletier, R., Lavoie, K. L., Gordon, J., Arsenault, A., Campbell, T. S., & Bacon, S. L. (2009). The role of mood disorders in exercise-induced cardiovascular reactivity. *Psychosomatic Medicine, 71,* 3, pp. 301-307, 0033-3174

Penninx, B. W., Leveille, S., Ferrucci, L., van Eijk, J. T., & Guralnik, J. M. (1999). Exploring the effect of depression on physical disability: longitudinal evidence from the established populations for epidemiologic studies of the elderly. *American Journal of Public Health, 89,* 9, pp. 1346-1352, 0090-0036

Persson, M. G., & Gustafsson, L. E. (1992). Ethanol can inhibit nitric oxide production. *European Journal of Pharmacology, 224,* 1, pp. 99-100, 0014-2999

Petersen, M., & Harper, A. (2004). Weight loss strategies--a change of focus is required. *Obesity Reviews, 5,* 4, pp. 239-240, 1467-7881

Petrie, J. R., Ueda, S., Webb, D. J., Elliott, H. L., & Connell, J. M. (1996). Endothelial nitric oxide production and insulin sensitivity. A physiological link with implications for pathogenesis of cardiovascular disease. *Circulation, 93,* 7, pp. 1331-1333, 0009-7322

Piko, B. F., & Keresztes, N. (2006). Physical activity, psychosocial health, and life goals among youth. *Journal of Community Health, 31,* 2, pp. 136-145, 0094-5145

Pine, D. S., Goldstein, R. B., Wolk, S., & Weissman, M. M. (2001). The association between childhood depression and adulthood body mass index. *Pediatrics, 107,* 5, pp. 1049-1056, 0079-0400

Plotsky, P. M., Owens, M. J., & Nemeroff, C. B. (1998). Psychoneuroendocrinology of depression. Hypothalamic-pituitary-adrenal axis. *Psychiatric Clinics of North America, 21,* 2, pp. 293-307, 0193-953X

Pretorius, M., Rosenbaum, D. A., Lefebvre, J., Vaughan, D. E., & Brown, N. J. (2002). Smoking impairs bradykinin-stimulated t-PA release. *Hypertension, 39,* 3, pp. 767-771, 0362-4323

Puranik, R., & Celermajer, D. S. (2003). Smoking and endothelial function. *Progress in Cardiovascular Diseases, 45,* 6, pp. 443-458, 0033-0620

Quyyumi, A. A. (2003). Prognostic value of endothelial function. *American Journal of Cardiology, 91,* 12A, pp. 19H-24H, 0002-9149

Rabkin, J. G., Charles, E., & Kass, F. (1983). Hypertension and DSM-III depression in psychiatric outpatients. *American Journal of Psychiatry, 140,* 8, pp. 1072-1074, 0002-953X

Rajagopalan, S., Brook, R., Rubenfire, M., Pitt, E., Young, E., & Pitt, B. (2001). Abnormal brachial artery flow-mediated vasodilation in young adults with major depression. *American Journal of Cardiology, 88,* 2, pp. 196-198, 0002-9149

Ramsdale, D. R., Faragher, E. B., Bray, C. L., Bennett, D. H., Ward, C., & Beton, D. C. (1985). Smoking and coronary artery disease assessed by routine coronary arteriography. *British Medical Journal Clinical Research Ed, 290,* 6463, pp. 197-200, 0267-0623

Reuben, D. B., Judd-Hamilton, L., Harris, T. B., & Seeman, T. E. (2003). The associations between physical activity and inflammatory markers in high-functioning older persons: MacArthur Studies of Successful Aging. *Journal of the American Geriatrics Society, 51,* 8, pp. 1125-1130, 0002-8614

Reusch, J. E., & Draznin, B. B. (2007). Atherosclerosis in diabetes and insulin resistance. *Diabetes, Obesity & Metabolism, 9,* 4, pp. 455-463, 1462-8902

Reynolds, R. M., & Walker, B. R. (2003). Human insulin resistance: the role of glucocorticoids. *Diabetes, Obesity & Metabolism, 5,* 1, pp. 5-12, 1462-8902

Richardson, L. P., Davis, R., Poulton, R., McCauley, E., Moffitt, T. E., Caspi, A., et al. (2003). A longitudinal evaluation of adolescent depression and adult obesity. *Archives of Pediatrics & Adolescent Medicine, 157,* 8, pp. 739-745, 1072-4710

Rinder, M. R., Spina, R. J., & Ehsani, A. A. (2000). Enhanced endothelium-dependent vasodilation in older endurance-trained men. *Journal of Applied Physiology, 88,* 2, pp. 761-766, 8750-7587

Rival, J., Riddle, J. M., & Stein, P. D. (1987). Effects of chronic smoking on platelet function. *Thrombosis Research, 45,* 1, pp. 75-85, 0049-3848

Roy, A., Pickar, D., De Jong, J., Karoum, F., & Linnoila, M. (1988). Norepinephrine and its metabolites in cerebrospinal fluid, plasma, and urine. Relationship to hypothalamic-pituitary-adrenal axis function in depression. *Archives of General Psychiatry, 45,* 9, pp. 849-857, 0003-990X

Rozanski, A., Blumenthal, J. A., & Kaplan, J. (1999). Impact of psychological factors on the pathogenesis of cardiovascular disease and implications for therapy. *Circulation, 99,* 16, pp. 2192-2217, 0009-7322

Rywik, T. M., Blackman, M. R., Yataco, A. R., Vaitkevicius, P. V., Zink, R. C., Cottrell, E. H., et al. (1999). Enhanced endothelial vasoreactivity in endurance-trained older men. *Journal of Applied Physiology, 87,* 6, pp. 2136-2142, 8750-7587

Sambola, A., Osende, J., Hathcock, J., Degen, M., Nemerson, Y., Fuster, V., et al. (2003). Role of risk factors in the modulation of tissue factor activity and blood thrombogenicity. *Circulation, 107,* 7, pp. 973-977, 0009-7322

Schatzberg, A. F., Rothschild, A. J., Bond, T. C., & Cole, J. O. (1984). The DST in psychotic depression: diagnostic and pathophysiologic implications. *Psychopharmacology Bulletin, 20,* 3, pp. 362-364, 0048-5764

Schwartz, M. A., & Lechene, C. (1992). Adhesion is required for protein kinase C-dependent activation of the Na+/H+ antiporter by platelet-derived growth factor. *Proceedings of the National Academy of Sciences of the United States of America, 89,* 13, pp. 6138-6141, 0027-8424

Senn, J. J., Klover, P. J., Nowak, I. A., & Mooney, R. A. (2002). Interleukin-6 induces cellular insulin resistance in hepatocytes. *Diabetes, 51,* 12, pp. 3391-3399, 0012-1797

Shen, Y., Rattan, V., Sultana, C., & Kalra, V. K. (1996). Cigarette smoke condensate-induced adhesion molecule expression and transendothelial migration of monocytes. *American Journal of Physiology, 270,* 5 Pt 2, pp. H1624-1633, 0002-9513

Sheps, D. S., Freedland, K. E., Golden, R. N., McMahon, R. P., Enhancing Recovery in Coronary Heart, D., & Sertraline Antidepressant Heart Attack, T. (2003). ENRICHD and SADHART: implications for future biobehavioral intervention efforts. *Psychosomatic medicine, 65,* 1, pp. 1-2, 0033-3174

Sherwood, A., Hinderliter, A. L., Watkins, L. L., Waugh, R. A., & Blumenthal, J. A. (2005). Impaired endothelial function in coronary heart disease patients with depressive symptomatology. *Journal of the American College of Cardiology, 46,* 4, pp. 656-659, 0735-1097

Smith, F. B., Lee, A. J., Fowkes, F. G., Price, J. F., Rumley, A., & Lowe, G. D. (1997). Hemostatic factors as predictors of ischemic heart disease and stroke in the Edinburgh Artery Study. *Arteriosclerosis, Thrombosis & Vascular Biology, 17,* 11, pp. 3321-3325, 1079-5642

Smith, J. K., Dykes, R., Douglas, J. E., Krishnaswamy, G., & Berk, S. (1999). Long-term exercise and atherogenic activity of blood mononuclear cells in persons at risk of developing ischemic heart disease. *Journal of the American Medical Association, 281,* 18, pp. 1722-1727, 0098-7484

Smoak, C. G., Burke, G. L., Webber, L. S., Harsha, D. W., Srinivasan, S. R., & Berenson, G. S. (1987). Relation of obesity to clustering of cardiovascular disease risk factors in children and young adults. The Bogalusa Heart Study. *American Journal of Epidemiology, 125,* 3, pp. 364-372, 0002-9262

Soardo, G., Donnini, D., Varutti, R., Moretti, M., Milocco, C., Basan, L., et al. (2005). Alcohol-induced endothelial changes are associated with oxidative stress and are rapidly reversed after withdrawal. *Alcoholism: Clinical & Experimental Research, 29,* 10, pp. 1889-1898, 0145-6008

Soodini, G. R., & Hamdy, O. (2004). Obesity and endothelial function. *Current Opinion in Endocrinology & Diabetes, 11,* 4, pp. 186-191, 1752-2978

Spyridopoulos, I., Wischhusen, J., Rabenstein, B., Mayer, P., Axel, D. I., Frohlich, K. U., et al. (2001). Alcohol enhances oxysterol-induced apoptosis in human endothelial cells by a calcium-dependent mechanism. *Arteriosclerosis, Thrombosis & Vascular Biology, 21,* 3, pp. 439-444, 1079-5642

Stanner, S. (Ed.). (2005). *Cardiovascular disease: diet, nutrition and emerging risk factors.* Blackwell Publishing, 9781405101660, Oxford

Steinberg, H. O., Chaker, H., Leaming, R., Johnson, A., Brechtel, G., & Baron, A. D. (1996). Obesity/insulin resistance is associated with endothelial dysfunction. Implications for the syndrome of insulin resistance. *Journal of Clinical Investigation, 97,* 11, pp. 2601-2610, 0021-9738

Steppan, C. M., Bailey, S. T., Bhat, S., Brown, E. J., Banerjee, R. R., Wright, C. M., et al. (2001). The hormone resistin links obesity to diabetes. *Nature, 409,* 6818, pp. 307-312, 0028-0836

Sun, H., & Mayhan, W. G. (2001). Temporal effect of alcohol consumption on reactivity of pial arterioles: role of oxygen radicals. *American Journal of Physiology - Heart & Circulatory Physiology, 280,* 3, pp. H992-H1001, 0363-6135

Swendsen, J. D., Merikangas, K. R., Canino, G. J., Kessler, R. C., Rubio-Stipec, M., & Angst, J. (1998). The comorbidity of alcoholism with anxiety and depressive disorders in four geographic communities. *Comprehensive Psychiatry, 39,* 4, pp. 176-184, 0010-440X

Tappia, P. S., Troughton, K. L., Langley-Evans, S. C., & Grimble, R. F. (1995). Cigarette smoking influences cytokine production and antioxidant defences. *Clinical Science, 88,* 4, pp. 485-489, 0143-5221

Tauchmanova, L., Rossi, R., Biondi, B., Pulcrano, M., Nuzzo, V., Palmieri, E. A., et al. (2002). Patients with subclinical Cushing's syndrome due to adrenal adenoma have increased cardiovascular risk. *Journal of Clinical Endocrinology & Metabolism, 87,* 11, pp. 4872-4878, 0021-972X

Thompson, P. D., Crouse, S. F., Goodpaster, B., Kelley, D., Moyna, N., & Pescatello, L. (2001). The acute versus the chronic response to exercise. *Medicine & Science in Sports & Exercise, 33,* 6 Suppl, pp. S438-445; discussion S452-433, 0195-9131

Timonen, M., Laakso, M., Jokelainen, J., Rajala, U., Meyer-Rochow, V. B., & Keinanen-Kiukaanniemi, S. (2005). Insulin resistance and depression: cross sectional study. *British Medical Journal, 330,* 7481, pp. 17-18, 1756-1833

Tisi, P. V., Hulse, M., Chulakadabba, A., Gosling, P., & Shearman, C. P. (1997). Exercise training for intermittent claudication: does it adversely affect biochemical markers of the exercise-induced inflammatory response? *European Journal of Vascular & Endovascular Surgery, 14,* 5, pp. 344-350, 1078-5884

Tracy, R. P., Psaty, B. M., Macy, E., Bovill, E. G., Cushman, M., Cornell, E. S., et al. (1997). Lifetime smoking exposure affects the association of C-reactive protein with cardiovascular disease risk factors and subclinical disease in healthy elderly subjects. *Arteriosclerosis, Thrombosis & Vascular Biology, 17,* 10, pp. 2167-2176, 1079-5642

Tsuji, H., Venditti, F. J., Jr., Manders, E. S., Evans, J. C., Larson, M. G., Feldman, C. L., et al. (1994). Reduced heart rate variability and mortality risk in an elderly cohort. The Framingham Heart Study. *Circulation, 90,* 2, pp. 878-883, 0009-7322

Tuomilehto, J., Lindstrom, J., Eriksson, J. G., Valle, T. T., Hamalainen, H., Ilanne-Parikka, P., et al. (2001). Prevention of type 2 diabetes mellitus by changes in lifestyle among subjects with impaired glucose tolerance. *New England Journal of Medicine, 344,* 18, pp. 1343-1350, 0028-4793

Uysal, K. T., Wiesbrock, S. M., Marino, M. W., & Hotamisligil, G. S. (1997). Protection from obesity-induced insulin resistance in mice lacking TNF-alpha function. *Nature, 389,* 6651, pp. 610-614, 0028-0836

Van der Kooy, K., van Hout, H., Marwijk, H., Marten, H., Stehouwer, C., & Beekman, A. (2007). Depression and the risk for cardiovascular diseases: systematic review and meta analysis. *International Journal of Geriatric Psychiatry, 22*, 7, pp. 613-626, 0885-6230

van Melle, J. P., de Jonge, P., Spijkerman, T. A., Tijssen, J. G., Ormel, J., van Veldhuisen, D. J., et al. (2004). Prognostic association of depression following myocardial infarction with mortality and cardiovascular events: a meta-analysis. *Psychosomatic Medicine, 66*, 6, pp. 814-822, 0033-3174

Veith, R. C., Best, J. D., & Halter, J. B. (1984). Dose-dependent suppression of norepinephrine appearance rate in plasma by clonidine in man. *Journal of Clinical Endocrinology & Metabolism, 59*, 1, pp. 151-155, 0021-972X

Vogelzangs, N., Beekman, A. T. F., Milaneschi, Y., Bandinelli, S., Ferrucci, L., & Penninx, B. W. J. H. (2010). Urinary cortisol and six-year risk of all-cause and cardiovascular mortality. *The Journal of clinical endocrinology and metabolism, 95*, 11, pp. 4959-4964, 0021-972X

Vogt, C. J., & Schmid-Schonbein, G. W. (2001). Microvascular endothelial cell death and rarefaction in the glucocorticoid-induced hypertensive rat. *Microcirculation, 8*, 2, pp. 129-139, 1073-9688

Walker, B. R., Soderberg, S., Lindahl, B., & Olsson, T. (2000). Independent effects of obesity and cortisol in predicting cardiovascular risk factors in men and women. *Journal of Internal Medicine, 247*, 2, pp. 198-204, 0954-6820

Wang, X. L., Tam, C., McCredie, R. M., & Wilcken, D. E. (1994). Determinants of severity of coronary artery disease in Australian men and women. *Circulation, 89*, 5, pp. 1974-1981, 0009-7322

Wannamethee, S. G., Lowe, G. D. O., Whincup, P. H., Rumley, A., Walker, M., & Lennon, L. (2002). Physical activity and hemostatic and inflammatory variables in elderly men. *Circulation, 105*, 15, pp. 1785-1790, 0009-7322

Watkins, L. L., & Grossman, P. (1999). Association of depressive symptoms with reduced baroreflex cardiac control in coronary artery disease. *American Heart Journal, 137*, 3, pp. 453-457, 0002-8703

Watts, G. F., O'Brien, S. F., Silvester, W., & Millar, J. A. (1996). Impaired endothelium-dependent and independent dilatation of forearm resistance arteries in men with diet-treated non-insulin-dependent diabetes: role of dyslipidaemia. *Clinical Science, 91*, 5, pp. 567-573, 0143-5221

Weiss, J. W., Mouttapa, M., Chou, C.-P., Nezami, E., Johnson, C., Palmer, P. H., et al. (2005). Hostility, depressive symptoms, and smoking in early adolescence. *Journal of Adolescence, 28*, 1, pp. 49-62, 0140-1971

Whitworth, J. A., Gordon, D., McLachlan-Troup, N., Scoggins, B. A., & Moulds, R. W. (1989). Dexamethasone suppression in essential hypertension: effects on cortisol and blood pressure. *Clinical & Experimental Hypertension - Part A, Theory & Practice, 11*, 2, pp. 323-335, 0730-0077

Whitworth, J. A., Mangos, G. J., & Kelly, J. J. (2000). Cushing, cortisol, and cardiovascular disease. *Hypertension, 36*, 5, pp. 912-916, 0194-911X

Whitworth, J. A., Saines, D., & Scoggins, B. A. (1984). Blood pressure and metabolic effects of cortisol and deoxycorticosterone in man. *Clinical & Experimental Hypertension - Part A, Theory & Practice, 6*, 4, pp. 795-809, 0730-0077

Whooley, M. A., de Jonge, P., Vittinghoff, E., Otte, C., Moos, R., Carney, R. M., et al. (2008). Depressive symptoms, health behaviors, and risk of cardiovascular events in patients with coronary heart disease. *Journal of the American Medical Association, 300*, 20, pp. 2379-2388, 0098-7484

Widlansky, M. E., Sesso, H. D., Rexrode, K. M., Manson, J. E., & Gaziano, J. M. (2004). Body Mass Index and Total and Cardiovascular Mortality in Men With a History of Cardiovascular Disease. *Archives of Internal Medicine, 164*, 21, pp. 2326-2332, 0003-9926

Williamson, P. M., Kelly, J. J., & Whitworth, J. A. (1996). Dose-response relationships and mineralocorticoid activity in cortisol-induced hypertension in humans. *Journal of Hypertension - Supplement, 14*, 5, pp. S37-41, 0952-1178

World Health Organization. (2007). Cardiovascular diseases fact sheet Retrieved February 18, 2008

Wyatt, R. J., Portnoy, B., Kupfer, D. J., Snyder, F., & Engelman, K. (1971). Resting plasma catecholamine concentrations in patients with depression and anxiety. *Archives of General Psychiatry, 24*, 1, pp. 65-70, 0003-990X

Xin, X., He, J., Frontini, M. G., Ogden, L. G., Motsamai, O. I., & Whelton, P. K. (2001). Effects of alcohol reduction on blood pressure: a meta-analysis of randomized controlled trials. *Hypertension, 38*, 5, pp. 1112-1117, 0194-911X

Yamauchi, T., Kamon, J., Waki, H., Terauchi, Y., Kubota, N., Hara, K., et al. (2001). The fat-derived hormone adiponectin reverses insulin resistance associated with both lipoatrophy and obesity. *Nature Medicine, 7*, 8, pp. 941-946, 1078-8956

Yeragani, V. K., Pohl, R., Jampala, V. C., Balon, R., Ramesh, C., & Srinivasan, K. (2000). Increased QT variability in patients with panic disorder and depression. *Psychiatry Research, 93*, 3, pp. 225-235, 0165-1781

Yokode, M., Kita, T., Arai, H., Kawai, C., Narumiya, S., & Fujiwara, M. (1988). Cholesteryl ester accumulation in macrophages incubated with low density lipoprotein pretreated with cigarette smoke extract. *Proceedings of the National Academy of Sciences of the United States of America, 85*, 7, pp. 2344-2348, 0027-8424

Yudkin, J. S., Kumari, M., Humphries, S. E., & Mohamed-Ali, V. (2000). Inflammation, obesity, stress and coronary heart disease: is interleukin-6 the link? *Atherosclerosis, 148*, 2, pp. 209-214, 0021-9150

Yusuf, S., Hawken, S., Ounpuu, S., Dans, T., Avezum, A., Lanas, F., et al. (2004). Effect of potentially modifiable risk factors associated with myocardial infarction in 52 countries (the INTERHEART study): case-control study. *Lancet, 364*, 9438, pp. 937-952, 0140-6736

Zeiher, A. M., Schachinger, V., & Minners, J. (1995). Long-term cigarette smoking impairs endothelium-dependent coronary arterial vasodilator function. *Circulation, 92*, 5, pp. 1094-1100, 0009-7322

Zilkens, R. R., Burke, V., Hodgson, J. M., Barden, A., Beilin, L. J., & Puddey, I. B. (2005). Red wine and beer elevate blood pressure in normotensive men. *Hypertension, 45*, 5, pp. 874-879, 0194-911X

Zilkens, R. R., Rich, L., Burke, V., Beilin, L. J., Watts, G. F., & Puddey, I. B. (2003). Effects of alcohol intake on endothelial function in men: a randomized controlled trial. *Journal of Hypertension, 21*, 1, pp. 97-103, 0194-911X

Part 2

Childhood and Adolescence

6

Mood Disorders in Childhood and Adolescence and Their Outcome in Adulthood

Ulf Engqvist
Mid Sweden University
Sweden

1. Introduction

In this chapter occurrence of mood disorders in a child and adolescent psychiatric population and among patients visiting a youth psychiatric clinic for young adults in a smaller county council and province study area in Sweden is described. It also describes the outcome in adulthood for patients having mood disorder in childhood and adolescence in terms of later general psychiatric care and diagnoses, mortality and criminality.

Mood disorders refer to a category of mental health problems that include all types of depression and bipolar disorder. They are defined by ICD-10 chapter V, block F30-F39 and described from reasons for admission and diagnoses given at a child and adolescent psychiatric clinic, and from the patients' own descriptions of their symptoms when they visited a youth psychiatric clinic for young adults.

1.1 Psychiatry in Sweden

Psychiatry is one of the most large-scale caring sectors in our time. It is probably the most changeable, and widely prejudice loaded sector in the health care system in Sweden. Among the most rapidly growing, controversy-ridden, and attention-attracting area of history over the past generation has been the history of psychiatry (Ottosson 2003)

Psychiatry is not at all a uniform concept. In Sweden, there are three psychiatric settings, general (adult) psychiatry, forensic psychiatry and child and adolescent psychiatry working with adult patients, children and young people up to 18 years and mentally disordered offenders, respectively (Levander et al. 2006).Without being a speciality there are also units for geriatric psychiatry in some hospitals. Particularly, in the research work delimitation of psychiatry is made with consideration to method and theoretical basis in for example dynamic psychiatry, social psychiatry and biological psychiatry (Ottosson 2009).

Psychiatry has a history dating back to medieval times when the church organized the first care of the sick in Sweden and the health care facilities belonged to the monasteries. In the 18th century the madhouses played a more important role. General hospitals for care of physical diseases were established but he mentally ill remained at the "madhouses" (Gunnarsson 2006). Round 1860 several large changes within psychiatry and psychiatric care in Sweden was effected. A new regulation was sanctioned, psychiatry became an established branch of learning at the medical seats of learning and more hospitals were built (Qvarsell 1982). In 1851 the Government laid the foundation for psychiatry as a medical speciality (Sjöström 1992).

Construction of new mental hospitals after the turn of the century took off and grew the following decades, until the 1960's as an expression of a general institutionalization trend. From 1900 to 1950 the number of beds at the mental hospitals increased from 4,600 to 25,000. In the beginning of the 1960s there were 31 free-standing mental hospitals in Sweden (Beckman 1984). In 1967, when the county councils took over the responsibility for mental health services from the Government the number of inpatient beds were approximately 36,000 most of them in mental hospitals.

In January 1 1995 the Community Mental Health Care reform (SOU 1992:73 1992) came into effect, a reform that all political parties supported. The reform made clear the liability for the municipality to provide for occupation and housing so that the psychiatric disabled patients could adapt themselves to the society. They were now seen as mental handicapped and placed under the same act as individuals with a bodily handicap, The Act on Support and Service (LSS) (SOU 2006:100 2006).

In 2001 only 5,565 beds for inpatient psychiatric care remained and 2006 the number was 4,600 beds. An increasing population and decreasing number of beds has lead to more individuals out of society (Socialstyrelsen 2003).

The national psychiatry coordinator (which was appointed as an investigator of psychiatry at the beginning of the 2000s) established that the Community Mental Health Care Reform from 1995 has implied an increased supply of housing and occupation in the municipalities and that the needs of the target group has been attended to in a better way than before. But the weak support in law for the reform, the lack of local political support and the lack of evidence based methods for the community-based support caused obstacles that to a great extent remains more than ten years later (SOU 2006:100 2006).

For a long time children and young people were not separated from adults in the psychiatric context. As a medical specialty child and adolescent psychiatry is altogether a 20th century matter but intelligence defects and psychoses among children had been paid attention to through earlier centuries. They were treated together with adults in the "madhouses" (Harding 1975).

Within paediatrics it was early observed during the latter part of the 19th century that also children could show nervous and mental diseases. Sweden was the first country in the world to set up special chairs on child diseases. The first psychiatric guidance unit was established in Stockholm in 1914. School psychiatry was also a form of early child psychiatry. The child custody authority in Stockholm established the Child Guidance Clinics Centrals in the beginning of the 1930s.

Professorships in child and adolescent psychiatry were established from 1958 onwards. In November 1960 the mental health care was subsidized by the State. Ten independent child and adolescent psychiatric clinics were active in Sweden at that time. In 1971 all county councils has an organization with in- and outpatient services for child and adolescent psychiatry. The State subsidy regulation remained to end of 1975 when the number of clinics was 30 (Karlén 1985; Schleimer 1999).

Three years after the Community Mental Health Care reform in 1995 another governmental report on child and adolescent psychiatry was published (SOU 1998:31 1998). That report established among other things that psychiatric care for youths with severe psychological disturbances such as eating disorders, autism, schizophrenia, severe obsessive-compulsive disorders and depression with suicide risk did not seem to be working in a satisfactory manner. Many proposals were given in the report but few were realized.

During the 1990s the inflow of admissions to child and adolescent psychiatry increased to a large extend. The waiting lists grow and it was not unusual with more than yearlong periods of waiting for care. This caused debate in media and in authorities (Nyberg, 2001, Schubert, 1999, SOU 1998:31, 1998, Wieselgren, 2007).

1.2 Psychiatric care in Jämtland county

1.2.1 General psychiatry

The history of general psychiatry in Jämtland County follows the history for the rest of the country. Frösö Hospital that was finished in 1912 was one of the large mental hospitals built in Sweden at that time. As late as the start of the 1980s there were more than 20 psychiatric wards with more than 1,000 inpatients. During the 1980s the psychiatry was divided in sectors.

At the start of the 1990s psychiatry was transferred from Frösö hospital to Östersunds hospital and became a psychiatric unit integrated with the physical care. Today there are five wards and less than 100 beds.

1.2.2 Child and adolescent psychiatry

As early as in 1944 Jämtland County Council came to a decision that child and youth care should be organized in the County and the activity started in August 1946 and was closed in the beginning of 1948 when the chief physician left his post. The work was conducted without a treatment ward but with the possibility to use three beds at the paediatric ward. Consulting hours existed at not less than nine different places in the County (Arbetsgruppen kring översyn av den barn- och ungdomspsykiatriska verksamheten i Jämtlands län 1984).

It took more than twenty years before child and adolescent psychiatry again was established in Jämtland. In September 1970 the activity started with one children's ward and one youth ward with seven beds each. The children's ward was transformed to a family ward in 1977 but that was closed one year later. From 1978 up to and until 1993 there was one ward with five beds. In the beginning the care was characterized by traditional nursing a successive change and development towards environment therapy and family therapy. In 1984 a treatment home was opened and it was closed ten years later.

The inpatient activities were replaced in the mid 1990s by an intensified systemic outpatient treatment at the same time as the treatment home and the ward were closed (Arbetsgruppen kring översyn av den barn- och ungdomspsykiatriska verksamheten i Jämtlands län 1984; Engqvist 1993). During the time for child and adolescent psychiatric care that this chapter comprises both outpatient and inpatient care at a ward and a treatment home was active.

1.2.3 The youth psychiatric clinic for young adults

In 1998 a survey was conducted with a view to comparing different occupational groups in child and adolescent psychiatry and general psychiatry. The aim was to compare the knowledge of children and youth, and staff perceptions of each other's activities and views on the youth group and cooperation in the in the various occupations within child and adolescent psychiatry and general psychiatry (Engqvist 2000).

The results showed that in general psychiatry was deficiencies in the child and youth literacy in general and that there were differences in views and approaches between the disciplines. Interest existed for cooperation on the youth group with a willingness to approach each other from this group's needs.

After an investigation work of a collaboration group, and when state funding became available, heads of child and adolescent psychiatry and general psychiatry applied for investment in outpatient psychiatric care for the age group 16-25 years. The youth psychiatric clinic officially started its operations in September 2002 and one month later, the activity began to take on young people. The clinic is jointly funded by child and adolescent psychiatry and general psychiatry.

1.3 Similarities and differences between general psychiatry and child and adolescent psychiatry

There are many similarities between general psychiatry and child and adolescent psychiatry. The two disciplines have operated in parallel during the last 100 years. Despite the similarities there is a lack of significant and reliable information regarding, for instance the relationship between psychiatric deviations and behavioural disturbances in childhood and psychiatric disorders and social adjustment problems and or criminality later in adulthood.

Information about the relationship between behavioural disturbances in childhood and criminality in later life has however developed further compared to information about psychological deviations during childhood and later psychiatric illness. There is also a virtual lack of information regarding possible early symptoms of mental disorders, and how they change during the process of maturing from childhood to adulthood.

It is easy to describe the difference between the two psychiatric settings. Child and adolescent psychiatry works with psychopathological states of the growing and maturing individual or personality where the psychopathological symptoms change in pace with maturity. General psychiatry, on the other hand, works with adult patients where the process of maturing has, if not ended, at least slowed down.

This means that psychiatrists working with adult patients can work with a system of classification of symptoms over time and states that are not as dependent on age and maturation.

In child and adolescent psychiatry, great individual differences and gender differences appear which is why symptoms and behavioural disturbances can actually have varying meaning even in children of the same age. A certain symptom can also form an adequate way of expression for children in a certain age, but be considered pathological in other ages.

1.4 Mood (affective) disorders in Sweden

The national public health report (the latest is from 2009) describes six national diseases[1]: Cardiovascular diseases; Mental illness; Tumour diseases, musculoskeletal diseases, allergies and Accident Injuries (Socialstyrelsen 2009).

In Sweden, clinical depression is a major national disease. One in four men and one in two women suffer at some time in the life of a deep depression according to the Swedish Lundby study (Rorsman et al. 1990). Clinical depression affects 6 percent of all teenagers each year (Olsson and von Knorring 1999)

According to the World Health Organization depression is common worldwide, affecting about 121 million people and depression is among the leading causes of disability worldwide. About one percent of the adult population in Sweden suffers from bipolar disorder. Depression is of great economic importance. An estimate for the entire Swedish population shows that the costs of depression in 2005 were 33 billion Swedish Kronor (over $ 5,000,000).

Treatment for depression and anxiety has become more common for young people. The proportion of 20-24 year olds admitted to hospital for depression doubled between 1997 and 2007, among both women and men. Among youths aged 16-19 years, it is now four times as common with such care. In the case of mental illness among young people, there are major differences between the sexes. Approximately twice as many women as men are cared for attempted suicide, suffer from anxiety and are treated with antidepressants. More than twice as many women are hospitalized for depression (Socialstyrelsen 2009).

Research shows that comorbidity is very high as regards the abuse of alcohol and depression. There is also evidence that alcohol abuse is not always detected when the depression is treated, and that depression is not recognized in people being treated for their addiction. A clear majority of young people who turned to Maria Ungdom[2] in Stockholm showed an extensive psychiatric disorder that often overshadowed the actual abuse problems (SOU 2006:100 2006).

1.5 Background

1.5.1 Previous longitudinal studies

In Sweden, we can proudly note that there are prospective and longitudinal studies describing child and adolescent psychiatry patient materials in different time periods, from 1928 until present, and this may be considered to be unique in the research field. Researchers in the Nordic countries, especially in Finland, have also the opportunity to use

---

[1] Term for diseases that have a high prevalence in the population. Usually chosen to express the prevalence as a percentage and considering diseases with a prevalence of 1%

[2] Specialized rehabilitation clinics for children and adolescents up to age 20 who have substance abuse problems or at risk of developing an addiction

registers that allow longitudinal studies of this kind. In the following sections different studies from Sweden and some Nordic studies are reviewed.

1.5.2 Descriptions of child and adolescent psychiatry patients

Child and adolescent psychiatry patients have been described from as early as 1928 when Alice Hellström started her description of psychopathic boys treated at the Mellansjö treatment home.

Hellström described all the children who came to Mellansjö treatment home 1928-1940 under a special research protocol. She followed then her patients long time after her retirement and until 1968. However, because of age she was not capable of completing her report.

The complete research material was bequeathed to the child and adolescent psychiatric clinic at the S:t Görans children's hospital in Stockholm, where Ingegärd Fried completed the work in her licentiate's dissertation (Fried 1995).

The next study of children treated by child psychiatry was the longitudinal prospective follow up of 2,164 child and adolescent guidance clinic patients (cared for 1953-1955) followed up to 1975. This group was originally studied by Curman and the study was later completed by Nylander (Curman and Nylander 1976; Nylander 1979).

1.5.3 Different Swedish risk groups

Various risk and control groups and children from the general population have been described with prospective longitudinal research from the perspective of several scientific disciplines (see Table 1). This provides a number of opportunities to discuss similarities and differences between patient groups, other groups at risk and children in general.

In addition to the studies named above there are also descriptions of child and adolescent psychiatry patients conducted with so called cross section methods.

In 1950 chief physician Svante Nycander at the Erica Foundation presented his thesis "Personlighetsutveckling på avvägar" (Personality Astray) where 308 boys and girls that were examined and treated at the Erica Foundation Curative Pedagogy Institute in Stockholm were described (Nycander 1950).

Ulf Otto, at the University of Lund, used retrospective longitudinal methods in his thesis from 1972 "Barns och ungdomars självmordshandlingar" (Suicidal acts by children and adolescents) (Otto 1972) to describe during a period of fifteen years the outcome for 1,727 boys and girls that at the end of the 1950s were treated for attempted suicide. The results of this study have been recovered also in later Swedish studies that indicate that children and youth's suicide acts are a "call for help" that needs treatment, support and help. However, attempted suicide in itself may be a "poor" predictor for later executed suicide. The high risk group that commits suicide before the age of 25 seems to be found among those who develop psychiatric illness or anti sociality and addiction (Engqvist and Rydelius 2006).

During pre-puberty and puberty behaviour problems and mental symptoms are common among children in general. When 222 common Stockholm boys were described during the 1950s it appeared that every tenth boy in the age of 8-16 years had phobias and compulsive

Sample	Discipline	Criminality	Alcoholism	Follow-up period
Psychopathic boys treated at the Mellansjö-treatment home	Child and adolescent psychiatry	41%	28%	1928-1968 (Fried 1995)
Delinquent boys treated at the Children's Village SKÅ	Child and adolescent psychiatry	67%	58%	1954-1973 (Andersson et al. 1976)
Young law-breakers from the general population	Child and adolescent psychiatry, Sociology, Psychology, Criminology	39%	46%	1960-1972 (SOU 1973:25 1973)
Teenage alcoholics	Child and adolescent psychiatry	42%	58%	1964-1985 1964-1977 (Rydelius 1985)
Adopted children with heredity for social problems /alcoholism	Child and adolescent psychiatry	14%	21%	1930-1972 (Bohman 1978)
Children of alcoholic fathers	Child and adolescent psychiatry	42%	35%	1958-1978 (Rydelius 1981)

Table 1. Different risk groups as described with longitudinal / prospective methodology.

acts. The result of the description of symptoms showed that every fourth schoolboy suffered from some kind of difficulty, nervous symptom or difficulty in adapting to the extent that specialist help or special treatment was needed. However, their prognosis was not decided by the load of symptoms or the behavioural disturbances. The 18 year follow up assessment showed that the prognosis for most of them was much better than was expected and it was not the load of symptoms but learning ability and home environment that was of importance for the future outcome (Andersson et al. 1976).

1.5.4 Studies on the Swedish general population

In addition, different risk and control groups and the general population have been studied with prospective longitudinal research methods from various disciplines during the second half of the 20th century.

The Lundby Study is a project exceptional and world leading by its prospective design (Ottosson 2003). International reviewers (MFR (Swedish medical research council) 2000) have stated that "the Lundby Study has produced prevalence, incidence and outcome data on depression and anxiety over many years, and is one of the few studies capable of producing reliable data on changes in incidence" (Nettelbladt 2009). It is a well-known longitudinal survey of the mental health in a total Swedish population during the time period from July 1, 1947, to July 1, 1997. The Lundby population consisting of 3,563 probands was investigated in 1947, 1957, and 1972. Sufficient information was available for 98–99% of the subjects. In 1997–2000 Per Nettelbladt and his research team carried through a fourth field investigation.

In a longitudinal research project, Nylander and Rydelius (Rydelius 1981) followed up the children of male alcoholics from the lowest social class. The results of these studies have shown that when they are still small, the children exhibit signs of mental disturbances and that when they become adults; the boys develop social maladjustment problems and addictions, and have a high rate of both somatic and psychiatric diseases. In comparison to their controls (matched for socio-economic class), the values found for these variables are consistently higher for the probands.

The Stockholm Metropolitan study is a follow-up of 15,000 individuals born in 1953 who were still living in the Stockholm Metropolitan area in 1963. The project was initiated by late professor Carl-Gunnar Janson and is now conducted by Sten-Åke Stenberg and Denny Vågerö (Stenberg and Vagero 2006). The study was primarily concerned with sociological issues of social mobility and organization, conformity and deviance, and marital patterns. This study has resulted in over 100 publications (Jansson 2006).

The IDA project is a prospective, longitudinal research program led by Professor David Magnusson from the outset in 1965, investigating three cohorts of subjects: all boys and girls who attended school in a municipality in mid-Sweden, and attended the third, sixth or eighth grade at the time of the first data collection. The main group, encompassing in excess of 1,400 individuals, are the participants in most of the research conducted in the program (Geisor 2007).

The Solna study is a birth-to-maturity study of 212 children born in Stockholm in the mid 1950s. The goal was to acquire a comprehensive picture of individual growth and development by charting the course of physical and psychological development and obtaining reasonably comprehensive life histories that would be useful in many areas of research. The information covers many issues from birth to mid-life. The database is large (Geisor 2008).

1.5.5 Studies in the Nordic countries

Thomsen followed up a total of 546 children and adolescents in Denmark, aged 5 to 15 years. They were inpatients in psychiatric hospitals throughout Denmark between 1970 and 1973, and followed up regarding later readmissions and mortality. Approximately one-third of the sample had at least one readmission after the age of 18 years. In total, 24 probands died during the study period. Eight individuals had committed suicide. The SMR was significantly increased (Thomsen 1996).

Thomsen also performed a register-based study of 485 children (0-15 years of age) admitted to a child psychiatric hospital from January 1, 1970, to December 31, 1972, who were followed up on December 31, 1986. They showed higher rates of admission to psychiatric hospitals in late adolescence or young adulthood than found in an age-standardized general population (Thomsen 1990).

Psychiatric morbidity, expressed as hospital admissions during a 30-year follow-up period, among 322 former child psychiatric patients in a register investigation was studied by Larsen in Denmark. They were admitted from 1949-1951, and were followed up as of December 31, 1980. By the variables employed, 37% of the sample was judged to have had a good overall outcome, with diagnosis being an inconsistent predictor of outcome (Larsen et al. 1990; Larsen 1991).

A Norwegian population of 1,276 former adolescent psychiatric inpatients was followed up 15 to 33 years after hospitalization by Kjelsberg by record linkage to the national registers of criminality (n=932), disability benefits (n=1,095) and causes of death (n=1,095). In total, 1,095 patients (53.7% males), representing 85.8% of the original sample could be traced. At follow-up, 52.0% of those investigated had engaged in criminal activity. It was shown that male sex, psychoactive substance use disorder, short hospital stay and poor impulse control remained strong and independent predictors of death (Kjelsberg and Dahl 1998, 1999; Kjelsberg et al. 1999; Kjelsberg 1999b, 1999a).

The Finnish "From a Boy to a Man study" has an objective to study associations between comorbid psychopathology and long-term outcomes in a large birth cohort sample from age 8 to early adulthood. The sample included long-term outcome data on 2,556 Finnish boys born in 1981 and the aim was to study the impact of early childhood psychopathology types and informant sources on young adult outcomes, based on data from a military registry of psychiatric diagnosis, a police registry on criminal and drug offences, and self-reported problems in late adolescence and early adulthood. The conclusions in this study were that the subjective suffering and long-term burden to society are especially high among children with comorbid conduct and internalizing problems in childhood. A major challenge for child and adolescent psychiatry, education, and social services is to develop effective intervention strategies focusing on these children (Sourander et al. 2007).

In a population-based birth cohort in Finland, with the aim of studying the continuity of psychopathology from the age of 3-12 years, children's emotional and behavioural problems were assessed at age 3 using the Child Behaviour Checklist 2-3, and at age 12 by parents using the Child Behaviour Checklist 11-18, and the Youth Self Report completed by the children. Behavioural ratings were obtained from 800 subjects at both time points and the results showed the complexity of pathways in psychopathology from early childhood to preadolescence. In particular, aggressive and destructive behaviour in very early childhood predicted later problems and require early recognition and possible intervention at an early age (Pihlakoski et al. 2006).

A Northern Finland 1966 birth cohort has been followed-up in different studies for 34 and 35 years regarding childhood precursors and risk factors of schizophrenia. The findings supported the longitudinal dimension and life span models of schizophrenia. No powerful risk factor, pre-morbid sign or a risk indicator was identified that was useful for the prediction of schizophrenia in the general population was identified (Isohanni et al. 2004; Isohanni et al. 2005; Isohanni et al. 2006; Jääskeläinen et al. 2008).

1.5.6 An important study on Swedish adolescents and depression

As stated below there has been a radical change in the past two decades in the understanding of depressive disorders arising early in life. Earlier psychiatrists did not see the current symptoms as depression the symptoms were instead associated to other things. A study that was important to increase knowledge about depression in adolescence was Gunilla Olsson's thesis from 1998 (Olsson 1998). At that time, more attention focused on mood disorders as evidenced by Figure 1 later.

Olsson's study engaged a total population of 16-17-year-old urban high-school students and 2,300 (93%) were screened for depression and previous suicide attempts. Adolescents with high depression scores in self-evaluation (12.3%) or reporting previous suicide attempts (2.4%) were diagnostically interviewed together with one control for each, matched for gender and educational program. After the interview self-ratings were completed regarding social network, family climate, and life events.

In her study Major depression was prevalent during the last year in 5.8% and during life time in 11.4%, 4 girls for every boy. A depression with remaining symptoms for a year or more was the most common type. Dysthymia without major depressive episodes was diagnosed in 1.1%, two girls for every boy. Short hypomanic episodes had been experienced by 13.2% of those with major depressive disorder.

Anxiety disorder was comorbid to depression in one half and conduct disorder in one fourth of the depressed adolescents. Alcohol was abused by 6.5% and used regularly by another 12%. Other drugs were used by 6.5% of depressed adolescents and not at all by controls. The depressed used tobacco twice as frequently as non-depressed.

Social network and family climate were compared within the originally matched pairs. Adolescents with long-lasting depressions had a smaller and unsatisfying social network. They also had experienced many stressful life events related to family adversities, while those with shorter depressive episodes had stress related to the peer group. Depressed adolescents with comorbid conduct disorder reported insufficient support from the close network and a more negative family climate.

1.6 Relationships between mental disorders in childhood and adulthood

In his opening speech at the 26th Berzelius symposium "Mental and Psychosocial Adaptation in Children - A Longitudinal and prospective Approach" (Linköping, May 1993) sir professor Michael Rutter discussed today's "current opinions" and level of attainment with 1960s and 1970s. This was an update of a lecture from 1972 about "Relationships between child and adult psychiatric disorders" (Rutter 1972).

Rutter stated that when the relationship was reviewed over the two last decades [54] there was a paucity of evidence on the adult outcome of mental disorders among children and the childhood antecedents of mental disorder in adulthood; it was concluded that most links over time became rather uncertain or non-specific. When the same topic was reconsidered a dozen years later, there were a few more research findings but the empirical base had not expanded greatly. The findings were used to consider both the conceptual issues and possible mediating mechanisms (Rutter 1995).

There is mounting evidence that many, if not most, lifetime psychiatric disorders' first appears in childhood or adolescence and that childhood maltreatment strongly predicts poor psychiatric and physical health outcomes in adulthood (Arnow 2004; Costello et al. 2005).

Most adult disorders should be reframed as extensions of juvenile disorders. In particular, juvenile conduct disorder is a priority prevention target for reducing psychiatric disorder in the adult population (Kim-Cohen et al. 2003).

Although, there has been a radical change over the last 1-2 decades in the understanding of depressive disorders arising early in life, Swedish longitudinal studies indicate that psychotic mood disorders are rare among child and adolescent psychiatric patients (Nylander 1979). Standardized methods of diagnostic assessment began to show that clinical conditions with phenomena quite similar to those seen with affective disorders in adult life were much more common in childhood and adolescence than had been hitherto appreciated. The scepticism that had dominated thinking about childhood depression for many years had in the 1990s been replaced by a huge upsurge of clinical and research interest in these disorders (Rutter 1995). The seriousness of the condition has been underlined by the substantial suicide risk (Harrington et al. 1994).

Adolescent depression carries an elevated risk of adult depression irrespective of comorbidity. Comorbid conduct disorder in childhood is associated with elevated rates of other psychiatric outcomes (Rutter and Maughan 1997). Ferguson and colleagues found that young people having early depression were at increased risk of a later adverse psychosocial outcome. There was a direct linkage in which early depression was associated with increased risk of later major depression and anxiety disorders. Linkages between early depression and other negative psychiatric outcomes appeared to reflect the effects of confounding factors (Fergusson and Woodward 2002).

Harrington and his colleagues found that depression in childhood or adolescence was associated with a fourfold increase, compared with a non depressed psychiatric group, in the risk of recurrence of major depressive disorder during early adult life. Even more important was the finding that the childhood depression group did not show any increased risk for any mental disorder other than affective conditions (Harrington et al. 1990).

Early risk factors for affective disorder exert effects by modifying person-environment relationships close to onset of adult symptoms. Sensitivity to life events may be transmitted from parents to offspring; psycho pathological continuity over the life-span may be explained in part by continuity of altered stress sensitivity (Van Os and Jones 1999).

Roza and colleagues suggest different developmental pathways for mood and anxiety disorders Anxiety disorders predominantly began in childhood and early adolescence whereas the incidence of mood disorders increased sharply in adolescence and young adulthood. A follow-up showed that mood disorders were predicted by high scores on the anxious / depressed scale and on the internalizing composite (withdrawn somatic complaints, and anxious / depressed). Anxiety disorders were significantly predicted by the social problems scale and the externalizing composite (delinquent behaviour and aggressive behaviour). The predictions based on problem behaviour remained stable during a 14-year period across adolescence and young adulthood (Roza et al. 2003).

2. Aims

The overall aim with this chapter was to describe and discuss the occurrence of mood disorders in some psychiatric in- and outpatient populations in a smaller county in Sweden.

A specific aim with the follow-up of 1,400 child and adolescent psychiatric patients was to describe their outcome in adulthood. The outcome is described in terms of psychiatric care in adulthood, the relationship between symptoms in childhood and adolescence and in adulthood, criminality and mortality and suicide.

The specific aim with the two questionnaire surveys conducted at the youth psychiatric clinic for young adults was to investigate the problems and symptoms the patients themselves stated that they had. It was also examined what expectations they had and what help they expected and if they felt they had received help for their problems.

3. The design of the studies

A longitudinal study was performed where child and adolescent psychiatric patients admitted 1975-1990 to inpatient or outpatient child and adolescent psychiatric care in Jämtland County, Sweden were followed until 2003. Data were collected from hospital records and official Swedish registers.

The homogeneity of the psychiatric care organization in the study area, the responsive study populations, the possibility of studying hospital records and the unique opportunity to use National registers allowed for the potential to carry out a longitudinal study, descriptive in its nature.

Two user surveys were conducted at a special youth psychiatric clinic for young adults 16-25 years of age in the same county and hospital organisation. In the first survey a questionnaire was answered when the patients visited the clinic for the first time and a second questionnaire was answered after the last contact with the clinic. In the second survey the patients answered a questionnaire when they had visited the clinic for the last time

4. Material and methods

4.1 The study area

Jämtland County is one of Sweden's 21 counties. It consists of the two provinces Härjedalen and Jämtland and minor parts of Ångermanland and Hälsingland. Jämtland is located in the western part of middle Sweden at the border to Norway. It represents 12% of Sweden's total land mass but only 1.5% of the population. From 1975 - 2003, the total population varied from 133,433 to 127,645 with a peak of 136,301 inhabitants in 1994.

Östersund is the only city in the county and the city is a major trade centre for the entire county. The county and the regional hospital in Östersund were well suited for this type of studies. In Jämtland County there were one child and adolescent psychiatric clinic and one general psychiatric clinic and a youth psychiatric clinic for young adults within the same health and hospital organization giving service to the whole population in Jämtland County.

4.2 Swedish registers

Sweden and the Nordic countries, especially Finland is unique concerning official registers. In this chapter several Swedish national registers were used. The parish registration required by the 1686 Church Ordinance laid the groundwork for future population statistics. Sweden began to keep population statistics in 1749, quite a unique phenomenon.

National registration is the fundamental registration of the Swedish population. In the national registration it is continuously registered who is living in the country and where they live. The register is based on personal identification numbers. For a long time it was conducted by the church through parish registration, but since July 1, 1991, the Swedish Tax Authority is responsible for the national registration (Statistiska Centralbyrån (SCB) 2006). Information from National registration; i.e. personal identification was used as a base for gathering information from all other official registers, which are described below.

From 1987 *the National Patient Register* covers all public inpatient care in Sweden. Statistics of diseases and surgical treatments of patients have a long history in Sweden. Data of this kind has been published for more than 100 years and was available during the entire 20th century. In the 1960s the National Board of Health and Welfare started to collect data on individual patients who had been treated as inpatients at public hospitals. The register built up at that time initially covered all patients treated in psychiatric care and around 16% of patients in somatic care, involving six of the 26 county councils in Sweden (Socialstyrelsen 2008a). This register has been used to collect data pertaining to inpatient psychiatric care and diagnoses.

The National Causes of Death Register has been used to establish time for death and causes of death. Along with the register death certificates have been examined. The National Causes of Death Register goes back to 1749 when a nation-wide report system first was introduced. The National Causes of Death Register is annually updated and based on personal identification numbers and includes all death certificates for the Swedish population. The National Swedish Board of Health and Welfare has been responsible for publication since 1994 (Socialstyrelsen 2008b).

Statistics of persons convicted of offences has continuously been recorded since 1972 by SCB. The Swedish NCCP (BRÅ) is since July 1, 1994, responsible for Sweden's official crime statistics. *The Register of Persons Convicted of Offences* has been used to obtain information regarding criminality.

4.3 Samples

4.3.1 The child and adolescent psychiatry patient group

The study was performed based on a total sample of former child and adolescent psychiatric patients. All 1,420 patients born between 1957 and 1976 that finalised child and adolescent psychiatric care in Jämtland County, Sweden, during the period 1975-1990 were initially considered for inclusion. Eight individuals not covered by the national registries and twelve who subsequently emigrated during the follows-up period were excluded.

The index group comprises 1,400 former child and adolescent psychiatry patients, including 285 in- and 1,115 outpatients, or 98.6% of the original population. The age groups were chosen on the basis of the fact that the oldest patients, born 1957 may have finalized their contact at child and adolescent psychiatry just before they turned 18 years; that is 1975. The youngest patients, those born 1976, may have made their first visit at child and adolescent psychiatry during their first year of life but may also have been readmitted later than 1990.

4.3.2 Patients at the youth psychiatric clinic for young adults

The two studies performed at the youth psychiatric clinic for young adults were both based on a total sample. All patients visiting the clinic from mid-November 2002 until 2004-12-31 were invited to participate in the first survey. A total of 219 questionnaires were received in connection with the first visit. One respondent submitted the form, but stated that he or she did not want to answer the questions and one has filed a blank form. The remaining 217 form the basis for this report. The loss at the first question time was 28.7 percent. A total of 126 questionnaires were received after the end of treatment. Three of them said they did not want to answer the questions. The remaining 123 form the basis for this report. The loss at the second follow-up time was 45.1 percent. The second survey included a total sample of those who completed their treatment in 2008, it comprises 85 submitted questionnaires and loss is not known.

All questionnaires were discussed with staff before shaping. They had the same design throughout the follow-ups. The questionnaires were not tested before they began to be used.

The problem descriptions were in the first survey made on the patient's first visit and they were asked to write freely in an open question which was: "Briefly describe your problems." The problems described were then categorized by the author. The categories were based on the categorization of problems and symptoms used by the Swedish Association for Child and Adolescent Psychiatry. The problem descriptions have not been transferred into diagnoses. Problem descriptions were very different. Some stated clearly their problems while others gave more vague descriptions of the problems they sought help for. Several patients provided a variety of problems. The number of problems will be more than the number of individuals.

In the second survey the description was made after treatment and the respondents had to mark their symptoms in a checkbox containing categories based on the first survey.

4.3.3 A patient group used for comparisons

It was also possible to use a previous Swedish follow up study for comparisons, a sample of 2,164 patients from Municipal Child Guidance Clinics in Stockholm (see Table 2) who were treated between 1953 and 1955 and followed over 20 years. Further when possible comparisons were also made to the general Swedish population.

	Stockholm		Jämtland County	
Primary Material	Number	Percent	Number	Percent
	2364		1420	
Excluded:				
Emigrated	100	4.2	12	0.8
Deceased during follow up	50	2.1	38	2.7
Unusable data	50	2.1	8	0.6

	Stockholm		Jämtland County	
Inpatient care	0	0	270	19.0
Less than 20 year follow up	0	0	484	34.1
Index group:	2164	91.5	608	42.8
Males	1417	65.5	325	53.5
Females	747	34.5	283	46.5

Table 2. Index groups for the Stockholm and Jämtland studies

4.4 Data collection

4.4.1 The child and adolescent psychiatric patient group

The study started in the mid 1990s when the index group was identified though an examination of registration books and hospital records at the child and adolescent psychiatric clinic.

The personal identification numbers were checked and information about place of current residence was obtained at Local Tax authorities. Information about a number of patients (75) could not be found at first due to incorrect or incomplete personal identification numbers and their information was searched and checked in birth records at the regional state archives.

When the index group was identified the hospital records at the child and adolescent psychiatric clinic were reassessed according to a protocol (based on previous empirical results from follow-ups of child and adolescent psychiatric patients).

The protocol entailed recording the following information from the child and adolescent psychiatric hospital records:

- Age at first admission to child and adolescent psychiatry and age at finishing the care and gender.
- Who referred the patient to the child and adolescent psychiatric clinic?
- Where or with whom they lived with during period of child and adolescent psychiatric care (family, foster family, institution, etc.).
- Problems with learning and/or behaviour problems at school.
- Information about inpatient or outpatient care at the child and adolescent psychiatric clinic.
- At the time of admission to the child and adolescent psychiatric clinic, family circumstances. If the patient's biological parents were living together, this would be described as a complete family and if not, as a split family.
- Any information about adoption.
- Reason for admission to the child and adolescent psychiatric clinic according to standards established by the Swedish Association for child and adolescent psychiatry.
- Diagnosis.

One-third of the outpatients were not given a formal diagnosis. Outpatients in the child and adolescent psychiatric clinic did not receive diagnoses when their problems were considered temporary due to growing and maturation.

The retrospective data from child and adolescent psychiatric hospital records were used as independent variables to the dependent variables mood disorder or not, deceased or alive, suicide or not suicide, convicted of offences or not and general psychiatric care or not.

4.4.2 Patients who visited the youth psychiatric clinic for young adults

In the first survey questionnaires were allocated out and were collected in connection with the youth's first visit. After the last visit the second questionnaire was sent out along with an addressed envelope. Distribution and collection of questionnaires was managed by the clinic secretary. In the second study patients answered the questionnaires in connection with the closing session.

4.5 Analysis of the data

The findings based on prospective data and answered questionnaires are descriptive in nature. All data analysis was performed using the SPSS for Windows, release 17.0 (SPSS Inc.) software.

The *chi-square* and *t-tests* were employed to analyze differences between categorical and continuous variables, respectively, with a P-value of < 0.05 being considered statistically significant in both cases.

Differences between proportions were analyzed utilizing a *two-by-two cross table* and *Student's t-test*. Although the t-test has earlier been debated as valid for making such comparisons, extensive studies have shown it to be applicable also in these respects, and, consequently, the Student's t-test has been widely and successfully used for analysis of proportional data (Uitenbroek 1997).

Binary logistic regression determined the effect of a set of variables on probability of a dependent variable — plus effect of individual variables. Binary logistic regression is a regression application for a dichotomous, dependent variable and when the independents are of any type. The logistic regression model is a non-linear transformation of the linear regression.

The Cox proportional-hazards regression model was used for survival analysis. The regression model is broadly applicable and the most widely used method of survival analysis. It offers the possibility of a multivariate comparison of hazard rates. Survival time was defined as (1) the interval between birth year and death or end of follow-up and as (2) the interval between diagnosis and death or end of follow-up.

Observed and expected numbers of deceased were calculated using *the prospective method* described by Hartz et al. (Hartz et al. 1983) and *the SMR method*. The difference between observed and expected numbers of deceased was tested using the z test variable, (Berry 1983) which we also applied in the SMR method:

$$z = \frac{D - E}{\sqrt{E}}$$

Where D denotes number of observed dead, E denotes expected number of dead, and z the test variable is asymptotic normally distributed (0.1). If the absolute value of z is larger than

1.96, then the hypothesis of equal mortality is rejected (the 5-percentage level). The limit for the 1-percentage level is 2.58, and for the 0.1-percentage level, 3.29. All-causes' SMR and sex-specific SMR were calculated for the entire country and for Jämtland County.

The relative risk or *the risk ratio (RR)* was presented with 95% CIs. The risk ratio takes on values between zero ('0') and infinity. One ('1') is the neutral value and means that there is no difference between the groups compared, close to zero or infinity means a large difference between the two groups on the variable concerned. A risk ratio larger than one means that group one has a larger proportion than group two.

The odds ratio (OR) is a way of comparing whether the probability of a certain event is the same for two groups. An odds ratio of 1 implies that the event is equally likely in both groups. An odds ratio greater than one implies that the event is more likely in the first group. An odds ratio less than one imply that the event is less likely in the first group. The odds ratio was presented with 95% CIs.

The rate ratio is most suited to study events in a constant domain while the denominator-i.e. the population at risk- is very large. The rate ratio was presented with 95% CIs (Wald).

Some questions in the questionnaires were designed to be answered on a scale graded from one to ten. These issues can be attributed to the categorical variables, also known as ordinal data. There is an order between the different categories. The central measure is *the median* and *quartile deviations* are specified.

4.6 Ethical considerations

Personal registers are an inalienable part of the research design for longitudinal studies as carried out in this longitudinal study. A comprehensive regulation of research on humans is of national and international nature. To the extent that personal data are processed in the framework of a research project the Personal Data Act (The Swedish Data Inspection Board 2008) that replaced the out-dated Swedish Data Act from 1973 is applicable. Some personal information is particularly sensitive nature. These are referred to in the Act as sensitive personal data. All health information is considered sensitive in the law, regardless of disease. Sensitive personal data may be processed for research purposes under the conditions set out in the Personal Data Act.

The data collection started long before the Act concerning the Ethical Review of Research Involving Humans came into force and in some cases before the Personal Data Act. The Ethical Review of Research Involving Humans includes, for example, research on living persons, research on the deceased, research on biological material from humans and research that involves the handling of sensitive personal data. Accompanying the new law an authority "*Centrala etikprövningsnämnden*" (the Central Ethical Review Board) and six regional councils was established.

The register study would be implemented within the general register license for health and medical service and the research ethical principles in force at each data collection were carefully followed. An initial application for ethical permission was sent 1995 and was approved by the ethical review committee of the Medical Faculty of Umeå University: Um document no. 95-051.

The research plan was later amended to include issues of crime and drug abuse. Following a request of the ethics committee, I received notice that the original license (Um document no. 95-051) could continue to apply to the register study on the child and adolescent psychiatric-patients and future psychiatric morbidity and mortality but that new ethical applications must be submitted for the parts of the project on crime and abuse. Additional applications were therefore sent in for ethical review. In 1999, ethical review committees of the Medical Faculty of Umeå University and the Regional Research Committee at Karolinska Institutet approved the expanded study on crime (Um document no. 99-023 and KI document no. 99-209). Every register search has also been approved by the authority responsible for the register.

The chief physician of the child and adolescence clinic and the general psychiatry clinic both respectively approved the reading of medical records within the general register license for health and medical service regularized in the Official Secrets Act.

Even though the study on patients at the youth psychiatric clinic for young adults was a continuous monitoring of what young people thought about the clinic and not intended as a scientific study, I have striven to follow the ethical principles regarding research involving human subjects stated by The Swedish Research Council (Vetenskapsrådet 2005) based on the WMA Declaration of Helsinki - Ethical Principles for Medical Research Involving Human Subjects.

The questionnaire contained no information about names, they were coded and included in a list that only a limited number of people had access to. The presentation of the results is also made in such a way that individuals are not to be identified even by informed individuals.

5. Results

5.1 Distribution of baseline information

5.1.1 The child and adolescent psychiatric-patient population at baseline

- The gender distribution was 733 female (52.4%) and 667 male (47.6%).
- Mean age at admission was 12.1 years of age (S.D. 4.0).
- The patients were referred to child and adolescent psychiatry by paediatricians or general practitioners (35%), by school or childcare personnel (22%), by social services (12%) or other authorities (2%) or else they by themselves and/or their parents sought help (29%).
- Inpatient care was given to 285 patients (20.4%).
- Over half of the study group (52.5%) lived in a split family.
- Nearly half of the patients (46.5%) had school problems noted in their records.

5.1.2 Patients in the youth psychiatric clinic for young adults at baseline

For those respondents who submitted questionnaires at one or two occasions in the first survey (n = 276) the gender balance was 222 women (80%) and 54 men (20%). 83 persons in the second survey reported gender and among them 74 were young women (89%) and nine young men (11%).

In the second study all 85 reported an age group. The majority (60%) were between 19 and 22 years of age, just over a quarter (26 percent) was found in the group 23-26 years of age and 14 percent were between 16 and 18 years of age.

In the first study the respondents also indicated their occupation. Nearly half of them (47.5%) studied, 19.8% worked, 17.5% were unemployed, 6.9% were put on the sick-list and 8.3% had mostly combinations of the aforementioned.

5.2 Mood disorders in childhood and adolescence

5.2.1 Mood disorders in the child and adolescent psychiatric-patient population

A total of 74 patients (5.3%) was admitted and or diagnosed with depression, depressed mood or sadness/mood disorder, more female than male (63,5% vs. 36.5%) ($p = 0.048$).

A lower proportion of them (40.5%) lived in a split family ($p = 0.034$) than those with other reasons for admittance and or diagnoses. Slightly less than half of them (45.1%) had problems in school noted in their hospital records.

Most of them (73%) were admitted to the child and adolescent psychiatric clinic when they were older than 13 years, which was the median age for admittance. The difference to others was strongly significant (≤ 0.001). Eleven patients (14.9%) were inpatients.

As shown in Table 3 60 patients (4.3%), more female 38 (63.3%) than male 22 (36.7%) were admitted to the child and adolescent psychiatric clinic due to depression, depressed mood or sadness.

The most frequent occurring reasons for admission to child and adolescent psychiatry are shown in Table 3 and diagnoses made at child and adolescent psychiatry are shown in Table 4.

Reason for admission to child and adolescent psychiatry	Frequency	Percentage
Behavioural disorder	292	21.0
Relationship problems	258	18.6
Anxiety	181	13.0
Somatic problems and eating disorders	138	9.9
Mental retardation and developmental problems	80	5.8
Suicide attempt, suicide thoughts	71	5.1
Depression , depressed mood, sadness	60	4.3
Reaction to stress	52	3.7
Enuresis or encopresis	47	3.4
Abused and/or neglected	38	2.7

Table 3. The ten most frequent occurring reasons for admission to child and adolescent psychiatry, n = 1389

As shown in Table 4 below 31 patients received a diagnosis in the block F30-F39 Mood [affective] disorders and just over half of them (17 or 54.8%) or had been admitted with depression, depressed mood or sadness as reason for admittance.

Diagnoses (block)	Frequency	Percentage
Z00-Z99 Factors influencing health status and contact with health services	321	35.6
F90-F98 Behavioural and emotional disorders with onset usually occurring in childhood and adolescence	196	21.8
F40-F48 Neurotic, stress-related and somatoform disorders	140	15.5
X60-X84 Intentional self-harm	48	5.3
F50-F59 Behavioural syndromes associated with physiological disturbances and physical factors	46	5.1
F80-F89 Disorders of psychological development	39	4.3
F30-F39 Mood [affective] disorders	31	3.4
F70-F79 Mental retardation	27	3.0
F10-F19 Mental and behavioural disorders due to psychoactive substance use	24	2.7
F20-F29 Schizophrenia, schizotypal and delusional disorders	20	2.2
F60-F69 Disorders of adult personality and behaviour	4	0.4
Other diagnoses	4	0.4
F99 Unspecified mental disorder	1	0.1

Table 4. Diagnoses (blocks) made at child and adolescent psychiatry according to ICD-10, n = 901

Other reasons for admission for those diagnosed within the block F30-F39 Mood [affective] disorders were anxiety (6), suicide attempt, suicide thoughts (2), behavioural disorder, somatic problems and eating disorders, conflicts with peers or bullying, conflicts in family, personality change and request for investigation from social services one respectively.

The mood disorder block included the following diagnoses:

- 31.4 Bipolar affective disorder, current episode severe depression without psychotic symptoms (6)
- 31.5 Bipolar affective disorder, current episode severe depression with psychotic symptoms (1)
- 32.9 Depressive episode, unspecified (24).

Two-thirds (39 patients or 65%) of those admitted to the child and adolescent psychiatric clinic with depression, depressed mood or sadness as reason for admittance were given a diagnosis at discharge, shown in Table 5:

Diagnoses (block)	Frequency	Percent
F30-F39 Mood [affective] disorders	17	43.6
F40-F48 Neurotic, stress-related and somatoform disorders	8	20.5
F90-F98 Behavioural and emotional disorders with onset usually occurring in childhood and adolescence	5	12.8
Z00-Z99 Factors influencing health status and contact with health services	5	12.8
F50-F59 Behavioural syndromes associated with physiological disturbances and physical factors	2	5.1
F20-F29 Schizophrenia, schizotypal and delusional disorders	1	2.6
X60-X84 Intentional self-harm	1	2.6

Table 5. Diagnoses made at the child and adolescent psychiatric clinic for patients admitted with depression, depressed mood or sadness as reason for admittance.

As shown in Figure 1 the proportion of patients admitted to the child and adolescence psychiatric clinic with depression, depressed mood or sadness as reason for admittance has increased significantly. Mean value over the whole period is 8.9 percent. The trend is also evident for patients in the study group (up to 2003) where the mean value is 4.3 percent.

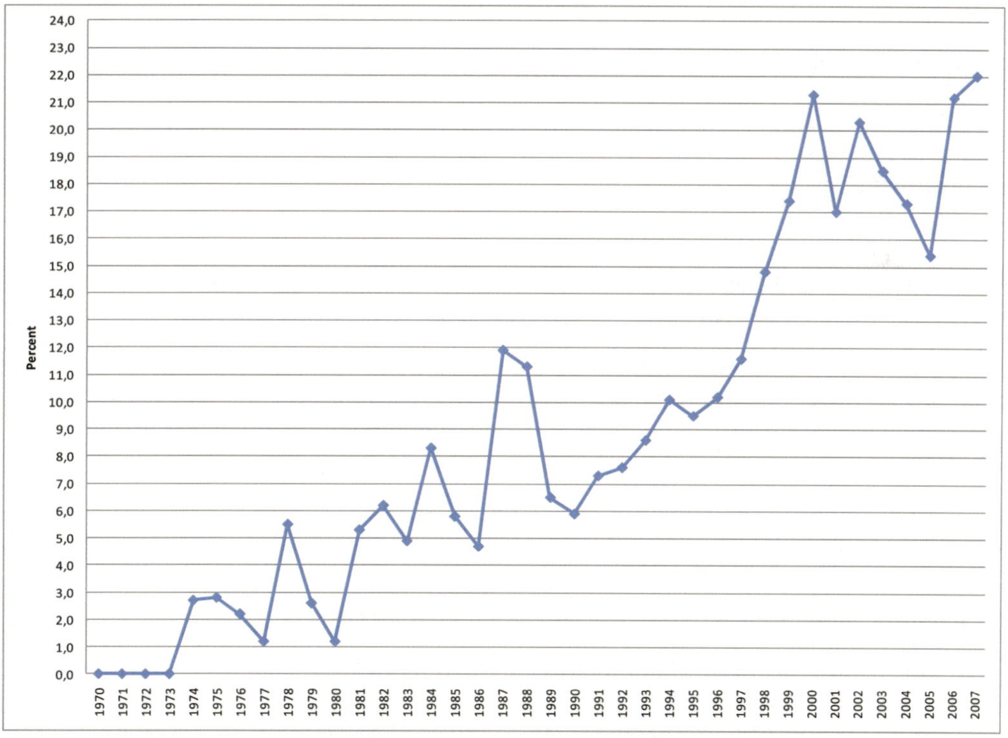

Fig. 1. Patients admitted to the child and adolescent psychiatric clinic with depression, depressed mood or sadness as the reason for admittance 1970 to 2007.

5.2.2 Mood disorders at the youth psychiatric clinic for young adults

As explained earlier the patients themselves described their problems and symptoms in an open question in the questionnaire in the first survey. Close to two hundred respondents reported 1.7 problems per person. Five problems account for 85% of the respondents answering the question and nearly half of the reported numbers of problems (49.7%).

In Table 6 the most frequent problem descriptions are listed.

Description of the problems that led to contact with the clinic	Frequency	Percentage of the reported number of problems (n = 338)	Percentage of respondents (n = 198)
Depression	58	17.2	29.3
Anxiety, phobia	51	15.1	25.8
Problems with self-confidence or self esteem	20	5.9	10.1
Crisis response	20	5.9	10.1
Relationship problems	19	5.6	9.6
Problems with food and eating	14	4.1	7.1
Family problems	11	3.3	5.6
Reaction to stress	9	2.7	4.5
Problems with sleep	9	2.7	4.5
Tired, worn	9	2.7	4.5
Victim of violence, maltreatment, assault	8	2.4	4.0
Suicidal thoughts	6	1.8	3.0

Table 6. Mood disorders described in the first survey by the patients themselves at the youth psychiatric clinic for young adults.

In the second survey 81 patients (95%) gave a description of the problems that led to contact with the clinic (see Table 7). They reported 129 different problems, 1.6 per person. Anxiety and Depression constitutes more than half of the problems. They were also often given together

Five problems account for 84% of the respondents answering the question and more than half of the reported numbers of problems (53%).

5.3 Patients' perception whether they received help for their problems

What did the patients think about the help they received? The patients in the first survey answered the question by marking on a scale of one to ten where one was "very poor" and 10 "very good". Only two respondents (1.6%) declined to respond. The lowest value was two and the maximum value of ten. The median value was 8.0 and the lower quartile, P_{25}, was 6.5 while the upper, P_{75}, was 9.5. None of the 121 respondents felt that the problem had become worse, one-third (33.1%) believed the problem remained unchanged and two-thirds (66.9%) thought it had improved or disappeared completely.

The second study had three possible answers: "yes", "to some extent" or "no". More than half of the 81 who responded, 50 or 62% said yes while 28 or 35% responded to some degree and three (4%) answered no.

Description of the problems that led to contact with the clinic	Frequency	Percentage of the reported number of problems (n = 129)	Percentage of respondents (n = 81)
Anxiety, phobia	22	17.1	27.2
Depression	20	15.5	24.7
Crisis response	10	7.8	12.3
Problems with self-confidence or self esteem	10	7.8	12.3
Suicidal thoughts	6	4.7	7.4
Victim of violence, maltreatment, assault	4	3.1	4.9
Parents with substance abuse	3	2.3	3.7
Family problems	3	2.3	3.7
Relationship problems	3	2.3	3.7
Self harm	3	2.3	3.7
Problems with alcohol, drugs	2	1.6	2.5
Problems with sleep	2	1.6	2.5
Obsessions, compulsions	2	1.6	2.5

Table 7. Mood disorders described in the second survey by the patients themselves at the youth psychiatric clinic for young adults.

5.4 Psychiatric care and mood disorders in adulthood for former child and adolescent psychiatric patients

5.4.1 Psychiatric care in adulthood

Information on later outcome in adulthood was available for the patient group comprising 1,400 patients cared for in child and adolescent psychiatry. At the end of the follow-up 531 patients or 37.9% of the child and adolescent psychiatric group had been treated in general psychiatry as adults, more female (62.3%) than male (37.7%) ($p < 0.001$). Mean age at first admission was 22.2 years of age (S.D. 5.2). More than half (56.5%) were inpatients in general psychiatry while 169 (31.8%) had been inpatients in child and adolescent psychiatry.

Risk factors for later general psychiatric care were: Psychosis symptoms as a reason for admission to child and adolescent psychiatry (RR 2.1); gender (female) (RR 1.9); older than 13 years at first admission to child and adolescent psychiatry; suicide attempt, suicide thoughts (RR 1.6) and anxiety (RR 1.3) as reason for admission and problems at school (RR

1.2). Depression, depressed mood, sadness as a reason for admittance was not a risk factor (p = 0.504), 25 patients (41.7%) were later admitted to general psychiatry.

5.4.2 Mood disorders in adulthood

The former child and adolescent patients were given diagnoses in general psychiatry shown in Table 8:

Diagnoses (block)	Frequency N = 524	Percent
F40-F48 Neurotic, stress-related and somatoform disorders	163	31.1
F30-F39 Mood [affective] disorders	77	14.7
F10-F19 Mental and behavioural disorders due to psychoactive substance use	58	11.1
Z00-Z99 Factors influencing health status and contact with health services	52	9.9
F60-F69 Disorders of adult personality and behaviour	50	9.5
F20-F29 Schizophrenia, schizotypal and delusional disorders	41	7.8
F50-F59 Behavioural syndromes associated with physiological disturbances and physical factors	18	3.4
X60-X84 Intentional self harm	17	3.2
F90-F98 Behavioural and emotional disorders with onset usually occurring in childhood and adolescence	14	2.7
F70-F79 Mental retardation	12	2.3
R00-R99 Symptoms, signs and abnormal clinical and laboratory	7	1.3
Other diagnoses	6	1.1
F00-F09 Organic, including symptomatic, mental disorders	4	0.8
F99 Unspecified mental disorder	3	0.6
F80-F89 Disorders of psychological development	2	0.4

Table 8. Diagnoses in general psychiatric care for patients earlier cared for in child and adolescent psychiatry

The 77 patients who received a mood disorder diagnosis in adulthood were give diagnoses in child and adolescent psychiatry according to Table 9. Their reasons for admission to child and adolescent psychiatry are shown in Table 10.

5.4.3 The relationship between symptoms in childhood and adolescence and in adulthood

Relationship between diagnoses made in child and adolescent psychiatry and later in general psychiatry is shown in Table 11. Overall one in five diagnostic blocks is equivalent (0.21).

Diagnoses (block)	Frequency N = 77	Percent
Z00-Z99 Factors influencing health status and contact with health services	18	23.4
F90-F98 Behavioural and emotional disorders with onset usually occurring in childhood and adolescence	17	22.1
F40-F48 Neurotic, stress-related and somatoform disorders	16	20.8
F30-F39 Mood [affective] disorders	7	9.1
X60-X84 Intentional self-harm	6	7.8
F70-F79 Mental retardation	4	5.2
F50-F59 Behavioural syndromes associated with physiological disturbances and physical factors	3	3.9
F80-F89 Disorders of psychological development	3	3.9
F10-F19 Mental and behavioural disorders due to psychoactive substance use	2	2.6
F20-F29 Schizophrenia, schizotypal and delusional disorders	1	1.3

Table 9. Diagnoses in child and adolescent psychiatric care for patients diagnosed within block F30-F39 Mood [affective] disorders in adulthood.

Reason for admission	Frequency N = 75	Percent
Anxiety	16	21.3
Behavioural disorder	11	14.7
Suicide attempt, suicide thoughts	10	13.3
Relationship problems	8	10.7
Mental retardation and developmental problems	7	9.3
Somatic problems and eating disorders	7	9.3
Depression , depressed mood, sadness	7	9.3
Reaction to stress	3	4.0
Enuresis or encopresis	2	2.7
Request for investigation	2	2.7
Confusion, personality change	1	1.3
Abused and/or neglected	1	1.3

Table 10. Reason for admission to the child and adolescent psychiatric clinic for patients diagnosed within the block F30-F39 Mood [affective] disorders in adulthood.

Diagnoses (block)	Childhood	Adulthood	Equivalent	
F20-F29 Schizophrenia, schizotypal and delusional disorders	20	41	12	0.60
F10-F19 Mental and behavioural disorders due to psychoactive substance use	17	60	8	0.47
F70-F79 Mental retardation	14	12	5	0.36
F50-F59 Behavioural syndromes associated with physiological disturbances and physical factors	31	54	11	0.35
F40-48 Neurotic, stress related and somatoform disorders	86	129	29	0.34
F30-39 Mood [affective] disorders	23	77	7	0.30

Table 11. The most consistent diagnoses from childhood and adolescence to adulthood.

5.4.4 Psychotic mood disorders incidence and early signs

By the end of the follow-up period 62 former child and adolescent psychiatric patients (26 males and 36 females, male/female ratio 1:1.4), 4.4% of the entire index group, had received an ICD-10 diagnosis of "F20-29: Schizophrenia, schizotypal and delusional disorders" (48 patients) and/or "F30-39: Psychotic mood disorders" (14 patients). The overall estimated incidence of first-episode psychosis per 10,000 person-years in the index group was 15.4 (17.1 for females and 13.7 for males).

The mean age at the time of the first onset of symptoms was 21.4 years (SD 6.4 years) and corresponding median age was 18 years. A majority of these (27 patients = 44%) were diagnosed between the age of 13 and 17 years, 17 patients (27%) between 18 and 25 years of age, 10 patients (16%) between the ages of 26 and 30 years, and the remaining 8 patients (13%) were older than 30 years of age when first diagnosed.

The 48 individuals diagnosed with schizophrenia were significantly younger (mean age 20.3 years; SD 5.2) at the time of the initial diagnosis of psychosis than were the 12 patients with psychotic mood disorders (mean age 26.8 years; SD 8.3) ($p = 0.018$).

The patients given diagnose of psychoses at an age of 25 years or older exhibited unspecific psychopathological symptoms, but did not exhibit any signs of a possible psychotic disorder during their child and adolescent psychiatric care. In adolescence acute psychoses were more common.

One-third (21 patients) of these individuals were given their initial diagnosis of psychosis during child and adolescent psychiatric care. Two of these 21 patients were not treated later for this disorder in general psychiatric care whereas the remaining 19 individuals were diagnosed with the same type of disorder as adults. The other 41 patients were not diagnosed as psychotic until they were cared for in general psychiatry. More than a third (15 patients) of them was given the diagnosis "F90-F98 Behavioural and emotional disorders with onset usually occurring in childhood and adolescence" during child and adolescent psychiatric care, eight individuals received the diagnosis "F40-F48 Neurotic, stress-related and somatoform disorders" and six patients were diagnosed with "Z00-Z99 Factors influencing health status and contact with health services."

Of the 21 patients given a diagnosis of psychosis in connection with child and adolescent psychiatric care, 19 had a psychosis diagnosis in both settings. 12 were placed in the same sub-category of "F20-29: Schizophrenia, schizotypal and delusional disorders" and one in the same sub-category of "F30-39: Psychotic mood disorders" at both time-points. Three patients with a child and adolescent psychiatric diagnosis in the sub-category of "F20-29: Schizophrenia, schizotypal and delusional disorders" were later categorized as "F30-39: Psychotic mood disorders" in adulthood. In contrast, three individuals treated during adolescence for "F30-39: Psychotic mood disorders" were later categorized in the sub-category of "F20-29: Schizophrenia, schizotypal and delusional disorders".

Changes in behaviour, including social isolation, refusal to go to school, loneliness and odd behaviour in general were the initial signs or symptoms most frequently observed prior or upon admission to child and adolescent psychiatric care. However, this was not the overall picture regarding the category of schizophrenia. Among the individuals diagnosed with schizophrenia or psychotic mood disorders, symptoms such as motor restlessness, obsessive rituals and poor sleep were equally common, being observed in 44% of the cases. Patients in both of these groups also frequently demonstrated anxiety and depression at the time of admission.

Of the three different groups of patients that could be discerned, the first and most distinct (Group I) included the 21 (34%) who exhibited signs and symptoms of psychosis in connection with child and adolescent psychiatric care and, consequently, received their first definitive diagnosis of psychosis as children. Among this group, 14 demonstrated obvious symptoms of a disorder at their initial contact with child and adolescent psychiatric care-givers, whereas the diagnosis for the 7 others was established only after an observation period of 1-4 years. Accordingly, the mean period of time that elapsed from first admission to child and adolescent psychiatric care until definitive diagnosis of a psychosis was 2.0 years, (SD 3.6).

A second group (Group II) of 15 patients (24%) also showed possible signs of psychosis during their child and adolescent psychiatric care, but were first diagnosed with such a disorder in connection with general psychiatric care, mostly at a relatively young age. Their diagnoses were established at a mean of 6.0 years, (SD 5.8) following first admission to child and adolescent psychiatric care.

The child and adolescent psychiatric records for the third group (Group. III) of 26 patients (42%) contained no notation of signs of psychosis and their definitive diagnosis of this disorder was made following completion of the child and adolescent psychiatric care. For these patients, the period from first admission to child and adolescent psychiatry to first diagnosed onset of psychosis was even longer; mean 12.4 years, SD 7.9. During child and adolescent psychiatric care, most of this group exhibited unspecific psychopathology, such as behavioural and emotional problems or problems with relationships.

Patients placed in the ICD-10 category "F20-29: Schizophrenia, schizotypal and delusional disorders" demonstrated more symptoms of psychosis at an earlier age than did those classified as "F30-39: Psychotic mood disorders" ($p = 0.019$).

5.5 Mortality and suicide

The findings in the study of 1,400 child and adolescent psychiatric patients showed that Swedish child and adolescent psychiatric patients are currently running an elevated risk of early death. Twenty-four males and 14 females died. The male/female ratio was 1.7:1. Twenty-eight of the patients (74% of the deceased) died before age 30. The all-causes standardized mortality ratio for the deceased was higher ($p < 0.001$) than the standardized mortality ratio for the general population of Sweden and for Jämtland County. The risk of dying was almost twice as high for young males as for young females and 50% of all deaths occurred more than ten years after child and adolescent psychiatric treatment.

None of the 74 patients who had been admitted to child and adolescent psychiatry due to depression, depressed mood or sadness or were given a mood disorder diagnosis had died at the end of the follow up.

Six patients died of somatic illnesses while the others died of unnatural causes. Suicide was the single most common cause of death (50%). Other unnatural causes were drug overdose, other accidents and suicidal intention unclear. Seven patients - all males who had experienced childhood histories of aggressive outbursts, difficulties controlling impulses, and troublesome psychosocial situations' - died in traffic accidents.

Nineteen patients (11 males and 8 females, male/female ratio 1.4:1.) committed suicide. When including an additional patient where a suicidal intention was unclear, this entails an approximate suicidal death rate of 14.3 per 1,000 persons in the cohort of former child and adolescent psychiatry patients. Males often used violent means to commit suicide while most of the female suicide victims died of intoxication.

Although suicide was the most common cause of death, only two of the 19, who later committed suicide, had been initially admitted for child and adolescent psychiatric clinical care because of attempted suicide. None of them had been admitted due to depression, depressed mood or sadness or were given a mood disorder diagnosis. Eleven of these 19 individuals exhibited obvious psychosocial risk factors related to their home environments at their first admission to a child and adolescent psychiatric unit. Two variables — problems at school and behavioural disorder — were the factors in the initial contacts that were found to be the most important for predicting premature death and suicide, irrespective of which statistical method was used.

5.6 Criminality development

Every third child and adolescent psychiatric patient treated between 1975 and 1990 (every second male and every fifth female) had been entered into the Register of Persons Convicted of Offences, which is a significantly higher rate than the general population. The male/female ratio was 2.3:1. The findings were compared to published results for patients from the Stockholm Child Guidance Clinics who were treated between 1953 and 1955 and followed over 20 years (Nylander 1979). The male patients in the present study were entered into the register for criminality at double the rate compared to the group of patients from the 1950s; and for girls, this rate increased sevenfold. Repeated criminality was common, a total of 315 (59%) reoffended.

It was shown that admission to child and adolescent psychiatry for behavioural disorder was the single most substantial risk factor for conviction of offences. Patients who had relationship problems also showed an increased risk. Both a split family and problems at school independently entailed nearly doubled risk later being entered into the Register of Persons Convicted of Offences.

Depression, depressed mood or sadness as a reason for admittance was not a risk factor. On the contrary, these patients were significantly less convicted of offences than others.

The 530 former child and adolescent psychiatric patients were entered into the Register of Persons Convicted of Offences for more than 2,000 incidents of committing crime, which included 81 types of crimes. Larceny, burglary, driving without a licence, motor vehicle theft, petty larceny, assault, drunk driving and narcotics were the most commonly registered crimes. The most serious crimes were robbery (10 cases), rape (6 cases), arson (3 cases), manslaughter (2 cases), and murder (1 case).

5.7 An overall outcome

A total of 835 child and adolescent psychiatric patients (59.6%) were found in one or more registers concerning general psychiatric care in adulthood, mortality, conviction of offences or compulsory care of addicts. Three-quarters appeared in one register. Only 3% were found in three registers. More male patients than female had this "poor" outcome.

The child and adolescent psychiatric patients that were found on these registers, in comparison to the patients not found on any of the registers, often had a split family, they had problems at school, they were older than 13 years of age at admission to child and adolescent psychiatry, more frequently they were inpatients at the child and adolescent psychiatric clinic, and behaviour disorder was a significant cause for admission to child and adolescent psychiatry resulting in a "poor" outcome. Mood disorders did not predict a significant "poor" outcome.

6. Discussion

6.1 Methodological considerations

One disadvantage of the present studies is that the population of Jämtland County cannot be considered to be representative of the entire Swedish population in all respects. Although comparison with an earlier longitudinal study of outpatients in Stockholm (Nylander 1979) as well as an unpublished comparison with child and adolescent psychiatric inpatients in the Stockholm metropolitan area reveals few significant differences, it should be kept in mind that the study group here came from a sparsely populated region.

The designs of the studies, using both retrospective and prospective longitudinal approaches, were well suited for the goals of the studies. The fact that Jämtland County had only one regional hospital with one child and adolescent psychiatric unit and one general psychiatric unit for the whole population made it well suited for a study of this kind. The study population is simple to describe and it is homogenous.

Furthermore, the primary information on child and adolescent psychiatric patients was obtained from psychiatric hospital records, which are in many respects not scientifically rigorous instruments of examination. Although the quality of these records was considered to be satisfactory, they were assessed employing a protocol chosen for the present study and, moreover, also contain information provided by parents, school personnel and other authorities. The protocols and the data used in this study have not been changed during the long observation time. However, no measures on an inter-reliability rate exist in regard to how the information from hospital files and registers were transformed into the protocols.

This was done only by one person and depending on his pre-understanding (conformability) of the available information which may be both an advantage and a limitation.

Although the follow-up period is relatively long for the 1,400 former child and adolescent psychiatric patients, they are still relatively young. At the end of the follow-up period, the youngest was 26 years old while the oldest was 46. Most likely additional clinical information of value for psychiatrists and child and adolescent psychiatrists will emerge if they are followed over an extended time period.

The weakness of the follow up of patients at the youth psychiatric clinic for young adults is particularly the categorization of their problems made by the author alone. No comparison with the staff description of the problems has been made.

The studies also lack, in common with most follow-ups of this nature, qualitative elements that can contribute to a more comprehensive picture of the respondents and their views on their problems and the psychiatric care.

6.2 The overall aim

The overall aim with this chapter was to describe and discuss the occurrence of mood disorders in some psychiatric in- and outpatient populations in a smaller county in Sweden.

The strength of these studies is the large amount of patients included. It has been possible to describe and discuss the occurrence of mood disorders and also to describe trends over time. It is clear that there has been a change over time. During the first third part of the follow-up of child and adolescent psychiatric patients (1970-1976) only eight patients were admitted to child and adolescent psychiatry with depression, depressed mood or sadness as reason for admittance. Mean value was 1.8% of all reasons for admittance. During the next seven years (1977-1983) the proportion had doubled to 3.7% (20 patients) and for the last third (1984-1990) it had risen to 7.9% (32 patients). The increase continued as evidenced by Figure 1. Men value for the period after the follow-up (1991-2007) was 14.7% with a peak in 2007 of 22 %. As shown in Table 6 and Table 7 depression constituted a quarter of the problems described by the patients themselves at the youth psychiatric clinic for young adults.

It can be assumed that there has been a change in description of child and adolescent symptoms and disorders over time. It is only in the past two decades, as knowledge of child and adolescent depression has been emerged. The symptoms were formerly not associated with depression, but instead to other things. They may be confused with common teenage problems or be hidden behind acting-out aggressive behaviour or irritability. In recent years there has been increasing awareness that many adults and children with AD/HD may also meet criteria for depression and bipolar disorder. Since this study was conducted before the wave of AD/HD diagnoses that were given to child and adolescent psychiatric patients more examinations of the material are required to investigate if comorbidity is the case here too.

The current findings are supported by the results of former Swedish longitudinal prospective studies of child and adolescent psychiatric patients that were carried out from the 1950s to 1980s (see Table 1).

6.3 The follow-up of 1,400 child and adolescent psychiatric patients

A specific aim with the follow-up of 1,400 child and adolescent psychiatric patients was to describe their outcome in adulthood. The outcome was described in terms of psychiatric care in adulthood, the relationship between symptoms in childhood and adolescence and in adulthood, criminality and mortality and suicide.

6.3.1 Psychiatric care in adulthood and the relationship between symptoms in childhood and adolescence and in adulthood

A total of 531 persons or 38% had been patients at a general psychiatry unit at end of follow-up. More girls than boys (male/female ratio 1:1.7) became patients in general psychiatry in adulthood. Girls also had more symptoms of depression, depressed mood or sadness.

Similar results have been found previous follow-ups of Swedish child and adolescent psychiatry patients (Nylander 1979) and from follows up other Swedish risk groups (Rydelius 1981) findings. This may depend on gender typical psychiatric symptoms, i.e., with anxiety, depression and somatic symptoms more commonly expressed by girls, and acting-out behaviour more common in boys. The symptoms more commonly expressed by girls may be easier for professionals to manage and in turn may lead to making it easier for girls to receive help within child and adolescent psychiatry, and may entail that they may more easily seek help even within general psychiatry. Acting-out behaviour on the other hand is often found disturbing by professionals, which is why the boys themselves may adopt a negative attitude towards child and adolescent psychiatry and to later general psychiatry contact. Another explanation for these gender differences may be that the profession perceives girl typical symptoms as psychiatry while boy typical symptoms are perceived as anti social behaviour. Due to this they are more often referred to social welfare, dependence care and institutions for treatment of offenders instead. Findings from the register on criminality support the latter explanation.

It appears that based on empirical findings during the past 70 years, psychotic disorders have been and continue to be relatively uncommon among patients admitted to child and adolescent psychiatric care in Sweden. The typical male Swedish child and adolescent psychiatric care patient is "a 10-year-old troublesome boy", while the typical female patient is "a 14-year-old depressed girl". Both come from families with psychosocial difficulties; have problems at school and risk later delinquency and/or alcohol and/or drug abuse. However, the rate of schizophrenia observed in the child and adolescent psychiatric patient group (3,4%) is threefold higher than in the general population, indicating that these individuals run an increased risk for developing severe chronic psychosis and that use of a specific treatment model for early psychosis among children and adolescents might be valuable (Engqvist and Rydelius 2006).

The present empirical findings indicate that psychotic disorders debut during the teen-age years and, moreover, that disorders in the ICD category "F20-F29: Schizophrenia, schizotypal and delusional disorders" are more common than those classified as "F30-39: Psychotic mood disorders". Clearly psychotic mood disorders are rare among children and adolescents.

There only seems to be a weak connection between diagnoses given to patients in child and adolescent psychiatry and later to the same patients in adult psychiatry. This can probably be explained by the different traditions of diagnosis within child and adolescent psychiatry and adult psychiatry during the current observation period. There can be another way to explain the weak connection, which also appears between certain child and adolescent psychiatric diagnoses and later psychiatric diagnoses. This applies particularly to crisis reactions, adjustment problems, and the age specific diagnoses. These groups of diagnoses can be related to children's maturity and development, and thereby so unspecified that, in different children they can have different meaning, and thus predict many different results in adulthood. Additional explanations may appear.

The observations indicate that disorderly behaviour and attention deficit disorders in childhood could be anxiety-related symptoms and that anxiety in childhood may be associated with mood disorders. The findings indicate a relationship between behaviour disturbances and later depressive disorder diagnoses. This suggests that behaviour disturbance and attention deficit disorder in childhood can be an expression for depressive internal feelings.

The current knowledge's Rutter (Rutter et al. 2006) discusses in reference to the relation between childhood symptoms and disorders in adulthood are similar to the findings appearing in this study.

6.3.2 Criminality

In this study 54% of the boys and 21 % of the girls were registered for criminality at the end of the follow up. Malfunctioning parental care and supervision, behaviour disorders and school problems were obvious risk factors. Those, who later developed criminal behaviour, also received more in-ward psychiatric care than other child and adolescent psychiatric patients.

Depression, depressed mood or sadness as a reason for admittance was not a risk factor. On the contrary, these patients were significantly less convicted of offences than others.

6.3.3 Mortality and suicide

The findings showed that the child and adolescent psychiatric patients have faced an elevated risk of early death, despite overall improvement in health that has occurred in Sweden in recent decades. Behavioural problems, problems at school, and crime were common, irrespective of the cause of death, whereas suicide attempts constituted a poor predictor for later suicides. It was surprising that none of the deceased belonged to the group that had been admitted and or diagnosed with depression, depressed mood or sadness/mood disorder.

Many studies of depressive disorders have stressed the importance of the mortality and morbidity associated with depression. The mortality risk for suicide in depressed patients is more than 20-fold greater than in the general population (Lépine and Briley 2011).

6.3.4 An overall outcome

The available evidence suggests that a poor adult outcome is most likely the result of early onset of conduct problems associated with hyperactivity / attention deficits and poor peer relations. It is also evident that the poor outcome is a function of persistent and pervasive conduct disturbances in childhood rather than a consequence of transient or isolated antisocial problems (Rutter 1995).

Using register data on their need of psychiatric care, the occurrence of criminality and mortality/suicides you can conclude that 60% of the former child and adolescent psychiatric patients had problems also as adults. This is an indication that they did not function well in adulthood which needs further discussion.

6.4 The two questionnaire surveys conducted at the youth psychiatric clinic for young adults

The specific aim with the two questionnaire surveys conducted at the youth psychiatric clinic for young adults was to investigate the problems and symptoms the patients themselves stated that they had. It was also examined what expectations they had and what help they expected and if they felt they had received help for their problems.

The large number of respondents is the greatest strength of the surveys at the youth psychiatric clinic for young adults. A large sample of patients participated and answered the questions asked. The aim has been fulfilled to a much greater degree than was expected when monitoring began.

The problems described by patients are perceived as large or very large and they have a significant impact both at work, study situation and relationships within and outside the family. Depression is one of the two most frequently described problems. There is a majority of women admitted to the youth psychiatric clinic for young adults and responding to the survey. As showed above mood disorders occur more frequently in women.

None of the respondents thought that the problem had become worse than it was on the first visit at the youth psychiatric clinic for young adults and the majority of the third who thought the problem was unchanged still thought that the problem was easier to live with. Two-thirds thought the problem gotten better or much better, which may be considered a good result considering the conditions at the first visit.

A summary conclusion of the two surveys is that the youth psychiatric clinic for young adults has filled a need for the target group it was intended for and that the activity during follow-up time really has been helpful for most of the patients

7. Conclusion

During the approximately 40 years that these surveys cover mood disorders have changed from being a rare phenomenon to be a national disease and one of the most common symptoms in psychiatric care for children and young people. This makes great demands on the psychiatric health care providers. More knowledge is needed about the early signs and on the presence of mood disorders together with other symptoms and disorders.

8. References

Andersson, M, Jonsson, G, and Kälvesten, A-L (1976), 'Hur går det för 50-talets Stockholmspojkar? En uppföljning av 222 vanliga skolpojkar och 100 Skåpojkar. [A follow-up of Stockholm boys of the 1950s]', (Stockholm: Monografier utgivna av Stockholms Kommunförvaltning, no 38 (summary in English)).

Arbetsgruppen kring översyn av den barn- och ungdomspsykiatriska verksamheten i Jämtlands län (1984), 'Översyn av den barn- och ungdomspsykiatriska verksamheten i Jämtlands län,(Östersund: Jämtlands läns landsting).

Arnow, BA (2004), 'Relationships between childhood maltreatment, adult health and psychiatric outcomes, and medical utilization', *J Clin Psychiatry,* 65 (Suppl 12), 10-5.

Beckman, Vanna (1984), *Sinnessjukhuset : bilder ur psykiatrins historia* (Stockholm: Norstedt).

Berry, G (1983), 'The analysis of mortality by the subject-years method', *Biometrics,* 39, 173-84.

Bohman, M. (1978), 'Some genetic aspects of alcoholism and criminality. A population of adoptees', *Arch Gen Psychiatry,* 35 (3), 269-76.

Costello, E.J. , Egger, H. , and Angold, A. (2005), '10-year research update review: the epidemiology of child and adolescent psychiatric disorders: I. Methods and public health burden.', *J Am Acad Child Adolesc Psychiatry.,* Oct;44 (10), 972-86.

Curman, H and Nylander, I (1976), 'A 10-Year Prospectiv Follow-up Study of 2268 Cases at the Child Guidance Clinics in Stockholm', *Acta Pædiatirica Scandinavia,* Suppl 260-76.

Engqvist, U (1993), 'Så tänkte man - hur blev det? Beskrivning av barn- och ungdomspsykiatrin i Jämtlands län 1992, åtta år efter översyn av verksamheten', (Östersund: Barn och ungdomspsykiatriska verksamheten i Jämtlands län).--- (2000), 'Barnpsykiatrins och vuxenpsykiatrins syn på varandras kunskaper och verksamhet [Child psychiatry's and general (adult) psychiatry's look at each others knowledge and work.]', *Socialmedicinsk Tidskrift,* 77 (4), 344-52.

Engqvist, U. and Rydelius, P. A. (2006), 'Death and suicide among former child and adolescent psychiatric patients', *BMC Psychiatry,* 6, 51.

Fergusson, DM and Woodward, LJ. (2002), 'Mental health, educational, and social role outcomes of adolescents with depression', *Arch Gen Psychiatry.,* Mar;59 (3), 225-31.

Fried, I (1995), 'Mellansjö school-home. Psychopathic children admitted 1928-1940, their social adaption over 30 years: a longitudinal prospective follow-up', *Acta Pædiatirica Scandinavia,* Suppl 408.

Geisor, Monika (2009), 'IDA: Individual Development and Adaptation', (updated April 7, 2007) <http://www.oru.se/templates/oruExtNormal____30650.aspx>, accessed February 21.--- (2009), 'The Solna study', (updated February 18, 2008) <http://www.oru.se/templates/oruExtNormal____30643.aspx>, accessed February 21.

Gunnarsson, Tove (2006), 'Psykiatrisk tvångsvård i Sverige–Från medeltida helgeandshus till LPT', *Tidskriften för Svensk Psykiatri. Tidskrift för Svenska Psykiatriska Föreningen och Svenska Barn- och Ungdomspsykiatriska Föreningen,* maj 2006.

Harding, Gösta (1975), *Tidig svensk psykiatri* (Lund: Berlingska Boktryckeriet).

Harrington, R, et al. (1994), 'Adult outcomes of childhood and adolescent depression: III. Links with suicidal behaviours.', *J Child Psychol Psychiatry.,* Oct;35 (7), 1309-19.

Harrington, R., et al. (1990), 'Adult outcomes of childhood and adolescent depression: I. Psychiatric status.', *Arch Gen Psychiatry,* May; 47 (5), 465-73.

Hartz, J.A., Giefer, E.E., and Hoffman, R.G. (1983), 'A comparison of two methods for calculating expected mortality', *Statistics in Medicine,* 2, 381-86.

Isohanni, M, et al. (2004), 'The persistence of developmental markers in childhood and adolescence and risk for schizophrenic psychoses in adult life. A 34-year follow-up of the Northern Finland 1966 birth cohort.', *Schizophr Res.*, Dec 1; 71 (2-3), 213-25.--- (2005), 'Predictors of schizophrenia: evidence from the Northern Finland 1966 Birth Cohort and other sources.', *Br J Psychiatry Suppl.*, Aug; 48, 4-7.--- (2006), 'Risk factors for schizophrenia. Follow-up data from the Northern Finland 1966 Birth Cohort Study.', *World Psychiatry.*, Oct; 5 (3), 168-71.

Jansson, Carl-Gunnar (2006), 'Commentary: The revitalization of the Stockholm Metropolitan Study ', *Int J Epidemiol*, 35 (3), 553.

Jääskeläinen, E, et al. (2008), 'Associations between early development and outcome in schizophrenia--A 35-year follow-up of the Northern Finland 1966 Birth Cohort.', *Schizophr Res.*, Feb;99 ((1-3)), 29-37. .

Karlén, Karl-Henrik (1985), 'Barn- och ungdomspsykiatrin i Sverige : dess historia, problem och patienter = [Child and youth psychiatry in Sweden] : [its history, problems and patients]', (Uppsala University).

Kim-Cohen, J, et al. (2003), 'Prior juvenile diagnoses in adults with mental disorder: developmental follow-back of a prospective-longitudinal cohort.', *Arch Gen Psychiatry.*, Jul;60 (7), 709-17.

Kjelsberg, E and Dahl, AA (1998), 'High delinqency, disability and mortality--a register study of former adolecent psychiatric in-patients.', *Acta Psychiatrica Scandinavia,* 1, 34-40.--- (1999), 'A long-term follow-up study of adolescent psychiatric in-patients. Part II. Predictors of delinquency.', *Acta Psychiatrica Scandinavia,* 4, 237-42.

Kjelsberg, E. (1999a), 'A long-term follow-up study of adolescent psychiatric in-patients. Part IV. Predictors of a non-negative outcome', *Acta Psychiatr Scand,* 99 (4), 247-51.--- (1999b), 'A long-term follow-up study of adolescent psychiatric in-patients. Part III. Predictors of disability', *Acta Psychiatr Scand,* 99 (4), 243-6.

Kjelsberg, E., Sandvik, L., and Dahl, A. A. (1999), 'A long-term follow-up study of adolescent psychiatric in-patients. Part I. Predictors of early death', *Acta Psychiatr Scand,* 99 (4), 231-6.

Larsen, F. W. (1991), 'A 30-year follow-up study of a child psychiatric clientele. II. Psychiatric morbidity', *Acta Psychiatr Scand,* 84 (1), 65-71.

Larsen, F. W., Dahl, V., and Hallum, E. (1990), 'A 30-year follow-up study of a child psychiatric clientele. I. Demographic description', *Acta Psychiatr Scand,* 81 (1), 39-45.

Lépine, JP and Briley, M (2011), 'The increasing burden of depression.', *Neuropsychiatr Dis Treat.* , 7 (Suppl 1), 3-7.

Levander, S, et al. (2006), *Psykiatri. En orienterande översikt [Psychiatry. An introductory overview]* (Lund: Studentlitteratur).

MFR (Swedish medical research council) (2000), *Swedish psychiatric research 2000.* (MFR-report, 7; Stockholm Medicinska forskningsrådet).

Nettelbladt, Per (2009), 'The Lundby Study – 50 years of psychiatric epidemiology', (updated January 21, 2008) <http://www.research.med.lu.se/en_projektdetaljer.php?Proj=398>, accessed February 21.

Nyberg, Lena (2001), 'Lång väntan för unga i kris', *Göteborgsposten.*

Nycander, G (1950), *Personlighetsutveckling på avvägar. Barnpsykiatriska studier.[Personality astray. Child Psychiatric Studies]* (Stockholm: Tidens Förlag).

Nylander, I. (1979), 'A 20-year prospective follow-up study of 2 164 cases at the child guidance clinics in Stockholm', *Acta Paediatr Scand Suppl*, 276, 1-45.

Olsson, G. I. and von Knorring, A. L. (1999), 'Adolescent depression: prevalence in Swedish high-school students', *Acta Psychiatr Scand*, 99 (5), 324-31.

Olsson, Gunilla (1998), 'Adolescent depression. Epidemiology, nosology, life stress and social network ', (Uppsala University).

Otto, U (1972), 'Suicidal acts by children and adolescents: A follow up study.', *Acta Psychiatr Scand*, Suppl 233, 1-123.

Ottosson, Jan-Otto (2003), *Psykiatrin i Sverige* (Stockholm: Natur och Kultur).--- 'Psykiatri', <http://www.ne.se/artikel/288196>, accessed Jan 09 2009.

Pihlakoski, L., et al. (2006), 'The continuity of psychopathology from early childhood to preadolescence: a prospective cohort study of 3-12-year-old children', *Eur Child Adolesc Psychiatry*, 15 (7), 409-17.

Qvarsell, R (1982), 'Ordning och behandling : psykiatri och sinnessjukvård i Sverige under 1800-talets första hälft [Order and treatment] : [psychiatry and the care and treatment of the mentally ill in Sweden during the first half of the 19th century]', (Umeå University).

Rorsman, B., et al. (1990), 'A prospective study of first-incidence depression. The Lundby study, 1957-72', *Br J Psychiatry*, 156, 336-42.

Roza, SJ, et al. (2003), 'Stable prediction of mood and anxiety disorders based on behavioral and emotional problems in childhood: a 14-year follow-up during childhood, adolescence, and young adulthood.', *Am J Psychiatry.*, Dec;160 (12), 2116-21.

Rutter, M (1972), 'Relationships between child and adult psychiatric disorders. Some research considerations.', *Acta Psychiatr Scand*, 48 (1), 3-21.--- (1995), 'Relationships between mental disorders in childhood and adulthood', *Acta Psychiatr Scand*, 91, 73-85.

Rutter, M and Maughan, B (1997), 'Psychosocial adversities in childhood and adult psychopathology', *J Personal Disord.*, Spring;11 (1).

Rutter, M, Kim-Cohen, J, and Maughan, B (2006), 'Continuities and discontinuities in psychopathology between childhood and adult life.', *J Child Psychol Psychiatry.*, 47 (3-4), 276-95.

Rydelius, P-A (1981), 'Children of Alcoholic Fathers. Their social adjustment and their health status over 20 years.', *Pædiatirica Scandinavia*, Suppl. 286.

Rydelius, P. A. (1985), 'Long-term prognosis for the young alcohol abuser.', in U Rydberg, C Alling, and J Engel (eds.), *Alcohol and the Developing Brain* (New York Raven Press Books cop.), 187-91.

Schleimer, Kari (1999), 'Child and adoelscent psychiatry in Sweden', in Remschmidt H and H van Engeland (eds.), *Child and Adolescent Psychiatry in Europe, Historical Development, Current Situation and Future Perspectives* (Darmstadt/New York: Steinkopff/Springer), 351-61.

Schubert, J. (1999), 'Chockökning av sökande till psykvård', *Dagens Nyheter*, p. A4-5.

Sjöström, Bengt (1992), 'Kliniken tar över dårskapen : om den moderna svenska psykiatrins framväxt', (Lund).

Socialstyrelsen (2003), 'Utredning av händelserna i Åkeshov och Gamla stan och dess möjliga samband med brister i bemötande och behandling inom den psykiatriska vården och socialtjänstens verksamhet [Analysis of incidents in Åkeshov and Old

Town and their possible connection with defectives in treatment within psychiatric care and social services actions] (S2003/5291/HS)', in Socialstyrelsen (ed.), (Stockholm).--- (2008), 'The National Patient Register', (updated 2006-03-13) <http://www.socialstyrelsen.se/en/Statistics/statsbysubject/The+Swedish+Hospital+Discharge+Register.htm>, accessed 2008-05-15.--- (2008), 'Dödsorsaksregistret [The National Cause of Death Register]', (updated May 13 2008) <http://www.socialstyrelsen.se/Statistik/statistik_amne/dodsorsaker/Dodsorsaksregistret.htm>, accessed May 30.--- (2009), *Folkhälsorapport 2009 [Health in Sweden: The National Public Health Report 2009]* (Västerås).

SOU 1973:25 (1973), 'Unga lagöverträdare III. Hem, uppfostran och kamratmiljö i belysning av intervju- och uppföljningsdata/Olofsson, B. [Young offenders III, Swedish Government official reports]', in Statens offentliga utredningar (ed.), *1956 års klientelundersökning rörande ungdomsbrottslingar* (Stockholm: Beckman).

SOU 1992:73 (1992), *Psykiatriutredningen. Välfärd och valfrihet - service, stöd och vård för psykiskt störda [Psychiatry official report. Welfare and freedom of choice - service, support and care for mentally disturbed. Swedish Government official reports]*, ed. Statens offentliga utredningar (Statens offentliga utredningar; Stockholm: Allmänna förlaget).

SOU 1998:31 (1998), 'Det gäller livet. Stöd och vård till barn och ungdomar med psykiska problem. Slutbetänkande av Barnpsykiatrikommittén. [It is a question of life. Support and care for children and adolescents with mental problems. Swedish Government official reports]', *Statens offentliga utredningar* (Stockholm: Statens offentliga utredningar).

SOU 2006:100 (2006), 'Ambition och ansvar. Nationell strategi för utveckling av samhällets insatser till personer med psykiska sjukdomar och funktionshinder [National strategy for development of society efforts for individuals with mental illness and impaired function] [Swedish Government official reports]', *Statens offentliga utredningar* (Stockholm).

Sourander, A., et al. (2007), 'Who is at greatest risk of adverse long-term outcomes? The Finnish From a Boy to a Man study', *J Am Acad Child Adolesc Psychiatry*, 46 (9), 1148-61.

Statistiska Centralbyrån (SCB) [Statistics Sweden] (2006), 'A brief history of Statistics Sweden'. <http://www.scb.se/templates/Listning2____44474.asp>, accessed January 31.

Stenberg, S. A. and Vagero, D. (2006), 'Cohort profile: the Stockholm birth cohort of 1953', *Int J Epidemiol*, 35 (3), 546-8.

The Swedish Data Inspection Board (2008), 'The Personal Data Act', accessed June 3.

Thomsen, P H (1996), 'A 22- to 25-year follow-up study of former child psychiatric patients: a register-based investigation of the course of psychiatric disorder and mortality in 546 Danish child psychiatric patients', *Acta Psychiatr Scand*, 1996: 94, 397-403.

Thomsen, P. H. (1990), 'The prognosis in early adulthood of child psychiatric patients: a case register study in Denmark', *Acta Psychiatr Scand*, 81 (1), 89-93.

Uitenbroek, D. (2008), 'SISA Binomial', <http://www.quantitativeskills.com/sisa/distributions/binomial.htm>, accessed August 06.

Van Os, J and Jones, PB (1999), 'Early risk factors and adult person--environment relationships in affective disorder.', *Psychol Med*, Sep;29 (5), 1055-67.

Vetenskapsrådet 'Forskningsetiska principer inom humanistisk-samhällsvetenskaplig forskning (Antagna av Humanistiskt samhällsvetenskapliga forskningsrådet i mars 1990) Senast ändrad: 6/26/2002', <http://195.17.252.28/vrshop_pdf/etikreglerhs.pdf>.

Wieselgren, Ing-Marie (2007), 'Tillgänglighet till insatser för att möta psykisk ohälsa hos barn och ungdomar.', in Sveriges kommuner och landsting (ed.), (Stockholm: Avdelningen för vård och omsorg. Sektionen för vård och socialtjänst).

7

Different Types of Childhood Adverse Experiences and Mood Disorders

Alessandra Alciati
Department of Psychiatry,
Luigi Sacco University Hospital, Milan
Italy

1. Introduction

A growing body of studies has suggested that adverse events or conditions, particularly when experienced early in life, are associated with an increased risk to develop mood disorders (Agid et al., 1999). The most salient forms of childhood adverse experiences are parental loss, as well as sexual, physical and emotional abuse.

The first studies were focused on a single event, such as parental death or sexual abuse, theorizing a unique effect of a particular adverse experience on a single type of mental disorders, most often depression (Tennant et al., 1980).

Recent studies have shown that multiple categories of retrospectively reported childhood adverse events often co-occur (Finkelhor et al., 2007) and, in many cases, they are non-specific in their associations with different mental disorders (Green et al., 2010). On this basis, the latest studies (Green et al., 2010) are aimed to assess various forms of abuse simultaneously and to examine multiple psychiatric outcomes in order to avoid overly narrow interpretations.

Although the demarcation lines between these various forms of childhood adverse experiences are not easily drawn, it is important to maintain the distinctions in order not to obscure a possible specific effect of any particular type of adverse event.

This approach has been supported by a recent study showing that, among 25 negative factors considered, late-life depression was significantly associated with only 8: verbal abuse from parents, mental cruelty, excessive punishment by parents, abuse by an adult outside the family, parental mental disorder, poverty, home conflict and excessive sharing of parental problems with children (Ritchie et al., 2009). In agreement with this result, Danese et al. (2009) have shown that different childhood adverse experiences do not necessarily overlap and exert independent effects on adult major depression risk.

This chapter summarizes the findings of researches from selected publications focused on single childhood adverse events and the risk to develop depression in adulthood. Then, it considers the neurobiological changes related to childhood adverse experiences and their relationship with the development of depression in adulthood.

2. Childhood adverse experiences and mood disorders

2.1 Childhood parental loss

An extensive literature, dating back to the work of Freud (1917), focused on childhood separation from parents due to death, illness, divorce or otherwise, as an antecedent to adult depression.

The death of a parent, which is experienced by 4% of children and adolescents in Western countries, is rated as one of the most stressful life events that a child can experience (Harrison & Harrington, 2001). Earlier reviews of the literature on this topic have not been consistent with each other: Crook & Eliot (1980) and Tennant et al. (1980) concluded that parental death in childhood has little effect on later development of depression, whereas Lloyd (1980) found that the childhood death of a parent increases the depressive risk by a factor of about 2 or 3. Reviews conducted in the late 80s concluded that, although there is no evidence that parental death is a significant risk factor for depression, separations, particularly those occurring in the context of family discord, seem to contribute to adult depression (Brown et al., 1986; Tennant, 1988). Studies based on patients who were hospitalized for severe depression and/or attempted suicide have found a higher prevalence of parental death among severe depressed subjects than in the general population (Kunugi et al., 1995; Munro, 1966).

Many studies have investigated the relationship between childhood parental loss and bipolar disorder, but their results are far from conclusive. Perris found that parental death in childhood was not more frequent in bipolar and unipolar patients than in the general population (Perris et al., 1986); furthermore, Furukawa et al. (1999) compared the rates of parental death and separation among bipolar patients and normal controls and found no statistically significant differences between the groups. On the contrary, other studies have shown that the loss of a parent during childhood significantly increases the likelihood of developing a bipolar disorder during adult life (Agid et al., 1999; Mortensen et al., 2003).

These inconsistencies are largely attributed to substantial methodological limitations, such as a sample consisting only of females or subjects who sought psychiatric services and, in older studies, the lack of symptom-based and standardized diagnostic criteria. Moreover, the failure to control for variables that could affect pathology (gender and current age of the participant, gender of the deceased, life events, socio-economic status) might have contributed to some inconsistencies of the findings.

Additional methodological problems, such as the differential death rates between men and women who are in the parenting age, make difficult to examine the effects of death by the sex of the deceased. Epidemiological data on parental death in childhood indicate that fathers die twice as frequently as mothers and, in addition, surviving fathers tend to be less willing to participate in research and to seek help when distressed (Gersten et al., 1991). Van Eederwegh et al. (1982) have found a link between the death of a father and severe depression in boys and older children, and Kendler & Baker (2007) reported that childhood parental loss was a more potent predictor of adult depressive episodes in males, compared to females. More recently, Jacobs & Bovasso (2009) have shown that, while maternal death was not a predictor of adult pathology, paternal death more than doubled the increased risk of major depression in adulthood. These results suggest that the father's death is a severe loss as the death of the mother. Financial complications may contribute to the effect of the father's death in childhood on adult depression, especially in a sample where the participants' childhood took place before 1960, when the discrepancies between men and women's income were greater than today (Jacobs & Bovasso, 2009).

It is conceivable that parental death could have a differential impact according to the age of the child because the level of emotional and cognitive development will influence the understanding of events and the response to the parental death. While some investigators have reported that the risk of developing a mood disorder following childhood parental loss is greater when the loss occurs during early childhood (Agid et al., 1999), others have not found such association (Kendler et al, 1990; Jacobs & Bovasso, 2009). Girls under the age of 11 years who lost their mother by separation or by death were reported to be at higher risk of depression in adult life (Brown et al., 1977), but the parental death increasing the risk for depression also applied to female subjects who lost parents between the age of 11 and 17 years (Barnes & Prosen, 1985). Bifulco et al. (1987) found that maternal loss doubled the risk for depression and anxiety in women and the effect was greater in those women who lost their mothers before the age of 6 years. In a meta-analysis of the literature, Patten (1991) found a significant association between maternal loss before the age of 11 years and depression in females.

Modes of parental death are associated with offspring long-term psychiatric morbidity severity. Recent evidence has shown that unnatural death of a parent during childhood and adolescence has a stronger association with psychiatric sequelae than sudden natural parental death (Wilcox et al., 2010). War-related loss of the father during childhood or adolescence strongly contributes to distress and disability in adulthood: 22% of bereaved children and adolescents and 11% of non-bereaved war survivors met the criteria for a major depressive episode a decade after the war (Morina, 2011). Children, whose parents died by suicide, are at greater risk for adverse outcomes, compared with offspring whose parents died from other causes (Brent, 2009). The association between bipolar disorders and parental suicide, suggested in earlier studies, has been supported by two investigations. Mortensen et al. (2003) found an increased risk for bipolar disorders among the children of parents who committed suicide, and Kessing et al. (2004) observed that an experience of maternal suicide has been associated with a history of mania or mixed episode upon first admission to psychiatric hospitals. No effect of maternal death due to a non-suicidal cause was found.

Few early studies have clearly differentiated between the psychological consequences of parental death and separation, referring to parental loss without making any distinction between the two experiences. However, the psychiatric outcome of the two types of childhood parental loss can be expected to be very different. Harris et al. (1986) made the distinction between loss by death or by separation in their sample and concluded that the rate of depression was almost the same among women with a loss of their mother by death (23%) and separation (21%). A scarce effect of childhood parental loss on adult depression has been found in a community sample, with no significant differences between the effect of death and separation (Tennant et al., 1982). Data from a community survey have shown that women, who were diagnosed with major depression, reported parental divorce or separation before the age of 17 years significantly more often (33%) than controls (14%), but without any significant effect of parental death (Harris et al., 1986). Similarly, several studies have suggested that childhood parental separation involves a greater risk of development of adult depression than the death of a parent. In a community sample, the association between prolonged separation from both parents and the increased risk of depression in women was stronger than for death or separation from one parent (Oakley Browne et al., 1995). Agid et al. (1999), in a case-control study, have found that the loss of a parent before the age of 17 years was significantly associated with major depression in adulthood; however, when the type of loss was categorized, the difference was significant only for permanent separation from a parent. In a study on high school students, adolescents who had experienced separation from parents had more anxious and depressive symptoms than those belonging to intact families, while the death of a parent was not associated with significantly different symptom scores. Separation from both parents was significantly more detrimental to the adolescent than separation from only one parent, probably because it reflects serious difficulties in parenthood (Canetti et al., 2000). The absence of one parent after divorce is very common and it does not always lead to psychological problems, unlike parental disharmony (Tennant et al., 1988).

These findings suggest that separation may be understood by the child as an event with more negative connotations than parental death. While separation may be perceived by the child as voluntarily made, even at the cost of distance from the child, death is an event that the parent could not avoid. Moreover, studies have highlighted a process of idealization of the parent after death, while such process has not been observed after separation from a parent. Adult psychiatric patients and normal controls perceive their parents already died in a more positive manner than the subjects whose parents are still alive (Richter et al., 1992).

2.2 Physical and sexual abuse

A Canadian province-wide survey of approximately ten thousand people showed that 27% of females reported having suffered either physical or sexual abuse or both during childhood (MacMillan et al., 1997). In a community sample of more than eight thousand subjects, a significant association between a history of physical or sexual abuse in childhood and major depression (followed or not by a manic episode) and a strong relationship between childhood physical abuse and mania have been found (Levitan et al., 1998). One study including almost 2000 women, who were seen in primary care practices, revealed that the subjects who reported childhood physical and/or sexual abuse had higher scores of depression and anxiety and had attempted suicide more frequently than women without a history of childhood abuse (McCauley et al., 1997). Hyun et al. (2000) found that a history of childhood physical and sexual abuse was significantly more frequent in bipolar than in unipolar adults. Wise et al. (2001) have shown, in a case-control study, that women who suffered physical abuse during childhood are three times more likely to experience adult depression than others. Women who were abused in childhood are four times more likely to develop syndromal major depression in adulthood than women who were not abused, with the magnitude of the abuse correlated with the severity of depression (Mullen et al., 1996). This result has been later replicated by a study showing that female subjects, who reported experiences of childhood physical or sexual abuse, had a three-to-four-fold increase in the lifetime prevalence of major depression (MacMillan et al., 2001).

Several studies have reported an association between mood disorders and childhood sexual abuse alone. Prevalence rates of sexual abuse in children range from 3.0% to 33.2% (Dube et al., 2005), but these values are likely to be underestimates because of the frequent underreporting of sexual abuse (Priebe & Svedin, 2008). Moreover, a significant variability in the percentage of documented survivors who recall the abuse as adults has been found (62%-81%) (Goodman et al., 2003). Molnar et al. (2001) examined the associations between childhood sexual abuse and a range of mood, anxiety and substance-related disorders in a nationally representative sample of nearly 6000 individuals in the United States. The percentage of women with lifetime depression was 39.3% among those reporting childhood sexual abuse, compared with 21.3% in the general population, as the rate reported by the National Comorbidity Survey (Kendler et al., 2000). Chronic childhood sexual abuse perpetrated by a close relative or other trusted acquaintance has more severe long-term consequences than isolated incidents committed by strangers. Among women treated for anxiety disorders and/or depression, the subjects with a history of childhood sexual abuse have a significantly higher load of all types of childhood adversities, compared to patients without any abuse, due to a greater number of family background risk factors. The lack of parental protection and an environment described as non-supportive, with conflict, violence and marital turbulence, were the most commonly family background risk factors associated with childhood sexual abuse (Gladstone et al., 1999).

Because most studies have used cross-sectional designs and relied to a large degree on retrospective self-report data, a method that risks having a range of retrospective biases, longitudinal studies are needed to provide information about the causal relationship between childhood sexual and/or physical abuse and mood disorders and to systematically evaluate developmental trajectories. The first prospective assessment of the risk of depression in a group of children with documented physical, sexual abuse and neglect and a matched comparison group, followed up in adulthood, has shown that childhood physical abuse was associated with an increased risk for lifetime major depression, in agreement with earlier cross-sectional studies (Widom et al., 2007). The same study has not found any relationship between childhood sexual abuse and lifetime or current major depression, but abused subjects reported more depressive symptoms than controls. Differences in the design of the studies (prospective longitudinal with documented cases of maltreatment versus cross-sectional and based on retrospective recall) may in part account for the discrepancy between these results and the larger extant literature. A systematic review of longitudinal studies of childhood maltreatment and psychiatric outcomes (between 2000 and 2008) found an association between childhood maltreatment and depression, PTSD, and suicide attempts (Gilbert, 2009). A recent review and meta-analysis of 37 longitudinal observational comparative studies found an association between a history of sexual abuse and a lifetime diagnosis of anxiety, depression, eating disorders, PTSD, sleep disorders, and suicide attempts that persisted regardless of the survivor's sex or age at which abuse occurred. Associations between sexual abuse and depression, eating disorders and post-traumatic stress disorder were strengthened by a history of rape (Chen et al., 2010).

There is growing evidence that childhood sexual and physical abuse leads to dysregulation of the hypothalamic pituitary-adrenal (HPA) axis in children (Cicchetti & Rogosch, 2001) that has enduring effects on cortisol responses to stress in adulthood (Heim et al., 2000). Women with severe physical or sexual abuse and neglect were more likely to exhibit endogenous depressive subtypes which have been associated with HPA axis dysregulation (Harkness & Monroe, 2002). Women with current depression and a history of childhood physical and/or sexual abuse had greater increases in plasma cortisol and ACTH in response to a laboratory psychological stress test (Heim et al., 2000). The role of childhood abuse has been further supported by the studies showing that patients with major depressive disorder and without a history of childhood abuse had a normal cortisol response to a psychological stress paradigm (Heim et al., 2000). Increased levels of cortisol during repeated childhood abuse, along with persistent sensitization of the HPA axis in adulthood, have been demonstrated to damage hippocampal neurons in adult women with major depressive disorder. Magnetic resonance imaging (MRI) has revealed reductions in hippocampus and amygdala (Schmahl, 2003) volumes as well as deficits in verbal declarative memory, measured with neuropsychological testing, in women who were sexually abused as children (Teicher et al., 2000). Evidence of the effects of traumatic stress in childhood on the hippocampus provides a possible neurophysiologic explanation for a phenomenon identified in studies of adults whose childhood abuse was documented, showing that their retrospective reports of childhood abuse underestimate the actual occurrence (Priebe & Svedin, 2008).

2.3 Emotional abuse

Emotional abuse encompasses several forms of childhood maltreatment, such as the witnessing of domestic violence and exposure to verbal aggression (Bernstein, 1997). It is not always recognized as a distinct form of maltreatment: some researchers have suggested that emotional abuse is inherent in all forms of maltreatment (Garbarino, 1986), while other investigators have demonstrated that it occurs independently of other types of abuse (Claussen & Crittenden, 1991). This second hypothesis has been supported by a study demonstrating that emotional abuse and neglect predicted adult psychopathology even after controlling the effects of other types of adverse experiences, such as physical and sexual abuse (Spertus et al., 2003).

Emotional abuse has been considered by some theorists as a non-specific risk factor for psychopathology in adulthood, while others have hypothesized that it may contribute to specific vulnerability to the development of depression. For example, Rose & Abramson (1992) hypothesized that childhood emotional abuse should be more likely to contribute to the development of a cognitive vulnerability to depression than physical or sexual abuse, because the depressive cognitive style is directly provided to the child by the abuser. In support of this theory, Gibb et al. (2003) reported that adult psychiatric outpatients with a history of childhood emotional abuse had a much higher rate of current depressive disorders than anxiety disorders. In contrast, reports of childhood physical and sexual abuse were equally strongly related to both depressive and anxiety disorders.

Several other studies support the relationship between emotional abuse and increased levels of depression (Briere & Runtz, 1988; Mullen et al., 1996; Rich et al., 1997) and suicidality (Briere & Runtz, 1988) in adulthood. Briere & Runtz (1988) demonstrated shared effects of multiple types of abuse in a sample of university women. However, even when these shared effects were statistically accounted for, paternal psychological abuse remained a significant predictor of anxiety, depression, interpersonal sensitivity, and dissociation in these women.

The impact of emotional abuse varied with the gender of abusers: children emotionally abused by their female caregivers were more prone to develop adult depressive symptoms, whereas subjects abused from a male family member had a negative impact on adult sexuality (Mullen et, 1996).

Teicher (2006) has demonstrated that the exposure to verbal abuse alone and to witnessing of domestic violence alone had moderately strong effects on depressive symptoms. The combined exposure to verbal abuse and witnessing of domestic violence had a greater additive negative effect, having the exposed subjects depression scores that were 2.8 times as high as those of the non-abused subjects. The effect of the combined exposure to verbal abuse and witnessing of domestic violence was greater than or equal to the effect of exposure to familial sexual abuse.

According to the attachment theory (Bowlby, 1982), it is hypothesized that the cognitive models developed on the basis of the negative pattern of interactions between the emotionally maltreating parent and the child provide a set of negative beliefs and expectations about the self and the others, centred on shame, vulnerability to harm and self-sacrifice, which contribute to the development of later psychiatric symptoms.

Although attempts have been made to define childhood verbal abuse, there is not yet an operational definition of it nor is there a consensus about the prevalence of childhood verbal abuse in the general population. Even though the exposure to verbal aggression has received little attention as a specific form of abuse, it may be common as 63% of American parents reported one or more instances of verbal aggression against their child (Vissing et al., 1991). Children who reported frequent verbal aggression exhibited higher rates of interpersonal problems than other children. A relationship between maternal verbal abuse during childhood and a higher risk of developing several personality disorders (borderline, narcissistic, obsessive-compulsive, and paranoid) has been demonstrated and it has remained significant after controlling for temperamental features and co-occurring psychiatric disorders, physical and sexual abuse, neglect and parental psychopathology (Johnson, 2001).

2.4 Emotional neglect

The Centers for Disease Control and Prevention (CDC) define neglect as "failure to provide for a child's basic physical, emotional, or educational needs or to protect a child from harm or potential harm" (Leeb et al., 2008).

Child neglect is the most prevalent, but least empirically studied, form of child maltreatment. Researches in this area are inherently difficult because neglected children may suffer both different subtypes of neglect and several associated adversities, such as physical or sexual abuse, witnessing of domestic violence, poverty, etc., that may confound the relationship between child neglect and adult development of depression or other psychiatric disturbances.

Emotional neglect has been defined as 'emotional unresponsiveness, unavailability and neglect characterized by lack of interaction between parent and child' (Glaser, 2002). It is generally characterized by parents who are emotionally and psychologically unavailable, detached, avoidant and unresponsive to their child's needs and desires.

The attachment between the mother and her infant is one of the most important developmental interactions in mammals. The infant maintains closeness to his caretakers through an attachment system consisting of emotive and behavioural response patterns facilitating the capacity for self-perception and perception of others, a process also known as mentalization (Bowlby, 1982). Frequent touching by the maternal caregiver is a biologic necessity for physical and psychological growth, as shown in infant rats and monkeys in which maternal deprivation results in persistent deficits of prefrontal executive function with impairment in social, behavioural and cognitive development (Black, 1998). Together with genetic predisposing factors, individuals with depressive manifestations in adulthood often show dysfunctional parental attachment. The study of Stansfeld et al. (2008) has demonstrated that, among the dysfunctional parental bonding, emotional neglect and mainly overprotection, characterized by parental intrusive behaviour, are strongly associated with depression. The same study has shown a correlation between attachment style and socio-demographic factors. Lower social class is robustly associated with emotional/material deprivation and low parental warmth. Less warm attachment relationships have been demonstrated to be a risk factor for the development of psychiatric symptoms in adulthood, particularly depressive symptoms. High parental warmth, which is more prevalent in upper socio-economic classes, is correlated with a decreased risk for insecure attachment styles.

The association between parental emotional neglect in childhood (assessed using the Parental Bonding Instrument (PBI) and the Childhood Experience of Care and Abuse (CECA) interview) and adult depression has been widely replicated (Parker et al., 1995; Hill et al., 2001; Bifulco et al., 2002). The result of a recent longitudinal study (Widom et al., 2007) has underlined the need to detect and to treat the long-term psychological sequelae of childhood neglect, showing that approximately one quarter of the neglected children in a sample of about 1200 subjects met the criteria for lifetime major depression and 15% for current depressive episode.

Despite this evidence, the developmental trajectories from emotional neglect to adulthood psychopathology are still poorly understood.

Childhood is a period of great vulnerability of the central nervous system to environmental factors. In the first years of life, children are most vulnerable to the effects of emotional neglect (Hildyard & Wolfe, 2002). From the postnatal period until the age of seven years, several processes (proliferation, migration, differentiation, synaptogenesis) affecting cognitive functions and emotional regulation take place (Keverne, 2004). Childhood adverse experiences are associated with abnormalities in brain development, particularly *Corpus Callosum* (CC) morphology, and neglect is the strongest experiential factor, accounting for a 15%-18% reduction in several *Corpus Callosum* regions (Teicher et al., 2004).

Neglected children may have difficulty in discriminating emotional expression (Fries & Pollak, 2004) and show various attention and social deficits (Turgeon & Nolin, 2004). The lack of emotional interaction during the crucial early period of development can result in poor emotional regulation that may be part of a cascade of adverse neurobiological events rendering a child vulnerable to the effects of later adverse experiences and triggering a vicious cycle towards adult depression.

3. Neurobiology of childhood adverse experiences and mood disorders

The exact neurobiological mechanisms through which childhood adverse experiences may increase the risk of developing depression are not yet known, but they may include sensitization to later life events, mediated by some combination of neurobiological changes. The possible impact of such adversity on both brain structure and function involves the activity of the hypothalamic-pituitary-adrenal (HPA) axis, specific cerebral areas and genetic factors.

3.1 Hypothalamic-pituitary-adrenal (HPA) axis response to childhood adverse experiences

It has been theorized that one neurobiological mechanism, which occurs as a consequence of childhood adverse events and results in effects that ultimately trigger depression after additional stress, is the increased activation of the hypothalamic-pituitary-adrenal (HPA) axis.

Several animal and human studies have shown that depression is associated with altered regulation of HPA axis activity, as indicated by elevated cortisol, disruption of circadian HPA rhythms, and failure to suppress cortisol levels following the administration of the synthetic steroid dexamethasone (Thase, 2002).

The hypothalamic-pituitary-adrenal (HPA) axis - which consists of the hypothalamus, the pituitary gland and the adrenal gland - is a critical component of the body's stress response system. Upon exposure to stress, neurons in the hypothalamic paraventricular nucleus (PVN) secrete corticotropin-releasing factor (CRF), which stimulates the production and release of adrenocorticotropin (ACTH) from the anterior pituitary region that, in turn, stimulates the release of glucocorticoids from the adrenal cortex. Glucocorticoids exert a negative feedback control on the HPA axis by the activity on mineralocorticoid and glucocorticoid receptors in the hippocampus, PVN, and the pituitary gland, in order to modulate responsiveness and to return the system to homeostasis (Jacobson & Sapolsky, 1991).

Laboratory animal studies have provided direct evidence that early stressful experiences influence the development of the hypothalamic-pituitary-adrenal (HPA) axis, leading to heightened stress reactivity that persists in adulthood. Early stress experimental paradigm, such as maternal separation in rats or adverse rearing conditions in non-human primates, produces long-lived hyperactivity of CRF neuronal systems as well as greater reactivity of the hypothalamic-pituitary-adrenal (HPA) axis to stress in adulthood (Ladd et al, 1996). Maternally separated adult rats also exhibit up to three-fold increases in ACTH and corticosterone responses to psychological stressors, when compared to control rats (Plotsky and Meaney, 1993).

Several clinical studies have documented the HPA axis involvement in man. Studies on the possible link between child maltreatment and HPA axis dysregulation in children have yielded inconsistent results. In a sample of prepubertal maltreated depressed, non-maltreated depressed and healthy control children, Kaufman et al. (1997) did not found any differences in cortisol measures, but the maltreated-depressed group exhibited elevated total, peak and net ACTH response in response to CRH, compared to the other groups of children. A study carried out on 175 maltreated children and 209 controls found no differences in morning or evening salivary cortisol measures between groups (Cicchetti and Rogosch, 2001). However, differences emerged when abuse subtype was considered. Children who had been both physically and sexually abused showed much higher morning cortisol levels, compared to the emotional abuse, neglect, physical abuse alone and control groups, with a positive correlation between the severity of sexual abuse and cortisol levels.

Among studies that examined the relationship between child maltreatment and HPA axis function later in life, those which used psychosocial stressors to examine the HPA axis response and negative feedback inhibition are particularly interesting. Maltreated women with and without current MDD exhibited higher plasma ACTH levels in response to a standard psychosocial stress than women with no history of maltreatment and a current diagnosis of major depression and healthy controls with no history of maltreatment (Heim et al., 2000).

Sustained glucocorticoid exposure interferes with the normal transcriptional mechanisms that control the expression of Brain derived neurotrophic factor (BDNF), a protein that promotes the survival of selected neuronal population, with adverse effects on hippocampal neurons, particularly in the CA3 region, and prefrontal cortex (PFC). In these cerebral regions, reduction in dendritic branching, loss of dendritic spines, impairment of neurogenesis, increased rate of neuronal death and atrophy have been observed. Such damage might progressively reduce the ability of hippocampus and prefrontal cortex to exert the inhibitory control over the HPA axis activity, resulting in greater exposure to glucocorticoids (Nestler et al., 2002), a condition that has shown to play a role in the development of adult depression.

Hyperactivity of the hypothalamic-pituitary-adrenal (HPA) axis in major depression is suggested by a large body of research, including basal and provoked measurements of plasma HPA axis hormone concentrations, imaging of pituitary and adrenal gland volume, cerebrospinal fluid (CSF) levels of corticotrophin-releasing hormone (CRH), and *post-mortem* measures of brain CRH receptor binding and CRH messenger ribonucleic acid (mRNA) levels (Heim et al., 2008).

3.2 Structural brain changes associated with childhood adverse experiences

Preclinical studies have shown an association between the prolonged exposure to glucocorticoids and the atrophy of brain regions involved in the regulation of HPA activity, such as the hippocampus (Sapolsky, 2000) and the medial prefrontal cortex (mPFC), in particular the anterior cingulate cortex (ACC) (Cerqueira et al., 2005).

A large body of research has investigated the relationship between childhood adverse events and changes in brain structures both in children who have experienced maltreatment and in adults reporting childhood adversity. The brain imaging studies included in this section are cross-sectional, therefore no conclusions can be drawn about the causal effect of maltreatment on the brain structures. We cannot exclude that the reported brain differences might represent a risk factor for exposure to maltreatment that, in turn, increases the risk of developing psychiatric symptoms or disorders.

3.2.1 Hippocampus

Several studies have been focused on the hippocampus, given its well-established role in regulating HPA activity, and the high density of glucocorticoid receptors expressed. The hippocampus plays a central role in declarative memory functions that are important in accurately identifying the signal of potential threat during stress situations and it is involved in fear responses.

MRI studies of children and adolescents with PTSD following maltreatment (Jackowski et al., 2009) have consistently failed to detect the pattern of reduced hippocampal volume that has been generally reported in adults who have experienced maltreatment as children (Vythilingam et al., 2002). One possible explanation to account for the discrepancy of findings from children and adults comes from the so-called "neurotoxicity hypothesis", which considers that years or decades of stress induced prolonged exposure to glucocorticoids could lead to a loss of hippocampal cells.

3.2.2 Amygdala

The amygdala plays a central role in differentiating threatening from non-threatening environmental stimuli on the basis of prototype matching to fear memories. The stimulation of amygdala in primate studies activates fear centers in the brain and results in behaviours consistent with anxiety, hyperarousal and hypervigilance. Results from human studies suggest that the amygdala is activated when reading threat words and viewing masked fearful faces (Whalen et al., 1998).

A recent meta-analysis did not find any significant differences in amygdala volume between children with maltreatment-related PTSD and non-maltreated children (Woon & Hedges, 2008). Later studies have reported increased amygdala volumes in children and adolescents who had been adopted after an experience of early institutionalization (Tottenham et al., 2010). The few studies that have examined the volume of amygdala in adults with a history of childhood maltreatment did not report any significant differences between maltreated and non-maltreated subjects (Andersen et al, 2008). A fMRI study has shown a strong positive correlation between physical abuse and right amygdala response to sad faces in a sample including 20 patients with depression and 16 healthy controls. The heightened amygdala response to sad faces was not a characteristic of individuals with depression, but rather of the subjects with a significant history of maltreatment (Grant et al., 2011).

3.2.3 Corpus callosum

The *corpus callosum* (CC) is the thickest band of myelinated fibres in the brain which connects, anatomically and functionally, the right and left hemispheres, allowing them to exchange information. Nerve fibre connections passing through the CC are fully formed before birth, with a rostral-caudal pattern of myelination that continues in young adulthood.

The majority of studies have shown a decrease in *corpus callosum* volume (particularly middle and posterior regions) in maltreated children and adolescents, as well in adults reporting childhood maltreatment. The *corpus callosum* size was more affected by early maltreatment in male than in female subjects (Teicher et al., 2004).

Reduction in the size of *corpus callosum* has been associated with a decrease in communication between the brain hemispheres as demonstrated, in adults with a history of childhood maltreatment, by the dramatic difference in hemispherical activation during the recall of neutral and disturbing memories, evaluated through evoked potentials (Schiffer et al., 1995).

3.2.4 Prefrontal cortex

Prefrontal cortex (PFC) consists of several related areas, including orbitofrontal cortex, anterior cingulate (ACC) and anterior prefrontal cortex. Prefrontal functions are related to action planning, decision making, working memory, and attention.

Studies comparing PFC volume of children with maltreatment-related post-traumatic stress disorder and non-maltreated children have yielded inconsistent results (McCrory et al., 2010).

In contrast to the studies on maltreated children, decreased PFC volume in adults with a history of childhood maltreatment has been a consistent finding both in non-clinical sample and in depressed subjects. Adult patients with major depressive disorder, who reported a history of childhood maltreatment, exhibited reduced volume of the rostral anterior cingulate cortex (ACC). This finding supports the hypothesis that the rostral ACC, like the hippocampus, might be vulnerable to prolonged glucocorticoid exposure due to chronic stress, which in turn may decrease its ability to regulate HPA activity in response to future stress, resulting in greater exposure to glucocorticoids (Treadway et al., 2009). Emotional abuse was associated with a reduction in left dorsal medial PFC, even in the absence of physical or sexual abuse in childhood. This volumetric change was independent of gender and it could not be attributed to current psychopathology, supporting the hypothesis that the observed brain differences might be associated with the experience of emotional abuse (van Harmelen et al., 2010).

3.3 Interaction between genetic variance and adverse events

Accumulating evidence supports the hypothesis of a gene-environment (GxE) interaction, in which specific polymorphisms exert genetic control of sensitivity to stressful experiences in early life, influencing the causal effect of environmental stressors on depression.

The original work by Caspi et al. (2003) demonstrated that individuals with one or two copies of the short (s) form of the serotonin transporter gene promoter region polymorphism 5-HTTLPR are at higher risk of developing depression in response to stressful life events than people who were homozygous for the long (l) allele. The 5-HTTLPR short allele appears to reduce *in vitro* the transcriptional activity of 5-HTTLPR, resulting in decreased expression of the serotonin transporter (5-HTT).

Emerging preclinical and clinical evidence suggests that the negative sequelae associated with early stress are not inevitable. The presence of positive supports is an important environmental factor in promoting resiliency in maltreated children, even in the presence of a genotype expected to confer vulnerability for psychiatric disorders. Maltreated children with the s/s genotype of 5-HTTLPR and no positive supports had depression scores that were twice as high as the non-maltreated comparison children with the same genotype. Nevertheless, the presence of positive supports reduced the risk associated with maltreatment and with the s/s genotype of 5-HTTLPR, so that maltreated children with this profile had only minimal increases in their depression scores (Kaufman et al., 2004).

GxE studies have been also focused on the role of genes involved in the physiological response to environmental stressors, via HPA axis, in moderating the risk to develop a psychiatric disorder following childhood adverse experiences.

Corticotropin-releasing hormone receptor 1 (CRHR1) is a G protein–coupled receptor localized in frontal cortical areas, forebrain, brainstem, amygdala, cerebellum, and anterior pituitary gland that plays a key role in the regulation of the HPA axis activity in response to stressful events, mediating the action of corticotropin-releasing hormone (CRH) on the pituitary gland to release adrenocorticotropic hormone (ACTH) that stimulates the production of cortisol in the adrenal cortex. Preclinical studies indicate that persistent hyperactivity of the HPA axis following developmental stress exposure is mediated, at least in part, by a hyperactive CRHR1 system (Lupien et al., 2009). Genetic variants in the corticotropin-releasing hormone receptor (CRHR1) gene polymorphisms appeared to moderate the effect of childhood abuse on the risk for adult depressive symptomatology (Bradley et al., 2008).

Recently, a study has provided preliminary evidence that gene by environment (G x E) interactions may play an important role in explaining the differential effectiveness of a given intervention. In 1-to-3-year-old children with externalizing problems, Bakermans-Kranenburg et al. (2008) found a moderating role for the dopamine D4 receptor (DRD4) in a video-feedback intervention designed to improve maternal sensitivity and discipline, showing that the intervention was effective primarily in those children with the DRD4 7-repeat polymorphism.

3.4 Epigenetics

The word "epigenetics" refers to processes by which environmental influences can regulate gene activity without altering the underlying DNA sequence. There is growing evidence that epigenetic mechanisms of gene regulation have been implicated in several psychiatric disorders such as depression, drug addiction and schizophrenia (Tsankova et al., 2006).

This epigenetic process is coordinated in large part through the control of chromatin structure. Chromatin is composed of nucleosomes that consist of ~147 base pairs of DNA wrapped around a core of histone proteins. These nucleosomes undergo a supercoiling process, which results in a highly compact structure that permits to control gene expression also by gating access of transcriptional activators to DNA. The structure of chromatin, and hence the access to DNA, is regulated by the direct methylation of DNA and post-translational modifications of histones, both of which can either promote or suppress gene transcription.

Methylation of DNA is a direct chemical modification of a cytosine, by adding a -CH3 group through a covalent bond. This process is associated with the suppression of gene transcription and, in cases of extensive DNA methylation, with the complete silencing of the associated gene. Modifications of histone proteins include acetylation, phosphorylation, methylation of histones and many other processes, with each modification either positively or negatively regulating the transcriptional activity of the underlying gene.

The regulation of gene expression has been also proposed as a potential mechanism that can mediate both vulnerability and resilience to environmental factors, explaining through which processes early environmental factors may produce long-lasting effects on HPA activity and neuronal function.

"Epigenetic" effects of maltreatment in brain areas, in which structural and functional changes have been observed in adults following maltreatment, have been reported by Roth et al. (2009), using a rat model of infant maltreatment by a caregiver. This study showed that early maltreatment produced persisting changes in methylation of BDNF DNA, through the lifespan to adulthood, leading to a reduced BDNF gene expression in the adult prefrontal cortex and hippocampus. The changes of BDNF DNA methylation have been demonstrated in the offspring of females that had been previously exposed to maltreatment as pups. This finding suggests the possibility of a trans-generational transmission of changes in gene expression associated with early maltreatment, even in a new generation of animals who had not been exposed to such environmental stressors.

It is well-documented that stress experienced during gestation causes brain and behavioural alterations in offspring that are comparable to those produced by postnatal adversity. Human infants of mothers with high levels of depression and anxiety during the third trimester have increased methylation of the glucocorticoid receptor gene promoter in cord blood cells (Oberlander et al., 2008).

In one of the few epigenetic studies in humans, differences in epigenetic regulation of hippocampal glucocorticoid receptor expression in suicide victims with a history of childhood abuse, as compared with either suicide victims with no childhood abuse or controls, have been observed (McGowan et al., 2009).

Epigenetic modulation of gene transcription has also been implicated in the long-term impact of positive caregiver experiences on adult rat stress responses: adult patterns of DNA methylation of the glucocorticoid receptor gene in the hippocampus, which plays a crucial role in mediating stress responses, are directly associated with the quality of maternal care received in infancy (Weaver et al., 2004). One important finding from this work is that a positive caregiving environment can reverse the epigenetic methylation changes associated with poor maternal care, highlighting the ongoing importance of positive caregiving in influencing the stress response at the biological level.

4. Conclusion

The empirical evidence concerning the association between childhood parental loss, due to death or separation, and adult mood disorders is inconsistent, probably for the important methodological limitations of the studies. On the contrary, past as well as recent findings converge on the conclusion that childhood maltreatment including sexual, physical, and emotional abuse as well as emotional neglect, are associated to the development of adult mood disorders.

Several animal and human studies have shown that childhood adversity produce long-lived hyperactivity of the hypothalamic-pituitary-adrenal (HPA) axis which lead to an over-sensitive stress response, that may be the basis of an etiological link to depression in adulthood.

The sustained cortisol response, which may cause brain damage leading to volumetric changes in several critical brain areas, such as hippocampus and prefrontal cortex, produces further disinhibition of HPA axis activity.

There are preliminary evidences of a gene-environment (GxE) interaction in which specific polymorphisms exert genetic control of sensitivity to stressful experiences in early life, influencing the effect of environmental stressors on the development of adult depression.

The epigenetic regulation of gene expression, a process that does not alter the underlying DNA sequence, has been also proposed as a potential mechanism that can mediate the vulnerability to early environmental factors, explaining through which processes they may produce long-lasting effects on HPA activity and neuronal function.

5. References

Agid, O., Shapira, B., Zislin, J., Ritsner, M., Hani, B., Murad, H., Troudart, T., Bloch, M., Heresco-Levy, U. & Lerer, B. (1999). Environment and vulnerability to major psychiatric illness: a case control study of early parental loss in major depression, bipolar disorder and schizophrenia. *Molecular Psychiatry* , Vol. 4, pp. 163-172. EISSN: 1476-5578

Andersen, S. L., Tomada, A., Vincow, E. S., Valente, E., Polcari, A. & Teicher, M. H. (2008). Preliminary evidence for sensitive periods in the effect of childhood sexual abuse on regional brain development. *Journal of Neuropsychiatry and Clinical Neuroscience,* Vol. 20, pp. 292–301, ISSN: 0895-0172.

Bakermans-Kranenburg, M. J., Van Ijzendoorn, M. H., Pijlman, F. T. A., Mesman, J.& Juffer, F. (2008). Experimental evidence for differential susceptibility: dopamine D4 receptor polymorphism (DRD4 VNTR) moderates intervention effects on toddlers' externalizing behavior in a randomized controlled trial. *Developmental Psychology,* Vol. 44, pp.293–300, EISSN: 1939-0599.

Barnes, G.E., & Prosen, H. (1985). Parental death and depression. *Journal of Abnormal Psychology*, Vol. 94, pp. 64–69, ISSN: 0021-843X

Bernstein, D.P., Ahluvalia, T., Pogge, D. & Handelsman, L. (1997). Validity of the childhood trauma questionnaire in an adolescent psychiatric population. *Journal of the American Academy of Child and Adolescent Psychiatry,* Vol. 36, pp. 340–348, ISSN: 0890-8567.

Bilfulco, A.T., Brown, G.W., & Harris, T.O. (1987) Childhood loss of parent, lack of adequate parental care and adult depression: A replication. *Journal of Affective Disorders*, Vol. 12, No. 2, pp. 115–128, ISSN: 0165-0327.

Bifulco, A., Moran, P., Baines, R., Bunn A. & Stanford K. (2002). Exploring psychological abuse in childhood: II association with other abuse and adult clinical depression, *Bulletin of the Menninger Clinic,* Vol. 66, pp. 241—258, ISSN: 0025-9284.

Black, I. B. (1998). Genes, brain, and mind: The evolution of cognition. *Neuron,* Vol. 20, pp.1073-1080, ISSN: 0896-6273.

Bowlby, J. (1982). *Attachment: Attachment and loss* (2nd ed.). NewYork: Basic Books.

Bradley, R.G., Binder, E.B., Epstein, M.P., Tang, Y., Nair, H.P., Liu, W., Gillespie, C.F., Berg, T., Evces, M., Newport, D.J., Stowe, Z.N., Heim, C.M., Nemeroff, C.B., Schwartz, A., Cubells. J.F. & Ressler KJ. (2008). Influence of child abuse on adult depression: moderation by the corticotropin-releasing hormone receptor gene. *Archives of General Psychiatry,* Vol. 65, No.2, pp. 190-200, ISSN 1538-3636.

Brent, D., Melhem, N., Donohoe, M.B. & Walker, M. (2009). The incidence and course of depression in bereaved youth 21 months after the loss of a parent to suicide, accident, or sudden natural death. *American Journal of Psychiatry,* Vol.166, No. 7, pp. 786–794, ISSN: ISSN 1535-7228.

Briere, J. & Runtz M. (1988). Multivariate correlates of childhood psychological and physical maltreatment among university women", *Child Abuse & Neglect,* Vol. 12, pp. 331-341, ISSN: 0145-2134.

Browne, A., & Finkelhor, D. (1986). Impact of child sexual abuse: a review of the research. *Psychological Bulletin*, Vol. 99, pp.66–77, ISSN: 0033-2909.

Brown, G.W., Harris, T. & Copeland, J,R. (1977). Depression and loss. *British Journal of Psychiatry*, Vol. 130, pp.1-18, ISSN: 1472-1465.

Canetti,L., Bachar,E., Bonne, O., Agid, O., Lerer, B., Kaplan De-Nour, A. & Shalev, A.Y. (2000). The Impact of Parental Death Versus Separation From Parents on the Mental Health of Israeli Adolescents. *Comprehensive Psychiatry*, Vol. 41, No. 5, pp. 360-368, ISSN: 0010-440X.

Caspi, A., Sugden, K., Moffitt, T.E., Taylor, A., Craig, I.W., Harrington, H., McClay, J., Mill, J., Martin, J., Braithwaite, A. & Poulton, R. (2003). Influence of life stress on depression: moderation by a polymorphism in the 5-HTT gene. *Science*, Vol.301, No. 5631, pp. 386-389, EISSN: 1095-9203.

Cerqueira, J.J., Cantania, C., Sotiropoulos, I., Schubert, M., Kalisch, R., Almeida, O.F, Auer, D.P. & Sousa, N. (2005) Corticosteroid status influences the volume of the rat cingulate cortex-a magnetic resonance imaging study. *Journal of Psychiatry Research*, Vol. 39, pp. 451–460, ISSN: 0022-3956.

Chen LP, Murad MH, Paras, ML, Colbenson, KM, Sattler, AL Goranson, E.N. Elamin, M.B.,Seime RJ, Shinoaki, G., Prokop, L.J. & Zirakzadeh, A. (2010). Sexual Abuse and Lifetime Diagnosis of Psychiatric Disorders: Systematic Review and Meta-analysis. *Mayo Clinic Proceedings*, Vol. 85, No. 7, pp. 618-629, EISSN: 1942-5546

Cicchetti, D. & Rogosch, F.A. (2001). Diverse patterns of neuroendocrine activity in maltreated children. *Development and Psychopathology*, Vol. 13, pp. 677–693, EISSN: 1469-2198.

Claussen, A.H. & Crittenden, P.M. (1991). Physical and psychological maltreatment: relations among types of maltreatment. *Child Abuse & Neglect*, Vol. 15, pp. 5–18, ISSN: 0145-2134.

Crook, T. & Eliot, J.(1980). Parental death during childhood and adult depression: a critical review of the literature. *Psychological Bulletin*, Vol. 87,pp. 252–259, ISSN: 0033-2909.

Danese, A., Moffitt, T.E., Harrington, H., Milne, B.J., Polanczyk, G., Pariante, C.M., Poulton, R. & Caspi A. (2009) Adverse childhood experiences and adult risk factors for age-related disease: depression, inflammation, and clustering of metabolic risk markers. *Archives of Pediatrics & Adolescent Medicine*, Vol. 163, No. 12, pp.1135-43, ISSN: 1072-4710.

Dube, S.R., Anda, R.F., Whitfield, C.L., Brown, D.W., Felitti, V.J., Dong, M. & Giles, W.H. (2005) Long-term consequences of childhood sexual abuse by gender of victim. *American Journal of Preventive Medicine*, Vol. 28, No. 8, pp. 430- 438, ISSN: 0749-3797.

Finkelhor, D., Ormrod, R.K. & Turner, H.A. (2007). Poly-victimization: A neglected component in child victimization. *Child Abuse & Neglect*, Vol. 31, No.1, pp. 7-26, ISSN 0145-2134.

Freud, Sigmund. (1917)"Mourning and Melancholia". (1963). Standard Edition of the Complete Psychological Works of Sigmund Freud, Vol. 14 (1914-1916) London: Hogarth, 237- 60.

Fries, A.B. & Pollak, S.D.(2004). Emotion understanding in post-institutionalized Eastern European children. Development and Psychopathology, Vol.16, No.2, pp. 355-69, EISSN: 1469-2198.

Furukawa, T.A., Ogura, A., Hirai, T., Fujihara, S., Kitamura, T., Takahashi, K. (1999) Early parental separation experiences among patients with bipolar disorder and major depression: a case-control study. *Journal of Affective Disorders* , Vol. 52, pp. 85-91, ISSN: 0165-0327.

Garbarino, J., Guttman, E. & Seeley, J. (1986). The Psychologically Battered Child. San Francisco, Jossey-Bass, ISBN: 0 642 274487.

Gersten, J. C., Beals, J., & Kallgren, C..A. (1991). Epidemiology and preventive interventions: parental death in childhood as a case example. *American Journal of Community Psychology, 19*, pp. 481-500, EISSN:1573-2770

Gibb, B.E., Butler, A.C. & Beck, J.S. (2003). Childhood abuse, depression, and anxiety in adult psychiatric outpatients. *Depression and Anxiety,* Vol. 17, pp. 226–228, ISSN: 1520-6394

Gilbert, R., Widom, C.S., Browne, K., Fergusson, D., Webb, E. & Janson, S. (2009). Burden and consequences of child maltreatment in high-income countries. *Lancet,* Vol. 373, pp. 68-81, ISSN: 0140-6736.

Gladstone, G., Parker, G., Wilhelm, K., Mitchell, P., & Austin, M.P. (1999). Characteristics of depressed patients who report child sexual abuse. *American Journal of Psychiatry,* Vol.156, pp. 431–437, ISSN: 1535-7228.

Glaser, D. (2002). Emotional abuse and neglect (psychological maltreatment): A conceptual framework. *Child Abuse and Neglect,* Vol. 26, pp. 697–714, ISSN: 0145-2134.

Goodman, G.S., Ghetti, S., Quas, J.A., Edelstein, R.S., Alexander, K.W., Redlich, A.D., Cordon, I.M. & Jones, D.P. (2003). A prospective study of memory for child sexual abuse: new findings relevant to the repressed-memory controversy. *Psychological Science, Vol.*14, pp. 113-118, EISSN: 1467-9280.

Grant, M. M., Cannistraci, C., Hollon, S.D., Gore, J. & Shelton, R. (2011).Childhood trauma history differentiates amygdala response to sad faces within MDD. *Journal of Psychiatry Research,* Vol. 45, pp.886–895, ISSN: 0022-3956.

Green, J.G., McLaughlin, K.A., Berglund, P.A., Gruber, M.J., Sampson, N.A., Zaslavsky, A.M., & Kessler, R.C. (2010) Childhood adversities and adult psychiatric disorders in the National Comorbidity Survey Replication I: associations with first onset of *DSM-IV* disorders. *Archives of General Psychiatry,* Vol. 67, No. 2, pp. 113-123, ISSN: 1538-3636.

Harkness, K.L. & Monroe, S.M. (2002). Childhood adversity and the endogenous versus nonendogenous distinction in women with major depression. *American Journal of Psychiatry,* Vol. 159, No. 3, pp. 387-93, ISSN: 1535-7228.

Harris, T., Brown, G.W. & Bifulco, A. (1986). Loss of parent in childhood and adult psychiatric disorder: the role of lack of adequate parental care. *Psychological Medicine,* Vol. 16, No. 3, pp. 641–659, EISSN: 1469-8978.

Harrison, L. & Harrington, R. (2001). Adolescent bereavement experiences: prevalence, association with depressive symptoms, and use of services. *Journal of Adolescence* , Vol. 24, No. 2, pp.159–169, ISSN: 0140-1971.

Heim, C., Newport, D.J., Heit, S., Graham, Y.P., Wilcox, M., Bonsall, R., Miller, A.H. & Nemeroff, C.B. (2000). Pituitaryadrenal and autonomic responses to stress in women after sexual and physical abuse in childhood. *JAMA,* Vol. 284, pp. 92–597, ISSN: 0098-7484.

Heim, C., Newport, D.J., Mletzko, T., Miller, A.H. & Nemeroff, C.B. (2008). The link between childhood trauma and depression: insights from HPA axis studies in humans. *Psychoneuroendocrinology,* Vol.33, No.6, pp.693-710,ISSN: 0306-4530.

Hildyard, K.L. & Wolfe, D.A. (2002). Child neglect: Developmental issues and outcomes. *Child Abuse and Neglect*, Vol.26, pp. 679–695, ISSN: 0145-2134.

Hill, J., Pickles, A., Burnside, E., Byatt, M., Rollinson, L., Davis R. & Harvey, K. (2001). Child sexual abuse, poor parental care and adult depression: evidence for different mechanisms. *British Journal of Psychiatry*, Vol. 179, pp. 104—109, ISSN 1472-1465.

Hyun, M., Friedman, S.D.& Dunner, D.L. (2000). Relationship of childhood physical and sexual abuse to adult bipolar disorder. Bipolar Disorders, Vol. 2, No.2,pp.131-135,ISSN: 1399-5618.

Jackowski, A.P., De Arau´ jo, C.M., De Lacerda, A.L.T., De Jesus Mari, J., & Kaufman, J. (2009). Neurostructural imaging findings in children with post-traumatic stress disorder: Brief review. *Psychiatry and Clinical Neurosciences*, Vol. 63, pp. 1–8, ISSN: 1440-1819

Jacobs, J.R. & Bovasso, G.B. (2009). Re-Examining the Long-Term Effects of Experiencing Parental Death in Childhood on Adult Psychopathology *The Journal of Nervous and Mental Disease,* Vol. 197, No. 1, pp. 24-27, ISSN: 1539-736X.

Jacobson, L., & Sapolsky, R. (1991).The role of the hippocampus in feedback regulation of the hypothalamic-pituitary-adrenocortical axis. *Endocrine Reviews,* Vol.12, No.2, pp.118-34, ISSN: 0163769X.

Johnson, J.G., Cohen, P., Smailes, E.M., Skodol, A.E., Brown, J. & Oldham, J.M. (2001). Childhood verbal abuse and risk for personality disorders during adolescence and early adulthood. *Comprehensive Psychiatry*, Vol. 42, pp.16–23 ISSN: 0010-440X.

Kaufman, J., Birmaher, B., Perel, J., Dahl, R.E., Moreci, P., Nelson, B., Wells, W. & Ryan, N.D. (1997). The corticotropinreleasing hormone challenge in depressed abused, depressed non abused, and normal control children. *Biological Psychiatry,* Vol. 42, pp. 669–679, ISSN: 0006-3223.

Kaufman, J., Yang, B.Z., Douglas-Palumberi, H., Houshyar, S., Lipschitz, D., Krystal, J.H. & Gelernter, J. (2004). Social supports and serotonin transporter gene moderate depression in maltreated children. *Proceedings of the National Academy of Science U S A,* Vol.101, No.49, pp.17316-17321, ISSN:1091-6490.

Kendler, K.S. (1990). Familial risk factors and the familial aggregation of psychiatric disorders. *Psychological Medicine,* Vol. 20, No. 2, pp. 311-319, EISSN: 1469-8978.

Kendler, K. S., Bulik, C. M., Silberg, J., Hettema, J. M., Myers, J., & Prescott, C. A. (2000). Childhood sexual abuse and adult psychiatric and substance use disorders in women: An epidemiological and co-twin control analysis. *Archives of General Psychiatry*, Vol.57, pp. 953–959, ISSN 1538-3636

Kendler, K.S. & Baker, J.H. (2007). Genetic influences on measures of the environment: a systematic review. *Psychological Medicine*, Vol. 37, No.5, pp. 615-26, EISSN: 1469-8978.

Keverne, E.B. (2004). Understanding well-being in the evolutionary context of brain development. *Philosophical Transactions* of the Royal Society of London. Series B, Biological Sciences, Vol. 359: pp.1349–1358. ISSN: 0962-8436.

Kessing, L.V., Agerbo, E. & Mortensen, P.B. (2004). Major stressful life events and other risk factors for first admission with mania. *Bipolar Disorders,* Vol. 6, pp. 122– 129,EISSN: 1399-5618.

Kunugi, H., Sugawara, N., Aoki, H., Nanko, S., Hirose, T., & Kazamatsuri, H. (1995). Early parental loss and depressive disorder in Japan. *European Archives of Psychiatry and Clinical Neuroscience,* 245, pp.109 –113, ISSN: 1433-8491.

Ladd, C.O., Owens, M.J. & Nemeroff, C.B.(1996). Persistent changes in corticotropin-releasing factor neuronal systems induced by maternal deprivation. *Endocrinology,* Vol.137, pp.1212–1218, EISSN: 1945-7170.

Leeb, R.T., Paulozzi, L., Melanson, C., Simon, T. & Arias, I. (2008). Uniform Definitions for Public Health and Recommended Data Elements. Version 10. Centers for Disease Control and Prevention, National Center for Injury Prevention and Control, Child Maltreatment Surveillance, Atlanta (GA).

Levitan, R.D., Parikh, S.V., Lesage, A.D., Hegadoren, K.M., Adams, M.,Kennedy, S.H. & Goering, P.N. (1998). Major depression in individuals with a history of childhood physical or sexual abuse: relationship to neurovegetative features, manic, and gender. *American Journal of Psychiatry,* Vol.155, No.12, pp. 1746-52, ISSN 1535-7228.

Lloyd, C.(1980). Life events and depressive disorder reviewed: I. events as predisposing factors. *Archives of General Psychiatry,* Vol. 37, pp.529–535, ISSN 1538-3636.

Lupien, S.J., McEwen, B.S., Gunnar, M.R., Heim, C. (2009). Effects of stress throughout the lifespan on the brain, behaviour and cognition. *Nature Review Neuroscience,* Vol. 10, pp. 434–445, ISSN: 1471-0048.

MacMillan, H.L., Fleming, J.E., Trocme, N., Boyle, M.H., Wong, M., Racine, Y.A., Beardslee, W.R. & Offord, D.R.(1997). Prevalence of child physical and sexual abuse in the community. *JAMA* , Vol.275, pp.131–135, ISSN: 0098-7484.

MacMillan HL, Fleming JE, Streiner DL, Lin E, Boyle MH, Jamieson E, Duku EK, Walsh CA, Wong MY, Beardslee WR. (2001). Childhood abuse and lifetime psychopathology in a community sample. American Journal of Psychiatry, Vol.155, No.12, pp. 1878-1883, ISSN 1535-7228

McCauley, J., Kern, D.E., Kolodner, K., Dill, L., Schroeder, A.F., DeChant, H.K., Schroeder AF, DeChant HK, Ryden J, Derogatis LR, Bass EB. (1997). Clinical characteristics of women with a history of childhood abuse: unhealed wounds. *JAMA,* Vol. 277, No. 17, pp.1362–1368, ISSN: 0098-7484.

McCrory, E., Stephane, A., De Brito, S.A. & Viding, E. (2010). Research Review: The neurobiology and genetics of maltreatment and adversity. *Journal of Child Psychology and Psychiatry,* Vol.51, No. 10, pp. 1079–1095, ISSN: 1469- 7619.

McGowan, P. O., Sasaki, A., D'alessio, A. C., Dymov, S., Labonté, B., Szyf, M., Turecki, G. & Meaney, M. J. (2009). Epigenetic regulation of the glucocorticoid receptor in human brain associates with childhood abuse. *Nature Neuroscience,* Vol. 12, pp. 342–348, EISSN: 1546- 1726.

Molnar, B.E., Buka, S.L. & Kessler,R.C. (2001). Child abuse and subsequent psychopathology: results from the National Comorbidity Survey. American Journal of Public Health , Vol. 91, pp. 753-760, EISSN: 0090-0036

Morina, N.., von Lersner, U. & Prigerson, H.G. (2011).War and bereavement: consequences for mental and physical distress. PLoS One, Vol.6, No. 7, e22140. Epub Jul 12, ISSN: 1932-6203.

Mortensen, P., Pedersen, C., Melbye, M., Mors, O. & Ewald H. (2003). Individual and familial risk factors for bipolar affective disorders in Denmark. *Archives of General Psychiatry,* Vol. 60 No.12, pp.1209–1215, ISSN:

Mullen, P. E., Martin, J. L., Anderson, J. C., Romans, S. E. & Herbison, G. P. (1996). The long-term impact of the physical, emotional, and sexual abuse of children: A community study. *Child Abuse & Neglect,* Vol.20, pp.7–21, ISSN: 0145-2134.

Munro, A. (1966) Parental deprivation in depressive patients. *British Journal of Psychiatry*, Vol. 112, pp. 443- 457, ISSN: 1472-1465.

Nestler, E.J., Barrot, M., DiLeone, R.J., Eisch, A.J., Gold, S.J., & Monteggia, L.M. (2002). Neurobiology of depression. *Neuron*, Vol.34, pp. 13–25, ISSN: 0896-6273.

Oakley Browne, M.A., Joyce, P.R., Wells, J.E., Bushnell, J.A. & Hornblow, A.R. (1995). Disruptions in childhood parental care as risk factors for major depression in adult women. *Australian and New Zealand Journal of Psychiatry*, Vol. 29, pp. 437-448, ISSN: 1440-1614.

Oberlander, T.F., Weinberg, J., Papsdorf, M., Grunau, R., Misri, S. & Devlin, A.M. (2008). Prenatal exposure to maternal depression, neonatal methylation of human glucocorticoid receptor gene (NR3C1) and infant cortisol stress responses. *Epigenetics*, Vol. 3, pp. 97–106, EISSN: 1559-2308

Parker, G., Hadzi-Pavlovic, D., Greenwald, S. & Weissman, M. (1995). Low parental care as a risk factor to lifetime depression in a community sample. *Journal of Affective Disorders*, Vol. 33, pp. 173—180 ISSN: 0165-0327

Patten, S.B.(1991).The loss of a parent during childhood as a risk factor for depression. *Canadian Journal of Psychiatry*, Vol. 36, No.10, pp. 706-711, ISSN: 0703-7437.

Perris, C., Holmgren, S., von Knorring, L., & Perris, H. (1986). Parental loss by death in the early childhood of depressed patients and of their healthy siblings. *British Journal of Psychiatry*, Vol. 148, pp.165–169. ISSN: 1472-1465.

Plotsky, P.M. & Meaney, M.J.(1993). Early, postnatal experience alters hypothalamic corticotropin-releasing factor (CRF) mRNA, median eminence CRF content and stress-induced release in adult rats. *Molecular Brain Research*, Vol.18, pp.195–200, ISSN: 0169-328X.

Priebe, G. & Svedin, C.G. (2008). Child sexual abuse is largely hidden from the adult society: an epidemiological study of adolescents' disclosures. *Child Abuse & Neglect*, Vol. 32, pp.1095-1108, ISSN: 0145-2134.

Rich, D. J., Gingerich, K. J. & Rosen, L. A. (1997). Childhood emotional abuse and associated psychopathy in college students. *Journal of College Student Psychotherapy*, Vol.11, pp.13–28, ISSN: 1540-4730.

Richter, J., Eisemann, M., Richter, G. & Perris, C. (1992). Determinants of recall of parental rearing behavior. Psychopathology, Vol.25, pp.120-127, ISSN: 0254-4962.

Ritchie, K., Jaussent, I., Stewart, R., Dupuy, A.M., Courtet, P., Ancelin, M.L., & Malafosse, A. (2009). Association of adverse childhood environment and 5-HTTLPR Genotype with late-life depression. *Journal of Clinical Psychiatry*, Vol. 70, No. 9, pp. 1281-8, ISSN: 1555-2101.

Rose, D.T. & Abramson, L.Y. (1992). Developmental predictors of depressive cognitive style: Research and theory. In: Cicchetti, D. & Toth, S., editors. Rochester Symposium of Developmental Psychopathology, Vol. IV. Rochester, NY: University of Rochester Press.pp. 323–349.

Roth, T. L., Lubin, F. D., Funk, A. J., & Sweatt, J. D. (2009). Lasting epigenetic influence of early-life adversity on the BDNF gene. *Biological Psychiatry*, Vol. 65, pp. 760–769, ISSN: 0006-3223

Sapolsky, R.M .(2000). Glucocorticoids and hippocampal atrophy in neuropsychiatric disorders. *Archives of General Psychiatry*, Vol. 57, pp. 925–935, ISSN 1538-3636.

Schiffer, F., Teicher, M.H. & Papanicolaou, AC. (1995). Evoked potential evidence for right brain activity during the recall of traumatic memories. *Journal of Neuropsychiatry and Clinical Neuroscience,* Vol. 7, No. 2, pp. 169-175, ISSN: 0895-0172.

Schmahl, C,G,, Vermetten, E., Elzinga, B.M. & Bremner, J.D. (2003). Magnetic resonance imaging of hippocampal and amygdale volume in women with childhood abuse and borderline personality disorder. *Psychiatry Research,* Vol. 122, pp. 193–198, ISSN: 0165-1781.

Spertus, I. L., Yehuda, R., Wong, C. M., Halligan, S., & Seremetis, S. V. (2003). Childhood emotional abuse and neglect as predictors of psychological and physical symptoms in women presenting to a primary care practice. *Child Abuse & Neglect,* Vol. *27,* pp. 1247–1258, ISSN: 0145-2134.

Stansfeld, S., Head, J., Bartley. M. & Fonagy, P. (2008). Social position, early deprivation and the development of attachment. *Social Psychiatry and Psychiatric Epidemiology,* Vol. 43, pp. 516–526, EISSN: 1433-9285.

Teicher, M. H. (2000) Wounds that time wouldn't heal: the neurobiology of childhood abuse. *Cerebrum,* Vol. 2, pp. 50–67, ISSN: 1524-6205.

Teicher, M.H., Dumont, N.L., Ito, Y., Vaituzis, C., Giedd, J.N. & Anderson, S.L. (2004). Childhood neglect is associated with reduced corpus callosum area. *Biological Psychiatry,* Vol. 56, pp. 80–85, ISSN: 0006-3223.

Teicher MH, Samson JA, Polcari A, McGreenery CE.(2006). Sticks, stones, and hurtful words: relative effects of various forms of childhood maltreatment. American Journal of Psychiatry, Vol 163, No. 6, pp. 993-1000, ISSN 1535-7228

Tennant, C., Bebbington, P., & Hurry, J. (1980). Parental death in childhood and risk of adult depressive disorders: a review. *Psychological Medicine,* 10(2), pp. 289–299, EISSN: 1469-8978.

Tennant, C., Bebbington, P. & Hurry, J. (1982). Social experiences in childhood and adult psychiatric morbidity: a multiple regression analysis. *Psychological Medicine,* Vol. 12, pp. 321-327, EISSN: 1469-8978.

Tennant, C. (1988). Parental loss in childhood: its effect in adult life. *Archives of General Psychiatry,* Vol. 45, pp.1045-1050, ISSN: 1538-3636.

Thase, M.E., Jindal, R. & Howland, R.H. (2002). Biological aspects of depression. In: Gotlib, I.H. & Hammen, C.L. eds. Handbook of Depression. New York: Guilford Press. pp. 192–218.

Tottenham, N., Hare, T. A., Quinn, B. T., Mccarry, T. W., Nurse, M., Gilhooly, T., Millner, A., Galvan, A., Davidson, M. C., Eigsti, I. M., Thomas, K. M., Freed, P. J., Booma, E. S., Gunnar, M. R., Altemus, M., Aronson, J. & Casey B. J. (2010). Prolonged institutional rearing is associated with atypically large amygdala volume and difficulties in emotion regulation. *Developmental Science,* Vol. 13, 46–61, ISSN: 1467-7687.

Treadway, M.T., Grant, M.M., Ding, Z., Hollon, S.D., Gore, J.C. & Shelton, R.C. (2009). Early adverse events, HPA activity and rostral anterior cingulate volume in MDD. *PLoS ONE,* Vol. 4, No. 3, e4887, ISSN: 1932-6203

Tsankova, N., Berton, O., Renthal, W., Kumar, A., Neve, R. & Nestler, E. (2006). Sustained hippocampal chromatin regulation in a mouse model of depression and antidepressant action. *Nature Neuroscience,* Vol. 9, pp. 519–525, EISSN: 1546- 1726.

Turgeon, M. & Nolin, P. (2004). Relationship between neglect and children's memory and verbal learning capacities. *Revue québécoise de psychologie,* Vol. 25, pp. 151–165, ISSN: 0225-9885.

Van Eederwegh, M. M., Bieri, M. D., Parrilla, R. H., & Clayton, P. J. (1982). The bereaved child. *British Journal of Psychiatry, 140,* pp. 23-29, ISSN 1472-1465.

van Harmelen, A.L., Van Tol, M.-J.,Van Der Wee, N. J. A., Veltman, D.J., Aleman, A., Spinhoven, P., Van Buchem, M. A., Zitman, F. G., Penninx, B. W. J. H. & Elzinga, B. M. (2010). Reduced medial prefrontal cortex volume in adults reporting childhood emotional maltreatment. *Biological Psychiatry,* Vol. 68, pp. 832–838, ISSN: 0006-3223.

Vissing, Y.M., Straus, M.A., Gelles, R.J. & Harrop, J.W. (1991). Verbal aggression by parents and psychosocial problems of children. *Child Abuse & Neglect,* Vol.15, pp. 223–238, ISSN: 0145-2134.

Vythilingam, M., Heim, C., Newport, J., Miller, A.H., Anderson, E., Bronen, R., Brummer, M., Staib, L., Vermetten, E., Charney, D.S., Nemeroff, C.B. & Bremner, J.D. (2002). Childhood trauma associated with smaller hippocampal volume in women with major depression. *American Journal of Psychiatry,* Vol. 159, pp. 2072–2080, ISSN 1535-7228.

Weaver, I.C.G., Cervoni, N., Champagne, F.A., D'Alessio, A.C., Sharma, S., Seckl, J.R, Dymov, S., Szyf, M. & Meaney, M.J. (2004). Epigenetic programming by maternal behavior. *Nature Neuroscience,* Vol. 7, pp.847–854, EISSN: 1546- 1726.

Whalen, P. J., Rauch, S. L., Etcoff, N. L., Mclnerney, S. C., Lee, M. B. & Jenike, M. A. (1998). Masked presentations of emotional facial expressions modulate amygdala activity without explicit knowledge. *Journal of Neuroscience,* Vol. 18, pp. 411-418, ISSN: 1529-2401.

Widom, C.S., Dumont, K. & Kzaja, S.J. (2007). A Prospective Investigation of Major Depressive Disorder and Comorbidity in Abused and Neglected Children Grown Up. *Archives of General Psychiatry,* Vol. 64, No. 1, pp. 49-56, ISSN 1538-3636.

Wilcox, H.C., Kuramoto, S.J., Lichtenstein, P., Langstrom, N., Brent, D.A. & Runeson, B. (2010). Psychiatric morbidity, violent crime, and suicide among children and adolescents exposed to parental death. *Journal of American Academy of Child & Adolescent Psychiatry,* Vol. 49 No. 5, pp. 514-523, ISSN: 0890-8567.

Wise, L.A., Zierler, S., Krieger, N. & Harlow, B.L.(2001), Adult onset of major depressive disorder in relation to early life violent victimisation: a case-control study. Lancet, Vol.358, No.9285,pp.881-887, ISSN: 0140-6736.

Woon, F.L. & Hedges, D.W. (2008). Hippocampal and amygdala volumes in children and adults with childhood maltreatment-related posttraumatic stress disorder: A meta-analysis. *Hippocampus,* Vol.18, pp. 729–736, ISSN: 1098-1063.

Part 3

Neurobiology

8

Bipolar Disorder: Diagnosis, Neuroanatomical and Biochemical Background

Kristina R. Semeniken[1,2] and Bertalan Dudás[2]
[1]*Department of Psychiatry, Millcreek Community Hospital, Erie, PA*
[2]*Neuroendocrine Organization Laboratory (NEO), Lake Erie College of Osteopathic Medicine (LECOM), Erie, PA*
USA

1. Introduction

Bipolar disorder is a mood disorder that is characterized by episodes of mania or hypomania that often alternate with episodes of depression. Bipolar disorder is also referred to as manic-depressive disorder and bipolar affective disorder. This potentially disabling mood disorder has a lifetime prevalence of 4% in the US population (Ketter, 2010). According to the National Institute of Mental Health (NIMH), bipolar disorder is classified as severe in 82.9% of adult patients, with 25 being the average age of onset. As is the case with the majority of psychiatric disorders, bipolar disorder tends to run in families (Smoller and Finn, 2003).

Bipolar disorder varies in its presentation and may be difficult to diagnose. Diagnosis requires the presence of a manic or a hypomanic episode; however, it is likely that the first few episodes may present as bouts of depression, particularly in younger patients (Saddock and Saddock, 2003). According to the Diagnostic and Statistical Manual of Mental Disorders fourth edition (DSM-IV-TR; American Psychiatric Association, 2000) a distinct period of abnormal mood for at least one week is required to meet diagnostic criteria.

There are two major subtypes of bipolar disorder: type I and type II. Type I requires the presence of at least one manic or mixed episode. Mania is categorized by euphoria, impulsivity, distractibility, racing thoughts, irritability, decreased need for sleep and grandiosity. Individuals with mania may stay up all night cleaning or working on flamboyant projects, or they may engage in large spending sprees, gambling or substance abuse. Manic individuals often exhibit pressured speech which may be loud, verbose, and intrusive in nature. The thought process in mania is often disorganized with flight of ideas and loosening of associations. Manic individuals often have poor judgment and insight and they are likely to give unreliable case histories (Saddock and Saddock, 2003).

Manic episodes may vary in severity, with the more severe cases involving psychosis. About 75% of manic individuals experience some form of perceptual disturbance (Saddock and Saddock, 2003). Psychotic features of manic individuals are characterized by the presence of hallucinations and or delusions. These features may be congruent with the mood and may revolve around possession of knowledge or power, disturbance in identity or belief in having relationship to a deity or famous person. Mood-incongruent psychotic symptoms do not focus on the themes mentioned above and are more likely to include thought insertion and delusions of being persecuted or controlled (Saddock and Saddock, 2003). Individuals suffering from bipolar disorder I may also experience mixed episodes. Mixed episodes often combine characteristics of a depressive episode with irritability, anxiety or inner tension, psychomotor agitation and racing thoughts (Koukopoulos and Koukopoulos, 1999; Ketter, 2010). Individuals experiencing mixed episodes are at highest risk for suicide, secondary to increased irritability and impulsivity combined with depressed mood (Goldberg et al., 1998; Akiskal and Benazzi, 2005).

Unlike bipolar I, bipolar disorder type II does not include full-blown manic episodes and it is often defined by episodes of hypomania combined with features of major depression and is considered to be less severe than bipolar I (Goodwin and Jamison, 1990). According to the DSM-IV-TR, hypomania is characterized by persistently elevated, irritable, or expansive mood lasting at least 4 days (as opposed to at least one week for mania). This mood must be distinct from the usual (nondepressed) mood. Although the mood disturbance in a hypomanic individual is observable by others, hypomanic episodes are not severe enough to cause marked impairment in social or occupational functioning, or to necessitate hospitalization (American Psychiatric Association, 2000). Additionally, psychotic features are absent from hypomania. Both mania and hypomania may be caused by comorbid states, such as hyperthyroidism or other medical conditions, hypothyroidism, electroconvulsive therapy (ECT) and medications such as antidepressants. In these cases, a diagnosis of bipolar disorder may not be established since diagnosis requires that the symptoms must not be due to a general medical condition or direct physiologic effects of a substance.

In bipolar disorder, episodes of mania or hypomania may alternate with episodes of depression. Episodes of depression are not exclusively limited to bipolar II and may also be present in bipolar I disorder. Depressive episodes involve either dysphoria (depressed mood) or anhedonia (loss of interest in pleasure). A major depressive episode is characterized by sadness or depressed mood, sleep disturbances (insomnia or hypersomnia), appetite disturbances, changes in body weight, suicidality, feelings of helplessness, hopelessness, worthlessness or guilt, loss of energy and loss of interest in activities (Saddock and Saddock, 2003). Individuals in a depressed episode may appear to be withdrawn and may exhibit slowing down of mental processes, known as psychomotor retardation. Not all depressed individuals admit to feeling depressed but they usually exhibit a negative outlook (Saddock and Saddock, 2003). During a depressive episode, patients may experience psychotic features such as hallucinations and or delusions. Patients in a depressed episode may also experience problems with memory and cognition. They may have problems concentrating or remembering recent events. Patients in a depressed episode are particularly at risk for suicide, with about 2/3 of them experiencing suicidal ideations (Saddock and Saddock, 2003). Comorbid states such as hypothyroidism and alcohol abuse may precipitate depression.

Bipolar disorder falls on a spectrum of severity. A less severe form is known as cyclothymia, which is characterized by cyclical mood swings that are less severe than full-blown mania or depression. Cyclothymia may involve hypomania but does not cause marked impairment in functioning and often involves a euphoric phase and a dysthymic phase. The DSM-IV-TR criteria for rapid cycling include at least four episodes within a 12 month period. Patients with rapid cycling are more likely to be female and to have depressive and hypomanic episodes (Saddock and Saddock, 2003). At times, the presentation of bipolar disorder may not fit into any specific category and therefore will be classified as unspecified.

Bipolar disorder is a progressive illness, in which frequency of episodes increases over time with subsequent decreased probability of treatment response (Berk et al., 2010). Individuals suffering from bipolar disorder have been shown to have altered reward processing (Pizzagalli et al., 2008). Bipolar manic patients were observed to produce more errors in a two-choice selection task, suggesting impairment in making decisions during times of uncertainty about the success of the outcome (Minassian et al., 2004). Response-reversal studies involving medicated euthymic children with bipolar disorder showed that these children were slower to learn variable stimulus-reward contingencies (Gorrindo et al., 2005). Additionally, manic bipolar patients were observed to make more unfavorable choices when presented with two choices of variable favorability (Murphy et al., 2001). Both symptomatic and euthymic bipolar patients have been noted to demonstrate a decreased bias response toward more frequently rewarded stimuli, with anhedonic patients showing greater impairments in reward learning (Pizzagalli et al., 2008). The implication is that bipolar patients may have impairment in the long-term integration of reinforcements and may struggle with adapting their behavior to alterations in reward (Pizzagalli et al., 2008).

2. Neuroanatomy and bipolar disorder

Regulation of mood is believed to involve primarily two major neuroanatomic circuits: the limbic-thalamic-cortical circuit and the limbic-striatal-pallidal-cortical circuit (Mayberg, 1997; Soares and Mann, 1997). The limbic-thalamic-cortical circuit is also involved in working memory and includes the medial-dorsal nucleus of the thalamus, the ventrolateral prefrontal cortex and the amygdala (Floresco et al., 1999; Parsey et al., 2006). The limbic-striatal-pallidal-cortical circuit is involved in emotional expression (Drevets et al., 2008). Dysfunction in either of these circuits may result in a mood disorder (Parsey et al., 2006). Neuroanatomical changes of patients affected by bipolar disorder have been associated with a dysfunction in the prefrontal cortex, basal ganglia, temporal and frontal lobes of the forebrain as well as parts of the limbic system including amygdala, thalamus and striatum (Cerullo et al., 2009). Patients with bipolar disorder have also been found to have abnormalities in the subgenual anterior cingulate cortex (sACC) (Shah et al., 2009; Rosen and Rich, 2010). More specifically, decreased glial cell volume, regional cerebral blood flow, and decreased glial cell number have been observed in this region in bipolar patients, as opposed to healthy controls (Drevets et al., 1997; Ongur et al., 1998). Increased limbic activity has been found to be associated with aberrant emotional processing while neurophysiological abnormalities in the prefrontal cortex have been associated with impaired executive function (Green et al., 2007). Dysfunction in the prefrontal cortex, basal ganglia and the limbic system plays a significant role in bipolar disorder due to the strong connection between these regions and the emotional and cognitive aspects of affect regulation (Papez J.W., 1937; Allman et al., 2001; Vogt BA, 2005).

The nature and location of the neuroanatomical dysfunction in bipolar patients often correlates with the resulting symptoms. Abnormalities in the frontosubcortical circuit, especially in the hippocampus and the prefrontal cortex, may explain the attention impairment observed in manic patients (Sax et al., 1999). Reduced neuronal and glial density and glial hypertrophy have been identified in the hippocampus and the dorsolateral prefrontal cortex of patients with bipolar disorder (Rajkowska et al., 2001). These observations may account for cell loss observed in brain imaging and post-mortem studies of bipolar patients (Brown et al., 2003). These deficits differ from reported observations of increased neuronal density noted in patients with schizophrenia. The specific pattern of cell loss described above appears to resemble the reduced cell density found in patients with major depressive disorder (Rajkowska et al., 2001; Brown et al., 2003). Patients with bipolar disorder have less reduction of hippocampal volume than schizophrenic patients (Radonic et al., 2011). The hippocampus, particularly the CA3 region, plays an important role in the formation of declarative memory, which requires separate encoding of elements of an event and their organization in relation to one another (Preston et al., 2005). Impairment in the circuitry involved in formation of this type of memory may be responsible for the observed impairment in spatial memory and reward processing that has been noted in bipolar patients.

Individuals with bipolar disorder appear to exhibit altered responses to emotional stimuli, with apparent dysfunction in the ventral-limbic regions including the ventrolateral prefrontal cortex, the orbitofrontal cortex, the subgenual cingulate cortex, the insula, amygdala and striatum (Wessa and Linke, 2009). Dysfunction in the fronto-striatal-limbic network, which includes the structures mentioned above, has been associated with aberrant emotional processing, particularly with regard to misreading facial expression of emotions (Rosen and Rich, 2010). Increased striatal activity has been reported in bipolar patients in response to potentially rewarding stimuli (Hassel et al., 2008). Additionally, decreased activity has been noted in the dorsal brain structures including right dorsolateral prefrontal cortex, posterior cingulate cortex, and the dorsal anterior cingulate cortex of patients with bipolar disorder following exposure to significant emotional stimuli (Hassel et al., 2008; Wessa and Linke, 2009). It appears that imbalance between the ventral-limbic network and the dorsal brain structures may be responsible for the emotional dysregulation observed in bipolar disorder (Wessa and Linke, 2009).

Functional magnetic resonance imaging (fMRI) studies have shown a dysfunction of subcortical pre-frontal networks (striato-thalamic) and associated regions of the limbic lobe (Strakowski et al., 2005). Decreased modulation of medial temporal and subcortical structures in the anterior limbic lobe (striatum, amygdala and thalamus) by prefrontal areas has been noted to appear early in the course of the illness (Strakowski et al., 2005). The amygdala, which plays a role in the regulation of mood, is noted to be decreased in volume in bipolar patients. This decrease appears to be more severe as the patients get older (Doty et al., 2008), suggesting a correlation between progression of bipolar disorder and increasing abnormalities in the brain. Individuals in the early stages of bipolar disorder were found to have changes in the volume of the white matter, ventricles, amygdala, caudate nucleus, subgenual prefrontal cortex and putamen (Hajek et al., 2005). Moreover, the use of lithium for mood stabilization in bipolar disorder has been shown to increase the size of the amygdala and hippocampus (Yucel et al., 2007; Foland et al., 2008; Yucel et al., 2008). Other abnormalities found in the brains of bipolar patients using neuroimaging techniques have been noted to appear with recurrent episodes of mood instability and affect dysregulation.

These abnormalities have been observed in the lateral ventricles, cerebellar vermis and other prefrontal regions such as the left anterior prefrontal cortex (Strakowski et al., 2005). It appears that although abnormalities may be present at the onset, the severity of these lesions progresses with the disease.

3. The role of oxidative stress

Oxidative stress has been shown to play a role in the pathophysiology of bipolar disorder (Gawryluk et al., 2011). Alterations in oxidative enzymes as well as increased lipid peroxidation products and reactive oxygen species have been detected in individuals with bipolar disorder and other psychiatric diseases (Steckert et al., 2010). Reduced levels of glutathione in the brain have been reported in bipolar disorder as well as schizophrenia (Fullerton et al., 2010; Dean et al., 2011; Magalhaes et al., 2011). Along with glutathione reductions, evidence of oxidative damage to mitochondrial proteins have been noted in the brains of patients (Andreazza et al., 2010; Gawryluk et al., 2011). As the primary antioxidant in the body, glutathione acts as a free radical scavenger and has been noted to be present in high levels in the brain (Choy et al., 2010; Gawryluk et al., 2011). Decreased glutathione levels increase susceptibility of cells to oxidative stress (Gawryluk et al., 2011). Glutathione depletion has been associated with cognitive deficits such as disruption in short-term recognition memory and depression in psychiatric illnesses including bipolar disorder and schizophrenia (Choy et al., 2010). Glutathione replacement appears to improve these deficits. Administration of N-acetyl-cysteine (NAC), which is a precursor of glutathione, has been associated with improved short-term spatial memory in rats and with depression in humans (Choy et al., 2010; Magalhaes et al., 2011). Moreover, antioxidant effects of mood stabilizers including lithium and valproic acid have been observed in animal models of oxidant-induced mania (Jornada et al., 2011). These findings further support the role of oxidative stress in bipolar disorders while emphasizing the protective effects of mood stabilizers with regard to their proposed antioxidant effects on the brain.

4. Cortisol and HPA axis

There is a general consensus that the hypothalamic-pituitary-adrenal (HPA) axis plays a significant role in the pathophysiology of bipolar disorders. HPA regulates stress responses via stress hormones, including cortisol, adrenocorticotropic hormone (ACTH), corticotropin-releasing hormone (CRH) and arginine vasopression (AVP). CRH is released by hypothalamic neurons into the portal circulation and stimulates the release of ACTH from the anterior pituitary. Studies have found an enhanced response to CRH in bipolar manic individuals as compared to controls (Vieta et al., 1999). CRH stimulation has been shown to result in dysregulation of ACTH and cortisol response in bipolar disorder, leading to HPA axis dysregulation and altered cortisol release (Daban et al., 2005). The severity of both manic and depressive symptoms appears to correlate with the severity of HPA dysfunction (Daban et al., 2005). Interestingly, the changes in CRH secretion appear before symptoms become evident in bipolar individuals (Daban et al., 2005).

Increased HPA axis activity has been associated with depression, mixed manic states and less commonly with true manic episodes (Garlow S, 1999; Varghese and Brown, 2001; Manji et al., 2003). Elevated daytime cortisol levels were observed in the offspring of parents with mood disorders, suggesting that cortisol excess may serve as a biomarker for susceptibility to developing bipolar disorder (Ellenbogen et al., 2011).

Since some of the morphological changes associated with chronic stress respond to mood stabilizers, chronic stress is believed to affect the development and progression of bipolar disorder (Wood et al., 2004). The complete evaluation of the extensive amount of data regarding the effects of chronic stress on brain morphology would certainly exceed the scope of the present chapter. Studies examining the human brain via MRI have noted decreased hippocampal volume in patients with post traumatic stress disorder (PTSD) and Cushing's syndrome, which are both conditions associated with excess cortisol levels (Manji et al., 2003). Of particular importance is the atrophy of CA3 hippocampal neurons which have been exposed to high levels of glucocorticoids (Sapolsky, 2000). Studies revealed that remodeling of apical dendrites on hippocampal CA3 pyramidal cells is mediated by adrenal steroids and excitatory amino acids (Wood et al., 2004). The threshold for cellular atrophy and death may be lowered by various pathological and physiological events including excitotoxicity from abnormal glutamatergic signaling enhancement and inhibition of glucose transport (Sapolsky, 2000). More importantly, these changes have been observed to be reversible and may even be prevented to some degree by pharmacological treatment with mood stabilizers (Wood et al., 2004).

5. Neurotransmitters involved in bipolar disorder

5.1 Norepinephrine

The dysfunctional behavior and mood dysregulation observed in mood disorders often involves abnormal neurotransmitter function. Monoamines are a class of neurotransmitters including dopamine, norepinephrine, epinephrine and serotonin. Neurochemical abnormalities involving monoamines have been observed in patients with bipolar disorder (Goodwin and Jamison, 1990; Fibiger, 1991; Willner et al., 1991; Schatzberg and Schildkrout, 1995; Manji H. and Potter W., 1997). Increased concentrations of norepinephrine have been found in the plasma and cerebrospinal fluid (CSF) of patients with bipolar mania (Goodwin and Jamison, 1990; Manji H. and Potter W., 1997). Increased norepinephrine turnover has also been noted in the thalamic and cortical areas in *post mortem* studies of individuals with bipolar disorder (Young et al., 1994; Vawter et al., 2000). On the other hand, norepinephrine was observed to be reduced to normal resting output in bipolar depressed patients (Manji and Lenox, 2000). *In vivo* studies have additionally revealed lower plasma levels of norepinephrine and its major metabolite, 3-methoxy-4-hydroxyphenylglycol (MHPG) in patients with bipolar depression as opposed to those with unipolar depression (Manji H. and Potter W., 1997). Urinary excretion of MHPG has also been noted to be higher in the manic state in comparison to the depressed state (Manji et al., 2003). The rate of norepinephrine turnover appears to be inversely proportional to the degree of depression, with higher turnover observed in more manic states. An explanation for this phenomenon may involve alteration in sensitivity of α_2 and β_2 adrenergic receptors in people with mood disorders (Schatzberg and Schildkrout, 1995). These observations suggest that bipolar individuals may be hypersensitive to fluctuations of norepinephrine levels, similarly to the noted hypersensitivity to acetylcholine in bipolar patients. Alterations in α_2 autoreceptor sensitivity may lead to increased activity of the α_2 autoreceptor, resulting in decreased release of norepinehprine, which subsequently affects mood (Manji and Lenox, 2000; Delgado and Moreno, 2000).

5.2 Dopamine

Dopamine is a catecholamine that has been shown to play a significant role in bipolar disorders (Cousins et al., 2009). Excess dopamine activity facilitates mania and delusional symptoms (Manji H. and Potter W., 1997; Serretti et al., 2001; Wightman and Robinson, 2002). Dopamine has also been proposed to play a role in the etiology of bipolar disorder. Manic states in healthy individuals that have received substances that increase dopaminergic activity (L-DOPA [the dopamine precursor L-3,4-dihydroxyphenylalanine], bromocryptine and amphetamine) have been observed (Murphy et al., 1971; Szczepankiewicz et al., 2011). Additional observations include the presence of manic symptoms following administration of high dose dopamine precursors to individuals with Parkinson's disease, with subsequent depressive state upon withdrawal of the dopamine precursor (Berk et al., 2007). Deficiency of dopamine has been suspected to play a role in the pathophysiology of depression. Reduced levels of homovanillic acid (HVA), a catecholamine metabolite that has been associated with dopamine, have consistently been found in the CSF of depressed and suicidal patients (Roy et al., 1992; Young et al., 1994; Manji H. and Potter W., 1997). These findings suggest that decreased neurotransmission of dopamine may correlate with depression and suicidality supporting the theory of catecholamine withdrawal in depressive states of bipolar disorder. It appears that dopamine excess correlates with manic states while dopamine depletion or withdrawal may precipitate depression. It has been suggested that the cyclical nature of bipolar disorder may result from potential downregulation of dopamine receptor sensitivity following excessive dopamine secretion (Berk et al., 2007).

Dopamine neurotransmission is primarily regulated by reuptake of dopamine from the synapse by the dopamine transporter (DAT) (Anand et al., 2011c). Dopamine binds to G protein-coupled receptors (D1, D2, D3, D4, D5), which then mediate its function in the nervous system (Beaulieu and Gainetdinov, 2011). The mechanism involving the DAT has been used as a marker for presynaptic dopamine function and has been found to be dysfunctional in individuals with bipolar disorder (Anand et al., 2011b). Significantly lower DAT availability has been found in the striatum of bipolar patients as opposed to healthy controls (Anand et al., 2011a). Animal studies revealed manic behavior traits such as increased risk-taking behavior in mice with reduced DAT function (Young et al., 2011), suggesting that reduced availability of the DAT results in subsequent increase in synaptic dopamine which therefore facilitates mania (Fig.1).

The pivotal role of dopamine in the pathomechanism of the bipolar disorders is supported by the observation that excessive dopamine levels observed in mania can be effectively treated with antipsychotic medications, whose mechanism involves dopamine blockade (Greenwood et al., 2001). Conversely bupropion (Wellbutrin), which binds to the dopamine transporter and also inhibits reuptake of norepinephrine, has been shown to be effective in treating bipolar depression (Sachs et al., 1994).

5.3 Serotonin

Abnormalities in the indolamine serotonin (5-HT) have also been observed in bipolar disorder. Studies examining hippocampal 5-HT_{1A}, 5-HT_{2A}, and 5-HT_{1B} mRNA and 5-HT_{1A} and 5-HT_{2A} receptor mRNA in the dorsolateral prefrontal cortex have revealed increased serotonin mRNA levels and decreased 5-HT_{2A} mRNA levels in the hippocampal formation

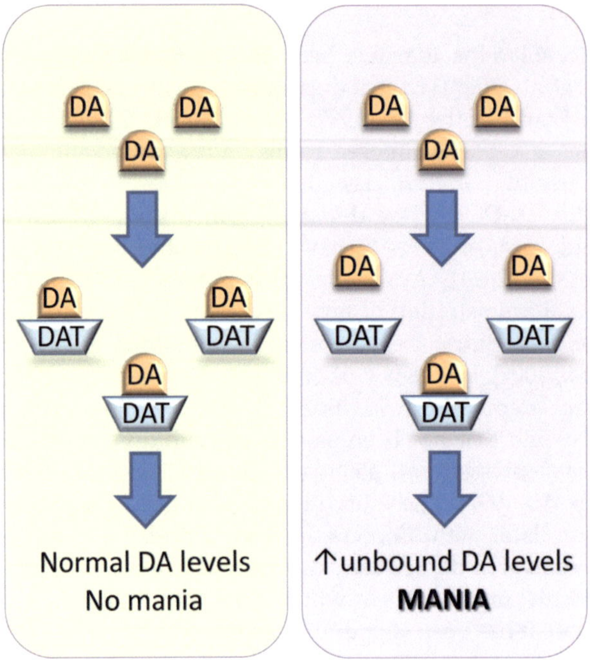

Fig. 1. The role of dopamine transporter (DAT) in the pathogenesis of mania.

of bipolar and schizophrenic patients (Lopez-Figueroa et al., 2004). Studies have found a correlation between major depressive episodes and altered serotonergic neurotransmission (Maes et al., 1995; Garlow S, 1999). Decreased radioligand binding to the serotonin transporter has been noted in the midbrain and platelets of individuals with depression (Garlow S, 1999). Reduction in central serotonergic activity has been found in bipolar individuals who were in the depressed phase, with similar findings in euthymic bipolar patients (Mahmood and Silverstone, 2001; Chou et al., 2010). Further studies reported decreased 5-HT_{1A} receptor binding in the hippocampus, amygdala and raphe in bipolar depressed patients, depressed relatives of bipolar patients as well as other individuals suffering from depression (Drevets et al., 1999). However, increased levels of 5-HT_{1A} were noted in depressed males with bipolar disorder (Sullivan et al., 2009). These findings indicate an increase in autoreceptor binding which may lead to decreased release of serotonin with a subsequent upregulation of postsynaptic 5-HT_{1A} receptors (Sullivan et al., 2009). Increased cortisol secretion has been proposed as one explanation for the decrease in 5-HT_{1A} receptor binding in depressive states, particularly since stimulation of corticosteroid receptors has been associated with inhibition of postsynaptic 5-HT_{1A} receptor mRNA expression in some regions of the brain (Manji et al., 2003). These data support the observation that the onset of bipolar symptoms and recurrence of manic and depressive episodes are more likely to occur during times of stress.

Models involving tryptophan depletion have further explored the role of serotonin in bipolar disorder. Tryptophan depletion lowers the levels of serotonin (Manji et al., 2003) and associated with recurrence of depression secondary to a reversed response to antidepressant medications in patients with a mood disorder but no onset or worsening of depression in nonmedicated depressed patients, healthy individuals without mental illness or in lithium-treated euthymic patients with bipolar disorder (Delgado et al., 1999; Hughes et al., 2000). Research has examined unaffected relatives of patients with bipolar disorder in order to investigate if sensitivity to cognitive effects of serotonin decrease might serve as a heritable marker or endophenotype for bipolar disorder (Manji et al., 2003). One double-blind crossover design study examined 20 unaffected relatives of bipolar patients and 19 control subjects undergoing acute tryptophan depletion (ATD). The study revealed that unaffected relatives of bipolar patients exhibited increased impulsivity and depressed mood following ATD as opposed to placebo (Sobczak et al., 2002). These symptoms were not observed in healthy controls. Additionally, unaffected relatives were found to have decreased number of lower affinity binding sites for imipramine and lower platelet concentrations of serotonin independent of ATD administration. Further studies revealed impairment in planning and memory in first degree healthy relatives of bipolar type I patients independent of ATD as well as impaired speed of information processing in first degree healthy relatives but not in healthy controls following ATD (Sobczak et al., 2002). These data suggest that impairments in serotonin metabolism observed in bipolar patients may be inherited.

A reduction in the serotonin metabolite 5-hydroxyindoleacetic acid (5-HIAA) was observed in patients with bipolar disorder as well as other mood disorders (Young et al 1994). This reduction was especially noted in individuals with aggression, impulsivity and suicide attempts (Manji et al., 2003). However, data regarding the 5-HIAA levels in manic versus depressed individuals is rather controversial. Some studies did not find a difference in CSF 5-HIAA levels between manic and depressed patients (Goodwin and Jamison, 1990), while others noted decreased 5-HIAA in the CSF in both manic and depressed individuals compared to controls as well as a significant decrease in CSF 5-HIAA accumulation in manic individuals as compared to their depressed counterparts and to controls (Goodwin and Jamison, 1990). Despite the inconsistent results, decreased CSF levels of the serotonin metabolite 5-HIAA appear to correlate with impulsive, aggressive behavior and mania, emphasizing the role of serotonin in the pathophysiology of bipolar disorder.

5.4 Glutamate

Abnormalities in glutamate and glutamate receptor functioning have been noted in patients with bipolar disorder. Glutamate is a major excitatory neurotransmitter affecting cognition, learning and memory (Shigeri et al., 2004). While glutamate mediates information affecting cellular survival, formation and destruction of synapses and brain development, it may be neurotoxic if not present in the right concentrations at appropriate locations and time in the nervous system (Danbolt, 2001). Stress-induced hippocampal changes in the brains of patients with bipolar disorder have been noted to include alterations in glutamate and its receptors. Patients with bipolar disorder have been found to have elevated lactate and glutamate levels (Dager et al., 2004). Elevated glutamate levels have been noted particularly in the dorsolateral prefrontal cortex of patients with acute mania (Michael et al., 2003). Elevated glutamate neurotransmission has been reported in bipolar patients and has been suggested as the reason for elevated glutamate levels in the cortex (Eastwood and Harrison, 2010). More specifically, increased expression of the vesicular glutamate transporter (VGluT1), which is expressed in glutaminergic neurons, has been found in the anterior cingulate cortex of bipolar patients (Eastwood and Harrison, 2010). VGluT1, netrin-G1 and its isoforms Gqc, G1d and G1f, as well as netrin-G2 are axon guidance and cellular adhesion molecules (Eastwood and Harrison, 2010). They participate in formation and maintenance of glutamatergic synaptic connections (Eastwood and Harrison, 2010). Netrin-G expression was also found to be elevated in the right anterior cingulate cortex of bipolar patients (Eastwood and Harrison, 2010). Netrin-G molecules have been known to affect the formation and plasticity of excitatory pathways (Eastwood and Harrison, 2010), suggesting that their increased expression may lead to increase in glutamate neurotransmission and possible neurotoxicity in the brains of bipolar patients.

5.5 GABA

Abnormal levels of gamma amino butyric acid (GABA) have been identified in bipolar disorder. GABA is an inhibitory neurotransmitter which modulates the activity of other neurotransmitters in the central nervous system (Vuoristo J, 2011). It is synthesized from glutamate, with the enzyme glutamate decarboxylase playing a key role in its synthesis (Bielau et al., 2007) (Fig.2). GABA plays a role in modulating hippocampal and cortical circuits and is involved in discriminative information processing, integrating sensory information and generating oscillatory rhythms in the corticolimbic system (Sakai et al., 2008). GABA participates in the inhibition of aggressive and impulsive behaviors, which are often prevalent in bipolar disorder, particularly in mixed and manic states (Sakai et al., 2008). GABA deficits as well as decreased inhibitory activity have been reported in the cortex of individuals with bipolar disorder (Duffy et al., 2000; Levinson et al., 2007). Aberrant modulation of GABA interneurons by glutamate has been noted to occur via the NR2A subunit of the N-methyl-D-aspartate (NMDA) receptor in the cerebral cortex (Woo et al., 2004). Decreased NR2A mRNA has been observed in bipolar patients, particularly in layer 2 of the anterior cingulate cortex (Woo et al., 2004). Decreased plasma levels of GABA have also been implicated in bipolar mania as well as in bipolar depression (Petty, 1995). Animal studies have revealed decreased GABA function in states of depression with subsequent improvement in mood following administration of GABA agonists (Petty, 1995). Interestingly, it has been reported that GABA levels do not normalize following treatment or remission of depression (Duffy et al., 2000). *Post mortem* studies examining GABA in humans noted significantly decreased protein in $GABA_b$ receptor subunits (BABBR1 and GABBR2) in the cerebellum of individuals with schizophrenia, bipolar disorder, autism and depression as compared to healthy controls (Fatemi et al., 2011). It appears that these deficits may be treatable as various classes of antidepressants have been reported to cause an up-regulation of $GABA_b$ receptors (Duffy et al., 2000). Additionally, mood stabilizing medications such as valproate, carbamazepine, gabapentin, and lithium carbonate have been noted to alter GABA levels, possibly decreasing aggressive and impulsive behavior by increasing availability and efficacy of GABA (Sakai et al., 2008).

5.6 Substance P

Substance P is a neuropeptide that is widely distributed throughout the central and peripheral nervous system (Pioro et al., 1990; Dudas and Merchenthaler, 2002; Lieb et al., 2003) and it has been shown to play a role in the pathomechanism of bipolar disorder (Lieb et al., 2003). Substance P regulates the release of acetylcholine in the cortex and co-localizes with dopamine in the midbrain and striatum, with GABA in the cortex, and with serotonin

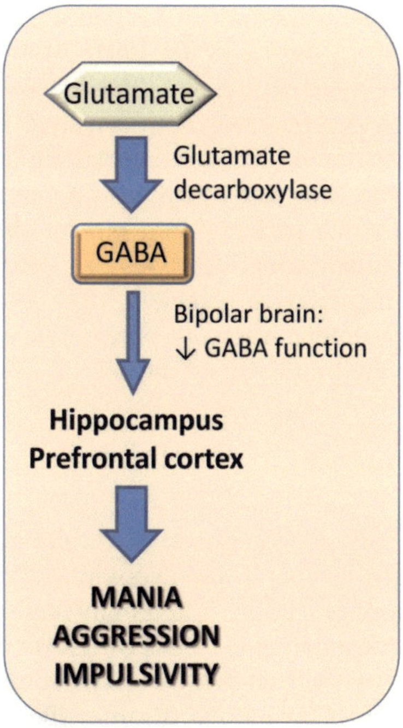

Fig. 2. Role of GABA in the pathogenesis of mania.

in the raphe nuclei (Jakab et al., 1997). Substance P, along with its receptor, the neurokinin 1 receptor (NK1R), is distributed in numerous areas of the brain including hippocampus, amygdala, frontal cortex and hypothalamus, and it appears to be involved in regulating responses to stress and fear (Herpfer and Lieb, 2005). Alterations of substance P content in these regions have been noted to influence the stress response and affective states (Lieb et al., 2003). Administration of substance P produces fear-related behaviors while blockade of NK1/2R or NK1/2/3R is associated with anxiolytic effects (Herpfer and Lieb, 2005). Moreover, an antidepressant effect has been noted when using substance P receptor antagonists, such as L-759274 and MK-869 (Mendlewicz et al., 2005). Substance P facilitates synthesis of interleukin-6 (IL-6), a cytokine, whose levels have been reported to be elevated in acute depressive states of bipolar disorder (Lieb et al., 2003). Inhibition of substance P as well as NK1R down-regulation has been noted following valproic acid administration (Lieb et al., 2003).

5.7 Acetylcholine

Acetycholine, along with cortisol, monoamines and amino acids has been suggested to play a role in the pathophysiology of bipolar disorder (Chen et al., 2010). Several studies reported catecholaminergic and cholinergic imbalance in bipolar patients. Administration of cholinomimetic agents and cholinesterase inhibitors reduced manic symptoms in bipolar manic patients (Davis et al., 1978; Fritze and Beckmann, 1988). Indeed, administration of the cholinergic agent physostigmine has been shown to result in decrease of manic symptoms in manic individuals as well as increase in depressive symptoms in non-manic individuals (Janowsky et al., 1974; Fritze and Beckmann, 1988). Animal models have noted exaggerated neuroendocrine and behavioral responses in response to administration of cholinergic agents (Overstreet et al., 1988; Janowsky et al., 1994). Moreover, rats bred for cholinergic sensitivity were noted to be more aggressive (Pucilowski et al., 1990). It appears that individuals with bipolar disorder may be more sensitive to the effects of acetylcholine. Studies involving humans have noted increased sensitivity to the effects of physostigmine administration in patients with affective disorders (Oppenheim et al., 1979; Risch et al., 1981). More specifically, euthymic patients with bipolar disorder were noted to become depressed after physostigmine administrations as compared to healthy controls (Oppenheim et al., 1979). Interestingly, previous data also reported increased serum epinephrine levels following administration of a cholinergic agent, suggesting that increased cholinergic tone may precipitate a subsequent increase in adrenergic tone (Janowsky et al., 1986).

6. The role of genes

According to the kindling theory, the likelihood of developing bipolar disorder increases in response to a series of stressful events in predisposed individuals (Saddock and Saddock, 2003). Studies involving adoption, twins and family reveal a genetic predisposition to developing bipolar disorder (Gershon et al., 1990; Muglia et al., 2002). A multitude of genes interacting with each other as well as with the environment is most likely lead to a multifactorial mode of inheritance for bipolar disorder (Berrettini W.H, 1999; Muglia et al., 2002). A significant association was reported between two single nucleotide polymorphisms (rs4952 and rs4953) in the human beta 3 nicotinic acetylcholoine receptor subunit gene (CHRNB3) on chromosome 8 and bipolar disorder (Hartz et al., 2011). Additionally, genetic linkage studies suggest a role for human alpha7 nicotinic acetylcholine receptor subunit gene (CHRNA7) on chromosome 15q13-q14 as one of the candidate genes for bipolar disorder (Hong et al., 2004). Interestingly, linkage to CHRNA7 has also been reported in schizophrenia and schizoaffective disorder, suggesting a similar genetic cause for these two disorders (Leonard et al., 2001; Martin et al., 2007). Other studies examining chromosome 15 have found a genetic association between bipolar disorder and GABA-A receptor ra5 subunit gene locus (GABRA5) on chromosome 15q11-q13 (Otani et al., 2005). Low GABA function has been proposed to be a heritable marker for susceptibility to developing a mood disorder and GABRA5 may be associated with bipolar disorder (Petty, 1995; Papadimitriou et al., 2001).

Variants in the interleukin-1 (IL-1) cluster on chromosome 2q13 has also been associated with increased risk for both schizophrenia and bipolar disorder (Papiol et al., 2008). Specific genetic polymorphisms of the IL-1B gene from the IL-1 gene cluster have been associated with grey matter deficits in the brain of bipolar disorder patients (Papiol et al., 2008). Altered or increased expression of proinflammatory genes has been reported in monocytes of individuals with bipolar disorder as well as their offspring (Padmos et al., 2008). This pattern of gene expression is a possible etiological cause for the increased oxidative state reported in bipolar disorder.

The short arm of chromosome 11 has been investigated for possible genetic factors leading to developing bipolar disorder. This particular region incorporates the tyrosine hydroxylase (TH) and the dopamine D4 receptor gene (DRD4) (Todd et al., 1996; Oruc et al., 1997; Furlong et al., 1999; Li et al., 1999; Serretti et al., 2001). Tyrosine hydroxylase plays a significant role in the formation of dopamine since it is the key and rate-limiting enzyme of the catecholamine synthesis converting tyrosine to DOPA, a precursor of dopamine (Muglia et al., 2002). DRD4 has been noted to be highly polymorphic, consisting of 2-10 tandem repeats in the third exon, encoding the receptor's longest loop (Asghari et al., 1995; Muglia et al., 2002). The gene for tyrosine hydroxylase contains a microsatellite known as HUMTH01, which has been proposed to play a regulatory role in its expression (Meloni et al., 1998). Studies investigating the short arm of chromosome 11 have been inconsistent, with many of them reporting lack of evidence for association with DRD4 and development of bipolar disorder (Byerley et al., 1992; Sidenberg et al., 1994; Smyth et al., 1996; Malafosse et al., 1997). Likewise, a positive and negative association has been observed between HUMTH01 and bipolar disorder (Serretti et al., 2000; Muglia et al., 2002). Studies examining a sample of nuclear families of bipolar patients and their biological parents reported a parent of origin effect (POE) for the DRD4 alleles 2- and 4- repeat alleles, which was inherited from the mother. This effect was noted to be located on chromosome 11p15.5, which is close to a group of imprinted genes (Muglia et al., 2002), indicating that genomic imprinting may be associated with bipolar disorder.

The dopamine transporter (DAT) gene plays a vital role in transmission of dopamine by mediating dopamine uptake into the synaptic terminal (Greenwood et al., 2001). Increased dopamine concentrations have been observed after administration of cocaine and amphetamine, which have been proposed to act at the DAT to inhibit reuptake of dopamine (Giros et al., 1992). Studies examining DAT for linkage to bipolar disorder in families from the Old Order Amish, Icelandic pedigree and the general North American population revealed a locus on chromosome 5p15.3 near the DAT gene, suggestive of autosomal dominant inheritance of bipolar disorder (Kelsoe et al., 1996; Homer et al., 1997). More specifically, data suggest that a functional variant of a 3′ noncoding sequence element may influence susceptibility for bipolar disorder (Greenwood et al., 2001). The significance of DAT in bipolar disorder is based on its involvement in regulating synaptic reuptake of dopamine (Kelsoe et al., 1996). Chromosomal aberration near the DAT gene locus in bipolar patients correlates with abnormalities in dopamine neurotransmission observed in bipolar disorder indicating a genetic predisposition to developing the disorder.

Dysregulation of the serotonin transporter and abnormalities in serotonin metabolism have also been noted in studies of patients with affective disorders. Serotonin transporter is involved in serotonin regulation and is the site of action for numerous antidepressants (Kelsoe et al., 1996). It is noteworthy that the serotonin transporter polymorphism (5-HTTLPR) gene as well as brain-derived neurotrophic factor (BDNF) polymorphism have been linked with suicidality in bipolar patients (Vincze et al., 2008). The 5-HTTLPR variant is the promoter region of the serotonin transporter gene, which has two alleles: the long 'l' allele and the short 's' allele (Neves et al., 2008). The short 's' allele of the 5-HTTLPR has been associated with significantly lower activation of the ventral anterior cingulate cortex, a dysfunction observed in bipolar patients (Shah et al., 2009). In addition to its link to bipolar disorder, 's' allele is particularly associated with suicidal behavior (Neves et al., 2008; Ferreira et al., 2009). Interestingly, mania and suicidality have been repeatedly observed in bipolar individuals following treatment with antidepressants (Ferreira et al., 2009). Since depression is often treated by selective serotonin reuptake inhibitors (SSRIs), understanding this mechanism can encourage more careful assessment and diagnosis of bipolar patients.

7. Concluding remarks

Bipolar disorder is a progressive illness which, if left untreated, results in cognitive decline. The rate and nature of the progression of the disease varies among individuals. The pathology of the bipolar disorder includes a variety of abnormalities in the neuroanatomy, endocrine homeostasis, neurotransmitter functions and genes of the patients.

Reduced hippocampal volume, dysfunction in the prefrontal cortex and limbic system observed in patients suffering in bipolar disorder emphasizes the importance of these brain regions in the development of the disease. The nature and extent of these neuroanatomical changes vary with the presentation, severity, and duration of the disease. While some anatomical abnormalities may be present prior to onset of disease, these lesions may become more severe with the progression of the disease. Conversely, certain abnormalities may be amendable to treatment. Additional factors involved in the development and progression of bipolar disorder include abnormal HPA axis function and oxidative stress. Abnormalities in neurotransmitter function, including dopamine, norepinephrine, serotonin, GABA, acetylcholine, glutamate and substance P have also been observed in bipolar disorder, offering the possibility of various avenues of pharmacologic interventions. Genetic variations found in bipolar patients indicate a predisposition in the development of the disease. The diversity of the factors involved in the pathomechanism of bipolar disorder may explain the wide variety of clinical manifestations and consequently may offer a broad range of options for successful treatment of this debilitating disease.

8. Acknowledgement

The authors are indebted to Dr. Michael Bradbury and Dr. Jack Caldwell for the critical evaluation of the manuscript.

9. References

Akiskal HS, Benazzi F (2005) Psychopathologic correlates of suicidal ideation in major depressive outpatients: is it all due to unrecognized (bipolar) depressive mixed states? Psychopathology 38:273-280

Allman JM, Hakeem A, Erwin JM, Nimchinsky E, Hof P (2001) The anterior cingulate cortex. The evolution of an interface between emotion and cognition. Ann N Y Acad Sci 935:107-117

American Psychiatric Association (2000) Diagnostic and Statistical Manual of Mental Disorders IV TR.

Anand A, Barkay G, Dzemidzic M, Albrecht D, Karne H, Zheng QH, Hutchins GD, Normandin MD, Yoder KK (2011c) Striatal dopamine transporter availability in unmedicated bipolar disorder. Bipolar Disord 13:406-413

Anand A, Barkay G, Dzemidzic M, Albrecht D, Karne H, Zheng QH, Hutchins GD, Normandin MD, Yoder KK (2011b) Striatal dopamine transporter availability in unmedicated bipolar disorder. Bipolar Disord 13:406-413

Anand A, Barkay G, Dzemidzic M, Albrecht D, Karne H, Zheng QH, Hutchins GD, Normandin MD, Yoder KK (2011a) Striatal dopamine transporter availability in unmedicated bipolar disorder. Bipolar Disord 13:406-413

Andreazza AC, Shao L, Wang JF, Young LT (2010) Mitochondrial complex I activity and oxidative damage to mitochondrial proteins in the prefrontal cortex of patients with bipolar disorder. Arch Gen Psychiatry 67:360-368

Asghari V, Sanyal S, Peterson A, Jovanovic V, Van Tol HH. Modulation of intracellular cyclic AMP levels by different human dopamine D4 receptor variants. Journal of Neurochemistry 65, 1157-1165. 1995.
Ref Type: Abstract

Beaulieu JM, Gainetdinov RR (2011) The physiology, signaling, and pharmacology of dopamine receptors. Pharmacol Rev 63:182-217

Berk M, Conus P, Kapczinski F, Andreazza AC, Yucel M, Wood SJ, Pantelis C, Malhi GS, Dodd S, Bechdolf A, Amminger GP, Hickie IB, McGorry PD (2010) From neuroprogression to neuroprotection: implications for clinical care. Med J Aust 193:S36-S40

Berk M, Dodd S, Kauer-Sant'anna M, Malhi GS, Bourin M, Kapczinski F, Norman T (2007) Dopamine dysregulation syndrome: implications for a dopamine hypothesis of bipolar disorder. Acta Psychiatr Scand Suppl:41-49

Berrettini W.H. On the future of genetic research in bipolar and schizophrenic syndromes. Neuropsychopharmacology 21, 1-2. 1999.
Ref Type: Abstract

Bielau H, Steiner J, Mawrin C, Trubner K, Brisch R, Meyer-Lotz G, Brodhun M, Dobrowolny H, Baumann B, Gos T, Bernstein HG, Bogerts B (2007) Dysregulation of GABAergic neurotransmission in mood disorders: a postmortem study. Ann N Y Acad Sci 1096:157-169

Brown CD, Wang JF, Young LT. Attenuation of N-methyl-C-aspartate-mediated cytoplasmic vacuolization in primary rat hippocampal neurons by mood stabilizers. Neuroscience 117[4], 949-955. 2003.
Ref Type: Abstract

Byerley W, Plaetke R, Hoff M, Jensen S, Holik J, Reimherr F, Mellon C, Wender P, O'Connell P, Leppert M (1992) Tyrosine hydroxylase gene not linked to manic-depression in seven of eight pedigrees. Hum Hered 42:259-263

Cerullo MA, Adler CM, DelBello MP, Strakowski SM (2009) The functional neuroanatomy of bipolar disorder. Int Rev Psychiatry 21:314-322

Chen J, Fang Y, Kemp DE, Calabrese JR, Gao K (2010) Switching to hypomania and mania: differential neurochemical, neuropsychological, and pharmacologic triggers and their mechanisms. Curr Psychiatry Rep 12:512-521

Chou YH, Wang SJ, Lin CL, Mao WC, Lee SM, Liao MH (2010) Decreased brain serotonin transporter binding in the euthymic state of bipolar I but not bipolar II disorder: a SPECT study. Bipolar Disord 12:312-318

Choy KH, Dean O, Berk M, Bush AI, van den BM (2010) Effects of N-acetyl-cysteine treatment on glutathione depletion and a short-term spatial memory deficit in 2-cyclohexene-1-one-treated rats. Eur J Pharmacol 649:224-228

Cousins DA, Butts K, Young AH (2009) The role of dopamine in bipolar disorder. Bipolar Disord 11:787-806

Daban C, Vieta E, Mackin P, Young AH (2005) Hypothalamic-pituitary-adrenal axis and bipolar disorder. Psychiatr Clin North Am 28:469-480

Dager SR, Friedman SD, Parow A, Demopulos C, Stoll AL, Lyoo IK, Dunner DL, Renshaw PF (2004) Brain metabolic alterations in medication-free patients with bipolar disorder. Arch Gen Psychiatry 61:450-458

Danbolt NC (2001) Glutamate uptake. Prog Neurobiol 65:1-105

Davis KL, Berger PA, Hollister LE, Defraites E (1978) Physostigmine in mania. Arch Gen Psychiatry 35:119-122

Dean OM, van den BM, Berk M, Copolov DL, Mavros C, Bush AI (2011) N-acetyl cysteine restores brain glutathione loss in combined 2-cyclohexene-1-one and d-amphetamine-treated rats: Relevance to schizophrenia and bipolar disorder. Neurosci Lett 499:149-153

Delgado PL, Miller HL, Salomon RM, Licinio J, Krystal JH, Moreno FA, Heninger GR, Charney DS (1999) Tryptophan-depletion challenge in depressed patients treated with desipramine or fluoxetine: implications for the role of serotonin in the mechanism of antidepressant action. Biol Psychiatry 46:212-220

Delgado PL, Moreno FA (2000) Role of norepinephrine in depression. J Clin Psychiatry 61 Suppl 1:5-12

Doty TJ, Payne ME, Steffens DC, Beyer JL, Krishnan KR, LaBar KS (2008) Age-dependent reduction of amygdala volume in bipolar disorder. Psychiatry Res 163:84-94

Drevets WC, Frank E, Price JC, Kupfer DJ, Holt D, Greer PJ, Huang Y, Gautier C, Mathis C (1999) PET imaging of serotonin 1A receptor binding in depression. Biol Psychiatry 46:1375-1387

Drevets WC, Price JL, Furey ML (2008) Brain structural and functional abnormalities in mood disorders: implications for neurocircuitry models of depression. Brain Struct Funct 213:93-118

Drevets WC, Price JL, Simpson JR, Jr., Todd RD, Reich T, Vannier M, Raichle ME (1997) Subgenual prefrontal cortex abnormalities in mood disorders. Nature 386:824-827

Dudas B, Merchenthaler I (2002) Close juxtapositions between LHRH immunoreactive neurons and substance P immunoreactive axons in the human diencephalon. J Clin Endocrinol Metab 87:2946-2953

Duffy A, Turecki G, Grof P, Cavazzoni P, Grof E, Joober R, Ahrens B, Berghofer A, Muller-Oerlinghausen B, Dvorakova M, Libigerova E, Vojtechovsky M, Zvolsky P, Nilsson A, Licht RW, Rasmussen NA, Schou M, Vestergaard P, Holzinger A, Schumann C, Thau K, Robertson C, Rouleau GA, Alda M (2000) Association and linkage studies of candidate genes involved in GABAergic neurotransmission in lithium-responsive bipolar disorder. J Psychiatry Neurosci 25:353-358

Eastwood SL, Harrison PJ (2010) Markers of glutamate synaptic transmission and plasticity are increased in the anterior cingulate cortex in bipolar disorder. Biol Psychiatry 67:1010-1016

Ellenbogen MA, Hodgins S, Linnen AM, Ostiguy CS (2011) Elevated daytime cortisol levels: a biomarker of subsequent major affective disorder? J Affect Disord 132:265-269

Fatemi SH, Folsom TD, Thuras PD (2011) Deficits in GABA(B) receptor system in schizophrenia and mood disorders: a postmortem study. Schizophr Res 128:37-43

Ferreira AA, Neves FS, da Rocha FF, Silva GS, Romano-Silva MA, Miranda DM, De ML, Correa H (2009) The role of 5-HTTLPR polymorphism in antidepressant-associated mania in bipolar disorder. J Affect Disord 112:267-272

Fibiger HC (1991) Dopaminergic-cholinergic interactions in the striatum. Jpn J Psychiatry Neurol 45:512

Floresco SB, Braaksma DN, Phillips AG (1999) Thalamic-cortical-striatal circuitry subserves working memory during delayed responding on a radial arm maze. J Neurosci 19:11061-11071

Foland LC, Altshuler LL, Sugar CA, Lee AD, Leow AD, Townsend J, Narr KL, Asuncion DM, Toga AW, Thompson PM (2008) Increased volume of the amygdala and hippocampus in bipolar patients treated with lithium. Neuroreport 19:221-224

Fritze J, Beckmann H (1988) The cholinergic agonist RS 86: a pharmacopsychological study. Neuropsychobiology 19:35-39

Fullerton JM, Tiwari Y, Agahi G, Heath A, Berk M, Mitchell PB, Schofield PR (2010) Assessing oxidative pathway genes as risk factors for bipolar disorder. Bipolar Disord 12:550-556

Furlong RA, Rubinsztein JS, Ho L, Walsh C, Coleman TA, Muir WJ, Paykel ES, Blackwood DH, Rubinsztein DC (1999) Analysis and metaanalysis of two polymorphisms within the tyrosine hydroxylase gene in bipolar and unipolar affective disorders. Am J Med Genet 88:88-94

Garlow S MDaNC (1999) The neurochemistry of mood disorders. In: Neurobiology of Mental Illness (Charney DS NEBB, ed), pp 348-364 New York: Oxford University Press.

Gawryluk JW, Wang JF, Andreazza AC, Shao L, Young LT (2011) Decreased levels of glutathione, the major brain antioxidant, in post-mortem prefrontal cortex from patients with psychiatric disorders. Int J Neuropsychopharmacol 14:123-130

Gershon ES, Martinez M, Goldin LR, Gejman PV. Genetic mapping of common diseases: the challenges of manic-depressive illness and schizophrenia. Trends in Genetics 6, 282-287. 1990.
Ref Type: Abstract

Giros B, el MS, Godinot N, Zheng K, Han H, Yang-Feng T, Caron MG (1992) Cloning, pharmacological characterization, and chromosome assignment of the human dopamine transporter. Mol Pharmacol 42:383-390

Goldberg JF, Garno JL, Leon AC, Kocsis JH, Portera L (1998) Association of recurrent suicidal ideation with nonremission from acute mixed mania. Am J Psychiatry 155:1753-1755

Goodwin FK, Jamison KR (1990) Manic Depressive Illness. New York: Oxford University Press.

Gorrindo T, Blair RJ, Budhani S, Dickstein DP, Pine DS, Leibenluft E (2005) Deficits on a probabilistic response-reversal task in patients with pediatric bipolar disorder. Am J Psychiatry 162:1975-1977

Green MJ, Cahill CM, Malhi GS (2007) The cognitive and neurophysiological basis of emotion dysregulation in bipolar disorder. J Affect Disord 103:29-42

Greenwood TA, Alexander M, Keck PE, McElroy S, Sadovnick AD, Remick RA, Kelsoe JR (2001) Evidence for linkage disequilibrium between the dopamine transporter and bipolar disorder. Am J Med Genet 105:145-151

Hajek T, Carrey N, Alda M (2005) Neuroanatomical abnormalities as risk factors for bipolar disorder. Bipolar Disord 7:393-403

Hartz SM, Lin P, Edenberg HJ, Xuei X, Rochberg N, Saccone S, Berrettini W, Nelson E, Nurnberger J, Bierut LJ, Rice JP (2011) Genetic association of bipolar disorder with the beta(3) nicotinic receptor subunit gene. Psychiatr Genet 21:77-84

Hassel S, Almeida JR, Kerr N, Nau S, Ladouceur CD, Fissell K, Kupfer DJ, Phillips ML (2008) Elevated striatal and decreased dorsolateral prefrontal cortical activity in response to emotional stimuli in euthymic bipolar disorder: no associations with psychotropic medication load. Bipolar Disord 10:916-927

Herpfer I, Lieb K (2005) Substance P receptor antagonists in psychiatry: rationale for development and therapeutic potential. CNS Drugs 19:275-293

Homer JP, Flodman PL, Spence MA (1997) Bipolar disorder: dominant or recessive on chromosome 5? Genet Epidemiol 14:647-651

Hong CJ, Lai IC, Liou LL, Tsai SJ (2004) Association study of the human partially duplicated alpha7 nicotinic acetylcholine receptor genetic variant with bipolar disorder. Neurosci Lett 355:69-72

Hughes JH, Dunne F, Young AH (2000) Effects of acute tryptophan depletion on mood and suicidal ideation in bipolar patients symptomatically stable on lithium. Br J Psychiatry 177:447-451

Jakab RL, Goldman-Rakic P, Leranth C (1997) Dual role of substance P/GABA axons in cortical neurotransmission: synaptic triads on pyramidal cell spines and basket-like innervation of layer II-III calbindin interneurons in primate prefrontal cortex. Cereb Cortex 7:359-373

Janowsky DS, el-Yousef MK, Davis JM (1974) Acetylcholine and depression. Psychosom Med 36:248-257

Janowsky DS, Overstreet DH, Nurnberger JI, Jr. (1994) Is cholinergic sensitivity a genetic marker for the affective disorders? Am J Med Genet 54:335-344

Janowsky DS, Risch SC, Ziegler MG, Gillin JC, Huey L, Rausch J (1986) Physostigmine-induced epinephrine release in patients with affective disorder. Am J Psychiatry 143:919-921

Jornada LK, Valvassori SS, Steckert AV, Moretti M, Mina F, Ferreira CL, Arent CO, Dal-Pizzol F, Quevedo J (2011) Lithium and valproate modulate antioxidant enzymes and prevent ouabain-induced oxidative damage in an animal model of mania. J Psychiatr Res 45:162-168

Kelsoe JR, Sadovnick AD, Kristbjarnarson H, Bergesch P, Mroczkowski-Parker Z, Drennan M, Rapaport MH, Flodman P, Spence MA, Remick RA (1996) Possible locus for bipolar disorder near the dopamine transporter on chromosome 5. Am J Med Genet 67:533-540

Ketter TA (2010) Diagnostic features, prevalence, and impact of bipolar disorder. J Clin Psychiatry 71:e14

Kirov G, Murphy KC, Arranz MJ, Jones I, McCandles F, Kunugi H, Murray RM, McGuffin P, Collier DA, Owen MJ, Craddock N (1998) Low activity allele of catechol-O-methyltransferase gene associated with rapid cycling bipolar disorder. Mol Psychiatry 3:342-345

Koukopoulos A, Koukopoulos A (1999) Agitated depression as a mixed state and the problem of melancholia. Psychiatr Clin North Am 22:547-564

Leonard S, Adler LE, Benhammou K, Berger R, Breese CR, Drebing C, Gault J, Lee MJ, Logel J, Olincy A, Ross RG, Stevens K, Sullivan B, Vianzon R, Virnich DE, Waldo M, Walton K, Freedman R (2001) Smoking and mental illness. Pharmacol Biochem Behav 70:561-570

Levinson AJ, Young LT, Fitzgerald PB, Daskalakis ZJ (2007) Cortical inhibitory dysfunction in bipolar disorder: a study using transcranial magnetic stimulation. J Clin Psychopharmacol 27:493-497

Li T, Liu X, Sham PC, Aitchison KJ, Cai G, Arranz MJ, Deng H, Liu J, Kirov G, Murray RM, Collier DA (1999) Association analysis between dopamine receptor genes and bipolar affective disorder. Psychiatry Res 86:193-201

Lieb K, Treffurth Y, Hamke M, Akundi RS, von KM, Fiebich BL (2003) Valproic acid inhibits substance P-induced activation of protein kinase C epsilon and expression of the substance P receptor. J Neurochem 86:69-76

Lopez-Figueroa AL, Norton CS, Lopez-Figueroa MO, rmellini-Dodel D, Burke S, Akil H, Lopez JF, Watson SJ (2004) Serotonin 5-HT1A, 5-HT1B, and 5-HT2A receptor mRNA expression in subjects with major depression, bipolar disorder, and schizophrenia. Biol Psychiatry 55:225-233

Maes M, Meltzer HY, D'Hondt P, Cosyns P, Blockx P (1995) Effects of serotonin precursors on the negative feedback effects of glucocorticoids on hypothalamic-pituitary-adrenal axis function in depression. Psychoneuroendocrinology 20:149-167

Magalhaes PV, Dean OM, Bush AI, Copolov DL, Malhi GS, Kohlmann K, Jeavons S, Schapkaitz I, nderson-Hunt M, Berk M (2011) N-acetyl cysteine add-on treatment for bipolar II disorder: a subgroup analysis of a randomized placebo-controlled trial. J Affect Disord 129:317-320

Mahmood T, Silverstone T (2001) Serotonin and bipolar disorder. J Affect Disord 66:1-11

Malafosse A, Leboyer M, d'Amato T, Amadeo S, Abbar M, Campion D, Canseil O, Castelnau D, Gheysen F, Granger B, Henrikson B, Poirier MF, Sabate O, Samolyk D, Feingold J, Mallet J (1997) Manic depressive illness and tyrosine hydroxylase gene: linkage heterogeneity and association. Neurobiol Dis 4:337-349

Manji H., Potter W. (1997) Monoaminergic mechanisms in bipolar disorder. In: Bipolar disorder: biological models and their clinical application (Young LT JR, ed), pp 1-40 New York: Decker.

Manji HK, Lenox RH (2000) Signaling: cellular insights into the pathophysiology of bipolar disorder. Biol Psychiatry 48:518-530

Manji HK, Quiroz JA, Payne JL, Singh J, Lopes BP, Viegas JS, Zarate CA (2003) The underlying neurobiology of bipolar disorder. World Psychiatry 2:136-146

Martin LF, Leonard S, Hall MH, Tregellas JR, Freedman R, Olincy A (2007) Sensory gating and alpha-7 nicotinic receptor gene allelic variants in schizoaffective disorder, bipolar type. Am J Med Genet B Neuropsychiatr Genet 144B:611-614

Mayberg HS (1997) Limbic-cortical dysregulation: a proposed model of depression. J Neuropsychiatry Clin Neurosci 9:471-481

Meloni R, Albanese V, Ravassard P, Treilhou F, Mallet J (1998) A tetranucleotide polymorphic microsatellite, located in the first intron of the tyrosine hydroxylase gene, acts as a transcription regulatory element in vitro. Hum Mol Genet 7:423-428

Mendlewicz J, Oswald P, Claes S, Massat I, Souery D, Van BC, Del-Favero J (2005) Patient-control association study of substance P-related genes in unipolar and bipolar affective disorders. Int J Neuropsychopharmacol 8:505-513

Michael N, Erfurth A, Ohrmann P, Gossling M, Arolt V, Heindel W, Pfleiderer B (2003) Acute mania is accompanied by elevated glutamate/glutamine levels within the left dorsolateral prefrontal cortex. Psychopharmacology (Berl) 168:344-346

Minassian A, Paulus MP, Perry W (2004) Increased sensitivity to error during decision-making in bipolar disorder patients with acute mania. J Affect Disord 82:203-208

Muglia P, Petronis A, Mundo E, Lander S, Cate T, Kennedy JL (2002) Dopamine D4 receptor and tyrosine hydroxylase genes in bipolar disorder: evidence for a role of DRD4. Mol Psychiatry 7:860-866

Murphy DL, Brodie HK, Goodwin FK, Bunney WE, Jr. (1971) Regular induction of hypomania by L-dopa in "bipolar" manic-depressive patients. Nature 229:135-136

Murphy FC, Rubinsztein JS, Michael A, Rogers RD, Robbins TW, Paykel ES, Sahakian BJ (2001) Decision-making cognition in mania and depression. Psychol Med 31:679-693

Neves FS, Silveira G, Romano-Silva MA, Malloy-Diniz L, Ferreira AA, De ML, Correa H (2008) Is the 5-HTTLPR polymorphism associated with bipolar disorder or with suicidal behavior of bipolar disorder patients? Am J Med Genet B Neuropsychiatr Genet 147B:114-116

Ongur D, Drevets WC, Price JL (1998) Glial reduction in the subgenual prefrontal cortex in mood disorders. Proc Natl Acad Sci U S A 95:13290-13295

Oppenheim G, Ebstein RP, Belmaker RH (1979) Effect of lithium on the physostigmine-induced behavioral syndrome and plasma cyclic GMP. J Psychiatr Res 15:133-138

Oruc L, Verheyen GR, Furac I, Jakovljevic M, Ivezic S, Raeymaekers P, Van BC (1997) Analysis of the tyrosine hydroxylase and dopamine D4 receptor genes in a Croatian sample of bipolar I and unipolar patients. Am J Med Genet 74:176-178

Otani K, Ujike H, Tanaka Y, Morita Y, Katsu T, Nomura A, Uchida N, Hamamura T, Fujiwara Y, Kuroda S (2005) The GABA type A receptor alpha5 subunit gene is associated with bipolar I disorder. Neurosci Lett 381:108-113

Overstreet DH, Russell RW, Crocker AD, Gillin JC, Janowsky DS (1988) Genetic and pharmacological models of cholinergic supersensitivity and affective disorders. Experientia 44:465-472

Padmos RC, Hillegers MH, Knijff EM, Vonk R, Bouvy A, Staal FJ, de RD, Kupka RW, Nolen WA, Drexhage HA (2008) A discriminating messenger RNA signature for bipolar disorder formed by an aberrant expression of inflammatory genes in monocytes. Arch Gen Psychiatry 65:395-407

Papadimitriou GN, Dikeos DG, Karadima G, Avramopoulos D, Daskalopoulou EG, Stefanis CN (2001) GABA-A receptor beta3 and alpha5 subunit gene cluster on chromosome 15q11-q13 and bipolar disorder: a genetic association study. Am J Med Genet 105:317-320

Papez J.W. A proposed mechanism of emotion. Archives of Neurological Psychiatry 34[4], 275-281. 1937.
Ref Type: Abstract

Papiol S, Molina V, Desco M, Rosa A, Reig S, Sanz J, Palomo T, Fananas L (2008) Gray matter deficits in bipolar disorder are associated with genetic variability at interleukin-1 beta gene (2q13). Genes Brain Behav 7:796-801

Parsey RV, Hastings RS, Oquendo MA, Huang YY, Simpson N, Arcement J, Huang Y, Ogden RT, Van Heertum RL, Arango V, Mann JJ (2006) Lower serotonin transporter binding potential in the human brain during major depressive episodes. Am J Psychiatry 163:52-58

Petty F (1995) GABA and mood disorders: a brief review and hypothesis. J Affect Disord 34:275-281

Pioro EP, Cuello AC, Mai JK (1990) Distribution of Substance P- and Enkephalin Immunoreactive Neurons and Fibers In The Human Nervous System. San Diego: Academic Press, Inc.

Pizzagalli DA, Goetz E, Ostacher M, Iosifescu DV, Perlis RH (2008) Euthymic patients with bipolar disorder show decreased reward learning in a probabilistic reward task. Biol Psychiatry 64:162-168

Preston AR, Shohamy D, Tamminga CA, Wagner AD (2005) Hippocampal function, declarative memory, and schizophrenia: anatomic and functional neuroimaging considerations. Curr Neurol Neurosci Rep 5:249-256

Pucilowski O, Eichelman B, Overstreet DH, Rezvani AH, Janowsky DS (1990) Enhanced affective aggression in genetically bred hypercholinergic rats. Neuropsychobiology 24:37-41

Radonic E, Rados M, Kalember P, Bajs-Janovic M, Folnegovic-Smalc V, Henigsberg N (2011) Comparison of hippocampal volumes in schizophrenia, schizoaffective and bipolar disorder. Coll Antropol 35 Suppl 1:249-252

Rajkowska G, Halaris A, Selemon LD (2001) Reductions in neuronal and glial density characterize the dorsolateral prefrontal cortex in bipolar disorder. Biol Psychiatry 49:741-752

Risch SC, Kalin NH, Janowsky DS (1981) Cholinergic challenges in affective illness: behavioral and neuroendocrine correlates. J Clin Psychopharmacol 1:186-192

Rosen HR, Rich BA (2010) Neurocognitive correlates of emotional stimulus processing in pediatric bipolar disorder: a review. Postgrad Med 122:94-104

Roy A, Karoum F, Pollack S (1992) Marked reduction in indexes of dopamine metabolism among patients with depression who attempt suicide. Arch Gen Psychiatry 49:447-450

Sachs GS, Lafer B, Stoll AL, Banov M, Thibault AB, Tohen M, Rosenbaum JF (1994) A double-blind trial of bupropion versus desipramine for bipolar depression. J Clin Psychiatry 55:391-393

Saddock BJ, Saddock VA (2003) Mood Disorders: Major Depression and Bipolar Disorder. In: Kaplan and Saddock's Synopsis of Psychiatry: Behavioral Sciences/Clinical Psychiatry, pp 535-572 Philadelphia: Lippincott Williams and Wilkins.

Sakai T, Oshima A, Nozaki Y, Ida I, Haga C, Akiyama H, Nakazato Y, Mikuni M (2008) Changes in density of calcium-binding-protein-immunoreactive GABAergic neurons in prefrontal cortex in schizophrenia and bipolar disorder. Neuropathology 28:143-150

Sapolsky RM (2000) Glucocorticoids and hippocampal atrophy in neuropsychiatric disorders. Arch Gen Psychiatry 57:925-935

Sax KW, Strakowski SM, Zimmerman ME, DelBello MP, Keck PE, Jr., Hawkins JM (1999) Frontosubcortical neuroanatomy and the continuous performance test in mania. Am J Psychiatry 156:139-141

Schatzberg AF, Schildkrout JJ (1995) Recent studies on norepinephrine systems in mood disorders. In: Psychopharmacology: the fourth generation of progress (Bloom FE KD, ed), pp 911-920 New York: Raven Press.

Serretti A, Lilli R, Lorenzi C, Lattuada E, Smeraldi E (2001) DRD4 exon 3 variants associated with delusional symptomatology in major psychoses: a study on 2,011 affected subjects. Am J Med Genet 105:283-290

Serretti A, Macciardi F, Cusin C, Lattuada E, Souery D, Lipp O, Mahieu B, Van BC, Blackwood D, Muir W, Aschauer HN, Heiden AM, Ackenheil M, Fuchshuber S, Raeymaekers P, Verheyen G, Kaneva R, Jablensky A, Papadimitriou GN, Dikeos DG, Stefanis CN, Smeraldi E, Mendlewicz J (2000) Linkage of mood disorders with D2, D3 and TH genes: a multicenter study. J Affect Disord 58:51-61

Shah MP, Wang F, Kalmar JH, Chepenik LG, Tie K, Pittman B, Jones MM, Constable RT, Gelernter J, Blumberg HP (2009) Role of variation in the serotonin transporter protein gene (SLC6A4) in trait disturbances in the ventral anterior cingulate in bipolar disorder. Neuropsychopharmacology 34:1301-1310

Shigeri Y, Seal RP, Shimamoto K (2004) Molecular pharmacology of glutamate transporters, EAATs and VGLUTs. Brain Res Brain Res Rev 45:250-265

Sidenberg DG, King N, Kennedy JL (1994) Analysis of new D4 dopamine receptor (DRD4) coding region variants and TH microsatellite in the Old Order Amish family (OOA110). Psychiatr Genet 4:95-99

Smoller JW, Finn CT (2003) Family, twin, and adoption studies of bipolar disorder. Am J Med Genet C Semin Med Genet 123C:48-58

Smyth C, Kalsi G, Brynjolfsson J, O'Neill J, Curtis D, Rifkin L, Moloney E, Murphy P, Sherrington R, Petursson H, Gurling H (1996) Further tests for linkage of bipolar affective disorder to the tyrosine hydroxylase gene locus on chromosome 11p15 in a new series of multiplex British affective disorder pedigrees. Am J Psychiatry 153:271-274

Soares JC, Mann JJ (1997) The functional neuroanatomy of mood disorders. J Psychiatr Res 31:393-432

Sobczak S, Riedel WJ, Booij I, an Het RM, Deutz NE, Honig A (2002) Cognition following acute tryptophan depletion: difference between first-degree relatives of bipolar disorder patients and matched healthy control volunteers. Psychol Med 32:503-515

Steckert AV, Valvassori SS, Moretti M, Dal-Pizzol F, Quevedo J (2010) Role of oxidative stress in the pathophysiology of bipolar disorder. Neurochem Res 35:1295-1301

Strakowski SM, DelBello MP, Adler CM (2005) The functional neuroanatomy of bipolar disorder: a review of neuroimaging findings. Mol Psychiatry 10:105-116

Sullivan GM, Ogden RT, Oquendo MA, Kumar JS, Simpson N, Huang YY, Mann JJ, Parsey RV (2009) Positron emission tomography quantification of serotonin-1A receptor binding in medication-free bipolar depression. Biol Psychiatry 66:223-230

Szczepankiewicz A, Dmitrzak-Weglarz M, Skibinska M, Slopien A, Leszczynska-Rodriewicz A, Czersky P, Hauser J. Study of dopamine receptors genes polymorphisms in bipolar patients with comorbid alcohol abuse. Alcohol and Alcoholism 42[2], 70-74. 2011.
Ref Type: Abstract

Todd RD, Lobos EA, Parsian A, Simpson S, DePaulo JR (1996) Manic-depressive illness and tyrosine hydroxylase markers. Bipolar Disorder Working Group. Lancet 347:1634

Varghese FP, Brown ES (2001) The Hypothalamic-Pituitary-Adrenal Axis in Major Depressive Disorder: A Brief Primer for Primary Care Physicians. Prim Care Companion J Clin Psychiatry 3:151-155

Vawter MP, Freed WJ, Kleinman JE (2000) Neuropathology of bipolar disorder. Biol Psychiatry 48:486-504

Vieta E, Martinez-De-Osaba MJ, Colom F, Martinez-Aran A, Benabarre A, Gasto C (1999) Enhanced corticotropin response to corticotropin-releasing hormone as a predictor of mania in euthymic bipolar patients. Psychol Med 29:971-978

Vincze I, Perroud N, Buresi C, Baud P, Bellivier F, Etain B, Fournier C, Karege F, Matthey ML, Preisig M, Leboyer M, Malafosse A (2008) Association between brain-derived neurotrophic factor gene and a severe form of bipolar disorder, but no interaction with the serotonin transporter gene. Bipolar Disord 10:580-587

Vogt BA. Pain and emotion interactions in subregions of the cingulate gyrus. National Review of Neuroscience 7, 533-544. 2005.
Ref Type: Abstract

Vuoristo J (2011) Genomic Organization of the Human GNAL Gene and Characterization of Two Novel Genes C18ORF2 and MPPE1 on Chromosome 18P11.2, a Susceptibility Region for Schizophrenia and Bipolar Disorder. In: Bipolar Disorder: University of Uolu.

Wessa M, Linke J (2009) Emotional processing in bipolar disorder: behavioural and neuroimaging findings. Int Rev Psychiatry 21:357-367

Wightman RM, Robinson DL (2002) Transient changes in mesolimbic dopamine and their association with 'reward'. J Neurochem 82:721-735

Willner P, Muscat R, Phillips G (1991) The role of dopamine in rewarded behavior: ability, insight, drive or incentive? Pol J Pharmacol Pharm 43:291-300

Woo TU, Walsh JP, Benes FM (2004) Density of glutamic acid decarboxylase 67 messenger RNA-containing neurons that express the N-methyl-D-aspartate receptor subunit NR2A in the anterior cingulate cortex in schizophrenia and bipolar disorder. Arch Gen Psychiatry 61:649-657

Wood GE, Young L.T., Reagan L.P., Chen E, McEwen BS. Stress-induced structural remodeling in hippocampus: Prevention by Lithium treatment. Proceedings in the National Academy of Sciences. 2004.
Ref Type: Abstract

Young JW, van EJ, Winstanley CA, Geyer MA (2011) Increased risk-taking behavior in dopamine transporter knockdown mice: further support for a mouse model of mania. J Psychopharmacol 25:934-943

Young LT, Warsh JJ, Kish SJ, Shannak K, Hornykeiwicz O (1994) Reduced brain 5-HT and elevated NE turnover and metabolites in bipolar affective disorder. Biol Psychiatry 35:121-127

Yucel K, McKinnon MC, Taylor VH, Macdonald K, Alda M, Young LT, MacQueen GM (2007) Bilateral hippocampal volume increases after long-term lithium treatment in patients with bipolar disorder: a longitudinal MRI study. Psychopharmacology (Berl) 195:357-367

Yucel K, Taylor VH, McKinnon MC, Macdonald K, Alda M, Young LT, MacQueen GM (2008) Bilateral hippocampal volume increase in patients with bipolar disorder and short-term lithium treatment. Neuropsychopharmacology 33:361-367

9

Neurotransmission in Mood Disorders

Zdeněk Fišar, Jana Hroudová and Jiří Raboch
Charles University in Prague and General University Hospital in Prague
Department of Psychiatry, First Faculty of Medicine, Prague
Czech Republic

1. Introduction

Mood disorders are characterized by depression, mania, or both. There are two groups of mood disorders, depressive disorders and bipolar disorder. Bipolar disorder is characterized by intermittent episodes of mania or hypomania and depressive episodes; rapid cycling, mixed states, and psychotic symptoms occur in some cases. Depression is a serious mental disorder that manifests with depressed mood, loss of interest or pleasure, feelings of guilt or low self-worth, disturbed sleep or appetite, low energy, and poor concentration. Mania is the opposite of depression; it is a state of abnormally elevated or irritable mood, arousal, and/or energy levels. Depression and mania are thought to be heterogeneous illnesses that can result from dysfunction of several neurotransmitter or metabolic systems.

The base of biological psychiatry is an assumption that human mind is connected with human body so that mental disorders (processes) are accompanied with biochemical changes, which can be measured. The second issue of biological psychiatry is a causality of all natural processes, including human mind. Molecular psychiatry studies disorders in human mind from the neurochemical, neurophysiological, neuroendocrine and genetic point of view mainly. It is postulated that mood disorders are caused by or associated with the disturbance of nervous signal transmission in the brain at the level of chemical synapses. On the molecular level, there is impaired neurotransmission mediated by neurotransmitters, their receptors and transporters, and by intracellular processes coupled to the activation of receptors for neurotransmitters and growth factors.

Because of difficult availability of human brain tissue for neurochemical measurements, psychotropic drugs effects are studied with the purpose to select those components of intracellular signalling pathways, which could be responsible for the therapeutic effects of tested drugs and consequently be related to origin of the disease. Results support the hypothesis that the treatment with antidepressants, mood stabilizers and even some antipsychotics leads to effects similar to neurotrophic and anti-inflammatory. Due to feedbacks inside the neuron the attention is devoted to pathways connected with receptors for monoamine neurotransmitters, receptors with inner calcium channel, and receptors for neurotrophins or Wnt glycoproteins. The role of neurochemical hypotheses of mood disorders is to suggest the relationship between symptoms of the disease, changes in signalling pathways and mechanisms of action of psychotropic drugs.

Understanding of neurochemistry of mood disorders as well as molecular mechanisms of action of antidepressants and mood stabilizers are necessary to declare valid molecular theory of the disorder as basis of effective diagnosis, prevention and treatment. However, sensitive and specific genetic, biochemical, physiological, neuroendocrine or other biological tests have not been so far developed to be capable for diagnosis of mood disorders and their subtypes or for prediction of efficacy of current pharmacotherapy. It seems that leading role in neurochemistry of mood disorders could be awarded to disturbed monoamine neurotransmission, dysfunction in energy metabolism of neurons, modulation of inflammatory pathway, and changes in activities of transcription factors, neurotrophic factors and other components involved in neuroplasticity and apoptosis.

Basic findings about signalling pathways included both in pathophysiology of mood disorders and in mechanisms of action of administered psychotropic drugs are summarized in this chapter. Finally, there are summarized advances in neurochemical hypotheses of mood disorders.

2. Synaptic signal transduction

Synaptic signal transduction is complex process by which a neurotransmitter (extracellular signalling molecule) is released in response to action potential or other stimuli and activates a specific membrane receptor that leads to alteration of intracellular molecules forming a cellular response. Signal transduction system consists of many components (neurotransmitters, specific receptors, ion channels, G proteins, effector enzymes, transporters and other membrane proteins, second messengers, protein kinases, phosphatases, transcription factors, neurotrophic factors etc.) and numberless interactions, interconnections and feedbacks among them. Structure and composition of lipid bilayer also plays an important role, because of the fact that most of membrane proteins require interaction with specific phospholipids or with cholesterol for their optimal function. Signalling pathways included in the pathogenesis of mood disorders are primarily activated mostly by serotonin (5-hydroxytryptamine, 5-HT), norepinephrine, dopamine, glutamate and γ- aminobutyric acid (GABA), i.e. they are connected with the processes in adenylate cyclase and phosphoinositide systems and with changes of intracellular ion concentrations, especially with calcium. Only selected components of signalling pathways are presented supposed to play a key role in pathophysiology and neurochemistry of mood disorders.

2.1 Neurotransmitters

A multitude of chemicals called neurotransmitters mediate intercellular communication in the nervous system. There are several groups of neurotransmitters: classical neurotransmitters, neuropeptides, endocannabinoids, nitric oxide (NO), and carbon monoxide (CO). Although they exhibit great diversity in many of their properties, most of them are stored in vesicles in nerve terminals (except for endocannabinoids, NO and CO) and are released to the extracellular space via processes requiring calcium ions. Their action is terminated by reuptake into presynaptic terminal or glia cells or by catabolism in extracellular space (e.g. in synaptic cleft) or in presynaptic terminal.

All classical neurotransmitters are synthesized in nerve terminals. The first molecule to be implicated as neurotransmitter was acetylcholine (ACh). ACh is synthesized from choline and acetyl-coenzyme A in the nerve endings; reaction is catalyzed by the enzyme choline acetyltransferase. ACh is rapidly degraded in synaptic cleft by the enzyme acetylcholinesterase.

There are three major amino acid neurotransmitters in the nervous system: GABA, glycine and glutamic acid. GABA and glycine are inhibitory neurotransmitters; glutamate and aspartate are excitatory neurotransmitters.

Monoamine neurotransmitters, such as dopamine, norepinephrine or serotonin are the most important neurotransmitters in pathophysiology of mood disorders and in mechanisms of action of antidepressants. Catecholamines (dopamine, norepinephrine and epinephrine) are synthesized from tyrosine. Dopamine is formed by tyrosine hydroxylase catalysed hydroxylation (rate limiting step) and decarboxylation of tyrosine. Norepinephrine is formed by hydroxylation of dopamine in presence of dopamine β-hydroxylase. Indolamines (serotonin and tryptamine) are synthesized from tryptophan by hydroxylation and decarboxylation; tryptophan hydroxylase mediated reaction is the rate-limiting step. Action both of catecholamines and indolamines on target cells is terminated much more slowly than that of acetylcholine; they are removed from synaptic cleft by reuptake. The major enzymes involved in the catabolism of catecholamines are monoamine oxidase (MAO) and catechol-*O*-methyltransferase (COMT). Serotonin is metabolized by MAO.

A great number of neuropeptides have been discovered and new neuropeptides being identified continually. All neuropeptides are synthesized in the cell body. The action of neuropeptides in the synaptic cleft is terminated by peptidases; there is no reuptake for neuropeptides. Neuropeptides usually act as co-transmitters and their transduction mechanism is coupled with G proteins.

The endocannabinoids are a family of lipid neurotransmitters that mediate retrograde signal from postsynaptic neurons to presynaptic ones (Fišar, 2009). Nitric oxide and carbon monoxide have specific properties from among neurotransmitters, since the NO or CO does not interact with membrane receptors but diffuse to target intracellular receptor (Hill et al., 2010; Snyder & Ferris, 2000). Abundant recent evidence favours a neurotransmitter/neuromodulator role for D-serine, because the D-serine, rather than glycine, is the endogenous ligand for *N*-methyl-D-aspartate (NMDA) receptors in many brain structures. D-serine is synthesized mainly in glial cells and it is released upon activation of glutamate receptors (Oliet & Mothet, 2009).

2.2 Growth factors

Growth factors are proteins that stimulate cellular growth, proliferation and differentiation, and promote cellular survival. They are essential for development and function of nervous system and have important function in neurotransmission. They are released from different cells; after interaction with membrane receptors changes in activity of intracellular enzymes occur leading to changes in gene expression and production of cellular molecules. Mechanism of their action is similar to action of neurotransmitters, but they are not released in response to membrane depolarization and to changes of intracellular calcium levels. Growth factors that promote the survival, development and function of neurons are known as neurotrophic factors. From the psychiatric point of view neurotrophins are the most important class of growth factors.

Neurotrophic factors act by preventing the neuron from initiation of programmed cell death (apoptosis); they induce differentiation of progenitor cells to form neurons. The term "neurotrophin" is reserved for four structurally related neurotrophic factors: nerve growth factor (NGF), brain-derived neurotrophic factor (BDNF), neurotrophin-3 (NT-3), and neurotrophin-4 (NT-4). Each of four neurotrophins activates one or more of three trk (tropomyosin-related kinase) receptors with intracellular tyrosine kinase activity called trkA, trkB and trkC. Furthermore, each neurotrophin can still bind (with low affinity) to neurotrophic receptor p75 (p75NTR), homologous to tumour necrosis factor (TNF) and without any tyrosine kinase activity.

BDNF has a key role in stress response and in action of antidepressants. Exposure to stress has been shown to decrease the expression of BDNF. Series of studies support the hypothesis that a reduction of BDNF could contribute to depression and that antidepressants mediate their therapeutic benefits by increasing levels of this factor in the hippocampus (Castrén et al., 2007; Castrén & Rantamäki, 2008; Duman & Monteggia, 2006).

2.3 Neurotransmitter receptors

Receptor is macromolecule specialized on transmission of information. It is defined as binding site with functional relationships. Receptor complex includes: 1. Specific binding site; 2. Internal ion channel or transduction element; 3. Effector system (ion channel or system of second messengers). Activation of receptors with internal ion channel leads to rapid change in membrane potential and to prompt cellular response. Activation of receptors coupled with G proteins leads to slower response through activation of effector system. Effector system includes G protein activated ion channels or enzymes; G protein activated enzymes generate second messengers, which activate protein kinases. Protein phosphorylation plays a significant role in a wide range of cellular processes. Transcription factors can be phosphorylated too, and phosphorylated transcription factors serve as third messengers, which activate gene expression.

Receptors are able to adapt their properties to increased or decreased activation. Changes in the density of receptors are known mechanism of their adaptation (receptor down-regulation or up-regulation). Additionally, response to receptor activation can be altered at unchanged density of receptors, because regulation of properties of receptors may consist of decreased or increased activity of post receptor events (desensitization or hypersensitivity).

2.4 Neurotransmitter transporters

Neurotransmitter transporters are necessary for synaptic transmission just as receptors, ion channels, G proteins and effectors. Generally, neurotransmitters are removed from synaptic cleft by enzymatic degradation or by active transport to presynaptic button or to surrounding glia cells. There are three main classes of membrane transporters: 1. Transporters dependent on sodium and chlorine transport serotonin, norepinephrine or dopamine back into presynaptic part; this enables that neurotransmitters may be stored in vesicles and repeatedly released in response to action potential; 2. Vesicular transporters carry neurotransmitters into synaptic vesicles; 3. Sodium dependent transporters are localized in the membrane of glia cells and transport neurotransmitters such as GABA, glutamate or aspartate into glia cells.

Functional polymorphisms in the promoter region of the serotonin transporter (SERT, 5-HTT) gene and *BDNF* gene were found to moderate the influence of stressful life events on depression (Aguilera et al., 2009; Caspi et al., 2003).

2.5 Postreceptor events

It is assumed that signal transduction mediated by receptors associated with G proteins and 2nd messenger systems are altered at mental disorders and during treatment with psychotropic drugs. After activation of receptor by first messenger G proteins are activated and activated G proteins activate effectors enzymes, such as adenylyl cyclase or phospholipase C, and second messengers are produced. Second messenger activates protein kinase of type A, C or calmodulin (CaM) dependent, which catalyses phosphorylation of cellular proteins and physiological response to receptor activation arises. Second messengers generated by enzymes activated by G proteins include cyclic adenosine monophosphate (cAMP), cyclic guanosine monophosphate (cGMP), inositol trisphosphate (IP_3), diacylglycerol (DAG), calcium (Ca^{2+}), metabolites of arachidonic acid and nitric oxide.

There are many neurotransmitters, many receptors, less G proteins and a few effector systems. However, there are many feedbacks at the cellular level and many cross-reactions at the intracellular level. Individual signalling systems interact together and form complex intracellular system, which enables neurons to compile signals from the different neurotransmitter systems, and which is involved in regulation of neuroplasticity and stress response. Reversible phosphorylation of proteins is probably the crucial molecular mechanism, which is used to realize the biological response to extracellular signals in target neurons. Attention is paid to cAMP dependent protein kinases (PKA), protein kinases C (PKC) activated by DAG, protein kinases dependent on calcium and calmodulin (CaMK), and protein kinases B (PKB, Akt). Transcription factor CREB, neurotrophin BDNF, glycogen synthase kinase-3 (GSK-3) and components regulating programmed cell death, especially family of Bcl-2 proteins and various mitochondrial factors (including MAO, enzymes of respiratory chain and mitochondrial DNA) are intensively studied, as well.

2.5.1 Monoamine oxidase

The enzyme monoamine oxidase (MAO; EC1.4.3.4) is a mitochondrial enzyme, which catalyzes oxidative deamination of biogenic and xenobiotic monoamines. It regulates the metabolic degradation of catecholamines and serotonin in neural and other target tissues. Major physiological role of intraneuronal MAO is to keep cytosolic monoamine concentrations very low. MAO exists in two isoforms that differ in substrate preference, inhibitor specificity, tissue and cell distribution, and immunological properties (Bach et al., 1988). The type A (MAO-A) metabolizes 5-HT and is sensitive to inhibition by low concentrations of clorgyline, whereas the type B (MAO-B) prefers benzylamine or 2-phenylethylamine as substrates and is sensitive to inhibition by low concentrations of *l*-deprenyl. Tyramine, tryptamine, dopamine, norepinephrine and epinephrine are equally well oxidized by both isoforms of MAO (Youdim et al., 2006). The high levels of both forms are found in the brain; MAO-B is found in dopamine-secreting neurons in the brain.

MAOs have an important role in brain development and function, and MAO inhibitors have a range of potential therapeutic uses (Ramsay & Gravestock, 2003). Generally, selective inhibitors of MAO-A and nonselective MAOIs seem to be effective in the treatment of patients with depression, panic disorder, and other anxiety disorders (Stahl & Felker, 2008). It is supposed that MAO-B inhibition may slow the course of various neurodegenerative disorders; thus, selective inhibitors of MAO-B may be efficacious in treatment of Parkinson's disease (Horstink et al., 2006) and possibly of Alzheimer's disease (Riederer et al., 2004).

2.5.2 Transcription factor CREB

The allocation of information and memory to specific cells and synapses within a neural network is modulated by many synaptic, cellular, and intercellular components and by mechanisms working at different time scales (Silva et al., 2009). The transcription factor CREB (cyclic adenosine monophosphate response element binding) regulates transcription of many genes and has a well-known role in the learning related to synaptic plasticity (Carlezon et al., 2005). CREB is also involved in antidepressant response (Chen et al., 2001). Increase in CREB function can enhance memory; however, cognitive performance can be disrupted under some circumstances. CREB activity has sometimes beneficial, sometimes detrimental roles, depending on the brain region involved. Therefore, alterations in CREB function do not produce uniform effects throughout the brain (Blendy, 2006; Carlezon et al., 2005; Tardito et al., 2006).

2.5.3 Bcl-2

Bcl-2 (acronym for B-cell CLL/lymphoma 2) protein family is included in the regulation of apoptotic cell death and is formed from members with both antiapoptotic (e.g. Bcl-2, Bcl-xL, Bcl-w) and proapoptotic (e.g. Bax, BAD, Bak) activities. There are a number of theories concerning how members of the Bcl-2 family exert their effects. E.g., proteins from Bcl-2 family together with other factors control the permeability of mitochondrial membranes (Chipuk & Green, 2008). The main changes in mitochondria during apoptosis are mitochondrial outer membrane permeabilization MOMP (regulated by members of Bcl-2 family) and depolarization of inner membrane. Bcl-2 itself is antiapoptotic; it reduces apoptosis by sequestration of caspases, by inhibition of release of mitochondrial apoptotic factors and by increasing of calcium uptake into the mitochondria. It is supposed that neurodegenerative diseases, including schizophrenia and mood disorders, may result from an abnormal ratio or function of pro- and antiapoptotic factors.

2.5.4 Glycogen synthase kinase 3

Glycogen synthase kinase 3 (GSK-3) is serine/threonine kinase that phosphorylates glycogen synthase and many other substrates such as transcription factors, enzymes and cytoskeletal elements. In human, it occurs in 2 isoforms, GSK-3α and GSK-3β; biological functions of β isoform have been more analyzed, it is widely present in the brain. Role of GSK-3 was found in different diseases, Alzheimer's disease, diabetes mellitus type 2, various carcinomas (Peineau et al., 2008), as well as in pathophysiology and treatment of bipolar disorder (Gould et al., 2007). Activity of GSK-3 is regulated by phosphorylation; phosphorylation on tyrosine residues increases the enzyme efficiency, phosphorylation of final serine has inhibitory effects. Furthermore, GSK-3 can be inactivated by stimuli coming from different signalling pathways, e.g. Wnt pathway (Fig. 5), phosphoinositide 3-kinase (PI3K) pathway, from PKA, PKC or from others. Main signalling pathways modulated by GSK-3 are insulin pathway, pathway of neurotrophic factors and Wnt pathway.

Different antidepressants, antipsychotics, amphetamines and growth factors have direct or indirect effects on GSK-3 and GSK-3 mediated signalling pathways. From the view of biological psychiatry, there is an important finding that GSK-3 is directly inhibited by lithium (by competition with Mg^{2+} ions) and indirectly inhibited with valproate; both of them are used in the treatment of bipolar disorder. GSK-3 inhibition is antiapoptotic, because of generally proapoptotic action of GSK-3. The hypothesis is tested that lithium (and other medications) can due to the GSK-3 inhibition induce cellular processes leading to changes in bioenergetics, neuroplasticity, neurogenesis, stability and survival of neurons (Gould & Manji, 2005).

3. Signalling pathways

There is relatively small amount of signalling pathways initiated by activation of membrane receptors, which leads to the majority of important intracellular physiological processes. Their modulations influence gene expression and the function of cellular proteins, the change of synaptic functions, neuroplasticity and the response to neurotransmitters, neuropeptides, neurohormones, glucocorticoids and other bioactive molecules. Understanding of these signalling pathways is essential for insight to pathogenesis and pathophysiology of mood disorders (Fišar & Hroudová, 2010).

3.1 Adenylate cyclase pathway

The mechanism of activation of the adenylate cyclase system is well known (Fig. 1). Neurotransmitter or other agonist binds to specific binding site of G protein-coupled receptor (GPCR) and activates G proteins. Main G proteins, regulating activation of adenylate cyclase (also known as adenylyl cyclase, AC), are G_s, G_q and G_i. Activated $G\alpha_s$ or $G\alpha_q$ subunits bind to AC and activate it directly; free $\beta\gamma$ complexes activate certain AC subtypes, as well. Activated AC catalyses conversion of ATP to cAMP; presence of Mg^{2+} ions is required and reaction is stopped by G protein inactivation. $G\alpha_i$ subunits decrease intracellular cAMP levels by inhibition of AC. Indirectly, AC activity can be modulated through the activation of phosphoinositide system, when protein kinases C (PKC) and calcineurin are activated. cAMP activates cAMP-dependent protein kinases (PKA). cAMP decomposition into 5′-AMP is catalyzed by phosphodiesterase (PDE) in the presence of Mg^{2+}. Activated PKA phosphorylates neuronal proteins; it leads to cross-connections of various signalling pathways and to different physiological effects (Sands & Palmer, 2008; Taylor et al., 2005). Phosphorylation (activation) of transcription factor CREB and consequent expression of neurotrophin BDNF seems to be important for therapeutic efficiency of antidepressants (Duman et al., 1997). Further, phosphorylation of nuclear factor κB (NFκB) is required for activation of transcription and interaction with CREB binding protein (CBP). NFκB obviously participates in survival of neurons (Hayden & Ghosh, 2004). PKA supports the neuronal survival by inhibition of proapoptotic factor BAD and by activation of NFAT (nuclear factor of activated T cells) protein (Wu et al., 2007). Importance of APC (anaphase-promoting complex) inactivation is also studied in processes of axon growth, neuronal survival and synaptic functions (Kim & Bonni, 2007).

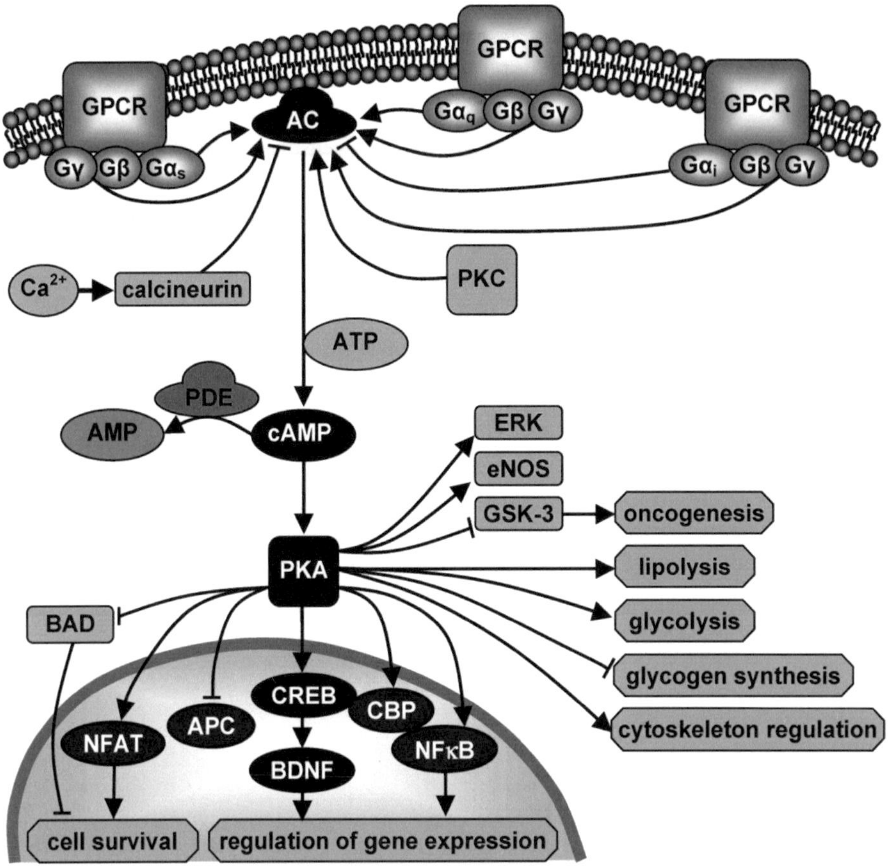

AC – adenylate cyclase (adenylyl cyclase); AMP – adenosine monophosphate; APC – anaphase-promoting complex; ATP – adenosine-5'-triphosphate; BAD – proapoptotic factor from Bcl-2 family (Bcl-2-associated death promoter); BDNF – brain-derived neurotrophic factor; cAMP – cyclic adenosine monophosphate; CBP – CREB binding protein; CREB protein – cAMP response element-binding protein; eNOS – endothelial nitric oxide synthase; ERK – extracellular signal-regulated kinase; GPCR – G protein-coupled receptor; GSK-3 – glycogen synthase kinase-3; NFκB – nuclear factor κB; NFAT – nuclear factor of activated T cells; PDE – phosphodiesterase; PKA – protein kinase A, cAMP dependent protein kinase; PKC – protein kinase C; ⟶ - activation; ⊣ - inhibition

Fig. 1. Adenylate cyclase pathway (Fišar & Hroudová, 2010). Detailed description in the text.

3.2 Guanylate cyclase pathway

Guanylate cyclase (GC, also known as guanylyl cyclase) catalyses production of cyclic guanosine monophosphate (cGMP) from guanosine triphosphate (GTP); presence of Mg^{2+} is required. There are two types of GCs, soluble (sGC) and membrane (Fig. 2). Membrane-bound form of GC is activated by peptide hormones; sGC is a receptor for nitric oxide (NO). The cGMP is a regulator of ion channels conductance, glycogenolysis and apoptosis; it also relaxes smooth muscles. cGMP degradation to 5′-GMP by phosphodiesterases (PDEs) is analogous to cAMP degradation. cGMP-dependent protein kinase (PKG) is activated by cGMP; PKG type I is expressed in specific brain regions.

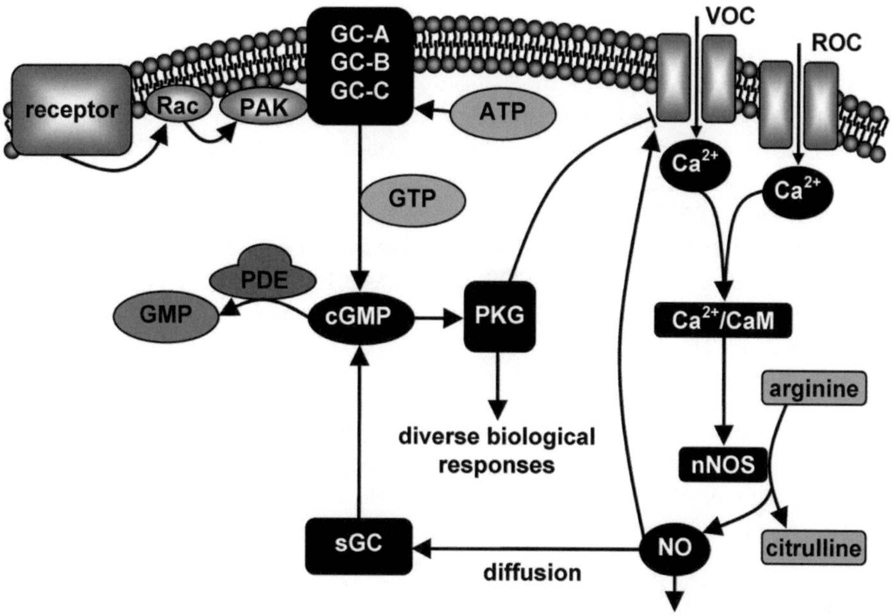

ATP - adenosine-5'-triphosphate; CaM - calmodulin; cGMP - cyclic guanosine monophosphate; GC - guanylate cyclase (guanylyl cyclase); GTP - guanosine-5′-triphosphate; NO - nitric oxide; nNOS - neuronal nitric oxide synthase; PAK - p21 activated kinase; PDE - phosphodiesterase; PKG - protein kinase G; Rac - small G protein with GTPase activity; ROC - receptor-operated channel; sGC - soluble guanylate cyclase; VOC - voltage-operated channel; → - activation; ⊣ - inhibition

Fig. 2. Guanylate cyclase pathway. Detailed description in the text.

3.2.1 Membrane guanylate cyclase

There are seven mammalian transmembrane GCs. GC-A and GC-B are natriuretic peptide receptors, GC-C can be activated by guanylin or uroguanylin (bacterial enterotoxins). The principal function of plasma membrane GCs in the CNS is modulation of sympathetic activity, inhibition of arginine vasopressin secretion, diminished salt appetite and water drinking, and effects on blood pressure/volume-regulating regions in the brain (Kuhn, 2003). Signalling pathway that provides a general mechanism for diverse signalling receptors to increase the cGMP concentrations was supposed in following sequence (Fig. 2): 1. Activation of membrane receptor, 2. Activation of Rac protein (small signalling G protein with GTPase activity) via guanine nucleotide-exchange factor, 3. Activation of p21 activated kinase (PAK) by Rac, 4. Activation of membrane GC by PAK, and 5. cGMP production (Guo et al., 2010).

3.2.2 Soluble guanylate cyclase and nitric oxide

In the brain, metastable free radical NO production is linked to the calcium (Ca^{2+}) influx during membrane depolarization through voltage-operated channels (VOC), e.g. L-type calcium channels, or after activation of receptor-operated channels (ROC), e.g. NMDA ionotropic glutamate receptors. Calmodulin (CaM) is activated following the increase of intracellular calcium and neuronal NO synthase (nNOS, NOS-1) is activated by Ca^{2+}/CaM;

nNOS catalyze conversion of arginine to NO and citrulline. NO regulates many kinds of ion channels, including voltage-gated (Jian et al., 2007) and ligand-gated Ca^{2+}-channels. NO functions as a neurotransmitter by diffusing through the membranes of postsynaptic cells, where it binds in sGC and activates this enzyme to convert GTP into the cGMP (Snyder & Ferris, 2000). Thus, increased cGMP production could be attributed to NO, establishing a role for NO in mediating actions of glutamate in the brain. Two major signalling mechanisms, namely cGMP pathway and *S*-nitrosylation, mediate the cellular effects of NO. It has been reported that cGMP-mediated processes occurs at low NO concentrations, whereas *S*-nitrosylation occurs when NO is available at levels ranging from physiological to pathophysiological conditions. However, there are many other biological activities affected by NO (Brown, 2010; Thomas et al., 2008).

NO-sensitive sGCs, PKGs, and cGMP-regulated PDEs have important functions as generators, effectors, and modulators of cGMP signals in the brain, respectively. The NO/sGC/cGMP/PKG pathway (Fig. 2) modulates long-term changes of synaptic activity in different brain regions. NO can modulate neuronal excitability and neurotransmitter release; thus, NO appears to play an important role in normal brain function and may have significant implications for the treatment of stress-related psychiatric disorders (Snyder & Ferris, 2000; Chiavegatto & Nelson, 2003). The human hippocampus contains a high density of NMDA receptors and neurons expressing nNOS suggesting that NMDA-NO transduction pathway can be involved in the pathogenesis of affective disorders and in the mechanism of action of antidepressants (Paul & Skolnick, 2003). It is well established that nNOS-derived NO inhibits neurogenesis (Zhu et al., 2006); moreover, several studies have indicated that nNOS inhibitors have antidepressant-like properties (Joca & Guimarães, 2006). Furthermore, several reports have demonstrated that increased plasma NO metabolites levels were associated with suicide attempts, especially in depressive patients (Kim et al., 2006; Lee et al., 2006). Activation of the NOS in the hippocampus can be involved in the pathogenesis of affective disorders, possibly triggered by effects of stress on hypothalamic-pituitary-adrenal (HPA) axis and mediated by impaired serotonin function.

3.3 Phosphoinositide pathway

Phosphoinositide pathway includes activation of phosphoinositide phospholipases C (PLCs), which participate in phosphatidylinositol bisphosphate (PIP_2) metabolism and lipid signalling pathways in a calcium-dependent manner (Fig. 3). PLCs are localized mostly in plasma membranes. PLC family contains 13 isoenzymes divided into six subfamilies. Neurotransmitter or agonist binds to receptor binding site and G proteins (mostly $G_{q/11}$, sometimes G_o or G_i) are activated. $G\alpha_{q/11}$ subunits activate (Ca^{2+} is required) PLC, mostly PLCβ isoform. Gβγ subunits activate only isoenzymes PLCβ2 and PLCβ3. PLCγ subtype is activated by both receptor (trk) and non-receptor tyrosine kinases, PLCδ by elevated calcium levels, PLCε by Ras and Rho GTPases, PLCη by Gβγ subunits (Stewart et al., 2007). Activated PLCs catalyze hydrolysis of phosphatidylinositol 4,5-bisphosphate (PIP_2) and give arise to the second messengers inositol trisphosphate (IP_3) and diacylglycerol (DAG). DAG activates (in the presence of Ca^{2+}) protein kinases type C (PKCs) and they phosphorylate many enzymes and other cellular proteins. IP_3 binds to intracellular receptors (IP_3R), leading to Ca^{2+} release from endoplasmic reticulum (ER). Ca^{2+} activates calmodulin (CaM), and Ca^{2+}/CaM-dependent protein kinases (CaMKs) phosphorylate a wide range of neuronal

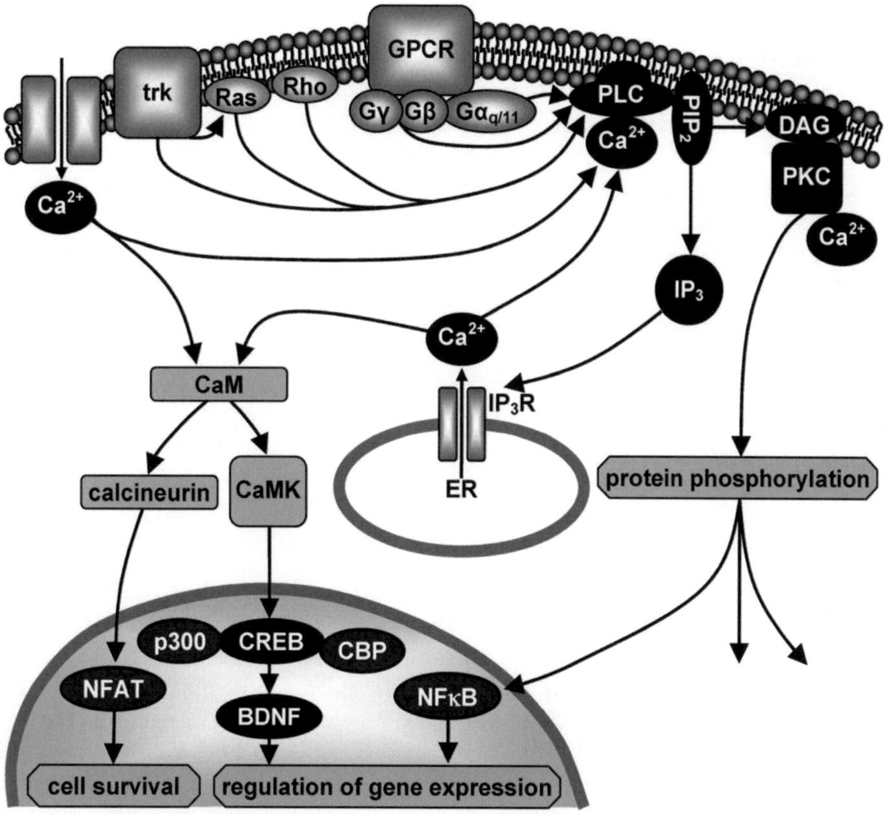

BDNF - brain-derived neurotrophic factor; CaM - calmodulin; CaMK - calcium and calmodulin-dependent protein kinase; CBP - CREB binding protein; CREB protein - cAMP response element-binding protein; DAG - diacylglycerol; ER - endoplasmic reticulum; GPCR - G protein-coupled receptors; IP_3 - inositol 1,4,5-trisphosphate; IP_3R - IP_3 receptor; NFκB - nuclear factor κB; NFAT - nuclear factor of activated T cells; p300 - transcriptional co-activating protein; PIP_2 - phosphatidylinositol 4,5-bisphosphate; PKC - protein kinase C; PLC - phosphoinositide phospholipase C, phospholipase C; Ras - small GTPase; Rho - small GTPase; trk - tropomyosin-related kinase; ⟶ - activation; ——┫ - inhibition

Fig. 3. Phosphoinositide pathway (Fišar & Hroudová, 2010). Detailed description in the text.

proteins, including transcription factor CREB (with CBP and p300 as coactivators). Calcineurin, which dephosphorylates (activates) NFAT (nuclear factor of activated T cells) transcription factor, is also activated by CaM.

Nervous tissue shows high PKC activities; there this enzyme system takes part in the regulation of ion channels, modulation of receptors, release of neurotransmitters, synaptic potentiation, synaptic depression, neuronal survival etc. Changes of activities and concentrations of PKC isoforms have been described in neurodegenerative diseases both acute (ischemia, trauma) and chronic (Alzheimer's disease, Parkinson's disease, amyotrophic lateral sclerosis) and by affective or psychotic disorders (Battaini, 2001; Pascale et al., 2007). A large range of biochemical data supports potential PKC involvement in pathophysiology of bipolar disorder and its treatment.

3.4 Calcium signalling pathway

Neurotransmitter, hormone or growth factor as well as depolarization of the membrane in excitable cells are responsible for change of intracellular calcium (Berridge et al., 2000). Intracellular calcium concentrations are regulated by transporters, which remove Ca^{2+} from the cell or store them in mitochondria and endoplasmic reticulum (Fig. 4). Intracellular calcium concentrations increase after the opening of Ca^{2+} channels in plasma membrane in the response to membrane depolarization (VOC, voltage-operated channel) or in response to activation of ionotropic receptors (ROC, receptor-operated channel; e.g. NMDA receptor) or in the response to activation of inositol trisphosphate receptor (IP_3R) in the membrane of endoplasmic reticulum (ER). Concentrations of cytosolic calcium are decreased both by transport from cell and by uptake into the mitochondria or endoplasmic reticulum. Plasma membrane Ca^{2+}-ATPases (PMCA), sodium-calcium exchanger (NCX) and mitochondrial calcium transporter (mCU) are responsible for the depletion of cytosolic calcium. Mitochondrial NCX in reverse mode of operation participates in the influx of Ca^{2+} ions into matrix in pathological conditions such as brain ischemia. Following Ca^{2+} uptake into the mitochondria, Ca^{2+} is slowly released by mitochondrial NCX (Kann & Kovács, 2007). Ca^{2+} can be released also from ER, by Ca^{2+}-activation of ryanodine receptor (RyR).

Increased cytoplasmic calcium activates calmodulin (CaM) and consequently both calcineurin and Ca^{2+}- and CaM-dependent protein kinases (CaMK) are activated. Cascade of calcium and CaM dependent protein kinases includes three kinases: kinase of CaM kinase (CaMKK) and CaM kinases CaMKI and CaMKIV which are activated by CaMKK. They occur frequently in the brain and T-lymphocytes. Nuclear CaMKIV regulates transcription through the phosphorylation of transcription factors, including CREB protein. Cross-connections of these protein kinases with other signalling pathways exist in the cytoplasm, e.g. with PKA, mitogen activated protein kinases (MAPK), and Akt (PKB). Consequently, CaMK are connected to the processes related to neuroplasticity and neuronal protection against apoptosis (Berridge et al., 2000; Miyamoto, 2006; Soderling, 1999).

Calcineurin dephosphorylates nuclear factor of activated T cells (NFAT), which afterwards enters into the nucleus and completes NFAT transcription complexes; these complexes regulate expression of growth factors, cytokines and other molecules essential for morphogenesis, development and function of neurons and other cells. Stimulus activating NFAT signalling pathway may originate from receptors with tyrosine kinase activity, ion channels, non-receptor tyrosine kinases, G protein-coupled receptors, and gap junctions (Wu et al., 2007). Other targets of calcium are adenylate cyclases (AC), some of them are activated others are inhibited. Further, Ca^{2+} stimulates some of cAMP phosphodiesterases (PDEs) resulting in changes of cAMP concentration; cAMP concentrations affect calcium levels through the activation of Ca^{2+}-channels or pumps by protein kinase A (PKA).

Activation of nitric oxide synthase (NOS) is an important role of Ca^{2+}. NOS enables NO production, activation of sGC and cGMP production; the cGMP influences activity of Ca^{2+} channels and pumps (feedback). Phosphoinositide-3-kinase (PI3K) interferes in the calcium pathways through the production of phosphatidylinositol 3,4,5-trisphosphate (PIP_3), which activates non-receptor tyrosine kinase Btk consequently activating PLCγ1, and IP_3 is produced. Increased Ca^{2+} concentrations also activate PLCδ which increases IP_3 concentrations. Conversely, Ca^{2+} can decrease IP_3 production by the activation of IP_3-kinase (IP_3K). Calcium activates tyrosine kinases enriched with proline (PYK2), which activate

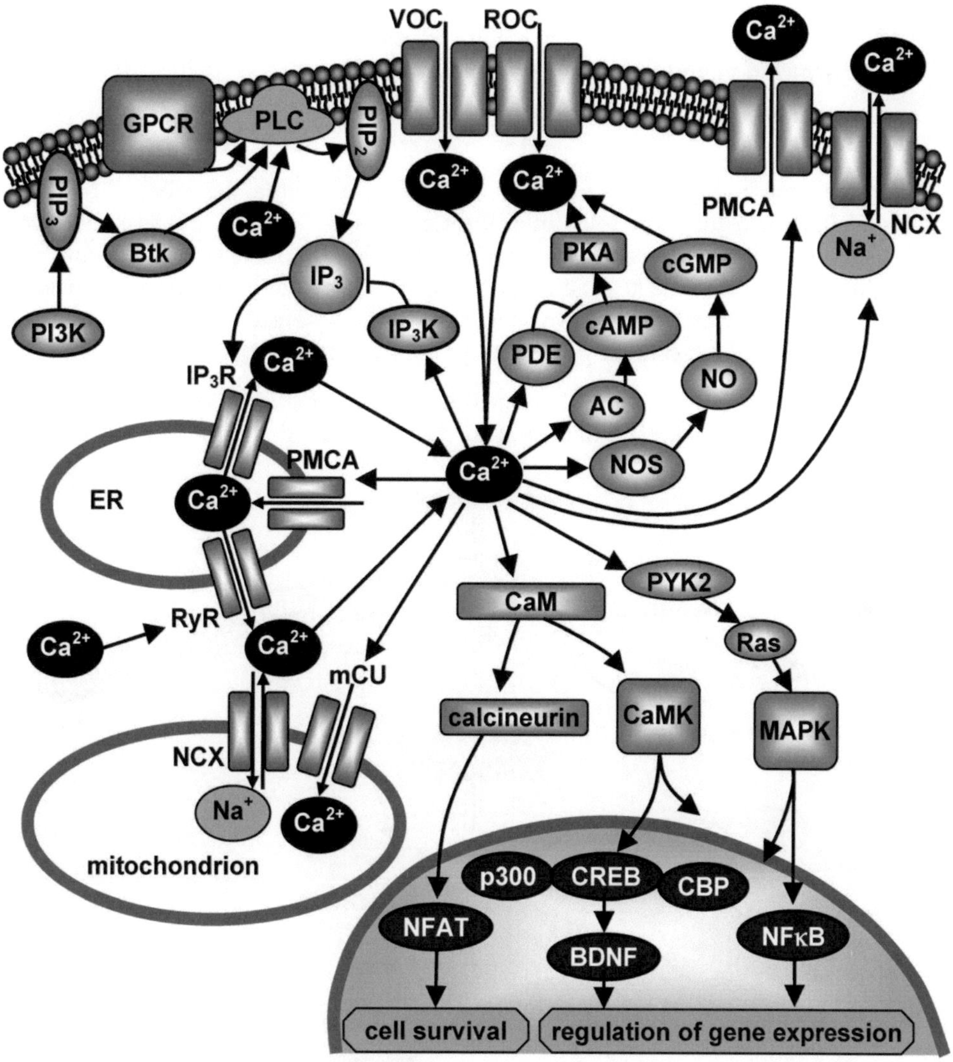

AC – adenylate cyclase (adenylyl cyclase); BDNF – brain-derived neurotrophic factor; Btk – Bruton tyrosine kinase; CaM – calmodulin; CaMK – calcium and calmodulin-dependent protein kinase; cAMP – cyclic adenosine monophosphate; CBP – CREB binding protein; CREB protein – cAMP response element-binding protein; cGMP – cyclic guanosine monophosphate; ER – endoplasmic reticulum; GPCR – G protein-coupled receptors; IP_3 – inositol 1,4,5-trisphosphate; IP_3K – IP_3 kinase; IP_3R – IP_3 receptor; MAPK – mitogen-activated protein kinase; mCU – mitochondrial Ca^{2+} uniporter; NCX – Na^+/Ca^{2+} exchanger; NFκB – nuclear factor κB; NFAT – nuclear factor of activated T cells; NO – nitric oxide; NOS – nitric oxide synthase; p300 – transcriptional co-activating protein; PDE – phosphodiesterase; PI3K – phosphoinositide 3-kinase; PIP_2 – phosphatidylinositol 4,5-bisphosphate; PIP_3 – phosphatidylinositol 3,4,5-trisphosphate; PKA – protein kinase A, cAMP dependent protein kinase; PLC – phosphoinositide phospholipase C, phospholipase C; PMCA – plasma membrane Ca^{2+} ATPase; PYK2 – tyrosine kinase enriched with proline; ROC – receptor-operated channel; RyR – ryanodine receptor; VOC – voltage-operated channel; ⟶ - activation; —┫- inhibition

Fig. 4. Calcium signalling pathway (Fišar & Hroudová, 2010). Detailed description in the text.

small GTPases Ras and mitogen-activated protein kinases (MAPKs). Protein kinases phosphorylate cellular proteins, which results in the cellular response to elevated Ca^{2+} concentrations, including induction of gene expression.

3.5 Wnt pathway

The Wnt signalling pathway activated by Wnt growth factors describes a complex of processes well known for their roles in embryogenesis and cancer, but also involved in normal physiological processes and in regulation of adult hippocampal neurogenesis (Clevers, 2006; Fuerer et al., 2008; Lie et al., 2005). Wnt signals are transmitted at least by three intracellular pathways; the most investigated one is canonical Wnt signalling pathway, which primarily modulates cellular processes during the cell development.

The canonical Wnt pathway (Fig. 5) describes a series of events occurring when Wnt glycoproteins bind to transmembrane receptor of the Frizzled family (WntR); membrane

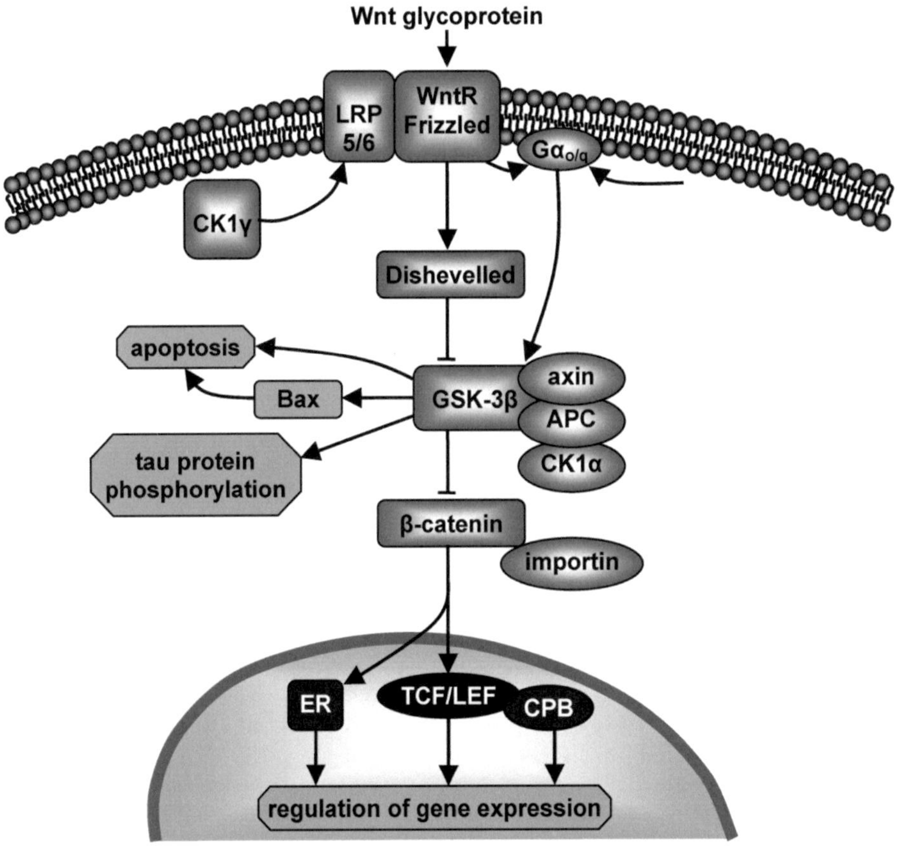

APC – adenomatous polyposis coli; Bax – proapoptotic factor from Bcl-2 family; CBP – CREB binding protein; CK – casein kinase; ER – estrogen receptor; GSK-3β – glycogen synthase kinase-3β; LEF – lymphoid-enhancing factor; LRP 5/6 – transmembrane protein; TCF – T cell factor; WntR – Wnt receptor (transmembrane receptor of the Frizzled family); → - activation; ⊣ - inhibition

Fig. 5. Canonical Wnt pathway (Fišar & Hroudová, 2010). Detailed description in the text.

protein LRP 5/6 phosphorylated by casein kinase 1γ (CK1γ) participates in the binding. Signal is transferred on the cytoplasmic protein Dishevelled that increases phosphorylation of glycogen synthase kinase-3β (GSK-3β). Therefore, complex of proteins that includes GSK-3β, axin, protein APC (adenomatous polyposis coli) and casein kinase 1α (CK1α) is inhibited, because phosphorylation of both of axin and β-catenin (mediated by GSK-3β) is decreased, and non-phosphorylated axin is degraded (Lee et al., 2003). It is suggested that heterotrimeric G proteins participate in disturbance of interaction GSK-3β with axin; some of these G proteins are probably associated with WntR (Liu et al., 2005). If the GSK-3β/axin complex is not inhibited, β-catenin is phosphorylated by CK1α and GSK-3β and β-catenin is degraded. If β-catenin is not degraded, it accumulates in cytoplasm and is transported by importin into the nucleus. Main target of β-catenin in the nucleus is transcription factor TCF (T cell factor) from LEF (lymphoid-enhancing factor) family; connection of TCF and β-catenin leads to activation of transcription factor. TCF, β-catenin and other transcription cofactors, such as CREB binding protein (CBP), activates the transcription of target genes. β-catenin activated by Wnt pathway interacts also with estrogen receptor (ER) and other nuclear receptors. Inhibition of GSK-3β links Wnt pathway also with apoptosis (through the activation of proapoptotic factor Bax from Bcl-2 family) and Alzheimer's disease (through tau protein phosphorylation). Recently, this pathway has been linked to therapeutic effects of lithium administered at bipolar disorders, probably through the activation of transcription factors by β-catenin.

3.6 Tyrosine kinase pathway

Tyrosine kinase pathway is activated by growth factors. Binding of neurotrophin to its trk receptor leads to three consequent processes (Fig. 6):

1. Receptor's autophosphorylation leads to phosphorylation of SHC protein, on which complex GRB2 (growth factor receptor bound protein-2)-SOS (son of sevenless protein) is attached to the membrane. SOS activates small G protein Ras and it attaches Raf kinase to the membrane. Activated Raf phosphorylates mitogen-activated protein kinase kinases MEK1, MEK2 or MEK5, and activated MEKs phosphorylate extracellular signal-regulated kinases (ERKs). G protein-coupled receptors are involved also in the ERK activation, namely by the activation of protein kinases PKA or PKC that activate Raf, consequently ERKs are activated. Activated ERK1 and ERK2 are transported into the nucleus and phosphorylate nuclear targets there, especially transcription factor Elk1 and kinase RSK (ribosomal protein S6 kinase). RSK has a principal role in regulation of gene transcription, because of its phosphorylation of a range of various factors. Phosphorylated Elk1 interacts with SRF (serum response factor) transcription factor, and transcription is initiated after the binding of a relevant response element in the area of c-Fos promoter. Phosphorylated RSK enables CREB phosphorylation; it binds to CREB binding protein (CBP) and to SRF/Elk1 complex, which leads to c-Fos transcription. As a result, there is regulation of immune and inflammatory processes, and control of cell growth and apoptosis. ERK5 is phosphorylated and activated by MEK5 (kinase of MAP kinase). In the contrast to ERK1 and ERK2, ERK5 is activated only by neurotrophins. ERK5 is included in cell survival through the MEF2 (myocyte enhancer factor 2) substrate (Wada & Penninger, 2004).
2. Phosphorylation of phospholipase Cγ (PLCγ) by trk receptor enables catalysis of phosphatidylinositol 4,5-bisphosphate (PIP_2) cleavage to diacylglycerol (DAG) and

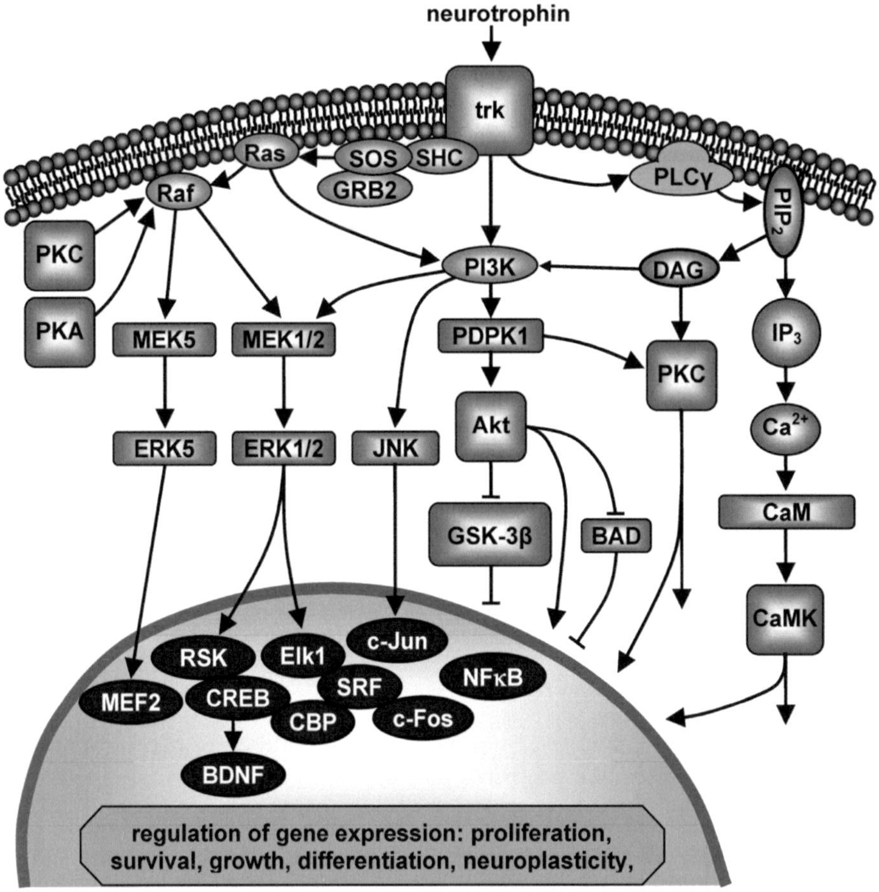

Akt - protein kinase B (PKB); BAD - proapoptotic factor (Bcl-2-associated death promoter); BDNF - brain-derived neurotrophic factor; CaM - calmodulin; CaMK - calcium and calmodulin-dependent protein kinase; CBP - CREB binding protein; c-Fos - transcription factor; c-Jun - transcription factor; CREB protein - cAMP response element-binding protein; DAG - diacylglycerol; Elk1 - transcription factor; ERK - extracellular signal-regulated kinase; GSK-3β - glycogen synthase kinase-3β; IP_3 - inositol 1,4,5-trisphosphate; JNK - c-Jun *N*-terminal kinase (stress-activated protein kinase); MEF2 - myocyte enhancer factor 2; MEK - mitogen-activated protein kinase kinase; NFκB - nuclear factor κB; PDPK1 - 3-phosphoinositide dependent protein kinase-1; PI3K - phosphoinositide 3-kinase; PIP_2 - phosphatidylinositol 4,5-bisphosphate; PKA - protein kinase A; PKC - protein kinase C; PLC - phosphoinositide phospholipase C; Raf - protein kinase ; Ras - small GTPase; RSK - ribosomal protein S6 kinase; SHC - protein with SH2 domain; SOS - son of sevenless protein; SRF - serum response factor; trk - tropomyosin-related kinase; ⟶ - activation; ⊣ - inhibition

Fig. 6. Tyrosin kinase pathway (Fišar & Hroudová, 2010). Detailed description in the text.

inositol trisphosphate (IP_3). DAG can activate phosphoinositide-3-kinase (PI3K) or various protein kinases C (PKCs). IP_3 releases calcium from intracellular stores and it activates calmodulin (CaM) and Ca^{2+}/CaM-dependent protein kinases (CaMKs).

3. Phosphatidylinositol 3,4,5-trisphosphate (PIP_3) is formed after the stimulation of PI3K heterodimers, which activates kinase PDPK1 (3-phosphoinositide dependent protein kinase-1) and therefore, protein kinase Akt (also known as PKB) is activated. Akt

supports survival and differentiation of neurons both by stimulation of transcription factors and by inhibition of proapoptotic factor BAD or glycogen synthase kinase-3β (GSK-3β). PI3Ks activate also PKCζ or they can indirectly activate protein kinases JNK (c-Jun *N*-terminal kinases), which are responsible for c-Jun phosphorylation and initiation of apoptosis (Cowan & Storey, 2003).

4. Neuroplasticity

Neuroplasticity is a fundamental mechanism of neuronal adaptation to environmental inputs. The term neuroplasticity (brain plasticity, cortical plasticity, cortical re-mapping) is used for description of either functional or structural changes of neurons and glial cells that occur in developing brain as well as in the adult brain in order to adjust to external or internal stimuli (Mesulam, 1999; Nestler et al., 2002). The most widely recognized forms of plasticity are learning, memory, and recovery from nervous system injury, which may happen through the change in the strength of connections among brain cells, by adding or removing connections, or by adding new cells.

Neuroplasticity is linked to the concept of synaptic pruning (neuronal pruning, neurostructural reassembly), defined as regulatory processes, which facilitate a productive change in neural structure by reducing the overall number of neurons or connections, leaving more efficient synaptic configurations; really, excess of neurons was observed in the human newborn mediodorsal thalamus compared with that of the adult (Abitz et al., 2007). New findings suggest that all areas of the brain are plastic even after childhood. Neuroplasticity in the adult brain includes changes of dendritic functions, reorganization of synapses, long-term potentiation (LTP), long-term depression (LTD), branching and sprouting of axons and dendrites, synaptogenesis, and neurogenesis (Fišar & Hroudová, 2010). Environmental changes could alter behaviour and cognition by adapting connections between neurons, and neurogenesis may be included in these processes. It is possible that there is a link between neurogenesis and learning-related changes in the brain. Adult neurogenesis in mammals is mainly restricted to the hippocampus and olfactory bulb; current research has revealed that other parts of the brain, the cerebellum included, may be also involved (Ponti et al., 2008). However, the reorganisation of the complex brain networks is not always beneficial for the individual. Maladaptive plasticity can be defined as behavioural loss or as development of disease symptoms resulting from plasticity changes in the adult brain (Draganski & May, 2008).

The development of new synapses, the activity dependent changes in the strength of existing synapses and the elimination of synapses have been proposed to form basis of synaptic plasticity. The synaptic plasticity could be the cellular basis of certain forms of learning and memory (Citri & Malenka, 2008). Synaptic plasticity has been studied in many brain regions, the most frequently in the hippocampus.

Recently, the concept was modified that synapse-specific forms of LTP and LTD at excitatory synapse can fully explain learning and experience-dependent plasticity. Intrinsic, inhibitory, and homeostatic plasticity were documented as additional forms of plasticity (Nelson & Turrigiano, 2008). A novel form of persistent synaptic plasticity was called metaplasticity (the plasticity of synaptic plasticity). Metaplasticity is induced by synaptic or cellular activity, but is not necessarily expressed as a change in the efficacy of normal synaptic transmission. Instead, it is manifested as a change in the ability to induce subsequent synaptic plasticity, such as LTP or LTD. Thus, metaplasticity is a higher-order form of synaptic plasticity (Abraham & Bear, 1996).

There is both postsynaptic and presynaptic plasticity. Postsynaptic plasticity involves changes in the number or sensitivity of postsynaptic receptors without any changes in the amount of neurotransmitter release. Basic mechanisms of postsynaptic plasticity are connected with activation of postsynaptic NMDA receptors followed by Ca^{2+} influx; calcium triggers following mechanisms contributing to synaptic plasticity in spines: 1. The regulation of channels and proteins involved in trafficking, cytoskeletal organization and protein synthesis; 2. Alterations of synaptic AMPA receptor properties, subunit composition and trafficking; 3. Actin reorganization and modulation of spine morphology; and 4. Initiation of local protein synthesis in spines and dendrites (Derkach et al., 2007).

Presynaptic plasticity generally translates into an increase or a decrease of neurotransmitter release (García-Junco-Clemente et al., 2005). Presynaptic plasticity has the potential to greatly influence all of the neurotransmitters release sites within a given axon, such that changes in the output of one inhibitory interneuron could modify the activity of many of its downstream target neurons (Tóth & McBain, 2000). Mechanisms of presynaptic LTP or LTD may be independent on NMDA receptors. It is hypothesized that various forms of presynaptic plasticity can operate in a manner fundamentally distinct from most postsynaptic forms of plasticity (García-Junco-Clemente et al., 2005; McBain & Kauer, 2009). A new class of presynaptic plasticity that requires signalling by endocannabinoids has been identified in several brain structures (Chevaleyre et al., 2006). At many of these synapses presynaptically expressed forms of LTD can coexist with postsynaptic forms of LTD mediated by internalization of AMPA receptors (Nelson & Turrigiano, 2008).

Learning and memory depend on long-lasting changes in synaptic strength. Long-term synaptic plasticity changes last from hours to weeks. These changes require induction of gene expression, production and insertion of new proteins. Activation of β-adrenoceptors can enhance LTP and facilitate long-term memory storage. Cyclic AMP/PKA and extracellular signal-regulated protein kinase cascades are important to express the long-lasting LTP in hippocampus, amygdala, and cortex (Pittenger & Duman, 2008). Transcription factor CREB is particularly important in modulation of synaptic plasticity. CREB is activated (phosphorylated) by PKA and other kinases upon synaptic stimulation during learning (Carlezon et al., 2005).

5. Stress, HPA axis and glutamate neurotoxicity

Stress may be defined as any environmental change, whether internal or external, that disturbs homeostasis (Leonard & Myint, 2009). Stress system is located in both the central nervous system and peripheral organs. Central functions of the stress response include facilitation of arousal, alertness, cognition, attention and aggression, inhibition of vegetative functions, and activation of counter-regulator feedback loops. Peripheral functions include increase of oxygenation, nutrition of the brain, hearth and skeletal muscles, increase of cardiovascular tone and respiration, increase of metabolism and detoxification, and activation of counter-regulatory feedback loops (Chrousos, 2009).

Stress activates both hypothalamic-pituitary-adrenal (HPA) axis and the sympathetic nervous system. The main central effectors of stress system include corticotrophin-releasing factor (CRF), arginine vasopressin (AVP) and norepinephrine. Chronic stress, as a result of the hypersecretion of cortisol, causes a decrease in serotonin turnover partly as a consequence of increased metabolism of tryptophan (Leonard & Myint, 2009). Increased activity of the HPA axis has been reported not only after stress but also in pregnancy and in many diseases, such as Cushing syndrome, depression, anorexia nervosa, obsessive-compulsive disorder, panic disorder, alcoholism, diabetes mellitus, metabolic syndrome, hyperthyroidism, etc. (Chrousos, 2009).

Stressors provoke the secretion of epinephrine and norepinephrine by the sympathetic nervous system, to induce the flight-or-fight response, and of the glucocorticoids by the adrenal gland. Catecholamine action involves activation of β-adrenoceptors and initiation of second messenger cascades in target cells within seconds; whereas, glucocorticoid's effects can take hours to emerge, as they involve transcriptional events. Brain areas involved in the stress response include the prefrontal cortex, the hippocampus, and the amygdala; they undergo stress-induced remodelling, which alters behavioural and physiological responses (McEwen, 2007). Many central aspects of stress response are modulated, and in some cases mediated, by glutamate neurotransmission in the prefrontal cortex (Moghaddam, 2002).

Stress and stress hormones produce both adaptive and maladaptive effects in the brain (McEwen, 2007). It is known that stress is important contributor to psychosocial and physical disorders. The relationship between stressful life events and development of mood disorders in vulnerable subjects has been long established (Aguilera et al., 2009; Caspi et al., 2003; Johnson, 2005; Kendler et al., 1999). Adverse childhood experiences have been described as one of the major environmental risk factors for depressive disorder.

Acute and chronic stress can have quite different effects on neuroplasticity. The mild stress for a few hours can enhance cognition by facilitating aspects of synaptic plasticity in the hippocampus; these effects are mediated by high-affinity corticosteroid receptors. In contrast, excessive glucocorticoid exposure in the hippocampus as a result of major and prolonged stress can be directly toxic to neurons, or can increase the neurotoxicity of various hippocampal insults (Lee et al., 2002; McEwen & Sapolsky, 1995; Sapolsky, 1996).

Stress produces a rapid increase in glutamate efflux in the prefrontal cortex and the hippocampus (Moghaddam, 2002). An excess of glutamate in the synapse leads to excess of cytosolic calcium, which produce overactivity of calcium-dependent enzymes and it leads to cytoskeletal degradation, protein malformation and oxygen radical generation. These processes can lead to atrophy or death of neurons (Atlante et al., 2001; Lipton, 1999). Different insults, such as hypoxia-ischemia, seizure and hypoglycaemia, all of them activate this pathway.

Neurons mobilize a variety of defences when are challenged with glutamatergic insults, e.g. removal of glutamate from the synapse, and of calcium from the cytoplasm, production of heat shock proteins (HSP), protective hyperpolarisation, and protective upregulation of antioxidant enzymes (Lee et al., 2002). The glia cells account for the majority of glutamate uptake.

Processes accompanying neuroplasticity are extremely energy-consuming and interfere with different intracellular pathways included in signal transduction or in apoptosis. Thus, it seems to be useful to study the mitochondrial dysfunctions in relation to neuroplasticity, mechanisms of stress response, pathophysiology of mood disorders, and mechanisms of action of psychotropic drugs, antidepressants and mood stabilizers included (Hroudová & Fišar, 2011). It was proposed that subtle deficits in the mitochondrial function likely play an important role in various facets of bipolar disorder, and that enhancing mitochondrial function may represent a critical component for the optimal treatment of the disorder (Quiroz et al., 2008).

There are close relationships between the mechanisms of synaptic plasticity, effects of stress, mechanisms of antidepressant action, and pathophysiology of depression (Pittenger & Duman, 2008). In order to explain these relationships, interactions among signal transduction pathways have been studied (Fišar & Hroudová, 2010; Hroudová & Fišar, 2011), including the role of mitochondria (Fig. 7).

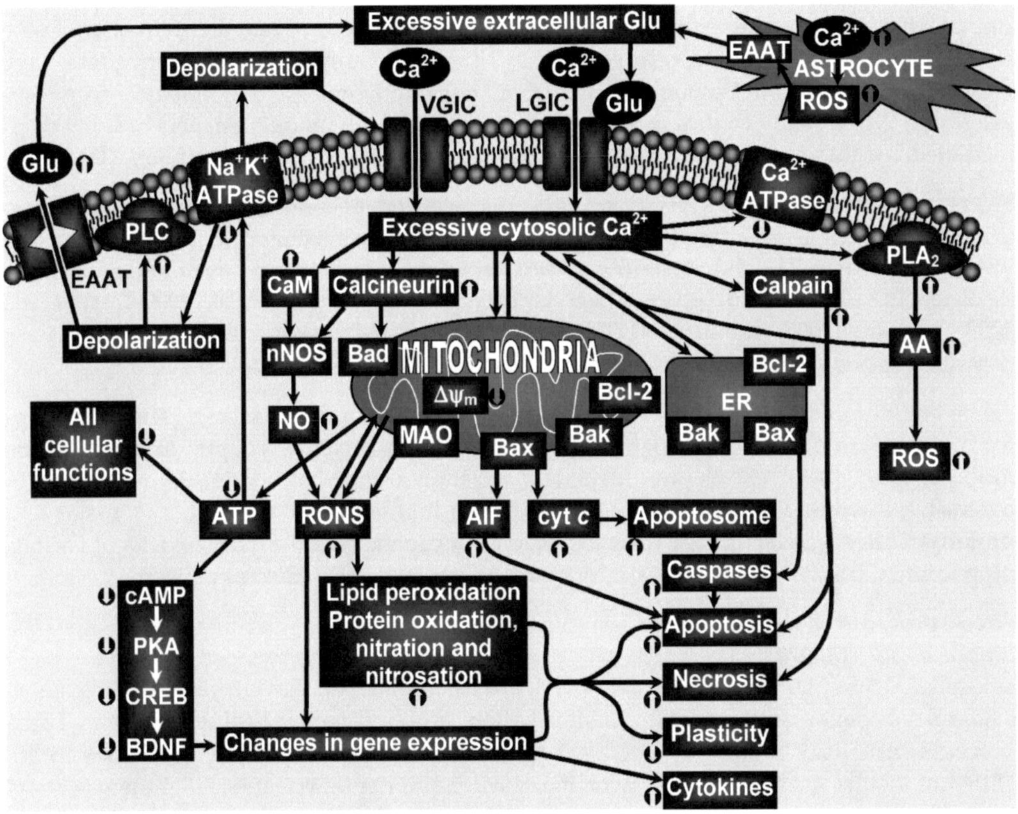

AA - arachidonic acid; ATP - adenosine-5'-triphosphate; Bax, Bak, BAD - proapoptotic factors of Bcl-2 family; Bcl-2 - antiapoptotic factor of Bcl-2 family; BDNF - brain-derived neurotrophic factor; CaM - calmodulin; cAMP - cyclic adenosine monophosphate; CREB - cAMP response element-binding; $\Delta\psi_m$ - potential on the inner mitochondrial membrane; EAAT - excitatory amino acid transporter; ER - endoplasmic reticulum; Glu - glutamate; MAO - monoamine oxidase; nNOS - neuronal nitric oxide synthase; NO - nitric oxide; PKA - protein kinase A; PLA_2 - phospholipase A_2; PLC - phospholipase C; ROS - reactive oxygen species; RONS - reactive oxygen and nitrogen species

Fig. 7. Principal mechanisms of glutamate neurotoxicity (Hroudová & Fišar, 2011). Detailed description in the text.

Principal mechanisms leading to neuronal impairment and cell death are composed of

1. Decreased adenosine triphosphate (ATP) production;
2. Increased production of reactive oxygen and nitrogen species (RONS);
3. Initiation of apoptotic processes;
4. Impaired calcium homeostasis.

Decrease of ATP production leads to impairment of ATP-dependent processes and therefore to changed cellular functions (Fig. 7). Insufficient function of $Na^{+}K^{+}$- ATPases leads to membrane depolarization. Membrane depolarization changes the functions of amino acids transporters and increases concentration of extracellular glutamate. Voltage-gated ion channels (VOC) and/or ligand-dependent calcium channels (ROC) are activated and mediate calcium influx. Intracellular calcium causes functional changes of amino acid transporters, enhances extracellular glutamate and extends neurotoxicity. Increased levels of synaptic glutamate can be mediated also by release of glutamate from astrocytes. The activation of ionotropic glutamate receptors leads to higher calcium influx into the cell, which is followed by increased activation of phospholipases, proteases, and endonucleases. For example, activation of phospholipase A_2 (PLA_2) by calcium releases membrane arachidonic acid (AA), which induces production of superoxide.

High intracellular calcium levels induce overload of mitochondrial calcium, leading to increase of reactive oxygen species (ROS) production and inhibition of ATP production. Activation of calcium-dependent protein phosphatases (e.g. calcineurin) causes translocation of proapoptotic factor BAD into the mitochondria and triggers apoptosis by sequestration of antiapoptotic factors Bcl-2 and Bcl-xL. Release of cytochrome *c* (cyt *c*) and other proapoptotic factors from the intermembrane space of mitochondria induces the formation of apoptosome, and consequently triggers activation of caspases and apoptosis. Apoptosis-inducing factor (AIF) is another factor released by mitochondria. Mitochondria in the brain are also a target of nitric oxide (NO) action.

6. Mechanisms of action of antidepressants and mood stabilizers

Current antidepressant medications are effective in about 60% of treated patients (Nelson, 1999; Papakostas et al., 2007) and significant therapeutic response is observed after several weeks of treatment. Thus, faster and more effective pharmacological treatments for depressive disorders are greatly needed.

Biochemical effects of antidepressants and mood stabilizers are studied with the aim to discover both molecular mechanism of their therapeutic efficiency and neurochemical nature of mood disorders. Both complex clinical pattern of mood disorders and adaptive changes in activity or availability of a large number of components of signalling pathways after long-term treatment with antidepressants is responsible for the fact, that definite molecular mechanisms responsible for therapeutic action of drugs are not known.

It seems that psychotropic drugs used in the therapy of mood disorders show neurotrophic or neuroprotective effects after long-term treatment. Thus, next to adenylate cyclase, guanylate cyclase, phosphoinositide and calcium systems, attention has been paid to tyrosine kinase pathway, Wnt pathway and inflammatory pathway.

6.1 Antidepressants

Antidepressants are psychiatric medication used to treatment of mood disorders, such as major depression, dysthymia, and anxiety disorders. Easily 41 drugs are currently used as antidepressants worldwide and many other drugs are administered as supportive therapy or holding course. At the level of chemical synapses, antidepressants usually act as serotonin or norepinephrine reuptake inhibitors, as inhibitors of the degradation of monoamine neurotransmitters, or as agonists or antagonists of their receptors. Therefore, the administration of antidepressants induces increased concentrations of norepinephrine and serotonin in the synaptic cleft. Consequently, increased activation or inhibition occurs in processes of intracellular signalling pathways, mainly in adenylate cyclase and phosphoinositide pathway.

Clinical effects of antidepressants are obviously caused by their ability to induce adaptive changes in neurotransmission, mainly serotonergic and noradrenergic. Changes in the availability of neurotransmitters and also in the density and sensitivity of their receptors and transporters are not sufficient to explain origin and course of the mood disorders neither the mechanisms of action of antidepressants and mood stabilizers. It was supposed that intracellular processes included in apoptotic, neurodegenerative, and inflammatory pathways are responsible for final therapeutic effects (Porcelli et al., 2011).

Antidepressants are currently classified according to their direct biochemical effects, which are well known in comparison with their long-term effects. The first antidepressants were monoamine oxidase inhibitors (MAOIs) and nonselective serotonin and/or norepinephrine reuptake inhibitors referred to as tricyclic antidepressants (TCAs). MAOIs act as antidepressants by blocking of enzyme that degrades monoamine neurotransmitters; TCAs act as antidepressants by blocking membrane transporters ensuring reuptake of 5-HT or NE, thus cause increased extracellular neurotransmitter concentrations. Next generations of antidepressants include selective serotonin reuptake inhibitors (SSRIs), norepinephrine reuptake inhibitors (NRIs), serotonin-norepinephrine reuptake inhibitors (SNRI), noradrenergic and specific serotonergic antidepressants (NaSSAs), serotonin antagonist and reuptake inhibitors (SARIs), norepinephrine-dopamine reuptake inhibitors (NDRIs), melatonin receptors agonist and selective serotonin antagonist (MASSA), sigma receptors agonist, and drugs directly affecting the neuroplasticity renewal. However, other classes of antidepressants may be suggested; e.g. antagonists of 5-HT_{2C} receptors may be classified as norepinephrine-dopamine disinhibitors (NDDIs), because they act by antagonizing receptors that normally acts to inhibit norepinephrine and dopamine release. Tianeptin was initially classified as selective serotonin reuptake enhancer (SSRE); however, its affinity to serotonin transporter is low and it was supposed that its therapeutic action is related to restore normal neuroplasticity in circumscribed limbic brain regions and to reverse stress-induced impairments in synaptic glutamate transmission (McEwen et al., 2010). It is of interest that new psychotropic drugs are multifunctional, i.e. agents with more than one putative therapeutic mechanism of action (Stahl, 2009).

6.1.1 Effect on glutamate system

Antidepressants affect learning and memory in animal models and enhance structural plasticity and hippocampal neurogenesis (Drzyzga et al., 2009; Kasper & McEwen, 2008;

Warner-Schmidt & Duman, 2006). Antidepressants can directly modulate glutamatergic neurotransmission through NMDA or AMPA receptors; it is likely that an intimate relationship exists between regulation of monoaminergic and glutamatergic neurotransmission and antidepressant effects (Paul & Skolnick, 2003). As mentioned above, tianeptine prevents or reverses stress-associated structural and cellular changes in the brain and normalizes disrupted glutamatergic neurotransmission in the hippocampus, the amygdala, and the cortex (Kasper & McEwen, 2008; McEwen et al., 2010). An inhibition of an excessive release of glutamate appears to be important for mechanisms of action of lamotrigine and riluzole (Zarate et al., 2006a). Robust and rapid antidepressant effect on individuals with treatment-resistant depression resulted from a single intravenous dose of ketamine (a non-competitive NMDA receptor antagonist and psychomimetic) (Diazgranados et al., 2010; Zarate et al., 2006b). These effects suggest that depressive symptoms can be improved by altering the action of glutamate (Krishnan & Nestler, 2008).

6.1.2 Effect on neuroplasticity and neurogenesis

Time-dependent therapeutic effects of antidepressants could be related to changes in gene expression in the brain. If a deficit in neuroplasticity is included in pathophysiology of depression, then it can be supposed that effects of stress on the mechanisms of neuroplasticity contribute to the genesis of depression and long-term antidepressant treatment affects the same mechanisms (Pittenger & Duman, 2008).

There are many evidences that antidepressants increase signalling pathways related to neuroplasticity by upregulation of cAMP/PKA/CREB cascade, by regulation of CaMKII activity and by upregulation of the MAPK cascade. The hypothesis that long-term antidepressant treatment enhances neuroplasticity is based on upregulation of expression of many neurotrophic factors, especially of BDNF, in the hippocampus and the prefrontal cortex (Bocchio-Chiavetto et al., 2006; Duman & Monteggia, 2006; Nibuya et al., 1996; Pittenger & Duman, 2008).

A neurotrophic model for the molecular mechanism of action of antidepressant treatments includes following steps: 1. Antidepressant treatment causes direct inhibition of serotonin and/or norepinephrine reuptake or breakdown, which is followed by elevation of extracellular levels of serotonin and/or norepinephrine; 2. Long-term treatment causes adaptive changes in the function and expression of serotonin and norepinephrine receptors, increase in the cAMP signal transduction and increase in expression of the transcription factor CREB; 3. Increased activity of the cAMP signal transduction cascade induces increased expression of BDNF and its receptor trkB; thus, neuroplasticity is improved and neuron survival and recovery is increased.

The results support the hypothesis that treatment by antidepressant lead to effects similar to neurotrophic. One of the target genes of pharmacotherapy is *BDNF* gene. BDNF supports processes implicated in neuronal plasticity and in renewal or improvement of neuronal connectivity. The renewal of synaptic connections and signalling pathways enables the normal function of neurotransmitters. It is suggested, that antidepressants can eliminate stress effects on the hippocampus and improve the symptoms of depression in this way. Due to the complexity of intracellular processes it is relatively difficult to establish this hypothesis. There is not large longitudinal study that might prove the ability of antidepressants to reverse atrophy of brain structures and prevent it.

The chronic administration of antidepressants also increases neurogenesis in hippocampal structures, i.e. it increases proliferation and survival of new neurons. Reverse effects of stress and antidepressant on the hippocampal neurogenesis indicate that changes in hippocampal neurogenesis can be significant in clinic syndrome of depression, although it is a complex disorder that targets more than one region of the brain. Coupling of hippocampal neurogenesis to pathophysiology of depression requires further research to be confirmed (Gass & Riva, 2007; Santarelli et al., 2003).

6.1.3 Effect on inflammatory pathway

Suggestion that the activation of immune responses and the release of inflammatory cytokines (including IL-1β, IL-6, TNF-α) may play a role in the pathophysiology of depression is supported by observation that there is association between antidepressant action and cytokine functioning. The most consistent finding has been that antidepressant treatment significantly reduces interleukin-6 (IL-6) plasma concentrations, which are elevated in depression (Janssen et al., 2010). Proposed mechanisms for the antidepressant effects on cytokine function include: 1. Increase of intracellular cAMP/PKA pathway by antidepressants leads to inhibition of NFκB, a transcription factor which promote proinflammatory gene expression; 2. Antidepressants and glucocorticoids may act via common intracellular signalling pathways; e.g. activated PKA is able to enhance glucocorticoid receptor-DNA binding (Pace & Miller, 2009); 3. Increased extracellular levels of serotonin after treatment with serotonergic antidepressants exhibit immunosuppressive effects (Kubera et al., 2000), and depleted intracellular 5-HT levels reduce cytokine levels through inhibition of mRNA expression (Maes, 2001); 4. Antidepressants may reverse effect of cytokines on depression of hippocampal neurogenesis (Koo & Duman, 2008). In conclusion, the inflammatory and neurodegenerative pathways might provide new targets for antidepressant development (Catena-Dell'Osso et al., 2011).

6.1.4 Direct, early and long-term effects

Sequence of biochemical events induced by antidepressants is crucial for discovery of molecular mechanisms associated with their therapeutic effects. Direct (immediate) biochemical effects of antidepressants leading to their therapeutic action include:

1. Inhibition of reuptake of monoamine neurotransmitters, i.e. inhibition of membrane transporters for serotonin, norepinephrine or dopamine.
2. Inhibition of metabolism of monoamine neurotransmitters, i.e. monoamine oxidase inhibition,
3. Receptor activation, e.g. postsynaptic serotonin receptors type 1A ($5\text{-}HT_{1A}$) or sigma receptors.
4. Blockade (antagonism) of monoamine receptors, e.g. postsynaptic serotonin receptors type 2A and 2C ($5\text{-}HT_{2A}$, $5\text{-}HT_{2C}$) and presynaptic α_2-adrenoceptors.
5. Inhibition or activation of several intracellular components of signalling pathways participant in neurotransmission.

Neurochemical events for antidepressant action classified into direct (immediate) effects, early (intermediate) effects, and long-term (delayed) effects are summarized in Fig. 8.

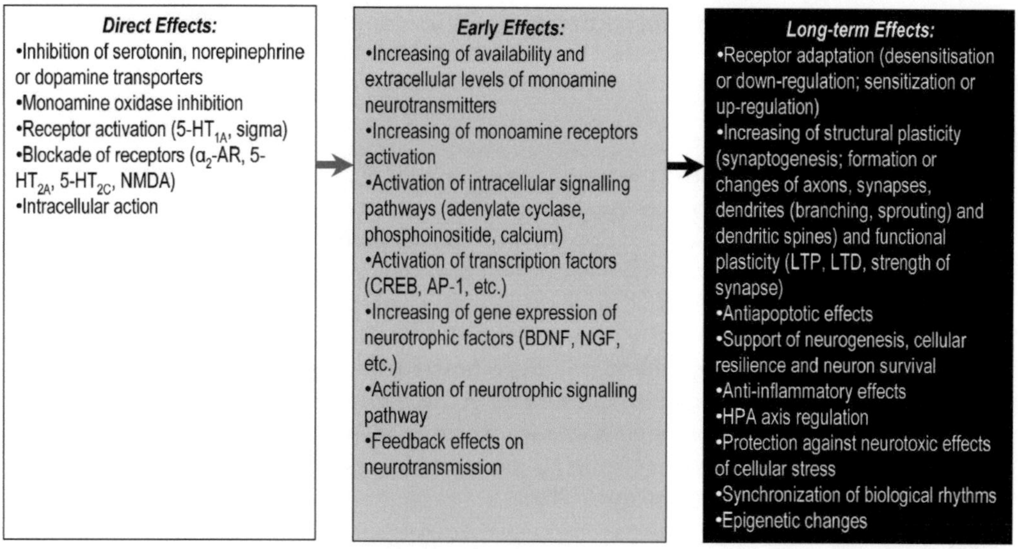

Fig. 8. Sequence of effects of antidepressants

Immediate biochemical effects of antidepressants are followed by early events, such as increasing of availability and extracellular levels of monoamine neurotransmitters, increasing of monoamine receptors activation, activation of intracellular signalling pathways, activation of transcription factors and increasing of gene expression of neurotrophic factors.

It is evident that adaptive changes in central neurotransmission are responsible for therapeutic effects of antidepressants. These adaptive processes and their regulatory mechanisms are studied on the neurochemical, neuroimmune, neuroendocrine, chronobiological, and genetic level primarily:

1. Neurochemical events include a) changes of neurotransmitter receptor density and sensitivity, which were not confirmed for *in vivo* treatment with many antidepressants and mood stabilizers on patients with affective disorders (Nikolaus et al., 2009; Richelson, 2003); b) antiapoptotic effects and changes in gene expression, neuroplasticity and neurogenesis, when stress-induced or depression-induced damage of brain cell functions and their interconnections are recovered after long-term treatment with antidepressants (Duman, 2009).
2. Neuroimmune approach is based on observations that a) major depressive disorder is an inflammatory disorder with an overproduction of proinflammatory cytokines (Maes et al., 2009); b) anti-inflammatory effects of antidepressants participate on their pharmacological effects (Janssen et al., 2010).
3. Neuroendocrine hypotheses suppose that therapeutic effects of antidepressants consist in regulation of HPA axis, which can be overactivated during depression (Nikisch, 2009).
4. Chronobiological hypotheses connect therapeutic effects of antidepressants with synchronization of biological rhythms disturbed in depression (Bunney & Potkin, 2008; Mendlewicz, 2009; Schulz & Steimer, 2009).

5. Genetic factors contribute for about 50% of the antidepressant response (Crisafulli et al., 2011). Epigenetic changes are studied both in relation to gene-environment interactions (G × E) (Caspi et al., 2003; Uher, 2008) and in animal models of stress, depression and antidepressant treatment (Schroeder et al., 2007; Tsankova et al., 2007; Yasuda et al., 2009).

6.2 Mood stabilizers

Mood stabilizers are psychiatric medication used in treatment of mood disorders, which are characterized by intense and sustained mood shifts (e.g. bipolar disorder). Most of mood stabilizers are anticonvulsants (valproate, carbamazepine, and lamotrigine), with an important exception of lithium, which is the oldest and the best known mood stabilizing drug. Some atypical antipsychotics (olanzapine, quetiapine, aripiprazole, risperidone, ziprasidone) have mood stabilizing effects, as well. It is also suggested that ω-3 fatty acids may have a mood stabilizing effect. It is hypothesized that ω-3 fatty acid deficiency may contribute to elevated phosphoinositide-PKC in neuropsychiatric illness (schizophrenia, bipolar disorder, major depressive disorder), because ω-3 fatty acids act as endogenous inhibitors of second messenger-regulated protein kinases (PKA, PKC, CaMK, MAPK) (McNamara et al., 2006; Mirnikjoo et al., 2001).

Mood stabilizers affect multiple sites in intracellular signalling pathways (Fig. 9). Main targets of mood stabilizers are neurotrophin BDNF, ERK pathway, and pathways modulated by GSK-3 or Bcl-2 (Einat & Manji, 2006; Gould & Manji, 2005; Shaltiel et al., 2007). Molecular and cellular targets of mood stabilizers include enzymes inhibited by lithium (inositol monophosphatase, inositol polyphosphate 1-phosphatase, GSK-3, fructose 1,6-bisphosphatase, bisphosphate nucleotidase, phosphoglucomutase), enzymes inhibited by valproate (succinate semialdehyde dehydrogenase, succinate semialdehyde reductase, histone deacetylase), targets of carbamazepine (sodium channels, adenosine receptors, adenylate cyclase), and components of signalling pathways regulated by multiple drugs (PKC, cAMP, arachidonic acid) (Gould et al., 2004). Furthermore, lithium and valproate reduce transport of *myo*-inositol into the cells, which leads to reduced PKC activity. Lithium and valproate increase Bcl-2 concentrations (Chen et al., 1999) and inhibit GSK-3 activity (lithium directly, valproate indirectly). Valproate activates MAPK signalling pathway and regulates stress proteins of endoplasmic reticulum (Bown et al., 2002). Through the effects on Bcl-2 and p53 (proapoptotic protein), lithium affects mitochondria by stabilization of membrane integrity and prevention of mitochondrial membrane permeability transition pores (MPTPs) opening; i.e. by regulating the key process in cell death leading to at least temporary loss of $\Delta\psi_m$, input of water into matrix and equilibration of ions concentrations.

Both lithium and valproate have neuroprotective effects based on protection from glutamatergic neurotoxicity by inactivation of NMDA receptors, on activation of cell survival factors such as phosphoinositide 3-kinase/protein kinase B pathway, and on induction of neurotrophic and neuroprotective proteins. Lithium protects against DNA damage, caspases activation, and apoptosis of neurons (Chuang, 2005). Increased concentrations of *N*-acetyl aspartate (NAA, marker of neuronal viability and functionality) in grey matter after the chronic lithium administration support its strong neuroprotective and neurotrophic effects in humans. Lithium also increases hippocampal neurogenesis.

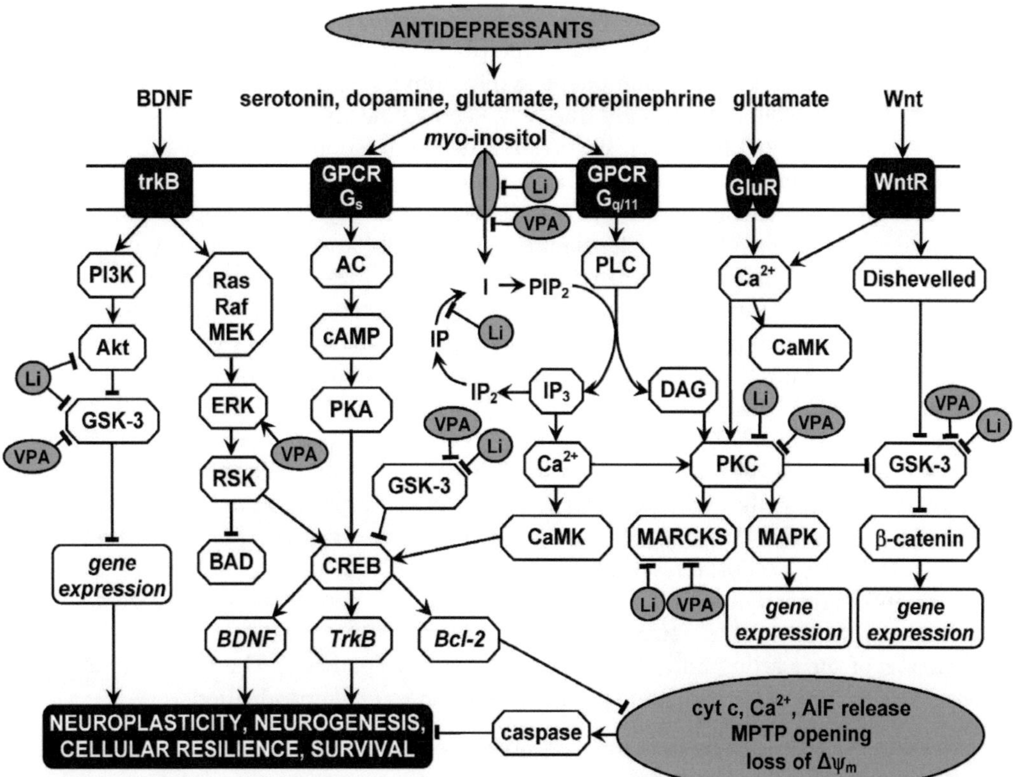

AC - adenylate cyclase (adenylyl cyclase); Akt - protein kinase B (PKB); BAD - proapoptotic factor (Bcl-2-associated death promoter); Bcl-2 - antiapoptotic factor of Bcl-2 (B-cell CLL/lymphoma 2) family; BDNF - brain-derived neurotrophic factor; CaM - calmodulin; CaMK - calcium and calmodulin-dependent protein kinase; cAMP - cyclic adenosine monophosphate; CREB protein - cAMP response element-binding protein; DAG - diacylglycerol; ERK - extracellular signal-regulated kinase; GluR - ionotropic glutamate receptor; GPCR - G protein-coupled receptors; GSK-3 - glycogen synthase kinase-3; I - inositol; IP - inositol monophosphate; IP_2 - inositol bisphosphate; IP_3 - inositol trisphosphate; MAPK - mitogen-activated protein kinase; MARCKS - myristoylated alanine-rich C-kinase substrate; MEK - mitogen-activated protein kinase kinase; MPTP - mitochondrial membrane permeability transition pore; PI3K - phosphoinositide 3-kinase; PIP_2 - phosphatidylinositol 4,5-bisphosphate; PKA - protein kinase A; PKC - protein kinase C; PLC - phosphoinositide phospholipase C; Raf - protein kinase ; Ras - small GTPase; RSK - ribosomal protein S6 kinase; trkB - neurotrophic tyrosine kinase, receptor, type 2; VPA - valproic acid; WntR - Wnt receptor (transmembrane receptor of the Frizzled family); ⟶ - activation; ⊣ - inhibition

Fig. 9. Neurotrophic and neuroprotective effects of antidepressants and mood stabilizers. Detailed description in the text.

7. Advances in neurochemical hypotheses of mood disorders

There are many hypotheses of mood disorders; however, current clinical, neurochemical, neurophysiological, genetic, and neuroimaging data are not sufficient for enunciation of theory of mood disorder. Thus, the ethiopathogenesis of major depressive disorder and other mood disorders is still unclear.

The neurochemical hypotheses of mood disorders were developed from neurotransmitter hypotheses to receptor and postreceptor hypotheses. Inadvertently induced depression by reserpine and effectiveness of first antidepressants from the class of TCA or MAOI led to the hypothesis that affective disorders are due to catecholamine and/or indolamine deficiency in the CNS (Coppen, 1967; Schildkraut, 1965). Single deficit of monoaminergic neurotransmission is insufficient to elucidation of neurochemistry of mood disorders. It was deduced that disposition of CNS characterized by disturbance in function of receptors for the key monoamine neurotransmitters is determining factor. Receptor malfunction may be caused by depletion of monoamine neurotransmitters, by abnormalities in the receptor density and sensitivity, or by problems with signal transduction on postreceptor level. It was formulated a large amount of receptor hypotheses. Classical norepinephrine receptor hypothesis of affective disorders supposed increased density of postsynaptic β-AR in depression (Banerjee et al., 1977; Vetulani & Sulser, 1975). Very elegant, although unconfirmed, was receptor catecholamine hypothesis (Richelson, 1990) that supposes that supersensitivity of catecholamine receptors in the presence of low levels of serotonin is the biochemical basis of depression. According to receptor hypotheses the common final result of chronic treatment by majority of antidepressants is the down-regulation or up-regulation of postsynaptic or presynaptic receptors. The delay of clinical response corresponds with these receptor alterations. However, problems in interpretation of molecular mechanisms of action of mood stabilizers and the introduction of new, more selective antidepressants led to new reflection upon the mechanisms of their action and to new neurochemical hypotheses. The most promising are recent investigations of the signalling pathways ensuring signal transduction.

Introduction of newer antidepressants sustained the monoamine deficiency hypothesis and the monoaminergic signalling pathways continues to be one of the key targets of new antidepressants. For decades, research focused mainly on the role of monoamines in mood disorders. However, data are inconclusive and the monoamine hypothesis seems to be insufficient to explain depression (Catena-Dell'Osso et al., 2011). E.g., *in vivo* neuroimaging studies on human brain neurotransmission provide little evidence about deficits in monoaminergic neurotransmission. Reports on positron emission tomography (PET) or single-photon emission computed tomography (SPECT) of neurotransmission in depressive disorders showed that there is usually no reliable relationship between the binding potential of the neuroreceptor or transporter and the clinical condition of depressed subject (Nikolaus et al., 2009; Smith & Jakobsen, 2009).

While dysfunctions within monoaminergic neurotransmitter systems are likely to play an important role in pathophysiology of mood disorders, it probably represents the downstream effects of more primary abnormalities in signal transduction. Thus, new theories about the pathophysiology of depression and the action of antidepressant treatment proposes that mood disorders are caused by structural or functional changes in particular molecules and signalling pathways in the brain, and that antidepressants function by counteracting these molecular changes. Neurodegeneration or impaired neurogenesis may be important mechanism of depression, and neurotrophins are key regulators of neurogenesis and neuroplasticity. Recovery of brain networks through increase of neuroplasticity induces antidepressant effect. It is supposed that structural and functional brain abnormalities in patients with depressive disorder may be associated with low levels of BDNF, abnormal function of hypothalamic-pituitary-adrenal (HPA) axis and glutamatergic toxicity (aan het Rot et al., 2009; Krishnan & Nestler, 2008; Mathew et al., 2008).

Research on the biological basis of mood disorders emphasises the changes of neural networks and synaptic plasticity. Evidence exists for impairment of neuroplasticity in major depression. Chronic stress is known to contribute both to development of major depression in vulnerable persons and to reduction of synaptic plasticity, induction of structural changes in dendrites, and impairment of neurogenesis (Pittenger & Duman, 2008). Mitochondria may be primary regulators of these processes, as they regulate not only neuronal survival and death, but also plasticity. There is a mounting evidence for the role of mitochondrial dysfunction in the pathophysiology and treatment of bipolar disorder (Quiroz et al., 2008). Attention is paid also to changes in inflammatory pathways, because monoaminergic systems, cellular energetics and inflammation are interrelated in many complex manners (Gardner & Boles, 2011).

7.1 Monoamine hypothesis

Classic monoamine hypothesis of depression proposed that depression might be produced by a serotonin or norepinephrine deficiency at functionally important receptor sites in the brain, i.e. that brain monoamine systems have a primary direct role in depression. Nevertheless, the monoamine hypothesis of depression in itself could neither explain the whole mechanism of action of antidepressants, nor could it provide an explanation for how monoamine loss occurs (Nestler et al., 2002). According to advanced monoamine theory (Meyer et al., 2006), serotonin or norepinephrine levels in the brain are regulated by monoamine oxidase type A (MAO-A) activity mainly, and severity of symptoms of depression is related to changes in the activity of monoamine transporters in specific brain regions. Thus, both MAO-A activity and density of transporters are included in the pathophysiology of affective disorders. The advanced monoamine hypothesis was supported by observation that during a major depressive episode, both MAO-A density is elevated, resulting in greater metabolism of monoamines in the brain (Meyer et al., 2009), and brain SERT binding is diminished (Selvaraj et al., 2011).

In order to test monoamine hypothesis, a series of studies was conducted to evaluate effects of monoamine depletion on depressive symptoms in depressed patients and in healthy controls. Relapse after serotonin depletion or catecholamine depletion was found to be specific to the type of antidepressant treatment and type of depletion. Serotonin or norepinephrine/dopamine depletion did not decrease mood in healthy controls and slightly lowered mood in healthy controls with a family history of major depressive disorder. In drug-free patients with major depressive disorder in remission, a moderate mood decrease was found for acute tryptophan depletion only. However, acute tryptophan depletion induced relapse in patients in remission who used serotonergic antidepressants (Delgado et al., 1999). Depletion studies failed to demonstrate a causal relation between serotonin and norepinephrine with depressive disorder (Cowen, 2008; Ruhé et al., 2007). The effects of acute tryptophan depletion on cognition in non-vulnerable participants are independent of mood changes (Mendelsohn et al., 2009). Even simultaneous disruption of serotonin and catecholamine systems didn't significantly alter mood in unmedicated depressed subjects (Berman et al., 2002). These findings forced a major revision of the classic monoamine hypothesis of depression. According to this revised monoamine theory of depression (aan het Rot et al., 2009; Heninger et al., 1996) monoamine systems are only modulating other brain neurobiological systems that have more primary role in depression.

7.2 Neurotrophic, neuroplasticity and neurogenesis hypotheses

The neurotrophic hypothesis of depression (Duman et al., 1997; Duman, 2002; Zarate et al., 2006a; Einat & Manji, 2006) supposed that vulnerability to depression can arise as a result of neuronal damage (Table 1), e.g. after chronic stress, long-term increased levels of glucocorticoids, hypoglycemia, ischemia, certain viral infections, effects of neurotoxins, etc. The therapeutic effects of antidepressants consist of increased function of the noradrenergic or serotonergic system. It leads to increased activity of transcription factor CREB, higher expression of neurotrophin BDNF and its receptor trkB, and consequently to increased neuronal plasticity and resumption of cellular functions.

Stimuli	*Mechanisms*
Chronic stress	Disturbed energetic metabolism
Hypoglycemia	Lack of growth factors
Ischemia	Excitotoxicity
Neurotoxins	Oxidative stress
Viral infections	Necrotic and apoptotic processes
γ-radiation, UV	Inflammatory processes

Table 1. Stimuli and mechanisms of neuron damage, which could be related to vulnerability to depression

Series of studies support the hypothesis that a reduction of BDNF could contribute to depression and that antidepressants mediate their therapeutic benefit by increasing levels of this factor in the hippocampus. A polymorphism in the *BDNF* gene has been associated with depression and bipolar disorder. BDNF levels have been found to be reduced in post-mortem brain samples and in the blood of depressed patients, and these reductions are reversible by successful antidepressant treatment (Castrén et al., 2007; Castrén & Rantamäki, 2008). In addition, the regulation of other growth factors may also play a role in the pathophysiology and the treatment of depression (Duman & Monteggia, 2006). Stress produces a sustained suppression of *BDNF* transcription through histone methylation, whereas antidepressants restore BDNF synthesis through histone acetylation (Tsankova et al., 2006). This suggests that chronic stress can cause long-lasting epigenetic changes that might be related to increased vulnerability to depression.

Local infusions of BDNF into specific brain regions showed to mimic antidepressant effects in behavioural models of depression. However, the loss of BDNF and signalling through the trkB receptor in broad forebrain regions *per se* is not sufficient to mediate depression-like behaviour (Duman & Monteggia, 2006). BDNF does not produce uniform effects throughout the brain. Infusion of BDNF into the ventral tegmental area might be related to the development of a depression-like phenotype; therefore, the role of BDNF in the ventral tegmental area - nucleus accumbens (VTA-NAc) pathway is opposite to the role of BDNF in the hippocampus (Eisch et al., 2003). It is hypothesized that BDNF signalling in the VTA-NAc is required for the establishment of associations with negative emotional stimuli; and under pathological conditions this signalling may establish abnormal associations leading to certain symptoms of depression (Nestler & Carlezon, 2006).

The neurogenesis (neurogenic) hypothesis of depression (Jacobs, 2002) proposes that depression may develop due to the decreased neurogenesis in hippocampus and antidepressants takes effect through the stimulation of neurogenesis. The first component of this hypothesis is not tenable, and the evidence for second component is conflicting (Sapolsky, 2004).

Behavioural data support the hypothesis that regulation of neuronal plasticity might play an important role in expression of symptoms of bipolar disorder. Major intracellular signalling pathways in bipolar disorder and neural plasticity include protein kinase C (PKC) and extracellular signal regulated kinase (ERK) cascades, glucocorticoid receptor modulation, glycogen synthase kinase-3 (GSK-3), and antiapoptotic Bcl-2 (B-cell CLL/lymphoma 2), AMPA (α-amino-3-hydroxyl-5-methyl-4-isoxazole propionic acid) receptors, and inositol homeostasis (Einat & Manji, 2006). The neuroplasticity hypothesis of depressive disorder (Pittenger & Duman, 2008) suggests that impaired mechanisms of neuroplasticity are core pathophysiological feature of the disorder, where chronic stress is an important causal factor in the development of this impairment, and where long-term treatment with antidepressants leads to modulation of impaired mechanisms of neuroplasticity. However, alterations in neuroplasticity do not produce the same effects in different brain regions.

Hypothesis of cellular plasticity (Kempermann & Kronenberg, 2003) relates the neurotrophic and the neurogenic hypothesis in the statement that depression can be generally caused by damaged cellular plasticity leading to inadequate relations between structure and function.

Molecular mechanisms leading to a disturbance of neuroplasticity are not known sufficiently. The bioenergetic and neurochemical model of bipolar disorder attempts to identify these mechanisms and focuses attention on mitochondrial dysfunctions (Kato & Kato, 2000; Stork & Renshaw, 2005).

7.3 Network hypothesis

According to the network hypothesis of depression, changes in levels of neurotrophins may not directly produce depression or an antidepressant effect, but neurotrophins may act as critical tools in the process whereby environmental conditions guide neuronal networks to better adapt to the environment (Castrén & Rantamäki, 2010).

The network hypothesis (Castrén, 2005) proposes that mood disorders reflect problems in information processing within particular neural networks in the brain; antidepressant drugs and other treatments, which alleviate depression, function by gradually improving information processing within these networks. Therefore, disorders of the nervous system, including depression, might represent disturbances in the activity-dependent information processing of the brain, rather than in the chemical balance of signalling molecules. However, some observations seem to be incompatible with the network hypothesis, e.g. rapid relapse of depressive symptoms in some patients, circadian variations in mood, and the effect of sleep deprivation on improvement of the mood of patients with depression. These rapid effects on mood can not be accounted for a gradual change in the structure of mood influencing neural networks.

The network hypothesis emphasizes the importance of processing of environmental information (such as social communication) in the recovery of the brain functions during treatment of mood disorders; so, treatment with psychotropic drug in itself is insufficient.

Findings that psychotherapy result in detectable changes in the brain (Etkin et al., 2005) and that combined use of psychotherapy and medication can lead to better treatment outcomes than the use of either mode of therapy alone (Elkin et al., 1989) are consistent with the network hypothesis.

7.4 Inflammatory and neurodegenerative hypothesis

The central nervous system, endocrine and immune systems use neurotransmitters, cytokines and hormones to communicate among them (Haddad et al., 2002). Now there is evidence that the activation of the immune system is associated with the symptoms of depression (Leonard & Myint, 2009; Catena-Dell'Osso et al., 2011). The inflammatory and neurodegenerative hypothesis of depression (Maes et al., 2009) supposes that depression is associated with both inflammatory processes, as well as with neurodegeneration and reduced neurogenesis. According to this hypothesis, enhanced neurodegeneration and impaired neurogenesis in depression are caused by inflammatory processes, related to the production of oxidative and nitrosative stress, tryptophan catabolites along the indoleamine-2,3-dioxygenase pathway, proinflammatory cytokines and lowered ω-3 polyunsaturated fatty acid status. Anti-inflammatory compounds should be able to counteract at least partly the enhanced neurodegeneration and decreased neurogenesis.

7.5 Mitochondrial dysfunction hypothesis

While dysfunctions within monoaminergic neurotransmitter systems are likely to play an important role in pathophysiology of mood disorders, it probably represents the downstream effects of more primary abnormalities in signal transduction. New theories about the pathophysiology of depression and mechanisms of action of antidepressants proposes that regulation of intracellular signalling pathways manifested by disturbed neuroplasticity plays a critical role in higher-order brain functions. Mitochondria may be primary regulators of these processes. It is supposed that mitochondrial dysfunctions are included in pathophysiology of mood disorders. Disturbances in activity of mitochondrial enzymes, effects of intracellular calcium and energy metabolism, damage of mitochondrial DNA, and action of psychotropics on mitochondria were observed in many neuropsychiatric illnesses, mood disorders included.

Subtle deficits in mitochondrial function likely play an important role in various aspects of bipolar disorder, and enhancing mitochondrial function may represent a critical component for the treatment of the disorder (Quiroz et al., 2008). Changes in cerebral concentrations of *N*-acetyl aspartate (NAA), glutamate/glutamine, choline-containing compounds, myo-inositol, lactate, phosphocreatine, phosphomonoesters, and intracellular pH in bipolar subjects were described (Yildiz-Yesiloglu & Ankerst, 2006). A hypothesis of mitochondrial dysfunction in bipolar disorder (Stork & Renshaw, 2005) was proposed and involved impaired oxidative phosphorylation, a resultant shift toward glycolytic energy production, a decrease in total energy production (decreased ATP production) and/or substrate availability, and changed concentrations of phosphomonoesters and altered phospholipid metabolism.

Neuronal calcium homeostasis and calcium signalling regulate multiple neuronal functions, including synaptic transmission, neuronal plasticity and survival. The idea that altered intracellular calcium signalling may be crucial for the molecular mechanisms leading to both schizophrenia and affective disorders was firstly suggested by Jimerson et al. (1979). Recently, disturbed calcium homeostasis has been studied in neurodegenerative (Wojda et al., 2008) and mood disorders. Mitochondrial DNA (mtDNA) mutations in the brain, associations of mtDNA polymorphisms and bipolar disorder and changes in gene expression related to mitochondria in the brain were observed (Kato, 2008). Calcium and mitochondrial dysfunction hypothesis of bipolar disorder offers that mtDNA polymorphisms/mutations or mtRNA deletions caused by nuclear gene mutations can cause mitochondrial dysregulation of calcium leading to symptoms of bipolar disorder (Kato & Kato, 2000; Kato, 2007, 2008).

Mitochondrial hypothesis of bipolar affective disorder corresponds both to the neurotrophic and neuroplasticity hypothesis and to the inflammatory and neurodegenerative hypothesis because of an important role of mitochondria in cell energetic, regulation of calcium signalling pathway, production of reactive oxygen species and apoptosis, i.e. in processes determining synaptic plasticity, damage, repairing, survival or death of neurons.

8. Conclusion

The understanding of signalling pathways and their interconnections on the intracellular level is necessary to understand normal and damaged functions of the CNS. The exact neurochemical processes included in mood disorders have not yet been sufficiently explored; the main attention has been focused on changes in monoaminergic neurotransmitter systems and neuroplasticity. It is suggested that the changes in cognitive functions, learning, memory and the emotions during depression are caused by damaged neuroplasticity in hippocampus, amygdala and cerebral cortex. Changes in effects of neurotransmitters, hormones and growth factors probably cause changes of neuroplasticity; processes leading to changed neuroplasticity can be initiated by mitochondrial dysfunctions.

The long-term administration of antidepressants leads to the effects similar to neurotrophic, as seen through the activation of transcription factors and increased gene expression of neurotrophins. Mood stabilizers, such as lithium and valproate, strongly activate also neurotrophic signal cascades and affect other signalling pathways. Intracellular processes among activation of receptor and increased gene expression of neurotrophins and other molecules have not yet been sufficiently clarified. Variability in these intracellular processes participates probably on interindividual differences of the response to treatment with antidepressant or in drug resistance.

On the basis of clinical analyses and findings about signalling pathway components affected by administration of psychotropic drugs, studies of following components seem to be perspective: neurotransmitter receptors and transporters (especially for serotonin and norepinephrine), monoamine oxidases, adenylate cyclase and phosphoinositide cascade and MAPK pathway, mitochondrial dysfunctions, reactive oxygen species, nitric oxide and cytoplasmic calcium. The role of membrane potential and membrane lipids in the proper function of neuronal membranes should not be forgotten.

Signal transduction is an amount of processes by which a cell converts one kind of signal into another. Neuronal signal transduction pathways involve neurotransmitter release, receptor activation and ordered sequence of biochemical reactions inside the cell. These processes can be rapid in the case of ion-channels opening (milliseconds), slower in the case of the activation of second messenger systems (minutes), and slow in the case of gene expression (hours, and even days). As signal propagates from a relatively small initial stimulus to a final large response, the signal cascade involves increasing number of proteins and other molecules (Fišar & Hroudová, 2010). Many feedbacks and cross reactions are included in this amplification of the signal. Determining factors in signal transduction are input signals and concentration, activity and subcellular localization of participating molecules. All these factors are determined by previous history of the brain, and it is probable that participate in interindividual differences in response to the same stimuli.

Regulation of critical intracellular signalling pathways plays a critical role in higher-order brain functions, which are altered in mood disorders, suggesting the involvement of dysfunctions of signalling pathways in the pathophysiology and the treatment of mood disorders (Gould et al., 2007). It can be concluded that both some structural deviations in neural networks and disturbances of signal transduction in certain neurons participate in development of mood disorders. It is assumed that novel biomarkers of mood disorders and/or predictors of efficiency of pharmacotherapy may be discovered on the basis of this research.

It is suggested that environmental stress and genetic risk variants interact with each other in a complex manner to alter neural circuitry and evoke illness. Considering the existence of very rapid distinct shifts of mood (e.g. ultra-ultra-rapid cycling in bipolar affective disorder) (Kramlinger & Post, 1996), developmental and structural changes in neural networks could be considered as necessary condition for vulnerability to the development of pathological states of mood, whereas disturbances in signal transduction pathways in chemical synapses could be related to the onset of specific symptoms of mood disorder. Distinct shifts in mood and activity occur both in "normal" and "disordered" man and can be induced both by endogenous and external stimuli. Therefore, occurrence of several symptoms, their severity and time duration depend on the proportion of dysfunction of brain homeostatic mechanisms in various brain areas. It may be supposed that both normal and pathological shifts of mood (e.g. ultradian cycling or rapid switching from episode of depression to mania) are associated with changes in expression or activity of specific compounds of signal transduction pathways related to monoaminergic and glutamatergic systems. Identification of compounds of signal transduction that are primarily responsible for shifts of mood remains incomplete. We suppose that mild dysfunction of some mitochondrial functions might be basis for homeostatic imbalance in synapses during episodes of depression, hypomania, mania or the appearance of mixed states.

Identification of biomarkers of mood disorders and factors capable of predicting the response to treatment with antidepressants has not been sufficiently successful (Balon, 1989; Joyce & Paykel, 1989; Fišar & Raboch, 2008). In accordance to actual neurochemical hypotheses of mood disorders, biological markers have been primarily found at the level of neurotransmitter concentrations, their metabolites or precursors. Subsequently, attention was shifted to the receptor systems, and since the 1990´s, intracellular processes have become of primary interest. There is evidence that growth factors, proinflammatory cytokines, endocrine factors, and metabolic markers contribute to the pathophysiology of major depressive disorder (Schmidt et al., 2011). The chance to find sensitive and specific biological predictors of antidepressant treatment has been increased, because of the introduction of new molecular biology methods. These methods enable better observation of cellular processes connected with the transduction of nervous signals in the brain. The choice of parameters, which should be studied as perspective biological markers of mood disorders, has been derived first of all from new findings of signalling pathways involved in neurotransmission and from above mentioned neurochemical hypotheses of mood disorders. Following parameters could play a principal role in findings of biological markers of mood disorders: energetic metabolism and mitochondrial dysfunctions, activities of protein kinases, transcription factors, neurotrophins, proapoptotic and antiapoptotic factors, caspases, polyunsaturated fatty acids, membrane potential and fluidity, cortisol, melatonin, homocysteine, cytokines, SERT, Na^+K^+-ATPases, ROS, NO, cytoplasmic Ca^{2+} and CaM, enzymes GSK-3, COMT, MAO and cytochrome P450. According to the complexity and connectivity of signalling pathways involved in ethiopathogenesis of mood disorders, number of chosen parameters is not final.

9. Acknowledgment

This work was supported by a Research project MSM0021620849 given by Ministry of Education, Youth and Sports of the Czech Republic and by grant No 41310 given by Grant Agency of Charles University.

10. References

aan het Rot, M., Mathew, SJ. & Charney DS. (2009). Neurobiological mechanisms in major depressive disorder. *CMAJ*, Vol. 180, No. 3, pp. 305-313

Abitz, M., Nielsen, RD., Jones, EG., Laursen, H., Graem, N. & Pakkenberg, B. (2007). Excess of neurons in the human newborn mediodorsal thalamus compared with that of the adult. *Cereb. Cortex*, Vol. 17, No. 11, pp. 2573-2578

Abraham, WC. & Bear, MF. (1996). Metaplasticity: the plasticity of synaptic plasticity. *Trends Neurosci.*, Vol. 19, No. 4, pp. 126-130

Aguilera, M., Arias, B., Wichers, M., Barrantes-Vidal, N., Moya, J., Villa, H., van Os, J., Ibáñez, MI., Ruipérez, MA., Ortet, G. & Fañanás L. (2009). Early adversity and 5-HTT/BDNF genes: new evidence of gene-environment interactions on depressive symptoms in a general population. *Psychol. Med.*, Vol. 39, No. 9, pp. 1425-1432

Atlante, A., Calissano, P., Bobba, A., Giannattasio, S., Marra, E. & Passarella, S. (2001). Glutamate neurotoxicity, oxidative stress and mitochondria. *FEBS Lett.*, Vol. 497, No. 1, pp. 1–5

Bach, AWJ., Lan, NC., Johnson, DL., Abell, CW., Bembenek, ME., Kwan, S-W., Seeburg, PH. & Shih, JC. (1988). cDNA cloning of human liver monoamine oxidase A and B: molecular basis of differences in enzymatic properties. *Proc. Natl. Acad. Sci. USA*, Vol. 85, No. 13, pp. 4934-4938

Balon, R. (1989). Biological predictors of antidepressant treatment outcome. *Clin. Neuropharmacol.*, Vol. 12, pp. 195-214

Banerjee, SP., Kung, LS., Riggi, SJ. & Chanda, SK. (1977). Development of β-adrenergic receptor subsensitivity by antidepressants. *Nature*, Vol. 268, pp. 455-456

Battaini, F. (2001). Protein kinase C isoforms as therapeutic targets in nervous system disease states. *Pharmacol. Res.*, Vol. 44, pp. 353-361

Berman, RM., Sanacora, G., Anand, A., Roach, LM., Fasula, MK., Finkelstein, CO., Wachen, RM., Oren, DA., Heninger, GR. & Charney, DS. (2002). Monoamine depletion in unmedicated depressed subjects. *Biol. Psychiatry*, Vol. 51, No. 6, pp. 469-473

Berridge, MJ., Lipp, P., Bootman, MD. (2000). The versatility and universality of calcium signalling. *Nat. Rev. Mol. Cell Biol.*, Vol. 1, pp. 11-21

Blendy, JA. (2006). The role of CREB in depression and antidepressant treatment. *Biol. Psychiatry*, Vol. 59, pp. 1144-1150

Bocchio-Chiavetto, L., Zanardini, R., Bortolomasi, M., Abate, M., Segala, M., Giacopuzzi, M., Riva, M. A., Marchina, E., Pasqualetti, P., Perez, J. & Gennarelli, M. (2006). Electroconvulsive Therapy (ECT) increases serum Brain Derived Neurotrophic Factor (BDNF) in drug resistant depressed patients. *Eur. Neuropsychopharmacol.*, Vol. 16, pp. 620-624

Bown, CD., Wang, JF., Chen, B. & Young, LT. (2002). Regulation of ER stress proteins by valproate: therapeutic implications. *Bipolar Disord.*, Vol. 4, pp. 145-151

Brown GC. (2010). Nitric oxide and neuronal death. *Nitric Oxide*, Vol. 23, No. 3, pp. 153-165

Bunney JN. & Potkin SG. (2008). Circadian abnormalities, molecular clock genes and chronobiological treatments in depression. *Br. Med. Bull.*, Vol. 86, pp. 23-32

Carlezon, WAJr., Duman, RS. & Nestler, EJ. (2005). The many faces of CREB. *Trends Neurosci.*, Vol. 28, No. 8, pp. 436-445

Caspi, A., Sugden, K., Moffitt, TE., Taylor, A., Craig, IW., Harrington, H., McClay, J., Mill, J., Martin, J., Braithwaite, A. & Poulton R. (2003). Influence of life stress on depression: moderation by a polymorphism in the 5-HTT gene. *Science*, Vol. 301, No. 5631, pp. 386-389

Castrén, E. (2005). Is mood chemistry? *Nat. Rev. Neurosci.*, Vol. 6, No. 3, pp. 241-246

Castrén, E., Võikar, V. & Rantamäki, T. (2007). Role of neurotrophic factors in depression. *Curr. Opin. Pharmacol.*, Vol. 7, No. 1, pp. 18-21

Castrén, E. & Rantamäki, T. (2008). Neurotrophins in depression and antidepressant effects. *Novartis Found. Symp.*, Vol. 289, pp. 43-52

Castrén, E. & Rantamäki, T. (2010). Role of brain-derived neurotrophic factor in the aetiology of depression: implications for pharmacological treatment. *CNS Drugs*, Vol. 24, No. 1, pp. 1-7

Catena-Dell'Osso, M., Bellantuono, C., Consoli, G., Baroni, S., Rotella, F. & Marazziti, D. (2011). Inflammatory and neurodegenerative pathways in depression: a new avenue for antidepressant development? *Curr. Med. Chem.*, Vol. 18, No. 2, pp. 245-255

Chen, AC., Shirayama, Y., Shin, KH., Neve, RL. & Duman, RS. (2001). Expression of the cAMP response element binding protein (CREB) in hippocampus produces an antidepressant effect. *Biol. Psychiatry*, Vol. 49, No. 9, pp. 753-762

Chen, G., Zeng, WZ., Yuan, PX., Huang, LD., Jiang, YM., Zhao, ZH. & Manji, HK. (1999). The mood-stabilizing agents lithium and valproate robustly increase the levels of the neuroprotective protein bcl-2 in the CNS. *J. Neurochem.*, Vol. 72, pp. 879-882

Chevaleyre, V., Takahashi, KA. & Castillo, PE. (2006). Endocannabinoid-mediated synaptic plasticity in the CNS. *Annu. Rev. Neurosci.*, Vol. 29, pp. 37-76

Chiavegatto, S. & Nelson, RJ. (2003). Interaction of nitric oxide and serotonin in aggressive behavior. *Horm. Behav.*, Vol. 44, No. 3, pp. 233-241

Chipuk, JE. & Green, DR. (2008). How do BCL-2 proteins induce mitochondrial outer membrane permeabilization? *Trends Cell Biol.*, Vol. 18, pp. 157-164

Chrousos, GP. (2009). Stress and disorders of the stress system. *Nat. Rev. Endocrinol.*, Vol. 5, No. 7, pp. 374-381

Chuang, DM. (2005). The antiapoptotic actions of mood stabilizers: molecular mechanisms and therapeutic potentials. *Ann. N. Y. Acad. Sci.*, Vol. 1053, pp. 195-204

Citri, A. & Malenka, RC. (2008). Synaptic plasticity: multiple forms, functions, and mechanisms. *Neuropsychopharmacology*, Vol. 33, No. 1, pp. 18-41

Clevers, H. (2006). Wnt/β-catenin signaling in development and disease. *Cell*, Vol. 127, pp. 469-480

Coppen, A. (1967). The biochemistry of affective disorders. *Br. J. Psychiatry*, Vol. 113, No. 504, pp. 1237-1264

Cowan, KJ. & Storey, KB. (2003). Mitogen-activated protein kinases: new signaling pathways functioning in cellular responses to environmental stress. *J. Exp. Biol.*, Vol. 206, pp. 1107-1115

Cowen, PJ. (2008). Serotonin and depression: pathophysiological mechanism or marketing myth? *Trends Pharmacol. Sci.*, Vol. 29, No. 9, pp. 433-436

Crisafulli, C., Fabbri, C., Porcelli, S., Drago, A., Spina, E., De Ronchi, D. & Serretti, A. (2011). Pharmacogenetics of antidepressants. *Front. Pharmacol.*, Vol. 2, Art. 6, pp. 1-21

Delgado, PL., Miller, HL., Salomon, RM., Licinio, J., Krystal, JH., Moreno, FA., Heninger, GR. & Charney, DS. (1999). Tryptophan-depletion challenge in depressed patients treated with desipramine or fluoxetine: implications for the role of serotonin in the mechanism of antidepressant action. *Biol. Psychiatry.*, Vol. 46, No. 2, pp. 212-220

Derkach, VA., Oh, MC., Guire, ES. & Soderling, TR. (2007). Regulatory mechanisms of AMPA receptors in synaptic plasticity. *Nat. Rev. Neurosci.*, Vol. 8, No. 2, pp. 101-113

Diazgranados, N., Ibrahim, L., Brutsche, NE., Newberg, A., Kronstein, P., Khalife, S., Kammerer, WA., Quezado, Z., Luckenbaugh, DA., Salvadore, G., Machado-Vieira, R., Manji, HK. & Zarate, CAJr. (2010). A randomized add-on trial of an *N*-methyl-D-aspartate antagonist in treatment-resistant bipolar depression. *Arch. Gen. Psychiatry*, Vol. 67, No. 8, pp. 793-802

Draganski, B. & May, A. (2008). Training-induced structural changes in the adult human brain. *Behav. Brain Res.*, Vol. 192, No. 1, pp. 137-142

Drzyzga, ŁR., Marcinowska, A. & Obuchowicz, E. (2009). Antiapoptotic and neurotrophic effects of antidepressants: a review of clinical and experimental studies. *Brain Res. Bull.*, Vol. 79, No. 5, pp. 248-257

Duman, RS., Heninger, GR. & Nestler, EJ. (1997). A molecular and cellular theory of depression. *Arch. Gen. Psychiatry*, Vol. 54, pp. 597-606

Duman, RS. (2002). Synaptic plasticity and mood disorders. *Mol. Psychiatry*, Vol. 7, pp. S29-S34

Duman, RS. & Monteggia, LM. (2006). A neurotrophic model for stress-related mood disorders. *Biol. Psychiatry.*, Vol. 59, No. 12, pp. 1116-1127

Duman, RS. (2009). Neuronal damage and protection in the pathophysiology and treatment of psychiatric illness: stress and depression. *Dialogues Clin. Neurosci.*, Vol. 11, No. 3, pp. 239-255

Einat, H. & Manji, HK. (2006). Cellular plasticity cascades: genes-to-behavior pathways in animal models of bipolar disorder. *Biol. Psychiatry*, Vol. 59, pp. 1160-1171

Eisch, AJ., Bolaños, CA., de Wit, J., Simonak, RD., Pudiak, CM., Barrot, M., Verhaagen, J. & Nestler, EJ. (2003). Brain-derived neurotrophic factor in the ventral midbrain-nucleus accumbens pathway: a role in depression. *Biol. Psychiatry.*, Vol. 54, No. 10, pp. 994–1005

Elkin, I., Shea, MT., Watkins, JT., Imber, SD., Sotsky, SM., Collins, JF., Glass, DR., Pilkonis, PA., Leber, WR., Docherty, JP., Fiester SJ. & Parloff MB. (1989). National Institute of Mental Health Treatment of Depression Collaborative Research Program. General effectiveness of treatments. *Arch. Gen. Psychiatry*, Vol. 46, No. 11, pp. 971-982

Etkin, A., Pittenger, C., Polan, HJ. & Kandel, ER. (2005). Toward a neurobiology of psychotherapy: basic science and clinical applications. *J. Neuropsychiatry Clin. Neurosci.*, Vol. 17, No. 2, pp. 145-158

Fišar, Z. & Raboch, J. (2008). Depression, antidepressants, and peripheral blood components. *Neuroendocrinol. Lett.*, Vol. 29, pp. 17-28

Fišar, Z. (2009). Phytocannabinoids and endocannabinoids. *Curr. Drug Abuse Rev.*, Vol. 2, No. 1, pp. 51-75

Fišar, Z. & Hroudová, J. (2010). Intracellular signalling pathways and mood disorders. *Folia Biol.*, Vol. 56, No. 4, pp. 135-148

Fuerer, C., Nusse, R. & Ten Berge, D. (2008). Wnt signalling in development and disease. *EMBO Rep.*, Vol. 9, pp. 134-138

García-Junco-Clemente, P., Linares-Clemente, P. & Fernández-Chacón, R. (2005). Active zones for presynaptic plasticity in the brain. *Mol. Psychiatry.*, Vol. 10, No. 2, pp. 185-200

Gardner, A. & Boles, RG. (2011). Beyond the serotonin hypothesis: mitochondria, inflammation and neurodegeneration in major depression and affective spectrum disorders. *Prog. Neuropsychopharmacol. Biol. Psychiatry*, Vol. 35, No. 3, pp. 730-743

Gass, P. & Riva, MA. (2007). CREB, neurogenesis and depression. *Bioessays*, Vol. 29, pp. 957-961

Gould, TD., Quiroz, JA., Singh, J., Zarate, CA. & Manji, HK. (2004). Emerging experimental therapeutics for bipolar disorder: insights from the molecular and cellular actions of current mood stabilizers. *Mol. Psychiatry*, Vol. 9, No. 8, pp. 734-755

Gould, TD. & Manji, HK. (2005). Glycogen synthase kinase-3: a putative molecular target for lithium mimetic drugs. *Neuropsychopharmacology*, Vol. 30, pp. 1223-1237.

Gould, TD., Picchini, AM., Einat, H. & Manji, HK. (2006). Targeting glycogen synthase kinase-3 in the CNS: implications for the development of new treatments for mood disorders. *Curr. Drug Targets*, Vol. 7, No. 11, pp. 1399-1409

Gould, TD., Dow, ER., O'Donnell, KC., Chen, G. & Manji, HK. (2007). Targeting signal transduction pathways in the treatment of mood disorders: recent insights into the relevance of the Wnt pathway. *CNS Neurol. Disord. Drug Targets*, Vol. 6, No. 3, pp. 193-204

Guo, D., Zhang, JJ. & Huang, XY. (2010). A new Rac/PAK/GC/cGMP signaling pathway. *Mol. Cell. Biochem.*, Vol. 334, No. 1-2, pp. 99-103

Haddad, JJ., Saadé, NE. & Safieh-Garabedian, B. (2002). Cytokines and neuro-immune-endocrine interactions: a role for the hypothalamic-pituitary-adrenal revolving axis. *J. Neuroimmunol.*, Vol. 133, No. 1-2, pp. 1-19

Hayden, MS. & Ghosh S. (2004). Signaling to NF-kappaB. *Genes Dev.*, Vol. 18, pp. 2195-2224

Heninger, GR., Delgado, PL. & Charney, DS. (1996). The revised monoamine theory of depression: a modulatory role for monoamines, based on new findings from monoamine depletion experiments in humans. *Pharmacopsychiatry*, Vol. 29, No. 1, pp. 2-11

Hill, BG., Dranka, BP., Bailey, SM., Lancaster, JRJr. & Darley-Usmar, VM. (2010). What part of NO don't you understand? Some answers to the cardinal questions in nitric oxide biology. *J. Biol. Chem.*, Vol. 285, No. 26, pp. 19699-19704

Horstink, M., Tolosa, E., Bonuccelli, U., Deuschl, G., Friedman, A., Kanovsky, P., Larsen, JP., Lees, A., Oertel, W., Poewe, W., Rascol, O. & Sampaio C; European Federation of Neurological Societies; Movement Disorder Society-European Section. (2006). Review of the therapeutic management of Parkinson's disease. Report of a joint task force of the European Federation of Neurological Societies and the Movement Disorder Society-European Section. Part I: early (uncomplicated) Parkinson's disease. *Eur. J. Neurol.*, Vol. 13, pp. 1170-1185

Hroudová, J. & Fišar, Z. (2011). Connectivity between mitochondrial functions and psychiatric disorders. *Psychiat. Clin. Neurosci.*, Vol. 65, No. 2, pp. 130-141

Jacobs, BL. (2002). Adult brain neurogenesis and depression. *Brain Behav. Immun.*, Vol. 16, pp. 602-609

Janssen, DG., Caniato, RN., Verster, JC. & Baune, BT. (2010). A psychoneuroimmunological review on cytokines involved in antidepressant treatment response. *Hum. Psychopharmacol.*, Vol. 25, No. 3, pp. 201-215

Jian, K., Chen, M., Cao, X., Zhu, XH., Fung, ML. & Gao, TM. (2007). Nitric oxide modulation of voltage-gated calcium current by S-nitrosylation and cGMP pathway in cultured rat hippocampal neurons. *Biochem. Biophys. Res. Commun.*, Vol. 359, No. 3, pp. 481-485

Jimerson, DC., Post, RM., Carman, JS., van Kammen, DP., Wood, JH., Goodwin, FK. & Bunney, WEJr. (1979). CSF calcium: clinical correlates in affective illness and schizophrenia. *Biol. Psychiatry*, Vol. 14, No 1, pp. 37-51

Joca, SR. & Guimarães, FS. (2006). Inhibition of neuronal nitric oxide synthase in the rat hippocampus induces antidepressant-like effects. *Psychopharmacology (Berl.)*, Vol. 185, No. 3, pp. 298-305

Johnson, SL. (2005). Life events in bipolar disorder: towards more specific models. *Clin. Psychol. Rev.*, Vol. 25, No. 8, pp. 1008-1027

Joyce, PR. & Paykel, ES. (1989). Predictors of drug response in depression. *Arch. Gen. Psychiatry*, Vol. 46, pp. 89-99

Kann, O. & Kovács, R. (2007). Mitochondria and neuronal activity. *Am. J. Physiol. Cell Physiol.*, Vol. 292, pp. C641-C657

Kasper, S. & McEwen, BS. (2008). Neurobiological and clinical effects of the antidepressant tianeptine. *CNS Drugs*, Vol. 22, No. 1, pp. 15-26

Kato, T. & Kato, N. (2000). Mitochondrial dysfunction in bipolar disorder. *Bipolar Disord.*, Vol. 2, No. 3 Pt 1, pp. 180-190

Kato, T. (2007). Mitochondrial dysfunction as the molecular basis of bipolar disorder: therapeutic implications. *CNS Drugs*, Vol. 21, No. 1, pp. 1-11

Kato, T. (2008). Role of mitochondrial DNA in calcium signaling abnormality in bipolar disorder. *Cell Calcium*, Vol. 44, No. 1, pp. 92-102

Kempermann, G. & Kronenberg, G. (2003). Depressed new neurons?—adult hippocampal neurogenesis and a cellular plasticity hypothesis of major depression. *Biol. Psychiatry*, Vol. 54, pp. 499-503

Kendler, KS., Karkowski, LM. & Prescott, CA. (1999). Causal relationship between stressful life events and the onset of major depression. *Am. J. Psychiatry*, Vol. 156, No. 6, pp. 837-841

Kim, AH. & Bonni, A. (2007). Thinking within the D box: initial identification of Cdh1-APC substrates in the nervous system. *Mol. Cell. Neurosci.*, Vol. 34, pp. 281-287

Kim, Y-K., Paik, J-W., Lee, S-W., Yoon, D., Han, C. & Lee, B-H. (2006). Increased plasma nitric oxide level associated with suicide attempt in depressive patients. *Prog. Neuropsychopharmacol. Biol. Psychiatry*, Vol. 30, No. 6, pp. 1091-1096

Koo, JW. & Duman, RS. (2008). IL-1β is an essential mediator of the antineurogenic and anhedonic effects of stress. *Proc. Natl. Acad. Sci. USA*, Vol. 105, No. 2, pp. 751-756

Kramlinger, KG. & Post, RM. (1996). Ultra-rapid and ultradian cycling in bipolar affective illness. *Br. J. Psychiatry*, Vol. 168, No. 3,pp. 314-323

Krishnan, V. & Nestler, EJ. (2008). The molecular neurobiology of depression. *Nature*, Vol. 455, No. 7215, pp. 894-902

Kubera, M., Kenis, G., Bosmans, E., Scharpé, S. & Maes, M. (2000). Effects of serotonin and serotonergic agonists and antagonists on the production of interferon-γ and interleukin-10. *Neuropsychopharmacology*, Vol. 23, No. 1, pp. 89-98

Kuhn, M. (2003). Structure, regulation, and function of mammalian membrane guanylyl cyclase receptors, with a focus on guanylyl cyclase-A. *Circ. Res.*, Vol. 93, pp.700-709

Lee, AL., Ogle, WO. & Sapolsky, RM. (2002). Stress and depression: possible links to neuron death in the hippocampus. *Bipolar Disord.*, Vol. 4, No. 2, pp. 117-128

Lee, BH., Lee, SW., Yoon, D., Lee, HJ., Yang, JC., Shim, SH., Kim, DH., Ryu, SH., Han, C. & Kim, YK. (2006). Increased plasma nitric oxide metabolites in suicide attempters. *Neuropsychobiology*, Vol. 53, No. 3, pp. 127-132

Lee, E., Salic, A., Krüger, R., Heinrich, R. & Kirschner, MW. (2003). The roles of APC and Axin derived from experimental and theoretical analysis of the Wnt pathway. *PLoS Biol.*, Vol. 1, pp. E10

Leonard, BE. & Myint, A. (2009). The psychoneuroimmunology of depression. *Hum. Psychopharmacol.*, Vol. 24, No. 3, pp. 165-175

Lie, DC., Colamarino, SA., Song, HJ., Désiré, L., Mira, H., Consiglio, A., Lein, ES., Jessberger, S., Lansford, H., Dearie, AR. & Gage, FH. (2005). Wnt signalling regulates adult hippocampal neurogenesis. *Nature*, Vol. 437, pp. 1370-1375

Lipton, P. (1999). Ischemic cell death in brain neurons. *Physiol. Rev.*, Vol. 79, No. 4, pp. 1431-1568

Liu, X., Rubin, JS. & Kimmel, AR. (2005). Rapid, Wnt-induced changes in GSK3β associations that regulate β-catenin stabilization are mediated by Gα proteins. *Curr. Biol.*, Vol. 15, pp. 1989-1997

Maes, M. (2001). The immunoregulatory effects of antidepressants. Hum. Psychopharmacol., Vol. 16, No. 1, pp. 95-103

Maes, M., Yirmyia, R., Noraberg, J., Brene, S., Hibbeln, J., Perini, G., Kubera, M., Bob, P., Lerer, B. & Maj, M. (2009). The inflammatory & neurodegenerative (I&ND) hypothesis of depression: leads for future research and new drug developments in depression. *Metab. Brain Dis.*, Vol. 24, No. 1, pp. 27-53

Mathew, SJ., Manji, HK. & Charney, DS. (2008). Novel drugs and therapeutic targets for severe mood disorders. *Neuropsychopharmacology*, Vol, 33, No. 9, pp. 2080-2092

McBain, CJ. & Kauer, JA. (2009). Presynaptic plasticity: targeted control of inhibitory networks. *Curr. Opin. Neurobiol.*, Vol. 19, No. 3, pp 254-262

McEwen, BS. & Sapolsky, RM. (1995). Stress and cognitive function. *Curr. Opin. Neurobiol.*, Vol. 5, No. 2, pp. 205-216

McEwen, BS. (2007). Physiology and neurobiology of stress and adaptation: central role of the brain. *Physiol. Rev.*, Vol. 87, No. 3, pp. 873-904

McEwen, BS., Chattarji, S., Diamond, DM., Jay, TM., Reagan, LP., Svenningsson, P. & Fuchs, E. (2010). The neurobiological properties of tianeptine (Stablon): from monoamine hypothesis to glutamatergic modulation. *Mol. Psychiatry*, Vol. 15, No. 3, pp. 237-249

McNamara, RK., Ostrander, M., Abplanalp, W., Richtand, NM., Benoit, SC. & Clegg, DJ. (2006). Modulation of phosphoinositide-protein kinase C signal transduction by omega-3 fatty acids: implications for the pathophysiology and treatment of recurrent neuropsychiatric illness. *Prostaglandins Leukot. Essent. Fatty Acids*, Vol. 75, No. 4-5, pp. 237-257

Mendelsohn, D., Riedel, WJ. & Sambeth, A. (2009). Effects of acute tryptophan depletion on memory, attention and executive functions: a systematic review. *Neurosci. Biobehav. Rev.*, Vol. 33, No. 6, pp. 926-952.

Mendlewicz, J. (2009). Disruption of the circadian timing systems: molecular mechanisms in mood disorders. *CNS Drugs*, Vol. 23, Suppl 2, pp. 15-26

Mesulam, MM. (1999). Neuroplasticity failure in Alzheimer's disease: bridging the gap between plaques and tangles. *Neuron*, Vol. 24, No. 3, pp. 521-529

Meyer, JH., Ginovart, N., Boovariwala, A., Sagrati, S., Hussey, D., Garcia, A., Young, T., Praschak-Rieder, N., Wilson, AA. & Houle, S. (2006). Elevated monoamine oxidase a levels in the brain: an explanation for the monoamine imbalance of major depression. *Arch. Gen. Psychiatry*, Vol. 63, pp. 1209-1216

Meyer, JH., Wilson, AA., Sagrati, S., Miler, L., Rusjan, P., Bloomfield, PM., Clark, M., Sacher, J., Voineskos, AN. & Houle S. (2009). Brain monoamine oxidase A binding in major depressive disorder: relationship to selective serotonin reuptake inhibitor treatment, recovery, and recurrence. *Arch. Gen. Psychiatry*, Vol. 66, No. 12, pp. 1304-1312

Mirnikjoo, B., Brown, SE., Kim, HF., Marangell, LB., Sweatt, JD. & Weeber, EJ. (2001). Protein kinase inhibition by ω-3 fatty acids. *J. Biol. Chem.*, Vol. 276, No. 14, pp. 10888-10896

Miyamoto, E. (2006). Molecular mechanism of neuronal plasticity: induction and maintenance of long-term potentiation in the hippocampus. *J. Pharmacol. Sci.*, Vol. 100, pp. 433-442

Moghaddam, B. (2002). Stress activation of glutamate neurotransmission in the prefrontal cortex: implications for dopamine-associated psychiatric disorders. *Biol. Psychiatry*, Vol. 51, No. 10, pp. 775-787

Nelson, JC. (1999). A review of the efficacy of serotonergic and noradrenergic reuptake inhibitors for treatment of major depression. *Biol. Psychiatry*, Vol. 46, pp. 1301-1308

Nelson, SB. & Turrigiano, GG. (2008). Strength through diversity. *Neuron*, Vol. 60, No. 3, pp. 477-482

Nestler, EJ., Barrot, M., DiLeone, RJ., Eisch, AJ., Gold, SJ. & Monteggia, LM. (2002). Neurobiology of depression. *Neuron*, Vol. 34, No. 1, pp. 13-25

Nestler, EJ. & Carlezon, WAJr. (2006). The mesolimbic dopamine reward circuit in depression. *Biol. Psychiatry*, Vol. 59, No. 12, pp. 1151-1159

Nibuya, M., Nestler, EJ. & Duman, RS. (1996). Chronic antidepressant administration increases the expression of cAMP response element binding protein (CREB) in rat hippocampus. *J. Neurosci.*, Vol. 16, pp. 2365-2372

Nikisch, G. (2009). Involvement and role of antidepressant drugs of the hypothalamic-pituitary-adrenal axis and glucocorticoid receptor function. *Neuro Endocrinol. Lett.*, Vol. 30, No. 1, pp. 11-16

Nikolaus, S., Antke, C. & Müller, HW. (2009). In vivo imaging of synaptic function in the central nervous system: II. Mental and affective disorders. *Behav. Brain Res.*, Vol. 204, No. 1, pp. 32-66

Oliet, SHR. & Mothet, J-P. (2009). Regulation of *N*-methyl-D-aspartate receptors by astrocytic D-serine. *Neuroscience*, Vol. 158, No. 1, pp. 275-283

Pace, TW. & Miller, AH. (2009). Cytokines and glucocorticoid receptor signaling. Relevance to major depression. *Ann. N. Y. Acad. Sci.*, Vol. 1179, pp. 86-105

Papakostas, GI., Thase, ME., Fava, M., Nelson, JC. & Shelton, RC. (2007). Are antidepressant drugs that combine serotonergic and noradrenergic mechanisms of action more effective than the selective serotonin reuptake inhibitors in treating major depressive disorder? A meta-analysis of studies of newer agents. *Biol. Psychiatry*, Vol. 62, pp. 1217-1227

Pascale, A., Amadio, M., Govoni, S. & Battaini, F. (2007). The aging brain, a key target for the future: the protein kinase C involvement. *Pharmacol. Res.*, Vol. 55, pp. 560-569

Paul, IA. & Skolnick, P. (2003). Glutamate and depression: clinical and preclinical studies. *Ann. N. Y. Acad. Sci.*, Vol. 1003, pp. 250-272

Peineau, S., Bradley, C., Taghibiglou, C., Doherty, A., Bortolotto, ZA., Wang, YT. & Collingridge, GL. (2008). The role of GSK-3 in synaptic plasticity. *Br. J. Pharmacol.*, Vol. 153, pp. S428-S437

Pittenger, C. & Duman, RS. (2008). Stress, depression, and neuroplasticity: a convergence of mechanisms. *Neuropsychopharmacology*, Vol. 33, No. 1, pp. 88-109

Ponti, G., Peretto, P. & Bonfanti, L. (2008). Genesis of neuronal and glial progenitors in the cerebellar cortex of peripuberal and adult rabbits. *PLoS ONE*, Vol. 3, No. 6, pp. e2366

Porcelli, S., Drago, A., Fabbri, C. & Serretti, A. (2011). Mechanisms of antidepressant action: An integrated dopaminergic perspective. *Prog. Neuropsychopharmacol. Biol. Psychiatry*, Vol. 35, No. 7, pp. 1532-1543

Quiroz, JA., Gray, NA., Kato, T. & Manji, HK. (2008). Mitochondrially mediated plasticity in the pathophysiology and treatment of bipolar disorder. *Neuropsychopharmacology*, Vol. 33, No. 11, pp. 2551-2565

Ramsay, RR. & Gravestock, MB. (2003). Monoamine oxidases: to inhibit or not to inhibit. *Mini Rev. Med. Chem.*, Vol. 3,pp. 129-136

Richelson, E. (1990). Antidepressants and brain neurochemistry. *Mayo Clin. Proc.*, Vol. 65, pp. 1227-1236

Richelson, E. (2003). Interactions of antidepressants with neurotransmitter transporters and receptors and their clinical relevance. *J. Clin. Psychiatry*, Vol. 64, Suppl 13, pp. 5-12

Riederer, P., Danielczyk, W. & Grünblatt, E. (2004). Monoamine oxidase-B inhibition in Alzheimer's disease. *NeuroToxicology*, Vol. 25, pp. 271-277

Ruhé, HG., Mason, NS. & Schene, AH. (2007). Mood is indirectly related to serotonin, norepinephrine and dopamine levels in humans: a meta-analysis of monoamine depletion studies. *Mol. Psychiatry*, Vol. 12, No. 4, pp. 331-359

Sands, WA. & Palmer, TM. (2008). Regulating gene transcription in response to cyclic AMP elevation. *Cell. Signal.*, Vol. 20, pp. 460-466

Santarelli, L., Saxe, M., Gross, C., Surget, A., Battaglia, F., Dulawa, S., Weisstaub, N., Lee, J., Duman, R., Arancio, O., Belzung, C. & Hen, R. (2003). Requirement of hippocampal neurogenesis for the behavioral effects of antidepressants. *Science,* Vol. 301, pp. 805-809

Sapolsky, RM. (1996). Stress, Glucocorticoids, and Damage to the Nervous System: The Current State of Confusion. *Stress,* Vol. 1, No. 1, pp. 1-19

Sapolsky, RM. (2004). Is impaired neurogenesis relevant to the affective symptoms of depression? *Biol. Psychiatry,* Vol. 56, No. 3, pp. 137-139

Schildkraut, JJ. (1965). The catecholamine hypothesis of affective disorders: a review of supporting evidence. *Am. J. Psychiatry,* Vol. 122, No. 5, pp. 509-522

Schroeder, FA., Lin, CL., Crusio, WE. & Akbarian, S. (2007). Antidepressant-like effects of the histone deacetylase inhibitor, sodium butyrate, in the mouse. *Biol. Psychiatry,* Vol. 62, No. 1, pp. 55-64

Schulz, P. & Steimer, T. (2009). Neurobiology of circadian systems. *CNS Drugs,* Vol. 23, Suppl 2, pp. 3-13

Selvaraj, S., Murthy, NV., Bhagwagar, Z., Bose, SK., Hinz, R., Grasby, PM. & Cowen, PJ. (2011). Diminished brain 5-HT transporter binding in major depression: a positron emission tomography study with [^{11}C]DASB. *Psychopharmacology (Berl.),* Vol. 213, No. 2-3, pp. 555-562

Shaltiel, G., Chen, G. & Manji, HK. (2007). Neurotrophic signaling cascades in the pathophysiology and treatment of bipolar disorder. *Curr. Opin. Pharmacol.,* Vol. 7, pp. 22-26

Silva, AJ., Zhou, Y., Rogerson, T., Shobe, J. & Balaji, J. (2009). Molecular and cellular approaches to memory allocation in neural circuits. *Science,* Vol. 326, No. 5951, pp. 391-395

Smith, DF. & Jakobsen, S. (2009). Molecular tools for assessing human depression by positron emission tomography. *Eur. Neuropsychopharmacol.,* Vol. 19, No. 9, pp. 611-628

Snyder, SH. & Ferris, CD. (2000). Novel neurotransmitters and their neuropsychiatric relevance. *Am. J. Psychiatry,* Vol. 157, No. 11, pp. 1738-1751

Soderling, TR. (1999). The Ca-calmodulin-dependent protein kinase cascade. *Trends Biochem. Sci.,* Vol. 24, pp. 232-236

Stahl, SM. & Felker, A. (2008). Monoamine oxidase inhibitors: a modern guide to an unrequited class of antidepressants. *CNS Spectr.,* Vol. 13, pp. 855-870

Stahl, SM. (2009). Multifunctional drugs: a novel concept for psychopharmacology. *CNS Spectr.,* Vol. 14, No. 2, pp. 71-73

Stork, C. & Renshaw, PF. (2005). Mitochondrial dysfunction in bipolar disorder: evidence from magnetic resonance spectroscopy research. *Mol. Psychiatry,* Vol. 10, No. 10, pp. 900-919

Tardito, D., Perez, J., Tiraboschi, E., Musazzi, L., Racagni, G. &, Popoli, M. (2006). Signaling pathways regulating gene expression, neuroplasticity, and neurotrophic mechanisms in the action of antidepressants: a critical overview. *Pharmacol. Rev.,* Vol. 58, No. 1, pp. 115-134

Taylor, SS., Kim, C., Vigil, D., Haste, NM., Yang, J., Wu, J. & Anand, GS. (2005). Dynamics of signaling by PKA. *Biochim. Biophys. Acta,* Vol. 1754, pp. 25-37

Thomas, DD., Ridnour, LA., Isenberg, JS., Flores-Santana, W., Switzer, CH., Donzelli, S., Hussain, P., Vecoli, C., Paolocci, N., Ambs, S., Colton, CA., Harris, CC., Roberts,

DD. & Wink, DA. (2008). The chemical biology of nitric oxide: implications in cellular signaling. *Free Radic. Biol. Med.*, Vol. 45, No. 1, pp. 18-31

Tóth, K. & McBain, CJ. (2000). Target-specific expression of pre- and postsynaptic mechanisms. *J. Physiol.*, Vol. 525.1, pp. 41-51

Tsankova, NM., Berton, O., Renthal, W., Kumar, A., Neve, RL. & Nestler, EJ. (2006). Sustained hippocampal chromatin regulation in a mouse model of depression and antidepressant action. *Nat. Neurosci.*, Vol. 9, pp. 519-525

Tsankova, N., Renthal, W., Kumar, A. & Nestler, EJ. (2007). Epigenetic regulation in psychiatric disorders. *Nat. Rev. Neurosci.*, Vol. 8, No. 5, pp. 355-367

Uher, R. (2008). The implications of gene-environment interactions in depression: will cause inform cure? *Mol. Psychiatry*, Vol. 13, No. 12, pp. 1070-1078

Vetulani, J. & Sulser, F. (1975). Action of various antidepressant treatments reduces reactivity of noradrenergic cyclic AMP-generating system in limbic forebrain. *Nature*, Vol. 257, No. 5526, pp. 495-496

Wada, T. & Penninger, JM. (2004). Mitogen-activated protein kinases in apoptosis regulation. *Oncogene*, Vol. 23, pp. 2838-2849

Warner-Schmidt, JL. & Duman, RS. (2006). Hippocampal neurogenesis: opposing effects of stress and antidepressant treatment. *Hippocampus*, Vol. 16, No. 3, pp. 239-249

Wojda, U., Salinska, E. & Kuznicki, J. (2008). Calcium ions in neuronal degeneration. *IUBMB Life*, Vol. 60, No. 9, pp. 575-590

Wu, H., Peisley, A., Graef, IA. & Crabtree, GR. (2007). NFAT signaling and the invention of vertebrates. *Trends Cell. Biol.*, Vol. 17, pp. 251-260

Yasuda, S., Liang, MH., Marinova, Z., Yahyavi, A. & Chuang, DM. (2009). The mood stabilizers lithium and valproate selectively activate the promoter IV of brain-derived neurotrophic factor in neurons. *Mol. Psychiatry*, Vol. 14, No. 1, pp. 51-59

Yildiz-Yesiloglu, A. & Ankerst, DP. (2006). Neurochemical alterations of the brain in bipolar disorder and their implications for pathophysiology: a systematic review of the in vivo proton magnetic resonance spectroscopy findings. *Prog. Neuropsychopharmacol. Biol. Psychiatry*, Vol. 30, No. 6, pp. 969-995

Youdim, MB., Edmondson, D. & Tipton, KF. (2006). The therapeutic potential of monoamine oxidase inhibitors. *Nat. Rev. Neurosci.*, Vol. 7, pp. 295-309

Zarate, CAJr., Singh, J. & Manji, HK. (2006a). Cellular plasticity cascades: targets for the development of novel therapeutics for bipolar disorder. *Biol. Psychiatry*, Vol. 59, No. 11, pp. 1006-1020

Zarate, CAJr., Singh, JB., Carlson, PJ., Brutsche, NE., Ameli, R., Luckenbaugh, DA., Charney, DS. & Manji, HK. (2006b). A randomized trial of an N-methyl-D-aspartate antagonist in treatment-resistant major depression. *Arch. Gen. Psychiatry*, Vol. 63, No. 8, pp. 856-864

Zhu, XJ., Hua, Y., Jiang, J., Zhou, QG., Luo, CX., Han, X., Lu, YM. & Zhu, DY. (2006). Neuronal nitric oxide synthase-derived nitric oxide inhibits neurogenesis in the adult dentate gyrus by down-regulating cyclic AMP response element binding protein phosphorylation. *Neuroscience*, Vol. 141, No. 2, pp. 827-836

10

Depression Viewed as a GABA/Glutamate Imbalance in the Central Nervous System

Joanna M. Wierońska[1], Agnieszka Pałucha-Poniewiera[1], Gabriel Nowak[1,2] and Andrzej Pilc[1,3]
[1]*Institute of Pharmacology, Polish Academy of Sciences, Kraków*
[2]*Jagiellonian University, Medical College, Department of Pharmacobiology, Kraków*
[3]*Jagiellonian University, Medical College, Faculty of Health Sciences, Kraków*
Poland

1. Introduction

Despite of the last few decades of investigations concerning the pathophysiology of affective disorders and their efficient treatment, the number of patients suffering from the diseases is growing every year. Compared to the data from the 1950s the percentage of people diagnosed with monopolar depression increased by nearly 7 times, and now it is estimated that more than 15% of the population suffers from the illness (Healy, 1999). Depressive episodes deeply influence the familiar and professional life of the patients, thus become both a clinical and social problem. The costs connected with the treatment are not only due to medical care, but also involve indirect wastes related to long-term disturbances in normal functioning. Statistical analysis reveal that depressive episodes became the most costly group of central nervous system (CNS) disorders in Europe and in the USA (in the recent decade they reached about 105 billion Euro and 83 Billion USD, respectively) (Andlin-Sobocki et al., 2005; Greenberg et al., 2003). Moreover, the problem of depression and mood disorders is hard to overestimate as only 1 out of 4 cases is either recognized or adequately treated, and approx. 15% of depressed people commit suicide (Brody et al., 1998).

The antidepressant pharmacotherapy presently used has been based on the modulation of monoaminergic neurotransmission (mainly noradrenergic and serotonergic). The accidentally discovered antidepressant activity of selective serotonin and noradrenalin reuptake inhibitors, as well as MAO inhibitors, laid at the grounds for the monoaminergic theory of depression (Kuhn, 1957; Loomer et al., 1957), which still dominates this field of research. The decreased level of monoamines in the depressed brain that was normalized after acute administration of antidepressant drugs (ADDs) became the main assumption of the hypothesis (Bunney et al., 1965; Lapin et al., 1969; Schildkraut, 1965). As the result of intensive studies, a number of antidepressant drugs were synthesized. However, most of them required chronic administration to evoke an antidepressant-like effect in humans. Therefore the increase in the serotonin and/or noradrenalin level in the brain observed immediately after the single administration of ADDs could not be responsible for the antidepressant effect observed only after several weeks of systemic treatment (for review see: Castren, 2005; Oswald et al., 1972). Furthermore, in nearly 40% of patients the antidepressant therapy presently conducted is not effective, thus confirming the limited role of monoamines in depression.

The further research concerning antidepressant therapy was focused on the adaptive changes observed after prolonged antidepressant treatment. The progress of research lead to the discovery of the new receptors, second messengers systems, protein kinases, transcription and trophic factors, genetic and epigenetic regulations (Duman et al., 1997; Nestler, 1998; Nestler et al., 2002; Shirayama et al., 2002; for recent review, see Millan, 2006). A number of new hypotheses were proposed, and the neurotrophic theory of the disease was probably the mostly recognized. It was based on the observed increase in the BDNF level after prolonged ADDs treatment (Altar, 1999; Nibuya et al., 1995), which was supposed to be involved in the hippocampal processes of plasticity and neurogenesis (Altar, 1999; Duman et al., 1997; Duman et al., 1998; Nestler, 1998; Duman et al., 2001; Warner-Schmidt & Duman, 2006).

However, despite all of the research, there are no drugs that are active after a single administration in depressed patients. Therefore, the new insight into the mechanism of action of antidepressants, based on anything beyond monoaminergic neurotransmission, is highly desired. Excitatory glutamate and inhibitory GABA seem to be good candidates for our consideration.

2. Amino acids neurotransmitters and depression

It has to be realized that glutamate constitutes 50-60% of all neurotransmission in the brain and the remaining 40-50% is GABAergic (Storm-Mathiesen and Iversen, 1979; Winfield et al., 1980; Winfield et al., 1981). Therefore 90-99% of neurons (depending on the source) are GABAergic or glutamatergic, and less than 10% is left for all the others monoamines, neuropeptides and neuroendocrine neuromodulators. The fundamental aspect of the proper functioning of the CNS is keeping the excitatory/inhibitory physiological balance (Altamura et al., 1993; Linden and Schoepp, 2006; Yildiz-Yesiloglu et al., 2006). Any disruptions within this balance may lead to a brain dysfunction reflected as a mental disorder. Viewing the pathophysiology of depression through the GABA-Glu interaction may bring new solutions concerning its effective treatment. Schematic representation of the twisted GABA/Glu balance in affective disorders is shown on the Figure 1.

The results of the preclinical and clinical investigations confirm that disrupted activity within these two amino acid neurotransmitters may lay at the grounds of depression. The other data shows that standard ADDs influence the inhibitory and excitatory neurotransmission, which contributes to their efficacy. Generally it can be concluded that antidepressant drug diminish glutamatergic neurotransmission (Tokarski et al. 2008) and increase GABAergic neurotransmission. Using these results as our basis it could be suggested that the direct pharmacological manipulation on GABA or Glu neurotransmission may become a faster and more efficient way to treat depression (Kendell et al., 2005).

A number of important review papers concerning the role of GABA and/or glutamate have been published lately, such as Cryan & Slatery, 2010; Drago et al., 2011; Froestl, 2010; Ghose et al., 2011; Hashimoto, 2011; Mitchell & Baker, 2010; Sanacora et al., 2011. In this review we decided to focus on issues which were not covered in the above mentioned reviews, mainly on interactions between GABAergic and glutamatergic systems with the involvement of

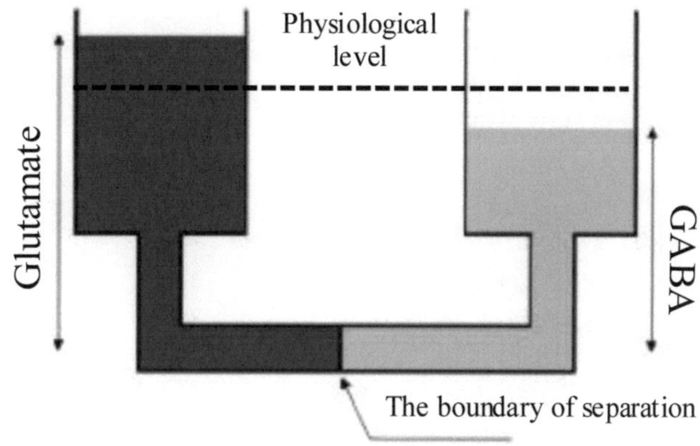

Fig. 1. The schematic representation of the Glutamate/GABA disbalance in the brain as an example of the communicating tubes. In the physiological conditions both liquids constitute a self-regulating system establishing the activity of both neurotransmitters on the same level. The disruption within the system leads to twist in the harmony, and the loss of the self-regulating properties. The overactivation of the glutamate and the decreased activity of GABA being responsible for mood disorders.

metabotropic receptors. This review is complementary to our recently published paper on the role of GABA and Glu interaction in anxiety (see Wieronska et at., 2011).

3. The role of the glutamatergic system in the treatment of depression

3.1 The role of the NMDA receptor in the mechanism of action of antidepressant drugs

The discovery and characterization of glutamatergic receptors opened a new possibilities to investigate the role of the glutamatergic system in the treatment of depression. The influence of the ADDs on the activity of glutamatergic receptors was studied on the one hand, whilst on the other hand the potential antidepressant effects of the ligands of those receptors were also investigated. Selected data are collected in Table 1.

The first reports on this subject concerned the NMDA receptor which belongs to the family of ionotropic receptor for the glutamate (together with AMPA and KA receptors). The receptor was described in details elsewhere (see: Danysz et al., 1998; Wieronska et al., 2011). The studies conducted in the 1990s revealed the antidepressant activity of the functional antagonists of the NMDA receptor, such as AP-7, MK-801, or ACPC in the forced swim test or tail suspension test (Trullas & Skolnick, 1990). The later studies carried out using those experimental procedures confirmed antidepressant activity of the other NMDA antagonists, such as CGP37849, CGP39551, memantine, eliprodil and zinc (Kroczka et al., 2001; Layer et al., 1995; Maj et al., 1992a; 1992b; Moryl et al., 1993; Przegalinski et al., 1997).

Despite of the evident antidepressant-like activity of NMDA receptor antagonists in simple screening tests, it was also shown that the compounds were active in commonly known models of depression. And thus MK-801, CGP37849, CGP40116 and ACPC (Papp & Moryl,

Author	Compound/dose	Species	Test
Maj et al., 1992a	MK-801 (0.3 mg/kg)	rats	Porsolt test
Maj et al., 1992b	CGP37849 (0.625-5 mg/kg) CGP39551	rats	Porsolt test
Moryl et al., 1993	Amantadine Memantine bifemelane	rats	Porsolt test
Papp &Moryl, 1994	MK-801 (0.3 mg/kg) CGP37849 (5 mg/kg) CGP40116 (5 mg/kg)	rats	Chronic mild stress
Papp&Moryl, 1995	ACPC (200)	rats	Chronic mild stress
Layer et al., 1995	Eliprodil (20-40 mg/kg)	mice	Porsolt test
Trullas &Skolnick, 1990	AP-7 ACPC MK-801	mice	Porsolt test Tail suspension test
Przegaliński et al., 1997	ACPC (200-400 mg/kg) CGP37849 (0.625-5 mg/kg)	rats	Porsolt test
Muhonen et al., 2008	Memantine (20mg/day)	human	Clinical study
Berman et al., 2000	Ketamine (0.5 mg/kg)	human	Clinical study
Zarate et al., 2004, 2005	Riluzol (100-200mg/day)	human	Clinical study
Zarate et al., 2006	Ketamine (0.5 mg/kg)	human	Clinical study
Nowak et al., 2003	Zinc (25mg/day)	human	Clinical study
Eby&Eby, 2006	Magnesium aspartate (125-300 mg/day)	human	Clinical study
Chouinard et al., 1990	Magnesium aspartate (40mg/day)	human	Clinical study
Siwek et al., 2009	Zinc (25mg/day)	human	Clinical study
Li et al., 2010	Ro 25-6981 (10 mg/kg)	rats	Chronic unpredictable stress
Preskorn et al., 2008	CP-101, 606	human	Clinical study

Table 1. Collected data concerning the antidepressant-like activity of the NMDA receptor antagonists (see also description in the text).

1994; Papp & Moryl, 1996) were active in chronic mild stress. Additionally, memantine and MK-801 were active in chronic unpredictable stress (Ossowska et al., 1997). Also, the antidepressant-like activity of MK-801 was evident in the olfactory bulbectomy model of depression (Redmond et al., 1997). Ro 25-6981, the subunit selective NR_{2B} antagonist, was found to exhibit rapid (24h) antidepressant-like effect in the FST (Maeng, 2008; Li et al., 2010).

The role of the NMDA receptor in the mechanism of action of ADDs was confirmed in the variety of biochemical (Nowak et al., 1993, 1996, 1998), electrophysiological (Bobula et al., 2003) and behavioural studies (Popik et al., 2000) in both rats and mice. All the results indicated a reduction of the NMDA receptor function after repeated administration of a variety of ADDs, as well as after electroconvulsive therapy. Receptor binding studies with the use of selective NMDA receptor binding sites radioligands revealed the decrease of glycine's affinity to its binding site (GLY_B) and to the abolishment of its ability to modulate glutamate binding sites in the NMDA receptor complex. The changes in mRNA synthesis and proteins of selected subunits of the NMDA receptor, and the changes in their mutual composition, were shown as being connected with the influence of the chronic ADDs administration on the NMDA receptor (Skolnick, 1999).

3.2 Glutamatergic theory of depression

The above noted results of the studies on the NMDA receptor have become the fundamental element of the glutamatergic theory of depression (Skolnick, 1999, 2009). The hypothesis was based on the results obtained in the studies on the role of trophic factors in depression (Duman, 1998; Skolnick, 1999). The brain-derived neurotrophic factor (BDNF) may decrease mRNAs for NR_{2A} and NR_{2C} NMDA receptor subunits, resulting in changes in their mutual composition and impartment in the NMDA receptor function in the matter resembling the effect of NMDA receptor antagonists (Brandoli et al., 1998). Therefore, the glutamatergic theory of depression states that the activation of BDNF, after ADDs administration, may lead to impartment of NMDA receptor function, which is commonly observed after administration of standard ADDs and NMDA receptor antagonists (Skolnick, 1999).

3.3 Antidepressant effects of NMDA receptor antagonists-clinical studies

The characteristic for ADDs behavioural effects obtained after administration of NMDA receptor antagonists in animal studies were confirmed in recent clinical trials. The antidepressant effect of memantine, a low-affinity, uncompetitive open channel NMDA receptor blocker, was observed in treatment-resistant patients with severe depression and comorbid alcohol dependence (Muhonen et al., 2008). However, another clinical study showed no antidepressant effect of memantine in patients suffering from major depression (Zarate et al., 2006). Thus, the clinical utility of memantine seems to be controversial.

Much more promising results were obtained with another NMDA channel blocker, ketamine. During initial studies, the drug was administered to patients with severe, long term, treatment-resistant depression, two weeks after cessation of standard pharmacotherapy (Berman et al., 2000; Zarate et al., 2006). Ketamine at a dose of 0.5 mg/kg was administered through intravenous infusion and the 21-degree Hamilton scale was taken to estimate the mental health of the patients. A statistically significant effect of ketamine was evident 24 hours after injection. The effect was similar or even greater than the effect of standard antidepressants such as venlafaxine, buprion or fluoxetine administered for the time period of 8 weeks. The effect of ketamine was observed until seven days after a single administration (Zarate et al., 2006). Another clinical study showed, that a single infusion of ketamine induced a rapid resolution of suicidal ideation in treatment-resistant patients suffering from major depression (DiazGranados et al., 2010). Moreover, a rapid antidepressant effect was observed after a single dose of ketamine in depressive patients with a family history of alcohol dependence (Phelps et al., 2009) and in electroconvulsive therapy-resistant patients (Ibrahim et al., 2011).

Another group of potential antidepressants are the divalent ions of zinc and magnesium, which are known to be NMDA receptor antagonists. The preliminary study by Nowak et al. (2003) showed the beneficial effects of zinc supplementation on standard antidepressant therapy with tricyclics and SSRIs in patients with unipolar depression. Then, a double blind, placebo- controlled study by Siwek et al. (2009) demonstrated that zinc supplementation enhanced the efficacy of imipramine and facilitated the treatment outcome in treatment-resistant depression. Several clinical studies indicated beneficial effects of magnesium supplementation in depression and depression-related or comorbid disorders, such as: major depression (Eby & Eby, 2006), mania (Pavlinac et al., 1979), bipolar disorder (Chouinard et al., 1990), and depression in elderly diabetic patients suffering from hypomagnesemia (Barragan-Rodriguez et al., 2008).

Clinical studies also showed the potential antidepressant activity of subtype-selective NMDA receptor modulators, which are supposed to be better and safer drugs than non-selective antagonists. Firstly, promising preclinical results showed the potential antidepressant-like effect of a selective antagonist of the NR_{2B} subunit, Ro 256981 (Maeng et al., 2008). Then, the clinical study of Preskorn et al. (Preskorn et al., 2008) confirmed the antidepressant effect of NR_{2B} subunit-specific NMDA receptor antagonist, CP-101,606 in a placebo controlled, double-blind study in patients with a treatment-refractory major depressive disorder.

This data strongly supports the hypothesis that glutamate receptors ligands may become more efficient antidepressants than the presently used pharmacotherapy.

The effectiveness of NMDA receptor antagonists is consistent with the notion of increased glutamatergic transmission in depression. However, the problem of profound adverse effects connected with the use of ketamine or the other NMDA receptor antagonists still remains unsolved; therefore the introduction of those compounds to the clinic requires further extensive preclinical and clinical studies.

3.4 Antidepressant effects of AMPA receptor ligands

The other member of the ionotropic glutamate receptors family, AMPA, was shown as being engaged in the mechanisms of antidepressant effects, too. A variety of animal studies indicate the potential antidepressant activity of the positive allosteric modulators (potentiators) of that receptor. Such effects were shown for LY392098, which was active in the forced swim test in mice and rats and in the tail suspension test in mice (Li et al., 2001, 2003). Furthermore, the compound induced an elevation of both the mRNA and BDNF protein level (Legutko et al., 2001), which stays in line with the trophic theory of disease. Another positive allosteric modulator of the AMPA receptor with antidepressant-like efficacy was LY451646. The potential antidepressant activity was shown in the forced swim test and in the tail suspension test (Bai et al., 2001). Moreover, LY451646 administration lead to the increase of neuronal proliferation in the hippocampus of the rat brain (Bai et al., 2003), and an increase of the level of BDNF mRNA and trkB, and to the enhancement of the BDNF protein synthesis (Maćkowiak et al., 2002). The potential antidepressant activity for AMPA potentiators in the preclinical studies was shown also for other compounds, such as CX546,

CX614, and LY404817 (Bai et al., 2001; Lauterborn et al., 2000). Additionally, the fundamental role of the AMPA receptor was also shown for ketamine activity, while the action of ketamine in the forced swim test and learned helplessness test was attenuated by AMPA antagonist, NBQX, indicating that AMPA activation is necessary for ketamine action in models of depression (Koike et al. 2011; Maeng et al. 2008).

3.5 The role of mGlu receptors in the mechanism of action of antidepressant drugs

The introduction of currently available NMDA receptor antagonists into the clinic is impossible because of the variety of profound adverse effects that can be triggered after their administration. Metabotropic glutamate receptors are a natural alternative target to influence the glutamatergic system. These receptors are responsible for the modulation, but not for the fast neuronal transmission (Nakanishi et al., 1992). The discovery of selective ligands of these receptors created a new possibilities in the therapy of variety of central nervous system disturbances, including psychiatric disorders.

Until now 8 subtypes of metabotropic glutamate receptors have been cloned. They were named from 1-8, according to the sequence homology, pharmacology and the second messenger system they activate. There can be several splice variants of selected subtypes of the receptors created by alternative splicing (Pin and Duvoisin, 1995; Pin et al., 1999), which makes them an established family of 21 subtypes known. At present mGlu receptors are divided into three different groups according to the sequence homology, pharmacology and the second messenger system they activate. And as such group I consists of $mGlu_1$ and $mGlu_5$ receptor subtypes, $mGlu_{2/3}$ receptors create the group II of mGlu receptors and the third group involve $mGlu_4$, $mGlu_6$, $mGlu_7$ and $mGlu_8$ receptors (Pin and Duvoisin, 1995; Pin et al., 1999; see: Wieronska et al., 2011).

3.5.1 The role of group I mGlu receptors in the mechanism of action of antidepressant drugs

Among all mGlu receptors, the greatest preclinical potential for possible antidepressant efficacy was shown for the ligands of the I group of mGlu receptors, especially the $mGlu_5$ subtype, which is localized in the brain structures known to be connected with emotional and motivational processes, such as the cortex, hippocampus and the amygdala (Romano et al., 1995; for review see Wieronska et al., 2010). The involvement of the NMDA receptor in the action of both compounds plays a substantial role, as a lot of evidence indicates that $mGlu_5$ receptors are physically and functionally linked with the NMDA receptor complex. The Homer family of proteins functions as a bridge between group I mGlu receptors and IP_3 receptors, as well as with Shank proteins, which are a part of the NMDA receptor-associated PSD-95 complex (Brakeman et al., 1997; Lujan et al., 1996; Tu et al., 1999; Xiao et al., 1998). The activation of the $mGlu_5$ receptor has been shown to potentiate NMDA receptor activity in the mechanism that requires G-protein activation (Attuci et al., 2001; Awad et al., 2000; Pisani et al., 2001), and antagonists of $mGlu_5$ receptors have been reported to decrease NMDA receptor activation (Doherty et al., 2000). Therefore, the inhibition of $mGlu_5$ receptors and antagonism towards the NMDA receptor evokes a similar effect in the brain. However, due to the indirect influence on the ion channel, side-effects typical for the channel blockers are not observed after MTEP or MPEP administration. The pathological changes within the receptor itself, and the malfunction of the $mGlu_5$/NMDA complex, may contribute to altered transmission in the CNS of depressed subjects. The observed decrease of the $mGlu_5$ receptor in the PFC of patients with depressive symptomatology (Deschwanden et al., 2011) supports this speculation, indicating on the impartment in the function of the receptor. The concomitant decrease of the PSD-95 enchoring protein (Feyissa et al., 2009) indicates on the dysfunction of the NMDA receptor, too. Although there is no experimental data on this subject, a hypothesis can be raised that $mGlu_5$/NMDA receptors lose their colocalization in depressive illness and that MPEP/MTEP administration restores the functionality of the complex.

It was shown that the blockade of $mGlu_5$ receptors may exert the effect similar to NMDA receptor antagonists e.g attenuation of the NMDA receptor function (Doherty et al., 2000). Therefore, to begin with, attempts to investigate the potential antidepressant-like effect of the antagonists of $mGlu_5$ receptors were undertaken. As shown in Table 2 the selective orthosteric antagonist for this receptor, MPEP (Gasparini et al., 1999) and its derivative MTEP (Cosford et al., 2003), were active in the forced swim test in both rats and mice, and in the tail suspension test in mice (Belozertseva et al., 2007; Li et al., 2006; Palucha et al., 2005; Pilc et al., 2002; Tatarczyńska et al., 2001). Moreover, in the olfactory bulbectomy model of depression it was shown that chronic administration of those substances evoked behavioural effects similar to those observed after the administration of ADDs administration (Palucha et al., 2005; Pilc et al., 2002; Wieronska et al., 2005). On the other hand repeated MPEP administration lead to an increase in the expression of BDNF mRNA

Author	Compound/dose	Species	Test
Tatarczyńska et al., 2001	MPEP (1-20mg/kg)	mice	Tail suspension
Pilc et al., 2002; Wierońska et al., 2002	MPEP (10 mg/kg)	rats	Olfactory bulbectomy model of depression
Belozertsewa et al., 2007	MTEP (5 and 10 mg/kg) MTEP (25mg/kg) MPEP (10 mg/kg) EMQMCM (1.25-10 mg/kg) EMQMCM (5 mg/kg)	rats mice mice rats mice	Porsolt test Tail suspension test Tail suspension test Porsolt test Tail suspension test
Pałucha et al., 2005	MTEP (0.3-30 mg/kg) MTEP (1 mg/kg)	mice rats	Tail suspension test Olfactory bulbectomy
Molina-Hernandez et al., 2008	MTEP (5 and 10 mg/kg) EMQMC (1.5 and 2 mg/kg)	rats rats	Porsolt test Porsolt test
Li et al., 2006	MPEP (10-30 mg/kg) MTEP (3-30 mg/kg)	mice mice	Porsolt test Porsolt test

Table 2. Collected data concerning the antidepressant-like activity of the group I mGlu receptors ligands. MPEP, MTEP-mGlu5 antagonists; EMQMC-mGlu1 antagonist.

in the hippocampus of the rat brain, which remains in line with the neurotrophic theory of depression (Legutko et al., 2006). Additional evidence suggesting the important role of the NMDA receptor in the underlying antidepressant mechanism of the $mGlu_5$ receptor antagonist comes from the studies of Pomierny-Chamioło et al. showing, that the anti-immobility action of MTEP in the Porsolt test was inhibited by NMDA administration (Pomierny-Chamioło et al., 2010).

The antidepressant effects could be evoked not only after administration of $mGlu_5$ receptor antagonists, but also after blockade of $mGlu_1$ receptor subtype, the second representative of group I mGlu receptors, as its antagonist EMQMCM was effective in the tail suspension and forced swim test in mice (Belozertseva et al., 2007). The role of the first group of mGlu receptors in depression and in the mechanism of action of ADDs is confirmed by experiments illustrating the changes in both the reactivity and expression of the $mGlu_1$ and $mGlu_5$ receptors after chronic ADDs treatment in both rats and mice (Pilc et al., 1998; Smiałowska et al., 2002; Zahorodna et al., 1999).

Unfortunatelly there is a limited amount of clinical data concerning antidepressant action of $mGlu_5$ receptor antagonists. Fenobam, discovered in 1978 as a nonbenzodiazepine anxiolytic (Itil et al., 1978), was in 2005 described as an $mGlu_5$ receptor antagonist with indication on anxiolytic activity (Porter et al., 2005). The antidepressant effects of fenobam were also reported (Lapierre & Oyewumi, 1982). The recent study of Berry-Kravis et al. show the positive results of fenobam administration in the treatment of fragile X syndrome, without any significant adverse reactions (Berry-Kravis et al., 2011), opening the possibility to evaluate the antidepressant potential of that agent or other new substances, such as ADX10059, which is effective in gastro-esophageal reflux disease (Zerbib et al.,2011) or in the treatment of migraine (Marin & Goadsby, 2010), with no major adverse effects.

The $mGlu_5$/NMDA receptor complexes were shown to be localized predominantly postsynaptically (Lujan et al. 1996) and in the hippocampus and prefrontal cortex known to be involved in depression those receptors are extensively expressed on GABAergic interneurons (van Hooft et al., 2000; Zhou et al., 1997). Similarly to the proposed earlier mechanism of anxiolytic-like action of MPEP or MTEP (Wieronska et al. 2011), the $mGlu_5$ antagonists would initiate a repertoire of changes between interneurons and pyramidal neurons to induce feedback inhibition of increased excitation in the brain.

As shown on Fig. 2 the inhibition of GABAergic neurotransmission by $mGlu_5$ receptor antagonists may occur at multiple sites: the inhibition of postsynaptic neurotransmission [Fig. 2 (1)] and the presynaptic inhibition of GABA release (Chu et al., 1998) [Fig.2 (2)] lead to the disinhibition of intermediate interneuron, which in turn inhibits the glutamatergic target neuronal element. The presynaptic localization of the $mGlu_5$ receptor in the medial prefrontal cortex (mPFC) was described in rare cases (Romano et al., 1995), and the activation of those receptors facilitates the release of glutamate (Thomas et al., 2000). Therefore the blockade of mGlu5-mediated presynaptic neurotransmission could have an inhibitory effect on glutamatergic neurotransmission [Fig. 2 (3)]. The studies of Marek & Zhang from 2008 stay in line with such a hypothesis, by showing the inhibition of DHPG-induced spontaneous EPSCs by MPEP (Marek & Zhang, 2008). As such, it can be supposed that the compound inhibits the activity of glutamate through pre- and postsynaptic binding sites at the interneurons and pyramidal neurons.

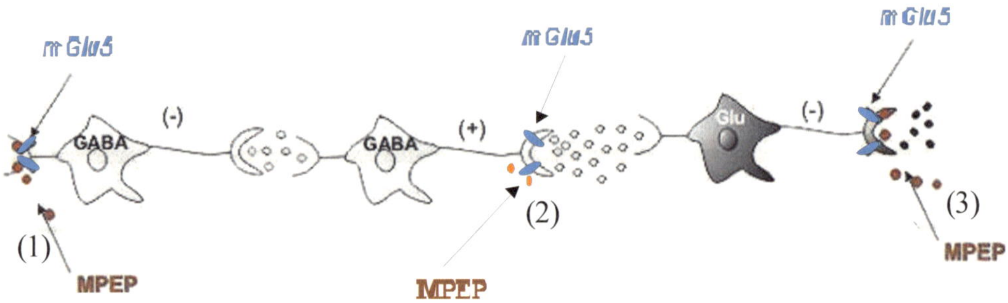

Fig. 2. The schematic representation of the mechanism of antidepressant-like action mediated by mGlu5 receptor antagonist(s) (see description in the text). Empty dots- GABA; black dots-Glu; (-)-inhibition; (+)-enhancement; the number of dots indicates the amount of neurotransmitter released

3.5.2 The role of group II mGlu receptors in the mechanism of action of antidepressant drugs

The investigations concerning the role of the II group of mGlu receptors was started with the orthosteric agonist of mGlu2/3 receptors, LY354740 (Helton et al., 1998). The compound was shown to be an effective anxiolytic, but not an antidepressant agent after peripheral administration in animal studies (Klodzinska et al., 1999). However, data on the antidepressant action of a novel $mGlu_2$ receptor potentiator THIIC (Fell et al., 2011) was recently reported, creating a hope for the antidepressant-like action for subtype selective agonists/PAMs.

As shown in Table 3 a number of experiments described the antidepressant-like activity of ligands of the $mGlu_{2/3}$ receptors. The best known selective and brain penetrating ligands of $mGlu_{2/3}$ receptors is MGS0039 (Nakazato et al., 2004). Another antagonist LY341495, is able to bind the third group of mGlu receptors too (Chung et al., 1997). Both compounds decreased the immobility time of animals in the Porsolt swim test or in the tail suspension test (Bespalov et al., 2008; Chaki et al., 2004). MGS0039 was shown to be active in the olfactory bulbectomy model of depression as well (Palucha-Poniewiera et al., 2010b). Moreover, the compound induced an increase in the neuronal proliferation in the rat hippocampus (Yoshimizu et al., 2004). The involvement of other than $mGlu_{2/3}$ receptor subtypes in the action of LY341495 cannot be ruled out, either as the compound is a mixed antagonist of both group II and group III receptors. The mechanism of the antidepressant-like action of both compounds was shown to be independent on the serotonergic system; the drugs were still active after the depletion of serotonin, and their action was not antagonized by serotonergic receptor antagonists ritanserin or WAY100635 (Pałucha-Poniewiera et al., 2010b). However, the activity of the compounds was antagonized by NBQX, further supporting the role of the AMPA receptor in the antidepressant action of $mGlu_{2/3}$ receptor ligands (Chaki et al., 2004; Karasawa et al., 2006; Kawashima et al., 2005; Palucha-Poniewiera et al., 2010b).

The possibility of involvement of the $mGlu_{2/3}$ receptor in depression was confirmed recently in the study of Feyissa et al showing the elevated level of the receptor in the prefrontal cortex of depressed suicide victims (Feyissa et al., 2010). These results correspond with the

Author	Compound/dose	Species	Test
Bespalov et al., 2008	LY341495 (0.3-3mg/kg)	mice	Porsolt test
Chaki et al., 2004	MGS0039 (0.3-3 mg/kg) LY341495 (0.1-3 mg/kg)	rats	Porsolt test
		rats	Porsolt test
Karasawa et al., 2005	MGS0039 (0.3-3 mg/kg)	mice	Tail suspension test
Yasuhara et al., 2006	7ao (MGS0039 prodrug, 3-10 mg/kg)	rats	Porsolt test
		mice	Tail suspension test
Yoshimizu et al., 2006	MGS0039 (10mg/kg)	rats	Learned helplessness
Pałucha-Poniewiera et al., 2010b	MGS0039 (1-3 mg/kg)	rats	Olfactory bulbectomy
Fell et al., 2011	THIIC (10 mg/kg)	mice	Porsolt test

Table 3. Collected data concerning the antidepressant-like activity of the group II mGlu receptors ligands. MGS0039-mGlu2/3 antagonist; LY341495-mixed group II/III antagonist; THIIC-mGlu2 PAM.

reports showing the antidepressant-like activity of the antagonists of the second group of mGlu receptors.

The localization of $mGlu_{2/3}$ receptors in the synaptic junction is both pre- and postsynaptic, and it seems that neither of them is predominant (Petralia et al., 1996). As presynaptic auto- or heteroreceptors, $mGlu_{2/3}$ receptors are located at perisynaptic sites of the synapse, often along axon terminals. Such a localization enables on one hand for the pharmacological regulation of the neurotransmitter release and, on the other hand, to make the regulation of the postsynaptic neuronal element. The $mGlu_{2/3}$ receptors are inhibitory in nature, being negatively coupled to adenylyl cyclase activity, therefore their blockade on postsynaptic membranes of glutamatergic pyramidal neurons may lead to an enhancement of glutamatergic transmission [Fig.3 (1)]. A similar effect would be observed after the inhibition of presynaptic autoreceptors, leading to an overflow of glutamate [Fig.3 (2)] and the activation of an inhibitory GABA-ergic neuron, which then inhibits the glutamatergic output element. Alternatively the blockade of the $mGlu_{2/3}$ heteroreceptors by the ligands cannot be excluded in their overall action in the brain. The antagonism towards $mGlu_{2/3}$ heteroreceptors localized on GABAergic nerve terminals would activate the release of inhibitory neurotransmission and thus contributing to the abolishment of the overexcitation and to the antidepressant-like effect of MGS0039, in a way similar to standard antidepressants, which were shown to elevate the level of inhibitory amino acid in the CNS [Fig. 3 (3)].

The involvement of the glial element also seems to be important, either, as $mGlu_3$ receptors are widely distributed on those non neuronal cells (Petralia et al., 1996). This part of action of the $mGlu_{2/3}$ antagonist may be especially important in the depression, as astrocytes were shown to contribute to the pathophysiology of the illness (for review see: Wierońska & Pilc, 2009).

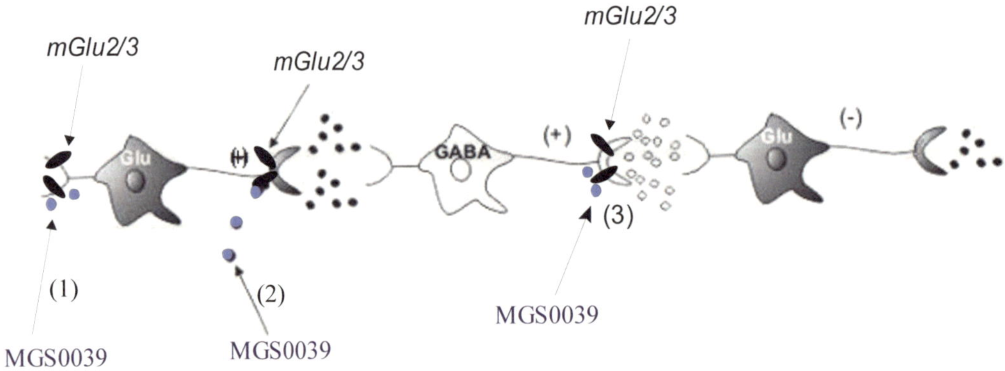

Fig. 3. The schematic representation of the mechanism of antidepressant-like action mediated by mGlu2/3 receptor antagonist(s) (see description in the text). Empty dots-GABA; black dots-Glu; (-)-inhibition; (+)-enhancement; the number of dots indicates the amount of neurotransmitter released

3.5.3 The role of group III mGlu receptors in the mechanism of action of antidepressant drugs

The third group of mGlu receptors is the largest and the least investigated. The lack of the selective and brain penetrating agents limited the investigations. Some results were obtained after the central administration of selected ligands such as ACPT-I and RS-PPG, which were shown to evoke a dose-dependent decrease in the immobility time after intrahippocampal administration (Palucha et al., 2004). However, the recent data obtained after peripheral administration of ACPT-I or LSP1-2111 did not indicate on any antidepressant-like efficacy (Stachowicz et al., 2009; Wieronska et al., 2010) of those preferential $mGlu_{4/8}$ receptor agonists (Acher et al., 1997; Beurrier et al., 2009). The $mGlu_7$ receptor was studied more extensively with the use of its positive modulator, AMN082. It was shown that the drug possessed antidepressant-like activity in the forced swim test in rats and mice, and in the tail suspension test in mice (Palucha et al., 2007, Table 4). Furthermore, the mechanism of the antidepressant action of the compound was shown to be serotonin-dependent, as it was absent in pCPA-treated animals and was inhibited by WAY100635, 5-HT_{1A} antagonist (Palucha-Poniewiera et al., 2010a). The involvement of $mGlu_7$ receptors in depression and in the mechanism of action of ADDs was confirmed by experiments showing the changes in the expression of the $mGlu_7$ receptor after standard antidepressant drugs, and in the olfactory bulbectomy model of depression (Wieronska et al., 2007, 2008).

Author	Compound/dose	Species	Test
Pałucha et al., 2007	AMN082 (3-6 mg/kg)	mice	Tail suspension test Porsolt test
Pałucha-Poniewiera et al., 2010	AMN082 (5-10 mg/kg)	rats	Porsolt test

Table 4. Collected data concerning the antidepressant-like activity of the group III mGlu receptors ligands. AMN082-mGlu7 positive modulator.

The receptor is expressed in the center of the synapse that is directly involved in the regulation of the neurotransmitter release, even in a very low concentrations of the neurotransmitter in the synaptic cleft at the site of the synaptic vesicle fusion (Shigemoto et al., 1996). The pyramidal neuron axon terminals expressing the $mGlu_7$ receptor were observed to predominantly form synapses with GABAergic interneurons (Shigemoto et al., 1996, 1997). Therefore, the final result of the pre-synaptic action of the activated mGlu7 receptor is the modulation of the postsynaptic GABAergic target interneuron [Fig.4 (1)]. This inhibition would cause the disinhibition of the other interneurons, targeting the glutamatergic network. Interestingly, $mGlu_7$ receptors are also expressed on some types of the interneuron population (e.g VIP positive) innervating $mGlu_{1\alpha}$-somatostatine postsynaptic interneurons [see: Fig.4 (2)] (Dalezios et al., 2002) and create a kind of GABA-GABA synaptic junction. The depression of the GABA release could lead to a disinhibition of postsynaptic interneuron and increased GABA release on their terminals [Fig.4 (3)], inhibiting the input zone to the pyramidal cells.

The pyramidal neurons expressing $mGlu_7$ on their terminals can form synapses with dendrites of the pyramidal cells in the prefrontal cortex (Samogyi et al., 2003). Therefore, AMN082, acting at presynaptic $mGlu_7$ receptors, may induce its antidepressant-like effect through the inhibition of the glutamate release [Fig. 4 (4)]. On the other hand, prefrontal pyramidal neurons have been shown as being inhibited by 5-HT via the activation of the inhibitory $5\text{-}HT_{1A}$ receptors [Fig.4 (5)] (Amargós-Bosch et al., 2004). Therefore, a selective blockade of $5\text{-}HT_{1A}$ receptors [Fig.4 (6)] may antagonize the inhibitory effects of 5-HT on pyramidal neurons, thus inducing an increased activity of these cells. This mechanism may account for a WAY100635-induced blockade of the antidepressant-like effect of AMN087 in the TST. This indicates that an interaction between group $mGlu_7$ receptors and $5\text{-}HT_{1A}$ receptors might be a general phenomenon involved in depression.

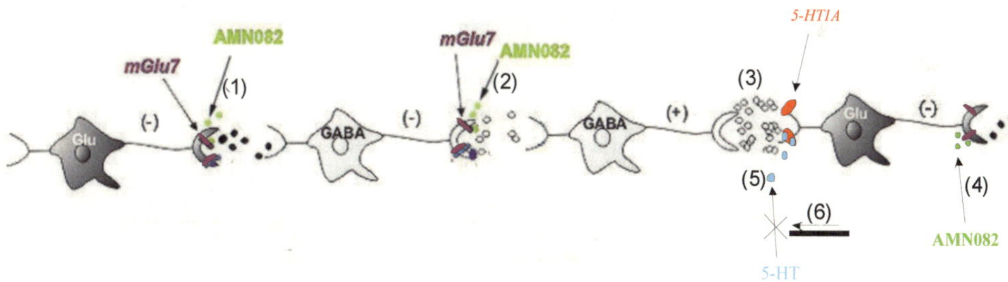

Fig. 4. Schematic representation of the mechanism of antidepressant-like action mediated by mGlu7 receptor positive allosteric modulator (see description in the text). Empty dots-GABA; black dots-Glu; (-)-inhibition; (+)-enhancement; the number of dots indicates the amount of neurotransmitter released

The mechanism of action of AMN082 involves not only GABAergic and glutamatergic neurotransmission. A major metabolite of the compound, Met-1, demonstrated a physiologically relevant transporter binding affinity at the serotonin transporter (SERT), dopamine transporter (DAT), and norepinephrine transporter (Sukoff Rizzo et al., 2011). Despite the involvement of the other mechanisms in the action of the $mGlu_7$ activator, the modulation of the receptor is still crucial and, as in the studies with $mGlu_7$ KO mice the antidepressant-like action of the ligand was shown to be receptor specific (Cryan et al., 2003; Palucha et al., 2007).

4. The role of the GABAergic system in the treatment of depression

γ-aminobutyric acid (GABA) is the main inhibitory amino acid in the brain and constitutes nearly 40% of all neurotransmission. In the properly functioning brain it stays in the physiological balance with excitatory glutamate. The aberrations within this balance may lead to mental and neurological disorders that could be treated alternatively by influencing the glutamatergic or GABAergic neurotransmission (Linden & Schoepp, 2006). Glutamatergic receptors ligands have been discussed widely in the previous chapters. However, the role of GABA cannot be omitted, as it constitutes a natural opposite force for the glutamate. Therefore it may be speculated, that influencing the GABAergic neurotransmission may lead to the normalization of glutamatergic activity.

4.1 GABAergic theory of depression

The GABAergic theory of depression is relatively young when compared to monoaminergic theory, and was raised in the 1990s. The first rationale for considering GABA in depression was justified with mixed GABAmimetics, acting on both $GABA_A$ and $GABA_B$ receptors. The first successful clinical trials were obtained with progabide in 1978 (Bartholini et al., 1978). The effect of the drug was described as similar to standard antidepressants (Bartholini et al., 1978). Later on another GABAergic agent with similar efficacy as progabide, named fengabine, was also tested in double blind clinical studies, with positive results (Carpenter et al., 2006; Magni et al., 1989; Nielsen et al., 1990).

Both compound were also active in standard animal models of depression, such as olfactory bulbectomy or learned helplessness (Lloyd et al., 1987 a, b).

Biochemical studies confirmed the involvement of GABAergic mechanisms in mood disorders, as it was hypothesized that antidepressant drugs may act through increasing the GABAergic tone. The up-regulation of $GABA_B$ receptors appeared to be the fundamental facet of antidepressant drug action (Pilc &Lloyd, 1984). Parallel to these observations it was shown that the level of GABA was decreased in the plasma of depressed patients and the level of GAD67, enzyme synthesizing GABA from glutamate, was lowered in the brains of those patients. Recent studies confirmed the importance of both GABA receptors in depression, suggesting $GABA_B$ neurophysiological deficits to be related to the pathophysiology of major depressive disorder (Fatemi et al., 2005; Guidotti et al., 2000).

4.2 The role of the $GABA_A$ receptor in the mechanism of action of antidepressant drugs

The role of the $GABA_A$ receptor is evident in the field of anxiety disorders, and since the benzodiazepines, $GABA_A$ receptor positive modulators, are the mostly effective and the best known anxiolytic drugs. However as there is no convincing data that the drugs are effective in major depression yet, this issue will be not discussed here.

4.3 The role of the $GABA_B$ receptor in the mechanism of action of antidepressant drugs

The involvement of the $GABA_B$ receptor in depression and antidepressant-like therapy seems to be more important than the previously mentioned $GABA_A$ ionotropic channel. The first report on this subject was released in 1984 (Pilc & Lloyd, 1984). Later on several other papers appeared, and stated that antidepressant drugs of all classes as well as electroconvulsive therapy caused the up-regulation of the $GABA_B$ receptor in the hippocampus and frontal cortex (Gray and Green, 1987; Lloyd et al., 1985; Pratt et al., 1993). By contrast, the down regulation of the receptor was described in animal models of depression, in particular in the olfactory bulbectomy and learned helplessness. Concomitantly, the GABA release was also shown to be decreased in those animals. However, $GABA_B$ binding sites are not changed in the brains of depressed suicide victims, when measured in the frontal and temporal cortex, and in the hippocampus (Arranz et al., 1992; Cross et al., 1988).

It is commonly known that the $GABA_B$ receptor constitutes of two subunits ($GABA_{B1}$ and $GABA_{B2}$). The $GABA_{B1}$ subunit is further represented by two spice variants $GABA_{B1A}$ and $GABA_{B1B}$. More detailed studies concerning the influence of antidepressant drugs on the $GABA_B$ receptor complex revealed that those drugs selectively up-regulated the $GABA_{B1A}$ subunit in the hippocampus, having no effect on the other subunits (Sands et al., 2004). The elevation of $GABA_B$ receptors was also observed in the frontal cortex and spinal cord. Simultaneously the receptor affinity was not changed (Sands et al., 2004).

The existence of $GABA_B$ has been known since 1981 (Hill & Bowery, 1981), but the receptor was cloned relatively recently, in 1997, as the last receptor from the family of major neurotransmitters (Kaupman et al., 1997).

The first selective agonist of the $GABA_B$ receptor, baclofen, was shown to induce some antidepressant-like activities in animal models detecting the antidepressant-like activity of drugs, such as the olfactory bulbectomy (Delini-Stula & Vassout, 1978). The results were not confirmed in later studies, as no activity of baclofen was observed in the forced swim test nor behavioural despair test (Borsini et al., 1986). Moreover, the drug was shown to attenuate the effect of standard antidepressants, such as desipramine, mianserin or imipramine in some tests of antidepressant-like activity (Nakagawa et al. 1996a, b, c). The newer compounds potentiating the activity of GABA at the $GABA_B$ receptors (positive allosteric modulator) were shown not to display such an activity in the forced swim test or in the tail suspension test. The lack of this activity was observed for GS39783 and other $GABA_B$ positive allosteric modulators (Mombereau et al., 2004a, b; Slattery et al., 2005).

The synthesis of the high-affinity phosphinic acid-derived antagonist of the $GABA_B$ receptor opened a new window in terms of GABA and depression (Froestl et al., 2004). The ligands were shown to possess a great antidepressant-like potential in animal models of depression (see Table 5). The first studies concerned CGP36742 in a learned helplessness model of depression (Nakagawa et al., 1996 a,b,c, 1999).

Later on the second available antagonist, CGP51176 was shown to be active in the forced swim test (Bittiger et al., 1996; Nowak et al. 2006). Similar results were observed for other antagonists such as CGP56433A and CGP55845A (Slattery et al., 2005). Concomitantly the drugs had no effects on the spontaneous locomotor activity. These studies strongly support the notion about the antidepressant-like properties of $GABA_B$ receptor antagonists. The

Author	Compound/dose	Species	Test
Delini-Stula&Vassout, 1978	Baclofen (3-10mg/kg)	rats	Olfactory-bulb-ablation induced muricide
Nowak et al., 2006	CGP36742 (10-30mg/kg) CGP51176 (3-30mg/kg)	mice rats	Porsolt test Olfactory bulbectomy Chronic mild stress
Nakagawa et al., 1999	CGP36742 (30-100mg/kg)	rats	Learned helplessness
Slattery et al., 2005	CGP56433A (1-10mg/kg) CGP55845A (3-10mg/kg)	rats	Porsolt test
Bartholini et al., 1978; Magni et al., 1989; Nielsen et al., 1990	progabide, fengabide	human	Clinical study

Table 5. Collected data concerning the antidepressant-like activity of the $GABA_B$ receptor ligands.

studies keep in line with results obtained in the experiments with the use of $GABA_B$ knockout animals, as an antidepressant-like phenotype was observed in mice lacking either B1 or B2 subunit of $GABA_B$ receptor (Mombereau et al., 2004a, b, 2005).

In 2006 the paper of Nowak et al. (Nowak et al., 2006) further confirmed the above mentioned studies showing positive results with CGP51176 and CGP36742 in the chronic mild stress model of depression and in the olfactory bulbectomy model of depression. These activity was observed after chronic treatment, but not acute administration. Moreover, the chronic treatment with CGP51176 induced an increase in $GABA_B$ receptor binding, similar to how it was observed after standard antidepressant drugs (Nowak et al., 2006). Therefore, $GABA_B$ receptor seems to be an interesting target in the search of novel antidepressants.

The mechanism of action of $GABA_B$ ligands, especially $GABA_B$ antagonists, seems to also involve other neurotransmitters systems. The studies of Slattery et al. 2005 revealed that the pattern of action of CGP55845, the $GABA_B$ receptor antagonist, in the modified Porsolt swim test was similar to the one observed after selective serotonin reuptake inhibitors. This differed from tricyclic antidepressants because the drug decreased immobility time and increased swimming, having no activity on climbing behavior (Slattery et al., 2005). Concomitantly, the decrease in immobility elicited by CGP56433A was abolished after *p*CPA pretreatment, corresponding to an attenuation of the increase in swimming time (Slattery et al., 2005). As the pCPA pretreatment induces nearly 90% of serotonin depletion, it was evident that the action of the $GABA_B$ antagonist is serotonin-dependent.

The mechanism of the antidepressant-like action of those compounds is mechanistically different from the one described for anxiolytic effects (Wieronska et al., 2011), as the anxiolysis is mediated through the stimulation of $GABA_B$ receptor, confirming the dissociation of the role of the $GABA_B$ receptors in depression and anxiety (Mombereau et al., 2004, Pilc & Nowak, 2006), similar to how it was observed in the case of $mGlu_{2/3}$ receptors.

$GABA_B$ receptors are expressed on nerve endings of pyramidal neurons exerting the inhibitory effect on glutamatergic transmission (Forti et al., 1997; Samulack et al., 1993), therefore their blockade will cause an overflow of glutamate [Fig. 5 (1)] leading to stimulation of the inhibitory GABAergic neuron. On the other hand the blockade of GABAergic autoreceptors will also lead to inhibition of the target glutamatergic neuron [Fig. 5 (2)].

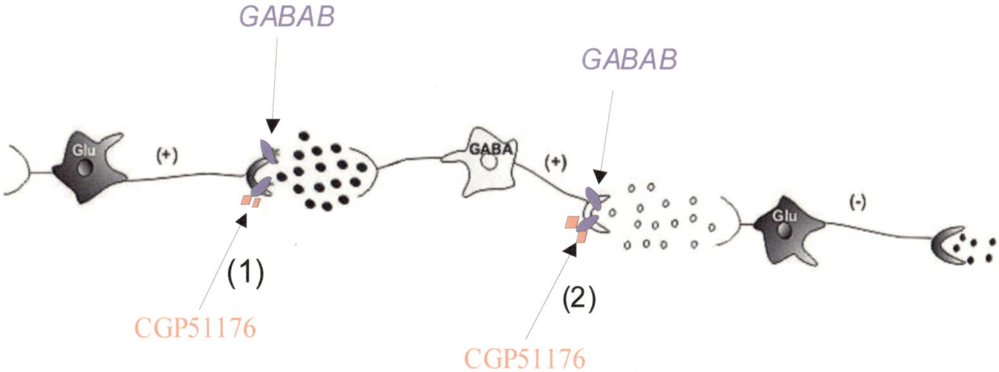

Fig. 5. Schematic representation of the mechanism of antidepressant-like action mediated by $GABA_B$ receptor antagonist(s) (see description in the text). empty dots- GABA; black dots- Glu; (-)-inhibition; (+)-enhancement; the number of dots indicates the amount of neurotransmitter released

Similar to the action of AMN082 described above, the antidepressant-like mechanism of action of $GABA_B$ antagonists occurs via an interaction with the serotonergic system (Slattery et al., 2005). However, determining the kind of receptors that are involved in the action of the compounds is still open for the investigations.

4.4 GABA receptors ligands and clinical studies

The clinical studies were started with progabide and fengabide, described earlier. As they were mixed $GABA_A/GABA_B$ mimetics, their action at particular receptor subtypes couldn't be estimated. More specific ligands acting at $GABA_B$ receptors were studied later on. The efficacy of $GABA_B$ antagonist, baclofen, and $GABA_A$ agonist, diazepam, was shown to be equal to amitryptiline in the treatment of affective disturbances in alcoholic patients (Krupitsky et al., 1993). Other studies showed that baclofen worsens the symptoms of depression (Post et al., 1991).

A study published in 2004 revealed the efficacy of SGS742 (CGP36742) in patients with mild cognitive impairment and opened the possibility for the compound to be investigated in humans (Froestl et al., 2004).

5. Conclusion

We described the mechanisms of the antidepressant-like efficacy of the ligands of metabotropic glutamatergic and GABAergic receptors in order to indicate, that the restoration of the GABA/Glu balance in the brain is an important part of their action. There are functional interactions between amino acids and monoamines (mainly serotonin), which may account for the behavioral effects observed. The unique pharmacology of the metabotropic receptors, their localization in key circuits involved in the pathophysiology of depression, and the promise of the subtle modulation of glutamatergic and GABAergic neurotransmission by regulating the transmitter release and/or acting at the postsynaptic neurons make these receptors intriguing targets for the development of novel medication against depression. Our deliberations further reinforce the hypothesis of a disrupted excitatory/inhibitory balance in the pathophysiology of MDD and its restoration after successful antidepressant treatment.

6. Acknowledgements

The work was supported by POIG. 01.01.02-12-004/09 "Depression-Methods-Therapy" grant, task 3.7.

7. References

Acher, FC. Tellier, FJ. Azerad, R. Brabet, IN. Fagni, & L. Pin, JP. (1997). Synthesis and pharmacological characterization of aminocyclopentanetricarboxylic acids: new tools to discriminate between metabotropic glutamate receptor subtypes. *J Med Chem.*, Vol. 40, No.19, (Sep 1997), pp.(3119-29).

Altamura, CA. Mauri, MC. Ferrara, A. Moro, AR. D'Andrea, G. & Zamberlan, F. (1993). Plasma and platelet excitatory amino acids in psychiatric disorders. *Am J Psychiatry*, Vol. 150, No. 11 (Nov 1993), pp.(1731-1733).

Altar, C A. (1999). Neurotrophins and depression. *Trends Pharmacol Sci.*, Vol. 20, No. 2 (Feb 1999), pp.(59-61).

Amargós-Bosch, M. Bortolozzi, A. Puig, MV. Serrats, J. Adell, A. Celada, P. Toth, M. Mengod, G. & Artigas, F. (2004). Co-expression and in vivo interaction of serotonin1A and serotonin2A receptors in pyramidal neurons of prefrontal cortex. *Cereb Cortex*, Vol. 14, No. 3, (Mar 2004), pp. (281-99).

Andlin-Sobocki, P., Jonsson, B., Wittchen, HU. & Olesen, J. (2005). Cost of disorders of the brain in Europe. *Eur J Neurol.* (Jun 2005), 12 Suppl 1, pp. (1-27).

Arranz, B. Cowburn, R. Eriksson, A. Vestling, & M. Marcusson, J. (1992). Gamma-aminobutyric acid-B (GABAB) binding sites in postmortem suicide brains. *Neuropsychobiology*, Vol. 26, No. 1-2, (1992), pp.(33-6).

Attucci, S. Carla, V. Mannaioni, G. & Moroni, F. (2001). Activation of type 5 metabotropic glutamate receptors enhances NMDA responses in mice cortical wedges. *Br J Pharmacol.*, Vol. 132, No. 4, (Feb 2001), pp. (799-806).

Awad, H. Hubert, GW. Smith, Y. Levey, AI. & Conn PJ. (2000). Activation of metabotropic glutamate receptor 5 has direct excitatory effects and potentiates NMDA receptor currents in neurons of the subthalamic nucleus. *J Neurosci.*, Vol. 20, No. 21, (Nov 2000), pp. (7871-7879).

Bai, F. Bergeron, M. & Nelson, DL. (2003). Chronic AMPA receptor potentiator (LY451646) treatment increases cell proliferation in adult rat hippocampus. *Neuropharmacology*, Vol. 44, No. 8, (Jun 2003), pp. (1013-1021).

Bai, F. Li, X., Clay, M. Lindstrom, T. & Skolnick, P. (2001). Intra- and interstrain differences in models of "behavioral despair". *Pharmacol Biochem Behav.*, Vol. 70, No. 2-3, (Oct-Nov 2001), pp. (187–192).

Barragan-Rodriguez, L. Rodriguez-Moran, M. & Guerrero-Romero, F. (2008). Efficacy and safety of oral magnesium supplementation in the treatment of depression in the elderly with type 2 diabetes: a randomized, equivalent trial. *Magnes Res* ., Vol. 21, No. 4, (Dec 2008), pp. (218-23).

Bartholini G. (1979). [Potential therapeutic activity of GABA-mimetic drugs in neuropsychiatry]. *Schweiz Arch Neurol Neurochir Psychiatr*, Vol. 125, No. 2, (1979), pp. (265-9).

Belozertseva, IV. Kos, T. Popik, P. Danysz, W. & Bespalov, AY. (2007). Antidepressant-like effects of mGluR1 and mGluR5 antagonists in the rat forced swim and the mouse tail suspension tests. *Eur Neuropsychopharmacol.*, Vol. 17, No. 3, (Feb 2007), pp. (172–179).

Berman, RM. Cappiello, A. Anand, A. Oren, DA. Heninger, GR. Charney, DS. & Krystal, JH. (2000). Antidepressant effects of ketamine in depressed patients. *Biol Psychiatry.*, Vol. 47, No. 4, (Feb 2000), pp. (351-354).

Berry-Kravis, E., Hessl, D., Coffey, S., Hervey, C. Schneider, A. Yuhas, J. Hutchison, J. Snape, M. Tranfaglia, M. Nguyen, DV. & Hagerman, R. (2009). A pilot open label, single dose trial of fenobam in adults with fragile X syndrome. *J Med Genet.*, Vol. 46, No. 4, (Apr 2009), pp. (266-71).

Bespalov, AY. van Gaalen, MM. Sukhotina, IA. Wicke, K. Mezler, M. Schoemaker, H. & Gross, G. Behavioral characterization of the mGlu group II/III receptor antagonist, LY-341495, in animal models of anxiety and depression. *Eur J Pharmacol.*, Vol. 592, No. 1-3, (Sep 2008), pp. (96-102).

Beurrier, C. Lopez, S. Révy, D. Selvam, C. Goudet, C. Lhérondel, M. Gubellini, P. Kerkerian-LeGoff, L. Acher, F. Pin, JP. & Amalric, M. (2009). Electrophysiological and behavioral evidence that modulation of metabotropic glutamate receptor 4 with a new agonist reverses experimental parkinsonism. *FASEB J.* Vol. 23, No. 10, (Oct 2009), pp. (3619-3628).

Bittiger, H. Froestl, W. Mickel, S. & Olpe, HR. (1993). GABAB receptor antagonists: from synthesis to therapeutic applications. *Trends Pharmacol Sci.*, Vol. 14, No. 11, (Nov 1993), pp. (391-4).

Bobula, B. Tokarski, K. & Hess, G. (2003). Repeated administration of antidepressants decreases field potentials in rat frontal cortex. *Neuroscience*, Vol. 120, No. 3, (2003), pp. (765-769).

Borsini, F. Evangelista, S. & Meli, A. (1986). Effect of GABAergic drugs in the behavioral 'despair' test in rats. *Eur J Pharmacol.*, Vol. 121, No. 2, (Feb 1986), pp. (265-268).

Brakeman, PR. Lanahan, AA. O'Brien, R. Roche, K. Barnes, CA. Huganir, RL. & Worley, PF. (1997). Homer: a protein that selectively binds metabotropic glutamate receptors. *Nature*, Vol. 386, No. 6622, (Mar 1997), pp. (284-288).

Brandoli, C. Sanna, A. De Bernardi, MA. Follesa, P. Brooker, G. & Mocchetti, I. (1998). Brain-derived neurotrophic factor and basic fibroblast growth factor downregulate NMDA receptor function in cerebellar granule cells. *J Neurosci.*, Vol. 18, No. 19, (Oct 1998), pp. (7953-7961).

Brody, DS. Hahn, S R. Spitzer RL. Kroenke, K. Linzer, M. deGruy, FV. & Williams, JB. (1998). Identifying patients with depression in the primary care setting: a more efficient method. *Arch Intern Med.*, Vol. 158, No. 22, (Dec 1998), pp. (2469-2475).

Bunney, WE. Jr. & Davis, JM. (1965). Norepinephrine in depressive reactions. A review. *Arch Gen.Psychiatry*, Vol. 13, No. 6, (Dec 1965), pp. (483-494).

Carpenter, LL. Schecter, JM. Tyrka, AR., Mello, AF. Mello, MF. Haggarty, R. & Price, LH. (2006). Open-label tiagabine monotherapy for major depressive disorder with anxiety. *J Clin Psychiatry,* Vol. 67, No. 1, (Jan 2006), pp. (66-71).

Castren, E. (2005). Is mood chemistry? *Nature Revies*, Vol. 6, No. 3, (Mar 2005), pp. (241-246).

Chaki, S. Yoshikawa, R. & Hirota, S. (2004). MGS0039: a potent and selective group II metabotropic glutamate receptor antagonist with antidepressant-like activity. *Neuropharmacology*, Vol. 46, No. 4, (Mar 2004), pp. (457-467).

Chouinard, G. Beauclair, L. Geiser, R. & Etienne, P. (1990). A pilot study of magnesium aspartate hydrochloride (Magnesiocard) as a mood stabilizer for rapid cycling bipolar affective disorder patients. *Prog Neuropsychopharmacol Biol Psychiatry*, Vol. 14, No. 2, (1990), pp. (171-80).

Chu, Z. & Hablitz, JJ. (1998). Activation of group I mGluRs increases spontaneous IPSC frequency in rat frontal cortex. *J Neurophysiol.*, Vol. 80, No. 2, (Aug 1998), pp. (621-627).

Chung, DS. Traynelis, SF. Murphy, TJ. & Conn, PJ. (1997). 4-Methylhomoibotenic acid activates a novel metabotropic glutamate receptor coupled to phosphoinositide hydrolysis. *J Pharmacol Exp Ther.*, Vol. 283, No. 2, (Nov 1997), pp. (742-749).

Cosford, ND. Roppe, J. Tehrani, L. Schweiger, EJ. Seiders, TJ. Chaudary, A. Rao, S. & Varney, MA. (2003). [3H]-methoxymethyl-MTEP and [3H]-methoxy-PEPy: potent and selective radioligands for the metabotropic glutamate subtype 5 (mGlu5) receptor. *Bioorg Med Chem Lett.*, Vol. 13, No. 3, (Feb 2003), pp. (351-354).

Cross, JA. Cheetham, SC. Crompton, MR. Katona, CL. & Horton, RW. (1988). Brain GABAB binding sites in depressed suicide victims. *Psychiatry Res.*, Vol. 26, No. 2, (Nov 1988), pp. (119-129).

Cryan, JF. Kelly, PH. Neijt, HC. Sansig, G. Flor, PJ. & van der Putten, H. (2003). Antidepressant and anxiolytic-like effects in mice lacking the group III metabotropic glutamate receptor mGluR7. *Eur J Neurosci.,* Vol. 17, No. 11, (Jun 2003), pp. (2409-2417).

Cryan, JF. Slattery, DA. (2010). GABAB receptors and depression. Current status. *Adv Pharmacol.*, Vol. 58, (2010), pp. (427-451).

Dalezios, Y. Luján, R. Shigemoto, R. Roberts, JD. & Somogyi, P. (2002). Enrichment of mGluR7a in the presynaptic active zones of GABAergic and non-GABAergic terminals on interneurons in the rat somatosensory cortex. *Cereb Cortex*, Vol. 12, No. 9, (Sep 2002), pp. (961-974).

Danysz, W. & Parsons, CG. (1998). Glycine and N-methyl-D-aspartate receptors: physiological significance and possible therapeutic applications. *Pharmacol Rev.*, Vol. 50, No. 4, (Dec 1998), pp. (597-664).

Delini-Stula, A. & Vassout, A. (1978). Influence of baclofen and GABA-minetic agents on spontaneous and olfactory-bulb-ablation-induced muricidal behaviour in the rat. *Arzneimittelforschung*, Vol. 28, No. 9, (1978), pp. (1508-1509).

Deschwanden, A. Karolewicz, B. Feyissa, AM. Treyer, V. Ametamey, SM. Johayem, A. Burger, C. Auberson, YP. Sovago, J. Stockmeier, CA. Buck, A. & Hasler, G. (2011). Reduced metabotropic glutamate receptor 5 density in major depression determined by [(11)C]ABP688 PET and postmortem study. *Am J Psychiatry*, Vol. 168, No. 7, (Jul 2011), pp. (727-734).

DiazGranados, N. Ibrahim, LA. Brutsche, NE. Ameli, R. Henter, ID. Luckenbaugh, DA. Machado-Vieira, R. & Zarate, CA Jr. (2010). Rapid resolution of suicidal ideation after a single infusion of an N-methyl-D-aspartate antagonist in patients with treatment-resistant major depressive disorder. *J Clin Psychiatry*, Vol. 71, No. 12, (Dec 2010), pp. (1605-1611).

Doherty, AJ. Palmer, MJ. Bortolotto, ZA. Hargreaves, A, Kingston, AE. Ornstein, PL. Schoepp, DD. Lodge, D. & Collingridge, GL. (2000). A novel, competitive mGlu(5) receptor antagonist (LY344545) blocks DHPG-induced potentiation of NMDA responses but not the induction of LTP in rat hippocampal slices. *Br J Pharmacol.*, Vol. 131, No. 2, (Sep 2000), pp. (239-244).

Drago, A., Crisafulli, C., Sidoti, A., & Serretti, A. (2011). The molecular interaction between the glutamatergic, noradrenergic, dopaminergic and serotoninergic systems informs a detailed genetic perspective on depressive phenotypes. *Prog Neurobiol.*, Vol. 94, No. 4, (Sep 2011), pp. (418-460).

Duman, RS. Nakagawa, S. & Malberg, J. (2001). Regulation of adult neurogenesis by antidepressant treatment. *Neuropsychopharmacology*, Vol. 25, No. 6, (Dec 2001), pp. (836-844).

Duman, RS. Heninger, GR. & Nestler, EJ. (1997). A molecular and cellular theory of depression. *Arch Gen.Psychiatry*, Vol. 54, No. 7, (Jul 1997), pp. (597-606).

Duman, RS. (1998). Novel therapeutic approaches beyond the serotonin receptor. *Biol Psychiatry*, Vol. 44, No. 5, (Sep 1998), pp. (324-335).

Eby, GA. & Eby, KL. (2006). Rapid recovery from major depression using magnesium treatment. *Med Hypotheses*, Vol. 67, No. 2, (Mar 2006), pp. (362-370).

Fatemi, SH. Stary, JM. Earle, JA. Araghi-Niknam, M. & Eagan, E. (2005). GABAergic dysfunction in schizophrenia and mood disorders as reflected by decreased levels of glutamic acid decarboxylase 65 and 67 kDa and Reelin proteins in cerebellum. *Schizophr Res.*, Vol. 72, No. 2-3, (Jan 2005), pp. (109-22).

Fell, MJ. Witkin, JM. Falcone, JF. Katner, JS. Perry, KW. Hart, J. Rorick-Kehn, L. Overshiner, CD. Rasmussen, K. Chaney, SF. Benvenga, MJ. Li, X. Marlow, DL. Thompson, LK. Luecke, SK. Wafford, KA. Seidel, WF. Edgar, DM. Quets, AT. Felder, CC. Wang, X. Heinz, BA. Nikolayev, A. Kuo, MS. Mayhugh, D. Khilevich, A. Zhang, D. Ebert, PJ. Eckstein, JA. Ackermann, BL. Swanson, SP. Catlow, JT. Dean, RA. Jackson, K. Tauscher-Wisniewski, S. Marek, GJ. Schkeryantz, JM. & Svensson, KA. (2011). N-(4-((2 - (trifluoromethyl) -3- hydroxy- 4-(isobutyryl) phenoxy)methyl)benzyl)- 1-methyl-1H-imidazole-4-carboxamide (THIIC), a novel metabotropic glutamate 2 potentiator with potential anxiolytic/antidepressant properties: in vivo profiling suggests a link between behavioral and central nervous system neurochemical changes. *J Pharmacol Exp Ther.*, Vol. 336, No. 1, (Jan 2011), pp. (165-177).

Feyissa, AM. Woolverton, WL. Miguel-Hidalgo, JJ. Wang, Z. Kyle, PB. Hasler, G. Stockmeier, CA. Iyo, AH. & Karolewicz, B. (2010). Elevated level of metabotropic glutamate receptor 2/3 in the prefrontal cortex in major depression. *Prog Neuropsychopharmacol Biol Psychiatry*, Vol. 34, No. 2, (Mar 2010), pp. (279-283).

Forti, M. & Michelson, HB. (1997). Novel glutamate- and GABA-independent synaptic depolarization in granule cells of guinea-pig hippocampus. *J Physiol.*, Vol. 504, No. 3, (Nov 1997), pp. (641-648).
Froestl, W. (2010). Chemistry and pharmacology of GABAB receptor ligands. *Adv Pharmacol.*, Vol. 58 (2010), pp. (19-62).
Froestl, W. Gallagher, M. Jenkins, H. Madrid, A. Melcher, T. Teichman, S. Mondadori, CG. & Pearlman, R. (2004). SGS742: the first GABA(B) receptor antagonist in clinical trials. *Biochem Pharmacol.*, Vol. 68, No. 8, (Oct 2004), pp. (1479-1487).
Gasparini, F. Lingenhöhl, K. Stoehr, N. Flor, PJ. Heinrich, M. Vranesic, I. Biollaz, M. Allgeier, H. Heckendorn, R. Urwyler, S. Varney, MA. Johnson, EC. Hess, SD. Rao, SP. Sacaan, AI. Santori, EM. Veliçelebi, G. & Kuhn, R. (1999). 2-Methyl-6-(phenylethynyl)-pyridine (MPEP), a potent, selective and systemically active mGlu5 receptor antagonist. *Neuropharmacology,* Vol. 38, No. 10, (Oct 1999), pp. (1493-1503).
Ghose, S. Winter, MK. McCarson, KE. Tamminga, CA. & Enna, SJ. (2011). The GABAB receptor as a target for antidepressant drug action. *Br J Pharmacol.*, Vol. 162, No. 1, (Jan 2011), pp. (1-17).
Gray, JA. & Green, AR. (1987). Increased GABAB receptor function in mouse frontal cortex after repeated administration of antidepressant drugs or electroconvulsive shocks. *Br J Pharmacol.*, Vol. 92, No. 2, (Oct 1987), pp. (357-362).
Greenberg, PE. Kessler, RC. Birnbaum, HG. Leong, SA. Lowe, SW. Berglund, PA. & Corey-Lisle, PK. (2003). The economic burden of depression in the United States: how did it change between 1990 and 2000? *J Clin Psychiatry,* Vol. 64, No. 12, (Dec 2003), pp. (1465-1475).
Guidotti, A. Auta, J. Davis, JM. Di-Giorgi-Gerevini, V. Dwivedi, Y. Grayson, DR. Impagnatiello, F. Pandey, G. Pesold, C. Sharma, R. Uzunov, D. & Costa, E. (2000). Decrease in reelin and glutamic acid decarboxylase67 (GAD67) expression in schizophrenia and bipolar disorder: a postmortem brain study. *Arch Gen Psychiatry,* Vol. 57, No. 11, (Nov 2000), pp. (1061-1069).
Hashimoto, K. (2011). The role of glutamate on the action of antidepressants. *Prog Neuropsychopharmacol Biol Psychiatry.*, Vol. 35, No. 7, (Aug 2011), pp. (1558-1568).
Healy, D. (1999). The three faces of the antidepressants: a critical commentary on the clinical-economic context of diagnosis. *J Nerv Ment Dis,* Vol. 187, No. 3, (Mar 1999), pp. (174-180).
Helton, DR. Tizzano, JP. Monn, JA. Schoepp, DD. & Kallman MJ. (1997). LY354740: a metabotropic glutamate receptor agonist which ameliorates symptoms of nicotine withdrawal in rats. *Neuropharmacology,* Vol. 36, No. 11-12, (Nov-Dec 1997), pp. (1511-1516).
Hill, DR. & Bowery, NG. 3H-baclofen and 3H-GABA bind to bicuculline-insensitive GABA B sites in rat brain. *Nature,* Vol. 290, No. 5802, (Mar 1981), pp. (149-152).
Ibrahim, L. Diazgranados, N. Luckenbaugh, DA. Machado-Vieira, R. Baumann, J. Mallinger, AG. & Zarate CA Jr. Rapid decrease in depressive symptoms with an N-methyl-d-aspartate antagonist in ECT-resistant major depression. *Prog Neuropsychopharmacol Biol Psychiatry.*, Vol. 35, No. 4, (Jun 2011), pp. (1155-1159).
Itil, TM. Seaman, PA. Huque, M. Mukhopadhyay, S. Blascutti, D. NQ, KT. & Ciccione, P E. (1978). The clinical and quantitative EEG effects and plasma levels of fenobam

(McN-3377) in subjects with anxiety: an open rising dose tolerance and efficacy study. *Curr Ther Res.*, Vol. 26, No. 6, (1978), pp. (708-724).

Karasawa, JI. Yoshimizu, T. Chaki, S. (2006). A metabotropic glutamate 2/3 receptor antagonist, MGS0039, increases extracellular dopamine levels in the nucleus accumbens shell. *Neurosci Lett.*, Vol. 393, No. 2-3, (Jan 2006), pp. (127–130).

Kawashima, N. Karasawa, J. Shimazaki, T. Chaki, S. Okuyama, S. Yasuhara, A. & Nakazato, A. (2005). Neuropharmacological profiles of antagonists of group II metabotropic glutamate receptors. *Neurosci Lett.*, Vol. 378, No. 3, (Apr 2005), pp. (131–134).

Kaupmann, K. Huggel, K. Heid, J. Flor, PJ. Bischoff, S. Mickel, SJ. McMaster, G. Angst, C. Bittiger, H. Froestl, W. & Bettler, B. (1997). Expression cloning of GABA(B) receptors uncovers similarity to metabotropic glutamate receptors. *Nature*, Vol. 386, No. 6622, (Mar 1997), pp. (239-246).

Kendell, SF. Krystal, JH. & Sanacora, G. (2005). GABA and glutamate systems as therapeutic targets in depression and mood disorders. *Expert Opin Ther Targets*, Vol. 9, No. 1, (Feb 2005), pp. (153-168).

Kłodzińska, A. Chojnacka-Wójcik, E. Pałucha, A. Brański, P. Popik, P. & Pilc, A. (1999). Potential anti-anxiety, anti-addictive effects of LY 354740, a selective group II glutamate metabotropic receptors agonist in animal models. *Neuropharmacology*, Vol. 38, No. 12, (Dec 1999), pp. (1831-1839).

Koike, H. Iijima, M. & Chaki, S. (2011). Involvement of the mammalian target of rapamycin signaling in the antidepressant-like effect of group II metabotropic glutamate receptor antagonists. *Neuropharmacology*, 2011 Aug 31.

Kroczka, B. Brański, P. Pałucha, A. Pilc, A. & Nowak, G. (2001). Antidepressant-like properties of zinc in rodent forced swim test. *Brain Res Bull*, Vol. 55, No. 2, (May 2001), pp. (297-300).

Krupitsky, EM. Burakov, AM. Ivanov, VB. Krandashova, GF. Lapin, IP. Grinenko, AJa. & Borodkin, YuS. (1993). Baclofen administration for the treatment of affective disorders in alcoholic patients. *Drug Alcohol Depend.*, Vol. 33, No. 2, (Sep 1993), pp. (157-163).

Kuhn, R. (1957). Uber die Behandlung depressiver Zustande mit einem Imidibenzylderivat. *Schweiz Med Wochenschr.*, Vol. 35/36, (1957), pp. (1135-1140).

Lapierre, YD. & Oyewumi, LK. (1982). Fenobam: Another Anxiolytic? *Current Therapeutic Research*, Vol. 31, No. 1, (1982), pp. (95-101).

Lapin, TP. & Oxenkrug, KF. (1969). Intensification of the central serotonergic processes as a possible determinant of the thymoleptic effect. *Lancet*, Vol. 1, (1969), pp. (132-136).

Lauterborn, JC. Lynch, G. Vanderklish, P. Arai, A. & Gall, CM. (2000). Positive modulation of AMPA receptors increases neutrophin expression by hippocampal and cortical neurons. *J Neurosci.*, Vol. 20, No. 1, (Jan 2000), pp. (8–21).

Layer, RT. Popik, P. Olds, T. & Skolnick, P. Antidepressant-like actions of the polyamine site NMDA antagonist, eliprodil (SL-82.0715). *Pharmacol Biochem Behav.*, Vol. 52, No. 3, (Nov 1995), pp. (621-627).

Legutko, B. Li, X. Skolnick, P. (2001). Regulation of BDNF expression in primary neuron culture by LY392098, a novel AMPA receptor potentiator. *Neuropharmacology*, Vol. 40, No. 8, (Jun 2001), pp. (1019–1027).

Legutko, B. Szewczyk, B. Pomierny-Chamioło, L. Nowak, G. & Pilc, A. (2006). Effect of MPEP treatment on brain-derived neurotrophic factor gene expression. *Pharmacol Rep.*, Vol. 58, No. 3, (May-Jun 2006), pp. (427–430).

Li, X. Need, AB. Baez, M. & Witkin, JM. (2006). Metabotropic glutamate 5 receptor antagonism is associated with antidepressant-like effects in mice. *J Pharmacol Exp Ther.*, Vol. 319, No. 1, (Oct 2006), pp. (254–259).

Li, X. Tizzano, JP. Griffey, K. Clay, M. Lindstrom, T. & Skolnick, P. (2001). Antidepressant-like actions of an AMPA receptor potentiator (LY392098). *Neuropharmacology*, Vol. 40, No. 8. (Jun 2001), pp. (1028–1033).

Li, X. Witkin, JM. Need, AB. & Skolnick, P. (2003). Enhancement of antidepressant potency by a potentiator of AMPA receptors. *Cell Mol Neurobiol.*, Vol. 23, No. 3, (Jun 2003), pp. (419–430).

Li, S. Gu, Y. Meng, B. Mei, B. & Li, F. (2010). The different effects of over-expressing murine NMDA receptor 2B subunit in the forebrain on conditioned taste aversion. *Brain Res.*, Vol. 1351, (Sep 2010), pp. 165-171.

Linden, AM. & Schoepp, DD. (2006). Metabotropic glutamate receptor targets for neuropsychiatric disorders. *Drug Discovery Today: Therapeutic Strategies*, Vol. 3, No. 4, (2006), pp. (507-517).

Lloyd, KG. Zivkovic, B. Sanger, D. Depoortere, H. & Bartholini, G. (1987). Fengabine, a novel antidepressant GABAergic agent. I. Activity in models for antidepressant drugs and psychopharmacological profile. *J Pharmacol Exp Ther.*, Vol. 242, No. 1, (Apr 1987), pp. (245-250).

Lloyd, KG. Morselli, PL. & Bartholini, G. (1987). GABA and affective disorders. *Med Biol.*, Vol. 65, No. 2-3, (1987), pp. (159-165).

Lloyd, KG. Thuret, F. & Pilc A. (1985). Upregulation of gamma-aminobutyric acid (GABA) B binding sites in rat frontal cortex: a common action of repeated administration of different classes of antidepressants and electroshock. *J Pharmacol Exp Ther.*, Vol. 235, No. 1, (Oct 1985), pp. (191-199).

Loomer, HP. Saunders, JC. & Kline, NS. (1957). A clinic and pharmacodynamic evaluation of iproniazid as a psychic energizer. *Psychiatry Res Rep.*, Vol. 8, (1957), pp. (129-141)

Lujan, R. Nusser, Z. Roberts, JD. Shigemoto, R. & Somogyi P. (1996). Perisynaptic location of metabotropic glutamate receptors mGluR1 and mGluR5 on dendrites and dendritic spines in the rat hippocampus. *Eur J Neurosci.*, Vol. 8, No. 7, (Jul 1996), pp. (1488-1500).

Maćkowiak, M. O'Neill, MJ. Hicks, CA. Bleakman, D. & Skolnick, P. (2002). An AMPA receptor potentiator modulates hippocampal expression of BDNF: an in vivo study. *Neuropharmacology*, Vol. 43, No. 1, (Jul 2002), pp. (1-10).

Maeng, S. Zarate, CA. Jr, Du, J. Schloesser, RJ. McCammon, J. Chen, G. & Manji, HK. (2008). Cellular mechanisms underlying the antidepressant effects of ketamine: role of alpha-amino-3-hydroxy-5-methylisoxazole-4-propionic acid receptors. *Biol Psychiatry*, Vol. 63, No. 4, (Feb 2008), pp. (349-352).

Magni, G. Garreau, M. Orofiamma, B. & Palminteri, R. (1989). Fengabine, a new GABAmimetic agent in the treatment of depressive disorders: an overview of six double-blind studies versus tricyclics. *Neuropsychobiology*, Vol. 20, No. 3, (1989), pp. (126-131).

Maj, J. Rogóż, Z. Skuza, G. & Sowińska, H. (1992a). Effects of MK-801 and antidepressant drugs in the forced swimming test in rats. *Eur Neuropsychopharmacol.*, Vol. 2, No. 1, (Mar 1992), pp. (37-41).

Maj, J. Rogóż, Z. Skuza, G. & Sowińska H. (1992b). The effect of CGP 37849 and CGP 39551, competitive NMDA receptor antagonists, in the forced swimming test. *Pol J Pharmacol Pharm.*, Vol. 44, No. 4, (Jul-Aug 1992), pp. (337-346).

Marek, GJ. & Zhang, C. (2008). Activation of metabotropic glutamate 5 (mGlu5) receptors induces spontaneous excitatory synaptic currents in layer V pyramidal cells of the rat prefrontal cortex. *Neurosci Lett.*, Vol. 442, No. 3, (Sep 2008), pp. (239-243).

Marin, JC. & Goadsby, PJ. (2010). Glutamatergic fine tuning with ADX-10059: a novel therapeutic approach for migraine? *Expert Opin Investig Drugs.*, Vol. 19, No. 4, (Apr 2010), pp. (555-561).

Michael, N. Erfurth, A. Ohrmann, P. Gössling, M. Arolt, V. Heindel, W. & Pfleiderer B. Acute mania is accompanied by elevated glutamate/glutamine levels within the left dorsolateral prefrontal cortex. *Psychopharmacology (Berl),* Vol. 168, No. 3, (Jul 2003), pp. (344-346).

Millan, MJ. (2006). Multi-target strategies for the improved treatment of depressive states: Conceptual foundations and neuronal substrates, drug discovery and therapeutic application. *Pharmacol Ther.*, Vol. 110, No. 2, (May 2006), pp. (135-370).

Molina-Hernández, M. Téllez-Alcántara, NP. Pérez-García, J. Olivera-Lopez, JI. & Jaramillo-Jaimes, MT. (2008). Desipramine or glutamate antagonists synergized the antidepressant-like actions of intra-nucleus accumbens infusions of minocycline in male Wistar rats. *Prog Neuropsychopharmacol Biol Psychiatry.*, Vol. 32, No. 2, (Oct 2008), pp. (1660-1666).

Mombereau, C. Kaupmann, K. Froestl, W. Sansig, G. van der Putten, H. & Cryan JF. (2004a). Genetic and pharmacological evidence of a role for GABA(B) receptors in the modulation of anxiety- and antidepressant-like behavior. *Neuropsychopharmacology,* Vol. 29, No. 6, (Jun 2004), pp. (1050-1062).

Mombereau, C. Kaupmann, K. van der Putten, H. & Cryan JF. (2004b). Altered response to benzodiazepine anxiolytics in mice lacking GABA B(1) receptors. *Eur J Pharmacol.*, Vol. 497, No. 1, (Aug 2004), pp. (119-20).

Mombereau, C. Kaupmann, K. Gassmann, M. Bettler, B. van der Putten, H. & Cryan JF. (2005). Altered anxiety and depression-related behaviour in mice lacking GABAB(2) receptor subunits. *Neuroreport*, Vol. 28, No. 3, (Feb 2005), pp. (307-310).

Moryl, E. Danysz, W. & Quack, G. (1993). Potential antidepressive properties of amantadine, memantine and bifemelane. *Pharmacol Toxicol.*, Vol. 72, No. 6, (Jun 1993), pp. (394-397).

Muhonen, LH. Lönnqvist, J. Juva, K. & Alho H. (2008). Double-blind, randomized comparison of memantine and escitalopram for the treatment of major depressive disorder comorbid with alcohol dependence. *J Clin Psychiatry.*, Vol. 69, No. 3, (Mar 2008), pp. (392-399).

Nakagawa, Y. Ishima, T. Ishibashi, Y. Tsuji, M. & Takashima, T. (1996a). Involvement of GABA(B) receptor systems in action of antidepressants .2. Baclofen attenuates the effect of desipramine whereas muscimol has no effect in learned helplessness paradigm in rats. *Brain Res., Vol.* 728, No. 2, (Jul 1996), pp. (225-230).

Nakagawa, Y. Ishima, T. Ishibashi, Y. Tsuji, M. & Takashima, T. (1996b). Involvement of GABA(B) receptor systems in experimental depression: Baclofen but not bicuculline exacerbates helplessness in rats. *Brain Res., Vol.* 741, No. 1-2, (Nov 1996), pp. (240-245).

Nakagawa, Y. Ishima, T. Ishibashi, Y. Yoshii, T. & Takashima, T. (1996c). Involvement of GABA(B) receptor systems in action of antidepressants: Baclofen but not bicuculline attenuates the effects of antidepressants on the forced swim test in rats. *Brain Res.,* Vol. 709, No. 2, (Feb 1996), pp. (215-220).

Nakagawa, Y. Sasaki, A. & Takashima, T. (1999). The GABA(B) receptor antagonist CGP36742 improves learned helplessness in rats. *Eur J Pharmacol.,* Vol. 381, No. 1, (Sep 1999), pp. (1-7).

Nakanishi, S. (1992). Molecular diversity of glutamate receptors and implications for brain function. *Science,* Vol. 258, No. 5082, (Oct 1992), pp. (597-603).

Nakazato, A. Sakagami, K. Yasuhara, A. Ohta, H. Yoshikawa, R. Itoh, M. Nakamura, M. & Chaki S. (2004). Synthesis, in vitro pharmacology, structure-activity relationships, and pharmacokinetics of 3-alkoxy-2-amino-6-fluorobicyclo[3.1.0]hexane-2,6-dicarboxylic acid derivatives as potent and selective group II metabotropic glutamate receptor antagonists. *J Med Chem.,* Vol. 47, No. 18, (Aug 2004), pp. (4570-4587).

Nestler, EJ. (1998). Antidepressant treatments in the 21st century. *Biol Psychiatry.,* Vol. 44, No. 7, (Oct 1998), pp. (526-533).

Nestler, EJ. Barrot, M. DiLeone, RJ. Eisch, AJ. Gold, SJ. & Monteggia, LM. (2002). Neurobiology of depression. *Neuron,* Vol. 34, No. 1, (Mar 2002), pp. (13-25).

Nibuya, M. Morinobu, S. & Duman, RS. (1995). Regulation of BDNF and trkB mRNA in rat brain by chronic electroconvulsive seizure and antidepressant drug treatments. *J Neurosci.,* Vol. 15, No. 11, (Nov 1995), pp. (7539-7547).

Nielsen, NP. Cesana, B. Zizolfi, S. Ascalone, V. Priore, P. & Morselli, PL. (1990). Therapeutic effects of fengabine, a new GABAergic agent, in depressed outpatients: a double-blind study versus clomipramine. *Acta Psychiatr Scand.,* Vol. 82, No. 5, (Nov 1990), pp. (366-371).

Nowak, G. Legutko, B. Skolnick, P. & Popik, P. (1998). Adaptation of cortical NMDA receptors by chronic treatment with specific serotonin reuptake inhibitors. *Eur J Pharmacol.,* Vol. 342, No. 2-3, (Jun 1998), pp. (367-370).

Nowak, G., Li, Y. & Paul, IA. (1996). Adaptation of cortical but not hippocampal NMDA receptors after chronic citalopram treatment. *Eur J Pharmacol.,* Vol. 295, No. 1, (Jan 1996), pp. (75-85).

Nowak, G. Trullas. R. Layer, RT. Skolnick, P. & Paul, IA. (1993). Adaptive changes in the N-methyl-D-aspartate receptor complex after chronic treatment with imipramine and 1-aminocyclopropanecarboxylic acid. *J Pharmacol Exp Ther.,* Vol. 265, No. 3, (Jun 1993), pp. (1380-1386).

Nowak, G. Siwek, M. Dudek, D., Zieba, A. & Pilc, A. (2003). Effect of zinc supplementation on antidepressant therapy in unipolar depression: a preliminary placebo-controlled study. *Pol J Pharmacol.,* Vol. 55, No. 6, (Nov-Dec 2003), pp. (1143-1147).

Nowak, G. Partyka, A. Pałucha, A. Szewczyk, B. Wierońska, JM. Dybała, M. Metz, M. Librowski, T. Froestl, W. Papp, M. & Pilc, A. (2006). Antidepressant-like activity of

CGP 36742 and CGP 51176, selective GABAB receptor antagonists, in rodents. *Br J Pharmacol.*, Vol. 149, No. 5, (Nov 2006), pp. (581-590).

Ossowska, G. Klenk-Majewska, B. & Szymczyk, G. (1997). The effect of NMDA antagonists on footshock-induced fighting behavior in chronically stressed rats. *J Physiol Pharmacol.*, Vol. 48, No. 1, (Mar 1997), pp. (127-135).

Oswald, I. Brezinova, V. & Dunleavy, DL. (1972). On the slowness of action of tricyclic antidepressant drugs. *Br J Psychiatry.*, Vol. 120, No. 559, (Jun 1972), pp. (673-677).

Pałucha, A. Brański, P. Szewczyk, B. Wierońska, JM. Kłak, K. & Pilc, A. (2005). Potential antidepressant-like effect of MTEP, a potent and highly selective mGluR5 antagonist. *Pharmacol Bioch Behav.*, Vol. 81, No. 4, (Aug 2005), pp. (901-906).

Pałucha, A. Kłak, K. Brański, P. van der Putten, H. Flor, PJ. & Pilc, A. (2007). Activation of the mGlu7 receptor elicits antidepressant-like effects in mice. *Psychopharmacology*, Vol. 194, No. 4, (Nov 2007), pp. (555-562).

Pałucha, A. Tatarczyńska, E. Brański, P. Szewczyk, B. Wierońska, JM. Kłak, K. Chojnacka-Wójcik, E. Nowak, G. & Pilc, A. (2004). Group III mGlu receptor agonists produce anxiolytic- and antidepressant-like effects after central administration in rats. *Neuropharmacology*, Vol. 46, No. 2, (Feb 2004), pp. (151-159).

Pałucha-Poniewiera, A. Brański, P. Lenda, T. & Pilc A. (2010a). The antidepressant-like action of metabotropic glutamate 7 receptor agonist N,N'-bis(diphenylmethyl)-1,2-ethanediamine (AMN082) is serotonin-dependent. *J Pharmacol Exp Ther.*, Vol. 334, No. 3, (Sep 2010), pp. (1066-1074).

Pałucha-Poniewiera, A. Wierońska, JM. Brański, P. Stachowicz, K. Chaki, S. & Pilc, A. (2010b). On the mechanism of the antidepressant-like action of group II mGlu receptor antagonist, MGS0039. *Psychopharmacology (Berl)*, Vol. 212, No. 4, (Dec 2010), pp. (523-535).

Papp, M. & Moryl, E. (1994). Antidepressant activity of non-competitive and competitive NMDA receptor antagonists in a chronic mild stress model of depression. *Eur J Pharmacol*, Vol. 263, No. 1-2, (Sep 1994), pp. (1-7).

Papp, M. & Moryl, E. (1996). Antidepressant-like effects of 1-aminocyclopropanecarboxylic acid and D-cycloserine in an animal model of depression. *Eur J Pharmacol.*, Vol. 316, No. 2, (Dec 1996), pp. (145-151).

Paul, IA. Layer, RT. Skolnick, P. & Nowak, G. (1993). Adaptation of the NMDA receptor in rat cortex following chronic electroconvulsive shock or imipramine. *Eur J Pharmacol.*, Vol. 247, No. 3, (Nov 1993), pp. (305-311).

Pavlinac, D. Langer, R. Lenhard, L. & Deftos L. (1979). Magnesium in affective disorders. *Biol Psychiatry*, Vol. 14, No. 4, (Aug 1979), pp. (657-661).

Petralia, RS. Wang, YX. Niedzielski, AS. & Wenthold, RJ. (1996). The metabotropic glutamate receptors, mGluR2 and mGluR3, show unique postsynaptic, presynapticand glial localizations. Neuroscience, Vol. 71, No. 4, (Apr 1996), pp. (949-76).

Phelps, LE. Brutsche, N. Moral, JR. Luckenbaugh, DA. Manji, HK. & Zarate CA Jr. Family history of alcohol dependence and initial antidepressant response to an N-methyl-D-aspartate antagonist. *Biol Psychiatry.*, Vol. 65, No. 2, (Jan 2009), pp. (181-184).

Pilc, A. Brański, P. Pałucha, A. Tokarski, K. & Bijak, M. (1998). Antidepressant treatment influences group I of glutamate metabotropic receptors in slices from hippocampal CA1 region. *Eur J Pharmacol.*, Vol. 349, No. 1, (May 1998), pp. (83-87).

Pilc, A. Kłodzińska, A. Brański, P. Nowak, G. Pałucha, A. Szewczyk, B. Tatarczyńska, E. Chojnacka-Wójcik, E. & Wierońska, JM. (2002). Multiple MPEP administrations evoke anxiolytic- and antidepressant-like effects in rats. *Neuropharmacology*, Vol. 43, No. 2, (Aug 2002), pp. (181-187).

Pilc, A. & Lloyd, KG. (1984). Chronic antidepressants and GABA "B" receptors: a GABA hypothesis of antidepressant drug action. *Life Sci.*, Vol. 35, No. 21, (Nov 1984), pp. (2149-2154).

Pilc, A. & Nowak, G. (2005). GABAergic hypotheses of anxiety and depression: focus on GABA-B receptors. *Drugs Today (Barc).* Vol. 41, No. 11, (Nov 2005), pp. (755-766).

Pin, JP. De Colle, C. Bessis, AS. & Acher, F. (1999). New perspectives for the development of selective metabotropic glutamate receptor ligands. *Eur J Pharmacol.*, Vol. 375, No. 1-3, (Jun 1999), pp. (277-294).

Pin, JP. & Duvoisin, R. (1995). The metabotropic glutamate receptors: structure and functions. *Neuropharmacology.*, (Jan 1995). Vol. 34, No. 1, pp:(1-26)

Pisani, A. Gubellini, P. Bonsi, P. Conquet, F. Picconi, B. Centonze, D. Bernardi, G. & Calabresi, P. (2001). Metabotropic glutamate receptor 5 mediates the potentiation of N-methyl-D-aspartate responses in medium spiny striatal neurons. *Neuroscience*, Vol. 106, (2001), pp. (579-587).

Pomierny-Chamioło, L. Poleszak, E. Pilc, A. & Nowak, G. (2010). NMDA but not AMPA glutamatergic receptors are involved in the antidepressant-like activity of MTEP during the forced swim test in mice. *Pharmacol Rep.*, Vol. 62, No. 6, (Nov-Dec 2010), pp. (1186-1190).

Popik, P. Wróbel, M. & Nowak, G. (2000). Chronic treatment with antidepressants affects glycine/NMDA receptor function: behavioral evidence. *Neuropharmacology*, Vol. 39, No. 12, (Sep 2000), pp. (2278-2287).

Porter, RH. Jaeschke, G. Spooren, W. Ballard, TM. Buttelmann, B. Kolczewski, S. Peters, JU. Prinssen, E. Wichmann, J. Vieira, E. Muhlemann, A. Gatti, S. Mutel, V. & Malherbe, P. (2005). Fenobam: a clinically validated nonbenzodiazepine anxiolytic is a potent, selective, and noncompetitive mGlu5 receptor antagonist with inverse agonist activity. *J Pharmacol Exp Ther.*, Vol. 315, No. 2, (Nov 2005), pp. (711-721).

Post, RM. Ketter, TA. Joffe, RT. & Kramlinger, KL. (1991). Lack of beneficial effects of l-baclofen in affective disorder. *Int Clin Psychopharmacol.*, Vol. 6, No. 4 (Winter 1991), pp. (197-207).

Pratt, GD. & Bowery, NG. (1993). Repeated administration of desipramine and a GABAB receptor antagonist, CGP 36742, discretely up-regulates GABAB receptor binding sites in rat frontal cortex. *Br J Pharmacol.*, Vol. 110, No. 2, (Oct 1993), pp. (724-735).

Preskorn, SH. Baker, B. Kolluri, S. Menniti, FS. Krams, M. & Landen, JW. (2008). An innovative design to establish proof of concept of the antidepressant effects of the NR2B subunit selective N-methyl-D-aspartate antagonist, CP-101,606, in patients with treatment-refractory major depressive disorder. *J Clin Psychopharmacol.*, Vol. 28, No. 6, (Dec 2008), pp. (631-637).

Przegaliński, E. Tatarczyńska, E. Dereń-Wesołek, A. & Chojnacka-Wójcik, E. (1997). Antidepressant-like effects of a partial agonist at strychnine-insensitive glycine receptors and a competitive NMDA receptor antagonist. *Neuropharmacology*, Vol. 36, No. 1, (Jan 1997), pp. (31-37).

Redmond, AM. Kelly, JP. & Leonard, BE. (1997). Behavioural and neurochemical effects of dizocilpine in the olfactory bulbectomized rat model of depression. *Pharmacol Biochem Behav.*, Vol. 58, No. 2, (Oct 1997), pp. (355-359).

Romano, C. Sesma, MA. Mcdonald, CT. Omalley, K. Vandenpol, AN. & Olney, JW. (1995). Distribution of metabotropic glutamate receptor mGluR5 immunoreactivity in rat brain. *J Comp Neurol.*, Vol. 355, No. 3, (May 1995), pp. (455–469).

Samulack, DD. & Lacaille, JC. (1993). Hyperpolarizing synaptic potentials evoked in CA1 pyramidal cells by glutamate stimulation of interneurons from the oriens/alveus border of rat hippocampal slices. II. Sensitivity to GABA antagonists. *Hippocampus*, Vol. 3, No. 3, (Jul 1993), pp. (345-358).

Sanacora, G. Gueorguieva, R. Epperson, CN. Wu, YT. Appel, M. Rothman, DL. Krystal, JH. & Mason, GF. (2004). Subtype-specific alterations of gamma-aminobutyric acid and glutamate in patients with major depression. *Arch Gen Psychiatry.*, Vol. 61, No. 7, (Jul 2004), pp. (705-713).

Sanacora, G. Treccani, G. & Popoli, M. (2011). Towards a glutamate hypothesis of depression An emerging frontier of neuropsychopharmacology for mood disorders. *Neuropharmacology*, 2011 Aug 3

Sands, SA. Reisman, SA. & Enna, SJ. (2004). Effect of antidepressants on GABA(B) receptor function and subunit expression in rat hippocampus. *Biochem Pharmacol.*, Vol. 68, No. 8, (Oct 2004), pp. (1489-1495).

Schildkraut, JJ. (1965). The catecholamine hypothesis of affective disorder: a review of supporting evidence. *Am J Psychiatry.*, Vol. 122, (1965), pp. (1032-1039).

Shigemoto, R. Kinoshita, A. Wada, E. Nomura, S. Ohishi, H. Takada, M. Flor, PJ. Neki, A. Abe, T. Nakanishi ,S. & Mizuno, N. (1997). Differential presynaptic localization of metabotropic glutamate receptor subtypes in the rat hippocampus. *J Neurosci.*, Vol. 17, No. 19, (Oct 1997), pp. (7503-22).

Shigemoto, R. Kulik, A. Roberts, JD. Ohishi, H. Nusser, Z. Kaneko, T. & Somogyi, P. (1996). Target-cell-specific concentration of a metabotropic glutamate receptor in the presynaptic active zone. *Nature*, Vol. 381, No. 6582, (Jun 1996), pp. (523-5).

Shirayama, Y. Chen, AC. Nakagawa, S. Russell, DS. & Duman, RS. (2002). Brain-derived neurotrophic factor produces antidepressant effects in behavioral models of depression. *J Neurosci.*, Vol. 22, No. 8, (Apr 2002), pp. (3251-3261).

Siwek, M. Dudek, D. Paul, IA. Sowa-Kućma, M. Zieba, A. Popik, P. Pilc, A. & Nowak, G. (2009). Zinc supplementation augments efficacy of imipramine in treatment resistant patients: a double blind, placebo-controlled study. *J Affect Disord.*, Vol. 118, No. 1-3, (Nov 2009), pp. (187-195).

Skolnick, P. (1999). Antidepressants for the new millennium. *Eur J Pharmacol.*, Vol. 375, No. 1-3, (Jun 1999), pp. (31-40).

Skolnick, P. Popik, P. & Trullas R. (2009). Glutamate-based antidepressants: 20 years on. *Trends Pharmacol Sci.* Vol. 30, No. 11, (Nov 2009), pp. (563-569).

Slattery, DA. Desrayaud, S. & Cryan, JF. GABAB receptor antagonist-mediated antidepressant-like behavior is serotonin-dependent. *J Pharmacol Exp Ther.*, Vol. 312, No. 1, (Jan 2005), pp. (290-296).

Somogyi, P. Dalezios, Y. Luján, R. Roberts, JD. Watanabe, M. & Shigemoto, R. (2003). High level of mGluR7 in the presynaptic active zones of select populations of GABAergic terminals innervating interneurons in the rat hippocampus. *Eur J Neurosci.*, Vol. 17, No. 12, (Jun 2003), pp. (2503-20).

Stachowicz, K. Kłodzińska, A. Palucha-Poniewiera, A. Schann, S. Neuville, P. & Pilc, A. (2009). The group III mGlu receptor agonist ACPT-I exerts anxiolytic-like but not antidepressant-like effects, mediated by the serotonergic and GABA-ergic systems. *Neuropharmacology*, Vol. 57, No. 3, (Sep 2009), pp. (227-34).

Storm-Mathisen, J. & Iversen, LL. (1979). Uptake of [3H]Glutamic acid in excitatory nerve endings: light and electron microscopic observations in the hippocampal formation of the rat. *Neuroscience*, Vol. 4, No. 9, (1979), pp. (1237-1253).

Śmiałowska, M. Szewczyk, B. Brański, P. Wierońska, JM. Pałucha, A. Bajkowska, M. & Pilc, A. (2002). Effect of chronic imipramine or electroconvulsive shock on the expression of mGluR1a and mGluR5a immunoreactivity in rat brain hippocampus. *Neuropharmacology*, Vol. 42, No. 8, (Jun 2002), pp. (1016-1023).

Tatarczyńska, E. Kłodzińska, A. Chojnacka-Wójcik, E. Pałucha, A. Gasparini, F. Kuhn, R. & Pilc, A. (2001). Potential anxiolytic- and antidepressant-like effects of MPEP, a potent, selective and systemically active mGlu5 receptor antagonist. *Br J Pharmacol.*, Vol. 132, No. 7, (Apr 2001), pp. (1423-1430).

Thomas, LS. Jane, DE. Harris, JR. & Croucher MJ. (2000). Metabotropic glutamate autoreceptors of the mGlu(5) subtype positively modulate neuronal glutamate release in the rat forebrain in vitro. *Neuropharmacology*, Vol. 39, No. 9, (Jul 2000), pp. (1554-1566).

Tokarski, K. Bobula, B. Wabno, J. & Hess, G. (2008). Repeated administration of imipramine attenuates glutamatergic transmission in rat frontal cortex. *Neuroscience*, Vol. 153, No. 3, (May 2008), pp. (789-795).

Trullas, R. & Skolnick, P. (1990). Functional antagonists at the NMDA receptor complex exhibit antidepressant action. *Eur J Pharmacol*, Vol. 185, No. 1, (Aug 1990), pp. (1-10).

Tu, JC. Xiao, B. Naisbitt, S. Yuan, JP. Petralia, RS. Brakeman, P. Doan, A. Aakalu, VA. Lanahan, AA. Sheng, M. & Worley, PF. (1999). Coupling of mGluR/Homer and PSD-95 complexes by the shank family of postsynaptic density proteins. *Neuron*, Vol. 23, No. 3, (Jul 1999), pp. (583–592).

Warner-Schmidt, JL. Duman, RS. (2006). Hippocampal neurogenesis: opposing effects of stress and antidepressant treatment. *Hippocampus*, Vol. 16, No. 3, (2006), pp. (239-349).

Wieronska, JM. Szewczyk, B. Branski, P. Palucha, A. & Pilc A. (2002). Antidepressant-like effect of MPEP, a potent, selective and systemically active mGlu5 receptor antagonist in the olfactory bulbectomized rats. *Amino Acids*, Vol. 23, No. 1-3, (2002), pp. (213-216).

Wierońska, JM. Kłak, K. Pałucha, A. Brański, P. & Pilc A. (2007). Citalopram influences mGlu7, but not mGlu4 receptors' expression in the rat brain hippocampus and cortex. *Brain Res.*, Vol. 1184, (Dec 2007), pp. (88-95).

Wierońska, JM. Legutko, B. Dudys, D. & Pilc A. (2008). Olfactory bulbectomy and amitriptyline treatment influences mGlu receptors expression in the mouse brain hippocampus. *Pharmacol Rep.*, Vol. 60, No. 6, (Nov-Dec 2008), pp. 844-855.

Wieronska, JM. Nowak, G. & Pilc, A. (2010a). Metabotropic approaches to anxiety. in Skolnick, P., (Ed). Milstones in drug therapy: Glutamate-based therapies for psychiatric disorders. Springer Basel AG, pp. 157-173.

Wierońska, JM. & Pilc, A. (2009). Metabotropic glutamate receptors in the tripartite synapse as a target for new psychotropic drugs. Neurochem Int., Vol. 55, No. 1-3, (Jul-Aug 2009), pp. (85-97).

Wierońska, JM. Stachowicz, K. Nowak, G. & Pilc, A. The Loss of Glutamate-GABA Harmony in Anxiety Disorders. In: InTech, Anxiety Disorders, 2011.

Wierońska, JM. Stachowicz, K. Pałucha-Poniewiera, A. Acher, F. Brański, P. & Pilc, A. (2010). Metabotropic glutamate receptor 4 novel agonist LSP1-2111 with anxiolytic, but not antidepressant-like activity, mediated by serotonergic and GABAergicsystems. Neuropharmacology, Vol. 59, No. 7-8, (Dec 2010), pp. (627-34).

Winfield, DA. Gatter, KC. & Powell, TP. (1980). An electron microscopic study of the types and proportions of neurons in the cortex of the motor and visual areas of the cat and rat. *Brain*, Vol. 103, No. 2, (Jun 1980), pp. (245-258).

Winfield, DA. Brooke, RN. Sloper, JJ. & Powell, TP. (1981). A combined Golgi-electron microscopic study of the synapses made by the proximal axon and recurrent collaterals of a pyramidal cell in the somatic sensory cortex of the monkey. *Neuroscience*, Vol. 6, No. 7, (1981), pp. (1217-1230).

van Hooft, JA. Giuffrida, R. Blatow, M. & Monyer, H. (2000). Differential expression of group I metabotropic glutamate receptors in functionally distinct hippocampal interneurons. *J Neurosci.*, Vol. 20, No. 10, (May 2000), pp. (3544-3551).

Xiao, B. Tu, JC. Petralia, RS. Yuan, JP. Doan, A. Breder, CD. Ruggiero, A. Lanahan, AA. Wenthold, RJ. & Worley, PF. (1998). Homer regulates the association of group 1 metabotropic glutamate receptors with multivalent complexes of homer-related, synaptic proteins. *Neuron*, Vol. 21, No. 4, (Oct 1998), pp. (707-716).

Yasuhara, A. Nakamura, M. Sakagami, K. Shimazaki, T. Yoshikawa, R. Chaki, S. Ohta, H, & Nakazato, A. (2006). Prodrugs of 3-(3,4-dichlorobenzyloxy)-2-amino-6-fluorobicyclo[3.1.0]hexane-2,6-dicarboxylic acid (MGS0039): a potent and orally active group II mGluR antagonist with antidepressant-like potential. *Bioorg Med Chem.*, Vol. 14, No. 12, (Jun 2006), pp. (4193-4207).

Yildiz-Yesiloglu, & A. Ankerst, DP. (2006). Review of 1H magnetic resonance spectroscopy findings in major depressive disorder: a meta-analysis. *Psychiatry Res*, Vol. 147, No. 1, (Jun 2006), pp. (1-25).

Yoshimizu, T. & Chaki, S. (2004). Increased cell proliferation in the adult mouse hippocampus following chronic administration of group II metabotropic glutamate receptor antagonist, MGS0039. *Biochem Biophys Res Commun*, Vol. 315, No. 2, (Mar 2004), pp. (493-496).

Yoshimizu, T. Shimazaki, T. Ito, A. & Chaki, S. (2006). An mGluR2/3 antagonist, MGS0039, exerts antidepressant and anxiolytic effects in behavioral models in rats. *Psychopharmacology*, Vol. 186, No. 4, (Jul 2006), pp. (587–593).

Zahorodna, A. & Bijak, M. (1999). An antidepressant-induced decrease in the responsiveness of hippocampal neurons to group I metabotropic glutamate receptor activation. *Eur J Pharmacol.*, Vol. 386, No. 2-3, (Dec 1999), pp. (173-179).

Zarate, CA. Jr, Payne, JL. Quiroz, J. Sporn, J. Denicoff, KK. Luckenbaugh, D. Charney, DS. & Manji, HK. (2004). An open-label trial of riluzole in patients with treatment-resistant major depression. *Am J Psychiatry.*, Vol. 161, No. 1, (Jan 2004), pp. (171-174).

Zarate, CA. Jr, Quiroz, JA. Singh, JB. Denicoff, KD. De Jesus, G. Luckenbaugh, DA. Charney, DS. & Manji HK. (2005). An open-label trial of the glutamate-modulating agent riluzole in combination with lithium for the treatment of bipolar depression. *Biol Psychiatry.*, Vol. 57, No. 4, (Feb 2005), pp. (430-432).

Zarate, CA. Jr, Singh, JB. Carlson, PJ. Brutsche, NE. Ameli, R. Luckenbaugh, DA. Charney, DS. & Manji, HK. (2006). A randomized trial of an N-methyl-D-aspartate antagonist in treatment-resistant major depression. *Arch Gen Psychiatry.*, Vol. 63, No. 8, (Aug 2006), pp. (856-864).

Zerbib, F. Bruley des Varannes, S. Roman, S. Tutuian, R. Galmiche, JP. Mion, F. Tack, J. Malfertheiner, P. & Keywood, C. (2011). Randomised clinical trial: effects of monotherapy with ADX10059, a mGluR5 inhibitor, on symptoms and reflux events in patients with gastro-oesophageal reflux disease. *Aliment Pharmacol Ther.*, Vol. 33, No. 8, (Apr 2011), pp. (911-921).

Zhou, FM. & Hablitz, JJ. (1997). Metabotropic glutamate receptor enhancement of spontaneous IPSCs in neocortical interneurons. *J Neurophysiol.*, Vol. 78, No. 5, (Nov 1997), pp. (2287-2295).

11

The Role of Blue Native/ SDS PAGE in Depression Research

Chunliang Xie, Ping Chen* and Songping Liang
Key Laboratory of Protein Chemistry and Developmental Biology of Education Committee, College of Life Sciences, Hunan Normal University, Changsha P. R. China

1. Introduction

Depression is common in medical patients and is associated with diminished health status and increased health care utilization (DiMatteo et al., 2000). It is characterized by low mood, low self-esteem, inappropriate guilt, thoughts of death and suicide, diminished concentration, loss of interest or pleasure in normally enjoyable activities, and disturbance of sleep and appetite. Approximately 3% of depressed patients in the United States commit suicide and around 60% of people who commit suicide have major depressive disorder (MDD) or another mood disorder (NIH). Over the past decade, evidence has accumulated to suggest that patients with depression were significantly impaired on tasks of attention, executive function, memory and psychomotor speed. These findings suggest deficits in cognitive function, attention, visual and verbal memory and learning, as vulnerability marker for major depressive disorder (Channon & Green, 1999; Gohier et al., 2009; Weingartner et al., 1981).

Synapse is the key structure for various brain functions such as learning and memory. The synapse consists of a pre-synaptic bouton and a post-synaptic spine separated by a synaptic cleft, and is connected via diverse types of cell adhesion molecules located in the opposing membranes of the pre- and post synapse. The fundamental role of the synapse in neurotransmission and plasticity is mediated by various kinds of neurotransmitter receptors and signal transduction proteins(Kaindl et al., 2008). Proteins responsible for the mechanisms of neurotransmitter release have been investigated in detail (Lang & Jahn, 2008; Murthy & De Camilli, 2003). The expression and modification of synaptic proteins is dynamically regulated, depending on the neuronal activity. This process is called synaptic plasticity, and is thought to be the cellular basis of learning and memory (Bagal et al., 2005; Maffei, 2011; Martin & Morris, 2002). Plasticity-induced changes can occur on both sides of the synapse by altering Ca^{2+}-dependent neurotransmitter release from the presynaptic nerve terminal and/or modifying the response to neurotransmitter in the postsynaptic neuron(Evans & Morgan, 2003). Abnormalities of synapse physiology have been implicated in a number of brain disorders, for example, mental retardation, drug addiction,

* Corresponding Author

neurodegenerative diseases and so on(Battaglia, 2011; Grant et al., 2009; Koob et al., 1998). Cognitive impairment suggests changes in the expression of numerous synaptic proteins in depressive disorder.

Although recent studies have identified several candidate genes and proteins associated with major depression, the roles of the resultant proteins remain unknown. The techniques of proteomics provide us with powerful tools to find the differentially expressed proteins in various conditions. Two-dimensional electrophoresis based on isoelectric focusing (IEF) in the first dimension and SDS-PAGE in the second dimension is the most commonly used method for protein separation and is a powerful proteomic technique to display differentially expressed proteins and posttranslational modifications in proteomic approaches. 2-DE can separate thousands of proteins based on their differences in charge and size(Gorg et al., 2000). However, isoelectric focusing is still not the method of choice for separation of hydrophobic membrane proteins, which represent at least 30% of all proteins coded in the genome, despite recent developments and systematic evaluation of detergents(Luche et al., 2003; Santoni et al., 2000). Several solutions have been developed to overcome the technical difficulties represented by the analysis of membrane proteins. For example, fractionation of cell lysates and biochemical enrichment has been reported to reduce pattern complexity on 2D gels and thus to improve visualization of low abundance proteins (Zhang et al., 2010). Alternatively isoelectric focusing can be eliminated (one-dimensional electrophoresis), modified, or judiciously replaced with a different separation technique such as blue native (BN)-PAGE. This technique allows the characterization of many proteins that cannot be resolved using conventional IEF SDS-PAGE(Wittig et al., 2006). Today, hundreds of proteins are identified by BN-PAGE which has had an enormous impact on the investigation of the respiratory chains and photosynthetic complexes(Schagger & von Jagow, 1991). When combined with SDS-PAGE, BN-PAGE allows an assignment of proteins to their protein complexes and display highly hydrophobic proteins in two dimensions. Recently, BN/SDS PAGE is used for the one-step isolation of microgram amounts of membrane protein complexes from different types of samples. Some publications demonstrate that BN-PAGE is also a valuable technique for the identification of physiological protein-protein interactions. The application of BN-PAGE will expand to the analysis of further receptors in the plasma membrane and in the membranes of cellular organelles, to the analysis of protein-chaperone interactions, and to the detection and analysis of antigen antibody interactions. This technology is beginning to be applied to clinical diagnostics of human mitochondrial disorders; to determine native masses and oligomeric states; to identify physiological protein–protein interactions and for studies of neurotransmitter assembly(Eubel et al., 2005; Nijtmans et al., 2002). This alternative 2D method was successfully used to screen for differentially expressed proteins and protein complexes in cystic fibrosis knock-out mice and salt-induced halotolerant alga, indicating that this approach is applicable to obtain a comprehensive map of the differentially expressed synaptic proteins occurring in the depression(Brouillard et al., 2005; Katz et al., 2007). We will discuss the capabilities of BN-PAGE with a focus on its potential use in depression research.

2. The potential role of blue native/SDS PAGE in depression research

The development of depression as well as recovery from depression is most likely accompanied by a change in protein or protein complexes expression profiles in synapse.

Mitochondria provide most of the energy for brain cells by the process of oxidative phosphorylation. Mitochondrial abnormalities and deficiencies in oxidative phosphorylation have been reported in individuals with schizophrenia (SZ), bipolar disorder (BD), and major depressive disorder (MDD) in proteomic studies(Lindholm et al., 1997; Rollins et al., 2009). Although significant progress has been made in our understanding of the depression, the molecular mechanisms underlying this disorder are still not clear. Novel technology such as blue native/SDS PAGE may offer new insights into this devastating illness because this technique not only allows the separation of protein sets for analysis but also provides information on their native interactions. Here, we will show the potential role of blue native/SDS PAGE as a tool to elucidate the differentially expressed proteins and protein complexes in synaptic plasma membrane and mitochondrial during depression and explore what could be done with this proteomic method to offer an insight to the pathogenesis of depression, biomarker establishment and drug development.

2.1 The working principle of BN-PAGE

Blue native polyacryamide gel electrophoresis (BN-PAGE) enables analysis of native (nondissociated) protein - protein interactions, particularly those involving hydrophobic membrane proteins. It was initially developed by Schagger and von Jagow to separate intact and functional mitochondrial membrane protein complexes responsible for oxidative phosphorylation(Schagger & von Jagow, 1991). It is used for the one-step isolation of microgram amounts of membrane protein complexes between 10 and 10 000 kDa from biological membranes and total cell or tissue homogenates. In principle, any nonionic detergent or mild anionic detergent, such as cholic acid derivatives, can be used for the solubilization of biological membranes for blue native PAGE, as long as the detergent can solubilize the desired protein and keep it in the native state. Mild neutral detergents are used for solubilisation of biological membranes, nonionic detergents such as digitonin, Triton X-100, dodecylmaltoside and so on are used for solubilisation the sample. To prevent re-aggregation of hydrophobic membrane proteins, the solubilization buffer contains a high concentration of aminocaproic acid; being a zwitterion, free aminocaproic acid does not migrate in the gel, thus avoiding issues of depletion from the proteins during electrophoresis. Following solubilization of biological membranes and centrifugation, the anionic dye Coomassie blue G-250 is added to the supernatant. G-250 can bind to membrane proteins because of its hydrophobic properties. This binding of a large number of negatively charged dye molecules to protein has several useful effects. (1) It shifts the isolectric point of the proteins to more negative values. As a result all proteins, even basic ones, migrate to the anode irrespective of their original isoelectric points upon electrophoresis at pH 7.5. (2) The excess negative charges on the surfaces of individual dye-associated proteins repel each other. The separation principle relies on binding of Coomassie blue G250 which provides negative charges to the surface of the protein. During migration to the anode, protein complexes are separated according to molecular mass and/or size and high resolution is obtained by the decreasing pore size of a polyacrylamide gradient gel. Thus a mixture of protein complexes of different native charges can be loaded at the top of a native gel, and be resolved according to differences in apparent native molecular weight. Native proteins and complexes migrate as blue bands through BN gels (Devreese et al., 2002; Lasserre et al., 2006; Schamel, 2008; Sunderhaus et al., 2007; Wang et al., 2007).

Protein complex assemblies that are retained during 1D BN-PAGE can be dissociated into the individual protein by applying an orthogonal SDS PAGE for the second dimension, thereby identifying the interaction partners and their stoichiometric ratio. For a second dimension, a lane of the BN-gel can be cut out and incubated in a buffer containing 1% SDS, 1% mercaptoethanol and then placed horizontally on the second dimension gel. This step ensures complete denaturation of the protein complexes necessary for the subsequent separation of their subunits. Loosely stuck in the pores of the gel, the subunits of the protein complexes remain at their position until they are forced electrophoretically into the second dimension gel. Due to the SDS used in the denaturation step (and residual Coomassie) the polypeptides are uniformly negatively charged and are separated according to their molecular weight in the gel. Subunits of a protein complex form a vertical row on the second dimension gel.

2.1.1 Detergents, stock solutions, and buffers

20% dodecyl-b-D-maltoside (wt/vol) (Fluka), dissolved in water; store 1-ml aliquots at –20 °C.

20% Triton X-100 (wt/vol) (Fluka), dissolved in water; store 1-ml aliquots at –20°C.

20% digitonin (wt/vol) (cat no. 37006, 450% purity, used without recrystallization) (Fluka), dissolved in water; store 0.1–1 ml aliquots at –20 °C.

49.5%T, 3%C Acrylamide: 24 g acrylamide, 0.75 g bisacrylamide / 50 ml H_2O

3 x Gel buffer: 150 mM BisTris-HCl, 1.5 M 6-aminohexanoic acid, pH 7.0, adjust pH to 7.0 with HCl at 4°C, store at 4°C.

50% (w/v) Glycerol, store at 4°C.

1 M imidazole/HCl, pH 7.0; store at 7 °C.

10 x Cathode buffer: 0.5 M Tricine, 75 mM imidazole, no need to adjust pH, store at 4°C.

5 x Anode buffer: 0.125 M imidazole, pH 7.0, adjust pH to 7.0 with HCl at 4°C, store at 4°C.

1 M 6-aminohexanoic acid, pH 7.0, adjust pH to 7.0 with HCl at 4°C, store at 4°C.

5% (wt/vol) stock of Coomassie blue G-250 (Serva) (suspend in 500 mM 6-aminohexanoic acid and store at 7 °C).

2.1.2 Prepare the sample: Isolation of synaptic plasma membrane from rat hippocampus

The rats were killed by decapitation, and the hippocampus were dissected quickly and stored on ice. The synaptic plasma membrane (SPM) fraction was isolated essentially as described. First, the hippocampus from 10 rats were combined and homogenized in solution A (0.32 M sucrose, 1 mM $NaHCO_3$, 1 mM $MgCl_2$, and 0.5 mM $CaCl_2$ containing protease inhibitor cocktail) at 4 °C using a hand-held disperser (IKA products, T8 ULTRA-TURRAX, Germany). A low speed (1400g for 10 min) pellet was obtained from the resultant homogenates and washed by resuspending the pellet in the solution A and then homogenized again. All the centrifugation in this experiment was carried out at 4 °C. The second centrifugation was performed at 710g for 15 min. The supernatants were pooled and centrifuged at 13 800g for 20 min. The resulting pellet was suspended in solution B (0.32 M sucrose and 1 mM $NaHCO_3$ containing protease inhibitor cocktail). The sucrose gradients, composed of suspended material, 0.85, 1.0, and 1.2 M sucrose were then centrifuged for 2 h at 82 500g. The bands between 1.0 and 1.2 M sucrose were collected and then diluted with solution B. The synaptosome was spun down at 32 800g for 20 min. The synaptosomal pellet was washed and resuspended in 12 mL of hypotonic solution containing 12 mM Tris-HCl (pH 8.1) and osmotic shock was carried out on ice for 45 min under gentle stirring. The resulting SPM was obtained after centrifugation at 20 000g for 30 min.

2.1.3 Blue native-PAGE and SDS-PAGE

BN-PAGE was performed according to the published protocols with minor modifications(Wittig et al., 2006). Briefly, a 4-13% gradient separation gel with a 3.5% stacking gel was poured and stored at 4 °C until further use.Composition of a sample gel and acrylamide gradient separation gel was shown in Table 1. The cathode buffer (7.5 mM Imidazole, 50 mM Tricine)containing 0.02% (w/v) Coomassie Brilliant Blue G250 and the anode buffer (25 mM Imidazole, pH 7.0) were chilled to 4 °C before use.

For BN-PAGE, 3 μL of 50% Glycerol and 3 μL of 5% G-250 were added to 30 μL of sample, which was then loaded onto the sample well. Electrophoresis was begun at 100 V at 4°C. After about 1 h, the cathode buffer was replaced by the same buffer containing 0.002% of G250, and the electrophoresis was continued with voltage at 200 V at 4 °C. The BN gel was fixed and stained with Coomassie Brilliant Blue G250. One typical gel separation was shown in Figure 1. For further separation in a second-dimensional SDS-PAGE, the excised lanes from BN-PAGE were denatured in SDS loading buffer for 2 h. The lanes were then

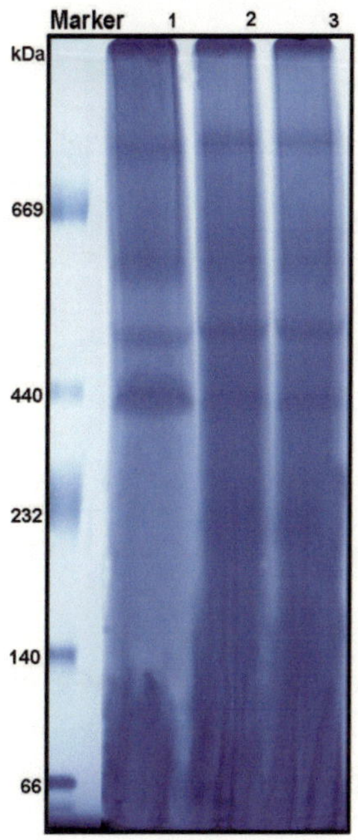

Fig. 1. The Blue Native PAGE of the synaptic plasma membrane fractions.

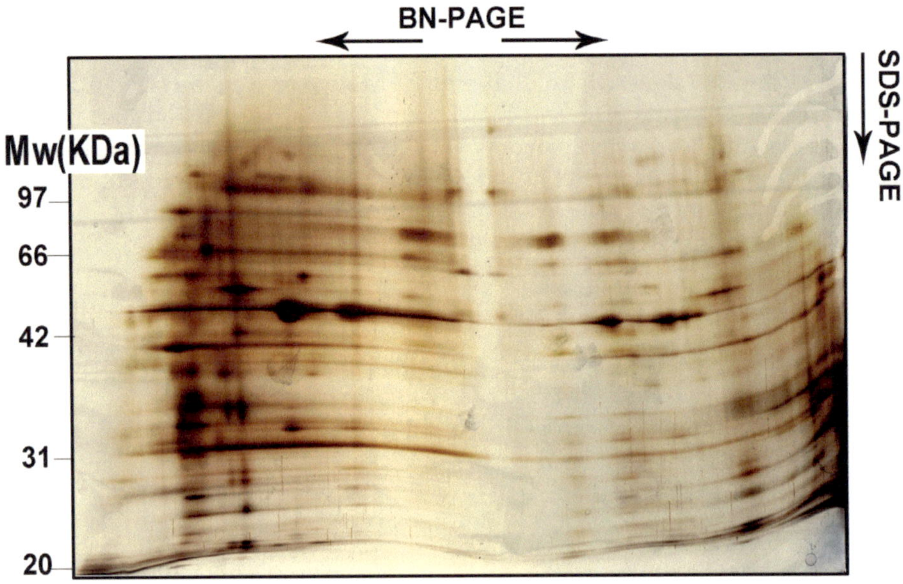

Fig. 2. The SDS-PAGE of the multiprotein complexes bands from Blue Native PAGE.

rinsed briefly with 20 mM Tris buffer. SDS-PAGE was performed using 7.5-12.5% acrylamide gradient gels. The excised lanes were then placed into the SDS-PAGE and sealed with hot agarose solution. Electrophoresis was performed at 20 mA until the front passed into the separation gel and then continued at 45 mA. Proteins were visualized using standard silver staining protocol. One typical gel separation was shown in Figure 2.

	Sample gel	Gradient separation gel	
	3.5% acrylamide	4% acrylamide	13% acrylamide
AB-3 mix	0.44 ml	1.5 ml	3.9 ml
Gel buffer 3×	2 ml	6 ml	5ml
Glycerol	-	-	3 g
Water	3.4 ml	10.4 ml	3 ml
Total volume	6 ml	18 ml	15 ml
10% APS	50 µl	100 µl	75 µl
TEMED	5 µl	10 µl	7.5 µl

Table 1. Composition of a sample gel and 4% and 13% acrylamide mixtures to prepare an acrylamide gradient gel.

2.2 Current biological applications of BN/SDS PAGE studies of mitochondrial and synaptic plasma membranes

2.2.1 Current biological applications of BN/SDS PAGE studies of mitochondrial

Mitochondria provide the main energy source for eukaryotic cells, oxidizing sugars and fats to generate ATP via oxidative phosphorylation (OXPHOS). This is accomplished by the respiratory chain, a series of multimeric enzyme complexes of the inner mitochondrial membrane comprising complex I (NADH-ubiquinone oxidoreductase,EC 1.6.5.3), complex II (succinate-ubiquinone oxidoreductase, EC 1.3.5.1), complex III (ubiquinol-ferricytochrome c oxidoreductase, EC 1.10.2.2), complex IV(cytochrome c oxidoreductase, EC 1.9.3.1), and complex V (F1F0 ATPase). Blue Native polyacrylamide gel electrophoresis (BN - PAGE), followed by immunodecoration, has proved to be invaluable for the analysis of individual respiratory complexes. Individual respiratory complexes also exist in larger supercomplexes, with the largest form (observed at 1700 kDa on BN - PAGE) estimated to contain four copies of complex IV, two copies of complex III, and one copy of complex I (Schagger & Pfeiffer, 2000). Two recent studies identified that complex I stability is reliant on an intact complex III, and this may relate to the association of these complexes within a supercomplex. In addition, destabilized supercomplexes resulting from cardiolipin remodeling defects were detected by BN - PAGE in patients with Barth syndrome, highlighting the importance of supercomplex stability for respiratory chain function (McKenzie et al., 2007; Schagger et al., 2004).

In humans, the respiratory complexes of the mitochondrial inner membrane contain approximately 90 different subunits, with 13 encoded by mtDNA. Assembly of these complexes requires subunit expression from both mitochondrial and nuclear genomes as well as the involvement of various assembly factors, chaperones, and protein translocation components required for targeting and folding of subunits at the inner membrane(Burger et al., 2003).

BN - PAGE has been used to examine the assembly of complex IV, identifying two assembly intermediates that contain mtDNA-encoded subunits. The efficiency of assembly of human respiratory complexes was monitored using radiolabeled, mitochondrially encoded subunits in conjunction with BN PAGE. Intermediate complexes containing newly synthesized mitochondrial DNA-encoded subunits could be observed for complexes I, III, and IV and human cytochrome b was detected as a monomer and as a component of a novel approximately 120 kDa intermediate complex at early chase times before being totally assembled into mature complex III. Furthermore, the researchers showed that BN PAGE was highly suited for the rapid detection of respiratory complex assembly defects in fibroblasts from patients with mitochondrial disease and, thus, has potential diagnostic applications(McKenzie et al., 2007; Wallace, 2005).

The structural organization of enzymes in biological membranes can be very complex, such as ATP synthase, which is usually isolated in monomeric form. However, it appears that ATP synthases are organized as more or less extended functional homooligomeric structures in the inner mitochondrial membrane. Blue native PAGE has been used to isolate ATP synthase as dimers or oligomers. This protein was first isolated as a dimer from yeast mitochondria, using TritonX-100 at very low detergent/protein ratio (Arnold et al., 1998). Dimeric and even oligomeric ATP synthases were later isolated using digitonin from yeast mitochondria. ATP synthases in the mass range 1.5-6 MDa have been isolated from mammalian mitochondria .BN PAGE was also used to detect the oligomycin-sensitive ATPase activity for monomeric and oligomeric ATP synthase and found that oligomerization/monomerization dynamics are not directly involved in regulating ATP synthase activity(Krause et al., 2005).

2.2.2 Current biological applications of BN/SDS PAGE studies of synaptic plasma membrane

Synaptic transmission involves an intricate network of synaptic proteins that forms the molecular machinery underlying transmitter release, activation of transmitter receptors, and signal transduction cascades. It is generally believed that neuronal activity-dependent change of synaptic efficacy is at the basis of learning and memory and is encoded by sequential molecular events at the synapse. Many neuropsychiatric and neurodegenerative diseases, such as Alzheimer's, are thought to involve altered expression of multiple structural and/or metabolic genes and proteins, and therefore are well-suited for proteomic analysis(Kim et al., 2004).

The study of other conditions, such as addiction and mood disorders that likely are secondary to altered expression of proteins involved in neurotransmission or neuroplasticity, can also take advantage of the power of global and narrow protein profiling that proteomics offers, for example, to examine the role of synaptic proteins in different disease states(K. W. Li & Smit, 2007). To date, most proteomic analyses of the brain either considered proteins located within different parts of the cell, or considered differentially expressed proteins in response to development or disease. Those studies lack information about the interaction pattern of these proteins, a very important aspect of the functions that the proteins perform. BN-PAGE offers great promise for the purification of intact protein complexes.

In our lab, BN /SDS-PAGE combined with tandem mass spectrometry, was used to screen multiprotein complexes in synaptic plasma membranes from rat hippocampus. 514 unique proteins were identified, of which 36% were integral membrane proteins. Several multiprotein complexes involved in membrane trafficking and synaptic vesicle exocytosis were identified, such as SNAREs. SNAREs play a role in the docking and fusion of synaptic vesicles to the active zone. The cores of SNARE complexes are composed of syntaxin, synaptobrevin, and SNAP-25. Besides SNAREs, we also found several known SNARE-associated proteins, such as the calcium channel alpha-2 delta-3 subunit, isoform 1 of the syntaxin-binding protein 1, and synaptotagmin-1(X. W. Li et al., 2009).

Most excitatory synapses in the brain use glutamate as a neurotransmitter. Fast signaling at these synapses occurs primarily by activation of amino-3-hydroxy-5-methyl-4-isoxazolepropionic acid (AMPA)-type and N-methyl-D-aspartate (NMDA)-type glutamate receptors. AMPAR channels comprise heterotetramers of subunits GluR1-4, and were reported to interact with cytoskeletal protein, BiP, AP-2 and NSF (Fukata et al., 2005; Vandenberghe et al., 2005). Three membranes containing AMPA-type glutamate receptor were identified from our analysis.

A potentially novel protein complex, involving syntaxin, synapsin I and Na^+/K^+ ATPase alpha-1, was further confirmed by co-immunoprecipitation and immunofluorescence staining. We found BN /SDS-PAGE was a powerful tool for the separation of hydrophobic membrane proteins and for the systematically identify synaptic plasma membrane protein complexes. To the best of our knowledge, this was the first synaptic plasma membrane proteome, and the largest data set determined by the BN PAGE strategy. It contributes promising starting points for further studies.

2.3 Future opportunities for BN in depression research

Brain is a high energy demand tissue and therefore more susceptible to reduction of aerobic metabolism. Even slight reduction in brain metabolism will impair judgment, memory and other higher brain function within seconds. Mitochondria provide most of the energy for brain cells by the process of oxidative phosphorylation. Mitochondrial abnormalities and deficiencies in oxidative phosphorylation have been reported in individuals with schizophrenia (SZ), bipolar disorder (BD), and major depressive disorder (MDD) in transcriptomic, proteomic, and metabolomic studies. The relationship between mitochondrial dysfunction and unipolar depression has been explored in several studies(Liu et al., 2011). Adult BALB/c mice were exposed to unpredictable chronic mild stress (UCMS) for different periods and differential 2D gel electrophoresis (DIGE) approach was employed to the brain tissue to find the differentially expressed proteins. DIGE analysis revealed marked alterations in the expression of proteins involved in energy metabolism, cytoskeleton, and response to cellular stress. Changes in expression of the protein, such as mitochondrial complex I, II, III, IV, V and enzymes relating to glycolysis were detected. Alterations of translational products linked to mitochondrial function, decreased gene expression for 6 of 13 mtDNA encoded transcripts, reduced of respiratory chain enzyme ratios and ATP production rates, and an increased prevalence of small mtDNA deletions (but not of the common 5 kb mtDNA deletion),were found in the depression in different areas of the brain. Impairment of complex I was seen in prefrontal cortex in all patients with bipolar disorder (Andreazza et al., 2010), and abnormalities of mitochondrial structure in the prefrontal cortex, fibroblasts and lymphocytes in another recent study (Cataldo et al., 2010).

Identifying the differentially expressed mitochondrial proteins is available in a rapidly growing subset of mitochondrial disease patients, yet this is still a minority. No systematic strategies for elucidating the mitochondrial proteome in depression patient or animal model have yet been reported. BN PAGE has been proven generally successful for the identification of integral membrane proteins and protein complexes from mitochondria and can maintain the enzyme activity of the complexes, so it will be powerful in research for identifying biomarker from mitochondrial in depression disorders.

Proteomic studies of hippocampus from a chronic stress rat model of MDD identified 27 differentially expressed proteins which participated in neurogenesis and oxidative metabolism. The authors claimed that cellular plasticity is a key issue for the understanding of the molecular mechanism of MDD (Mu et al., 2007). Protein is the actual effector of biological functions and, therefore, proteomic analysis is a powerful way to deepen our knowledge of the molecular mechanisms leading to depression or promoting recovery. Proteome analyses of HAB/LAB mice model led to the identification of glyoxalase-I as a protein marker downregulated in multiple brain areas and an alternate enolase phosphatase isoform in HAB mice (Mu et al., 2007).

Synaptic transmission involves fusion of neurotransmitter-containing vesicles with the plasma membrane and activation of postsynaptic receptors. In nerve terminals, the arrival of the depolarizing wave triggers the opening of presynaptic calcium channels, producing the calcium influx that induces the fusion of docked synaptic vesicles at the active zones and elicit release of glutamate-containing vesicles, which activate postsynaptic receptors that can be generally described as having two broad functions. The receptors that mediate the postsynaptic depolarizations that are responsible for initiating the action potential are Na^+-permeable receptors known as AMPA receptors and those that activate signalling and plasticity mechanisms are NMDA receptors and mGluRs (metabotropic glutamate receptors)(Becherer & Rettig, 2006; Schoch & Gundelfinger, 2006; Sheng & Kim, 2002). Synaptosomes are isolated nerve endings of brain tissue and is greatly enriched in proteins involved in synaptic plasticity. Thus synaptosomes are an ideal tool to analyze the global expression of proteins at synaptic sites where many biological targets of antidepressants are located. A few proteomic studies have been carried out on characterization of the synaptic proteome in animal models of depression. A proteomic analysis of synaptosomes obtained from hippocampus (HPC) and prefrontal/frontal cortex (P/FC) of LH rats or their controls, nonlearned helpless (NLH) rats. 2-DE was employed to identify proteins whose expression is differentially regulated in LH vs. NLH rats at synaptic level. Only 2 synaptic proteins (synapsin and synuclein) were identified in the research. The majority of the identified proteins is mitochondrial and signal proteins (Mu et al., 2007). In our study, we found that blue native PAGE was a powerful tool for the separation of synaptic plasma membrane proteins. So we can combine BN/SDS-PAGE with mass spectrometry to identify the synaptosome protein expression patterns of synaptosome in depression.

Although experimental animal, genomic and proteomic studies are providing some insight into the pathophysiological processes that may be occurring in the depression, at present the molecular correlates underpinning these abnormalities are not fully understood. Proteomics analysis by BN/SDS PAGE combined with tandem mass spectrometry allows simultaneous separation and identification of many hundreds of proteins and protein complexes located in mitochondria and synaptic plasma membrane and is proving to be an effective method to identify the molecular changes associated with depression.

3. Conclusion

This chapter has presented a new proteomic approach to identify the molecular changes associated with depression. BN/SDS PAGE together with other techniques allows the large-scale study and analysis of plasma membrane protein complexes and identifies differentially expressed proteins and protein complexes in large-scale, which serves as a powerful and promising mechanism for extracting useful knowledge and reaching interesting biological conclusions. This technique is enriched with new applications which focus on detecting previously unknown protein functions and relations in depression, which can offer significant experimental knowledge to experts in drug design and clinical applications. BN-PAGE might be the ideal tool to begin study of new protein complexes in brain and search for the differentially expressed proteins and protein complexes in various diseases and gain a deeper understanding of the unique aspects of protein-protein interactions in depression cellular processes.

4. Acknowledgment

This work was supported by a grant from National 973 Project of China (2007CB914203), and National Natural Science Foundation of China (30770437 and 81070353).

5. References

Andreazza, A.C., Shao, L., Wang, J.F. & Young, L.T. (2010). Mitochondrial complex I activity and oxidative damage to mitochondrial proteins in the prefrontal cortex of patients with bipolar disorder. *Arch Gen Psychiatry.*Vol. 67, No. 4, pp. 360-368, ISSN 1538-3636

Bagal, A.A., Kao, J.P., Tang, C.M. & Thompson, S.M. (2005). Long-term potentiation of exogenous glutamate responses at single dendritic spines. *Proc Natl Acad Sci U S A.*Vol. 102, No. 40, pp. 14434-14439, ISSN 0027-8424

Battaglia, A. (2011). Sensory impairment in mental retardation: a potential role for NGF. *Arch Ital Biol.* Vol. 149, No. 2, pp. 193-203, ISSN 0003-9829

Becherer, U. & Rettig, J. (2006). Vesicle pools, docking, priming, and release. *Cell Tissue Res.*Vol. 326, No. 2, pp. 393-407, ISSN 0302-766X

Brouillard, F., Bensalem, N., Hinzpeter, A., Tondelier, D., Trudel, S., Gruber, A.D., Ollero, M. & Edelman, A. (2005). Blue native/SDS-PAGE analysis reveals reduced expression of the mClCA3 protein in cystic fibrosis knock-out mice. *Mol. Cell. Proteomics.* Vol. 4, No. 11, pp. 1762-1775, ISSN 1535-9476

Burger, G., Gray, M.W. & Lang, B.F. (2003). Mitochondrial genomes: anything goes. *Trends Genet.*Vol. 19, No. 12, pp. 709-716, ISSN 0168-9525

Cataldo, A.M., McPhie, D.L., Lange, N.T., Punzell, S., Elmiligy, S., Ye, N.Z., Froimowitz, M.P., Hassinger, L.C., Menesale, E.B., Sargent, L.W., Logan, D.J., Carpenter, A.E. & Cohen, B.M. (2010). Abnormalities in mitochondrial structure in cells from patients with bipolar disorder. *Am J Pathol.*Vol. 177, No. 2, pp. 575-585, ISSN 1525-2191

Channon, S. & Green, P.S. (1999). Executive function in depression: the role of performance strategies in aiding depressed and non-depressed participants. *J Neurol Neurosurg Psychiatry.* Vol. 66, No. 2, pp. 162-171, ISSN 0022-3050

Devreese, B., Vanrobaeys, F., Smet, J., Van Beeumen, J. & Van Coster, R. (2002). Mass spectrometric identification of mitochondrial oxidative phosphorylation subunits separated by two-dimensional blue-native polyacrylamide gel electrophoresis. *Electrophoresis.* Vol. 23, No. 15, pp. 2525-2533, ISSN 0173-0835

DiMatteo, M.R., Lepper, H.S. & Croghan, T.W. (2000). Depression is a risk factor for noncompliance with medical treatment: meta-analysis of the effects of anxiety and depression on patient adherence. *Arch Intern Med.* Vol. 160, No. 14, pp. 2101-2107, ISSN 0003-9926

Eubel, H., Braun, H.P. & Millar, A.H. (2005). Blue-native PAGE in plants: a tool in analysis of protein-protein interactions. *Plant Methods.*Vol. 1, No. 1, pp. 11, ISSN 1746-4811

Evans, G.J. & Morgan, A. (2003). Regulation of the exocytotic machinery by cAMP-dependent protein kinase: implications for presynaptic plasticity. *Biochem Soc Trans.*Vol. 31, No. Pt 4, pp. 824-827, ISSN 0300-5127

Fukata, Y., Tzingounis, A.V., Trinidad, J.C., Fukata, M., Burlingame, A.L., Nicoll, R.A. & Bredt, D.S. (2005). Molecular constituents of neuronal AMPA receptors. *J Cell Biol.*Vol. 169, No. 3, pp. 399-404, ISSN 0021-9525

Gohier, B., Ferracci, L., Surguladze, S.A., Lawrence, E., El Hage, W., Kefi, M.Z., Allain, P., Garre, J.B. & Le Gall, D. (2009). Cognitive inhibition and working memory in unipolar depression. *J Affect Disord.*Vol. 116, No. 1-2, pp. 100-105, ISSN 1573-2517

Gorg, A., Obermaier, C., Boguth, G., Harder, A., Scheibe, B., Wildgruber, R. & Weiss, W. (2000). The current state of two-dimensional electrophoresis with immobilized pH gradients. *Electrophoresis.*Vol. 21, No. 6, pp. 1037-1053, ISSN 0173-0835

Grant, J.E., Bradshaw, A.D., Schwacke, J.H., Baicu, C.F., Zile, M.R. & Schey, K.L. (2009). Quantification of Protein Expression Changes in the Aging Left Ventricle ofRattus norvegicus. *Journal of Proteome Research.*Vol. 8, No. 9, pp. 4252-4263, ISSN 1535-3893

Kaindl, A.M., Koppelstaetter, A., Nebrich, G., Stuwe, J., Sifringer, M., Zabel, C., Klose, J. & Ikonomidou, C. (2008).. Brief Alteration of NMDA or $GABA_A$ Receptor-mediated Neurotransmission Has Long Term Effects on the Developing Cerebral Cortex. *Mol. Cell. Proteomics* Vol. 7, No. 12, pp. 2293-2310.

Katz, A., Waridel, P., Shevchenko, A. & Pick, U. (2007). Salt-induced Changes in the Plasma Membrane Proteome of the Halotolerant Alga Dunaliella salina as Revealed by Blue Native Gel Electrophoresis and Nano-LC-MS/MS Analysis. *Mol. Cell. Proteomics.* Vol. 6, No. 9, pp. 1459-1472.

Kim, S.I., Voshol, H., van Oostrum, J., Hastings, T.G., Cascio, M. & Glucksman, M.J. (2004). Neuroproteomics: expression profiling of the brain's proteomes in health and disease. *Neurochem Res.* Vol. 29, No. 6, pp. 1317-1331, ISSN 0364-3190

Koob, G.F., Sanna, P.P. & Bloom, F.E. (1998). Neuroscience of addiction. *Neuron.*Vol. 21, No. 3, pp. 467-476, ISSN 0896-6273

Krause, F., Reifschneider, N.H., Goto, S. & Dencher, N.A. (2005). Active oligomeric ATP synthases in mammalian mitochondria. Biochem Biophys Res Commun.Vol. 329, No. 2, pp. 583-590, ISSN 0006-291X

Lang, T. & Jahn, R. (2008). Core proteins of the secretory machinery. *Handb Exp Pharmacol.*Vol. No. 184, pp. 107-127, ISSN 0171-2004

Lasserre, J.P., Beyne, E., Pyndiah, S., Lapaillerie, D., Claverol, S. & Bonneu, M. (2006). A complexomic study of Escherichia coli using two-dimensional blue native/SDS polyacrylamide gel electrophoresis. *Electrophoresis.*Vol. 27, No. 16, pp. 3306-3321, ISSN 0173-0835

Li, K.W. & Smit, A.B. (2007). Proteomics of brain synapses and molecular dissection of synaptic subdomains. *Proteomics Clin Appl.*Vol. 1, No. 11, pp. 1476-1484, ISSN 1862-8346

Li, X., Xie, C., Jin, Q., Liu, M., He, Q., Cao, R., Lin, Y., Li, J., Li, Y., Chen, P. & Liang, S. (2009). Proteomic Screen for Multiprotein Complexes in Synaptic Plasma Membrane from Rat Hippocampus by Blue Native Gel Electrophoresis and Tandem Mass Spectrometry. *Journal of Proteome Research.* Vol. 8, No. 7, pp. 3475-3486, ISSN 1535-3893

Lindholm, E., Cavelier, L., Howell, W.M., Eriksson, I., Jalonen, P., Adolfsson, R., Blackwood, D.H., Muir, W.J., Brookes, A.J., Gyllensten, U. & Jazin, E.E. (1997). Mitochondrial sequence variants in patients with schizophrenia. *Eur J Hum Genet.* Vol. 5, No. 6, pp. 406-412, ISSN 1018-4813

Liu, Y., Yang, N., Hao, W., Zhao, Q., Ying, T., Liu, S., Li, Q., Liang, Y., Wang, T., Dong, Y., Ji, C. & Zuo, P. (2011). Dynamic proteomic analysis of protein expression profiles in whole brain of Balb/c mice subjected to unpredictable chronic mild stress: implications for depressive disorders and future therapies. *Neurochem Int.* Vol. 58, No. 8, pp. 904-913, ISSN 1872-9754

Luche, S., Santoni, V. & Rabilloud, T. (2003). Evaluation of nonionic and zwitterionic detergents as membrane protein solubilizers in two-dimensional electrophoresis. *Proteomics.*Vol. 3, No. 3, pp. 249-253, ISSN 1615-9853

Maffei, A. (2011). The many forms and functions of long term plasticity at GABAergic synapses. *Neural Plast.*Vol. 2011, No., pp. 254724, ISSN 1687-5443

Martin, S.J. & Morris, R.G. (2002). New life in an old idea: the synaptic plasticity and memory hypothesis revisited. *Hippocampus.* Vol. 12, No. 5, pp. 609-636, ISSN 1050-9631

McKenzie, M., Lazarou, M., Thorburn, D.R. & Ryan, M.T. (2007). Analysis of mitochondrial subunit assembly into respiratory chain complexes using Blue Native polyacrylamide gel electrophoresis. *Anal Biochem.* Vol. 364, No. 2, pp. 128-137, ISSN 0003-2697

Mu, J., Xie, P., Yang, Z.S., Yang, D.L., Lv, F.J., Luo, T.Y. & Li, Y. (2007). Neurogenesis and major depression: implications from proteomic analyses of hippocampal proteins in a rat depression model. *Neurosci Lett.*Vol. 416, No. 3, pp. 252-256, ISSN 0304-3940

Murthy, V.N. & De Camilli, P. (2003). Cell biology of the presynaptic terminal. *Annu Rev Neurosci.* Vol. 26, No., pp. 701-728, ISSN 0147-006X

Nijtmans, L.G., Henderson, N.S. & Holt, I.J. (2002). Blue Native electrophoresis to study mitochondrial and other protein complexes. *Methods.* Vol. 26, No. 4, pp. 327-334, ISSN 1046-2023

Rollins, B., Martin, M.V., Sequeira, P.A., Moon, E.A., Morgan, L.Z., Watson, S.J., Schatzberg, A., Akil, H., Myers, R.M., Jones, E.G., Wallace, D.C., Bunney, W.E. & Vawter, M.P. (2009). Mitochondrial variants in schizophrenia, bipolar disorder, and major depressive disorder. *PLoS ONE.*Vol. 4, No. 3, pp. e4913, ISSN 1932-6203

Santoni, V., Molloy, M. & Rabilloud, T. (2000). Membrane proteins and proteomics: un amour impossible? *Electrophoresis.* Vol. 21, No. 6, pp. 1054-1070, ISSN 0173-0835

Schagger, H., de Coo, R., Bauer, M.F., Hofmann, S., Godinot, C. & Brandt, U. (2004). Significance of respirasomes for the assembly/stability of human respiratory chain complex I. *J Biol Chem.* Vol. 279, No. 35, pp. 36349-36353, ISSN 0021-9258

Schagger, H. & Pfeiffer, K. (2000). Supercomplexes in the respiratory chains of yeast and mammalian mitochondria. *EMBO J.* Vol. 19, No. 8, pp. 1777-1783, ISSN 0261-4189

Schagger, H. & von Jagow, G. (1991). Blue native electrophoresis for isolation of membrane protein complexes in enzymatically active form. *Anal Biochem.* Vol. 199, No. 2, pp. 223-231, ISSN 0003-2697

Schamel, W.W. (2008). Two-dimensional blue native polyacrylamide gel electrophoresis. *Curr Protoc Cell Biol.*Vol. Chapter 6, No., pp. Unit 6 10, ISSN 1934-2616

Schoch, S. & Gundelfinger, E.D. (2006). Molecular organization of the presynaptic active zone. *Cell Tissue Res.* Vol. 326, No. 2, pp. 379-391, ISSN 0302-766X

Sheng, M. & Kim, M.J. (2002). Postsynaptic signaling and plasticity mechanisms. *Science.* Vol. 298, No. 5594, pp. 776-780, ISSN 1095-9203

Sunderhaus, S., Eubel, H. & Braun, H.P. (2007). Two-dimensional blue native/blue native polyacrylamide gel electrophoresis for the characterization of mitochondrial protein complexes and supercomplexes. *Methods Mol Biol.*Vol. 372, No., pp. 315-324, ISSN 1064-3745

Vandenberghe, W., Nicoll, R.A. & Bredt, D.S. (2005). Stargazin is an AMPA receptor auxiliary subunit. *Proc Natl Acad Sci U S A.* Vol. 102, No. 2, pp. 485-490, ISSN 0027-8424

Wallace, D.C. (2005). A mitochondrial paradigm of metabolic and degenerative diseases, aging, and cancer: a dawn for evolutionary medicine. *Annu Rev Genet.*Vol. 39, No., pp. 359-407, ISSN 0066-4197

Wang, Z.J., Xu, X.P., Fan, K.Q., Jia, C.J. & Yang, K.Q. (2007). Sample preparation for two-dimensional blue native/SDS polyacrylamide gel electrophoresis in the identification of Streptomyces coelicolor cytoplasmic protein complexes. *J Biochem Biophys Methods.* Vol. 70, No. 4, pp. 565-572, ISSN 0165-022X

Weingartner, H., Cohen, R.M., Murphy, D.L., Martello, J. & Gerdt, C. (1981). Cognitive processes in depression. *Arch Gen Psychiatry.* Vol. 38, No. 1, pp. 42-47, ISSN 0003-990X

Wittig, I., Braun, H.P. & Schagger, H. (2006). Blue native PAGE. *Nat Protoc.* Vol. 1, No. 1, pp. 418-428, ISSN 1750-2799

Zhang, L., Peng, X., Zhang, Z., Feng, Y., Jia, X., Shi, Y., Yang, H., Zhang, X., Liu, L., Yin, L. & Yuan, Z. (2010). Subcellular proteome analysis unraveled annexin A2 related to immune liver fibrosis. *J Cell Biochem.*Vol. 110, No. 1, pp. 219-228, ISSN 1097-4644

Part 4

Treatment

12

Mood Disorders in the Puerperium and the Role of the Midwife: Study on Improvement of Midwives' Knowledge About Post-Natal Depression After an Educational Intervention

Ana Polona Mivšek and Teja Zakšek
Faculty of Health Ljubljana, Midwifery Department
Slovenia

1. Introduction

Family, not only in terms of cohabitation of two adults, but in term of intragenerational symbiosys, is formed with the birth of a child (Musek, 1995). Pregnancy and birth are one of the major transitions in the life of a family members; they can promote family's evolution (Nastran Ule, 1993) however if they represent adaptational stress that family members are unable to cope with, can have harmful effect, resulting in the mental crisis of individuals (Zavrl, 1999); especially the mother. Being a mother nowadays is harder that ever before. Parenthood must be squeezed in the rifts between the strata of social life - policy, economy, industry, trade ect. and parenthood is considered of low value in comparison with those cathegories. Child represents some sort of private luxury. Women, often professionally engaged, therefore expect the task of motherhood in a strategic spirit – they carefully plan the pregnancy, are determined to deal with the motherhood succesfully and to return as soon as possible into the 'normal state', this is life as before the pregnancy (Kitzinger, 1994; Gatrell, 2005). As they became aware that parenthood is a life style and that child affect all aspects of their life that is going to change forever, they can be surprised, or even feel traped. This feelings, beside the newborn care that can be stressful because of repetitive and monotonous everyday activities, can provoke in women feelings of loss – loss of professional identity, previous lifestyle, autonomy ect. (Mauthner, 1999; Beck C.T., 1995a). Because of high (and sometimes unrealistic) expectations that they developed in pregnancy for themselves as mothers, they can suffer the feelings of guilt and inadequacy because they failed to become 'perfect mother' and have 'innapropriate' feelings towards the child. They must realize that the perfect mother does not exist (Kitzinger, 1994); it is just an arteficial product of the society, created by the ilusions of the media with the reminders of praised traditional role of women. Self-esteem and self-confidence are exposed to severe stress during postpartum. A lot of women pass this transition unharmed, but some of them can not cope mentally with all the changes; they can respond with feelings of dissapointment, anger, guilt or even depression. It is ironic that family, social institution, primary constructed for reproduction, is at most at strike just with the birth of the child (Welburn, 1980).

2. Background

Mothers' postnatal mental illness affects all family members and Post-Natal Depression [PND] can be therefore regarded as a problem of the society. It is therefore crucial to prevent it, or at least diagnose and treat it as soon as possible. Interventions must be family centered. Midwives who care for the women postnatally can contribute a lot to the well-being of the family. To understand the concept of PND that was tested in the study, described in this chapter later on, first the literature review was done about main features of postnatal mental health disorders.

2.1 History of postnatal mood disorders

The first description of postnatal mental health disorders appeared in ancient times, when Hippocrates wrote about a woman who became 'unstable' after the delivery of twins (Ličina & Radovanovič, 1995). While the doctors in ancient times presumed that lochia and milk went into the mother's head and caused mental disorders, in the Middle Ages the predominant thinking was that possession with demons caused these disorders. In 16th century Protestantism, when everything was seen from the religious perspective, motherhood was considered to be a means of salvation from original sin (Littlewood & McHugh, 1997) and so mental instability after labour was an expected consequence.

In the 19th century Holm stated that infection was the main cause of postnatal disturbances, Magnan emphasised hereditary factors, while Regis believed that depression arose in response to an autointoxication by an unknown substance produced by the body on its own (as cited in Brochington et al., 1990). In the 1950s psychoanalysts considered motherhood to be the mainstay of women's roles (Leathers et al., 1997) and thought depression was caused by a suppression of sexuality or inappropriate psychosexual development (Cox, 1986). A Parisian doctor, Louis Victor Marcé, was the first person to describe postnatal mood disorders, stating that emotionally stable women could be prone to episodes of instability after giving birth (Littlewood & McHugh, 1997). However, his approach was psychiatric and it took until the 1950s for PND to be described as a non-psychotic disorder for the first time (Stewart & Henshaw, 2002). In the 1960s, a biological basis for PND was revealed (Leathers et al., 1997). On the basis of the biological background, which supported a hormonal or hereditary cause of PND, Römer developed a theory that assumed that the endogen predisposition which birth triggers is the reason for mental disorders in the postnatal period, and Graver also sought explanations for PND in endocrine changes (as cited in Brochington et al., 1990).

In the inter-war years in the United Kingdom (UK) the Infanticide act (1932) was adopted. It provided extenuating circumstances for "women who were responsible for the death of their newly born babies (up to a one year), as they had not fully recovered from the effect of lactation and for this reason the balance of their mind was disturbed" (Littlewood & McHugh, 1997, p. 15). Women were considered to be totally under the influence of their hormones and biology. It was not until the feminist movement that psychosocial aspects of postnatal mood disorders began to be recognised. In the 1970s and 1980s women's roles were in transition and more importance was being given to their roles as mothers (Leathers et al., 1997).

This brief history shows how the varied and controversial opinions have influenced thinking about PND and made it difficult to define the aetiology of PND.

2.2 Postnatal mood disorders in different cultures

It was believed for a long time that only women from western societies suffered from PND and that postnatal mood disorders were defined by culture (DeAngelis, 1997). However, conditions with similar symptoms have also been identified in other countries (Cox, 1986). Some studies have found the same prevalence of PND in different societies (Huang & Mathers, 2001), however, in general, European and Australian women appear to have lower levels of PND than women in the United States of America (USA). Women from Asia and South Africa have been identified as being most at risk (Affonso et al., 2000).

Culture influences the perception of childbearing. It defines the woman's own perception and desire for support (Stuchbery et al., 1998) and also the level of expressed dissatisfaction if the help provided does not match her needs. The expression of complaints is more frequent in some cultures, while in others women are taught to suppress dissatisfaction and personal problems (Hearn et al., 1998; Stuchbery et al., 1998). The support needed also varies in type - some societies perceive emotional support as helpful, while others prefer practical support. Culture also defines the support clients expect from the care-giving staff (Simkin, 1996). The culture of a society also affects its predominant view of mental illnesses, and the social stigma of postnatal mood disorders may prevent women from expressing negative or ambivalent feelings.

Many societies have definite rules about how a woman should behave in the first month after the birth of a child. These rules also define the appropriate attitude of the community towards puerperal women. Extra attention is often given to a mother by her family during the six-week period after the birth (Cox, 1986). In Japan, for example, there is a custom called »Stagaeri-Bunben« that provides the mother with practical support and marks her rite of passage to parenthood (Okano et al., 1998). It is believed that rituals lower the levels of distress and depression in the postnatal period (Stuchbery et al., 1998), but only when the 'passenger' believes in the efficacy of the rituals (Littlewood & McHugh, 1997).

It is possible that a lack of universally accepted rituals and traditions around childbearing in western societies (Stuchbery et al., 1998) partially contributes to the phenomenon of PND. The ethic of individualism, competition, mobility, economical narrowness and the medicalization of childbirth are not supportive of new mothers (LoCicero et al., 1997). In modern society a return to normal domestic functioning is expected of women soon after their arrival home from the maternity hospital and the fact that a woman is still physically and psychologically very vulnerable at that time is not considered (Cox, 1986). The culture breeds expectations of motherhood that in current circumstances become unrealistic.

Today women miss helping hands and someone close to talk to, which in the past the extended family provided, but now, with the predominance of the nuclear family, this support has gone. Furthermore, motherhood is not as instinctive as is often suggested, but it is a highly skilled job that someone can learn only from observing and role modelling. This is hard in a modern society where breastfeeding and other activities in connection with motherhood are often closed to outsiders or hidden from the public eye. In developing countries motherhood is valued differently to western communities where being a mother is equated with low status, reinforces isolation and lacks recognition, all of which in turn provoke feelings of worthlessness in the mother (Welburn, 1980). It is not surprising that in this context, women understand motherhood through the context of loss (Mauthner, 1999;

Beck C.T., 1995a). These conditions are not specific to the individual, but are a socially structured issue (Thurtle, 1995).

2.3 Classification of postnatal mood disorders

Until recently, postnatal mood disorders were not considered to be mental disorders in their own right and so did not warrant separate categorisation (Littlewood & McHugh, 1997). This was also obvious from the International Classification of Diseases [ICD] 10 (World Health Organisation [WHO], 1992). The WHO considers postnatal mood disorders to be clinically identical to affective disorders in general and classifies them in the ICD 10 according to the syndrome, rather than aetiologically (Murray, 1996). In ICD 10 the general code for mental health disorders in the postnatal period is F53, and subtypes mark the severity and the form of the depression.

Another classification is given in the Diagnostic and Statistical Manual of mental disorders [DSM] IV (American Psychiatric Association [APA], 1994). This has a special categorisation for postnatal onset, but it still does not describe postnatal mood disorders as a separate entity (LoCicero et al., 1997; Sand Sobey, 2002; Bozoky & Corwin, 2002). However, the codes and terms are said to be compatible with ICD 10 (Benner-Carson, 2000). Additionaly Cox (2004) claims that this classification is limited because it permits the classification of mental and behavioural disorders associated with the puerperium only if they onset within six weeks and if they cannot be classified elsewhere.

Experts have agreed that both classifications should be changed in order to correctly address the phenomenon of PND (Paykel, 2000), but first it must be agreed whether PND is even an illness in its own right. It is not clear, whether postnatal forms of mental disorders are different from those in other life periods (Pajntar, 1994). Some experts claim that PND is different from conditions that occur at other times in a woman's life (Dalton & Herxheimer, 1997; Pitt, 1968). That is, severe PND, when compared with general conditions of severe depression, has more significant self-accusing behaviour, psychomotor agitation, suicidal tendencies, auditory hallucinations and disorientation (Dean & Kendall, 1981).

It is known that rates of depression in the postnatal period are not significantly greater than those of depression at other times in women's lives (Hanusa et al., 2008). Thus it could be thought that a postnatal mood disorder is just a depressive state that coincidentally occurs at the time of childbearing. Some authors have observed that there is little difference between the symptoms of PND and general depression but with the symptoms appearing to be less severe in postnatal cases (Whiffen & Gotlib, 1993).

Clinicians and researchers divide mood disorders after the childbirth into three categories (Murray, 1996):

- postnatal blues;
- PND that can range from mild to moderate or severe (Brockington, 2000) and
- puerperal psychosis (Clarke-Akalanue & Myles, 2002).

There is still a question as to whether these are three separate conditions or a single condition with different levels of severity (Brochington, 1992). The line between postnatal severe depression with suicidal tendency and puerperal psychosis is very hard to define.

Postnatal blues is defined as a transitory mood swing that affects approximately 66% (estimates range from 50 to 80%) of women after the birth (Murray, 1996; Clarke-Akalanue & Myles, 2002; Hewitt et al.,2009). The peak occurrence is around the fourth day after the birth and the unstable mood usually resolves by the tenth day (Wisner et al., 2002). Sometimes postnatal blues last only for a few hours (Boath & Henshaw, 2001). The postnatal blues is thought to be the result of changes in a woman's hormonal balance after the birth. Initial euphoria converts to emotional lability with weepiness, lack of will, inability to concentrate, irritability (Ličina & Radovanovič, 1995; Leathers, 1997), sadness, anxiety and confusion (Wisner et al., 2002). If it ends spontaneously within approximately ten days does not leave any consequences (Hirst & Moutier, 2010). In cases where it lasts longer, approximately 20% of women progress to PND (Campbell et al., 1992; Sand Sobey, 2002). That is why it is important to support women who experience this minor mood disorder in order to prevent more serious conditions. Otherwise it requires no treatment and is considered to be a normal occurence (Raphael-Leff, 1991).

Puerperal or postpartum psychosis is perhaps the most serious of the postnatal mood disorders. Postpartum psychosis is the most severe and uncommon form of postnatal affective illness, with rates of 0.2% of new mothers (Jones & Craddock , 2001) but some researchers estimate the incidence to be as high as 1-2% (Heron et al.,2008). The clinical onset is rapid, with symptoms presenting as early as the first 48 to 72 hours postpartum, and the majority of episodes developing within the first 2 weeks after delivery and can last from six weeks to six months (Cox, 1986; CRAG, 1996). Symptoms include hallucinations, agitation (Leathers et al., 1997), hypomania or mania (Brockington, 2000), delusions, confusion (Boath & Henshaw, 2001; Wisner et al., 2002) and suicidal tendencies with thoughts of infanticide, as the child is often included in the mother's delusional world (Raphael-Leff, 1991). Patients are out of touch with reality and have a lack of insight into their disturbed state. Puerperal psychosis is easy to recognise because it is very disturbing (Cox, 1986) with bizarre behaviour and has a sudden and dramatic onset (Buist, 1997). It often resembles manic depression (Gutteridge, 2001) where mood swings from the initial manic phase are followed by depression. If it manifests as a unipolar disorder, it is often depressive with psychotic features (Stowe & Nemeroff, 1995). Follow-up studies have shown that the majority of women with puerperal psychosis meet criteria for bipolar disorder (Robling et al., 2000).

Postnatal depression is a disorder that comes somewhere in between the postnatal blues and puerperal psychosis. It is a later, more prolonged and more serious condition than the postnatal blues (McIntosh, 1993), and as it is not as sudden and obvious as puerperal psychosis, it is hard to detect. The study reported in this article refers only to this form of postnatal mood disorder, as it is not as obvious as psychosis, but is much more dangerous than the blues and is often a neglected area.

2.4 Postnatal depression

Postantal depression is the most common complication of childbearing and as such represents a considerable public health problem affecting women and their families. It is a later, more prolonged and more serious condition than the postnatal blues (McIntosh, 1993), and as it is not as sudden and obvious as puerperal psychosis, it is hard to detect.

2.4.1 Definition, onset, duration and incidence

APA (1994) defines PND as postnatal onset of major depression, which begins 4-6 weeks postnatally and can last up to a year. It is not puerperal psychosis, which has an onset in the first postnatal week, or maternity blues, which begins 3-5 days postnatally and lasts up to 14 days (p.386). In their later definition APA (2000) states that post-natal depression is a non-psychotic depressive illness that occurs in the first month postpartum. It is defined in DSM IV Text Revision [DSM-IV-TR] as a major depressive episode that begins within one month of delivery (APA, 2000). There is no specific definition of PND and the features are similar to a general major depressive episode, therefore the same diagnostic criteria are ussualy adopted. These are dysphoric mood, loss or decrease of interest/pleasure, accompanied with at least four biological depressive symptoms, that are present every day, most of the day, for more than two weeks (Scott, 1997; Wisner et al., 2002). Three of these symptoms should be present from the seven listed in the ICD 10 classification or four from the eight symptoms listed by DSM IV (Clement & Elliott, 1999).

As definition of PND, also the duration and the onset of PND are not clearly defined. Researchers, differently from the official classifications, ascertain the period of onset to be between 6 and 12 weeks postnatally (Yonkers et al., 2001; Wisner et al., 2002). Most studies evaluate women for symptoms of PND during between 2 and 12 weeks postpartum. For many women, this time period may not be enough long, as depressive symptoms may appear and linger much later than 12 weeks postpartum (Hewitt & Gilbody, 2009). Buist et al. (1999) state that routinuous screening for depression during the perinatal period is a strategy that provides an opportunity to access large numbers of women and facilitate pathways to best care. The warning signs of PND are elated feelings in late pregnancy (Littlewood & McHugh, 1997) or in the first five days after labour, when about 10% of women experience certain kinds of 'highs' (Glover et al., 1994; Lane et al., 1997). Kesser et al. (2003) say that residual depressive symptoms are common with 50% of mothers remaining clinically depressed at 6 months postpartum and 25% of mothers with untreated PND remaining clinically depressed even after the first year (Holden, 1991). Some researchers claim that most cases of PND spontaneously resolve in three months (Watson et al., 1984; Cox et al., 1987), some state six months (Beck C.T., 1995a; Cooper & Murray, 1998; Yonkers et al., 2001), while other researchers think PND steadily remits only in one year (Noh, 1987; Beck C.T., 1995b). However, it is estimated that approximately 25% of mothers with PND will still be depressed one year after childbirth (Yonkers et al., 2001) if they are untreated (Cox, 1986; DeAngelis, 1997). Women who have suffered from PND are also twice as likely to experience future episodes of depression over a 5-year period (Cooper & Murray, 1995). In severe cases, spontaneous remission is less likely (Stowe et al., 1997).

The average duration of a postpartum episode of depression (without treatment) is seven months (O'hara et al., 2000). Episode can be repeated after the discontinuation of medication, and the risk increases with the number of previous episodes. Therefore, long-term treatment for the prevention of recurrence should be considered for women who have had three or more episodes of severe depression (Wisner, 2002).

Longitudinal and epidemiological studies have yielded varying PND prevalence rates ranging from 3% to more than 28% of women in the first year following delivery Letourneau et al., 2006, Brockington, 2000). A meta-analysis of 59 studies reported an overall prevalence of PND of 13% (O'Hara & Swain, 1996). Beck C.T. & Gable (2001) found 12% prevalence of severe depression and 19% of minor. In their meta-analysis of 59 studies, O'Hara & Swain (1996) found the prevalence of PND to be 13%. Estimates of the risk of a woman having a subsequent episode of PND are also inconsistent. They range from 25% (Wisner et al., 2001; Wisner et al., 2002) to 50-62% (Llewellyn et al., 1997).

2.4.2 Aetiology

It is still not known exactly what triggers the outbreak of mental disturbances in the postnatal period. Many reasons have been suggested throughout history and are associated with the medical knowledge and trends of that period. The literature regarding the aetiology of PND is inconclusive and many researchers support the theory of a synergistic effect of a number of factors (Cox, 1986; Brockington, 2000; LoCicero et al., 2001; Clarke-Akalanue & Myles, 2002; Skočir, 2004): the combination of hormones, genetics, environment, biochemistry, psychodynamics and behaviour (Dietch & Bunney, 2002; Clement & Elliott, 1999).

Beck C.T. (2002) in her meta-analysis identified 13 major risk factors for PND: marital status, marital satisfaction, social support, socio-economic status, self-esteem, antenatal depression or anxiety, history of previous depression and maternity blues, unplanned or unwanted pregnancy, life stress, child care stress and infant temperament. The quantity of these predictors calls into question their usefulness at forecasting impending development of PND (Wylie et al., 2011). In a meta-analysis of 59 studies that included in total 12 810 participants, O'Hara & Swain (1996) concluded that the prevalence rate of PND based on predictors was 13%.

The onset of PND can be seen from different perspective. Among health workers, the *medical model* is the dominant perspective. It views PND as an illness initiated by biological factors, and that is why the preferred treatment is usually pharmacological (Beck C.T., 2002). Social factors that influence the woman's life are not considered. The woman is considered a passive person, under the influence of her hormones (Mauthner, 1998). *Feminists,* on the other hand, consider PND to be a social construct, developed from a loss of control and reinforced by feminine helplessness. They do not consider PND to be an illness but more a social construct (Mauthner, 1998; Beck C.T., 2002). They argue that society makes standards for motherhood that are impossible to achieve (Lazzare, 1997). In *attachment theory* the focus is on the woman's unmet wishes for support by her partner (Beck C.T., 2002). Pregnancy and the postnatal period are potentially stressful periods of life and partners rely on each other's support. If they are securely attached, they will react to each other's needs for support. However, if there are differences between the partners, the mother can perceive there to be a lack of support and this can lead to PND (Whiffen & Johnson, 1998). The *interpersonal theory* stresses that people are the constructs of society and that personality is determined by interpersonal interactions (Beck C.T., 2002). People protect themselves from the negative reactions of others through different protective mechanisms (Sullivan, 2003). Interpersonal therapy (IPT) can therefore successfully improve interpersonal relationships by identifying the real cause of PND and alleviating it (Stuart & O'Hara, 1995). People who support the *self-labelling theory* of PND see the depression as being the result of a discrepancy between a mother's feelings and society's expectations regarding motherhood. Mothers who are well socialized and are well aware of the social norms of motherhood, find their ambivalent or negative feelings towards the child inappropriate and label themselves as inappropriate mothers. Self-labelling is thought to be a private activity (Beck C.T., 2002) that identifies the discrepancy between real feelings and feeling rules (Thoits, 1985).

Wylie et al. (2011) say that because women are individuals, a healthcare professional would be required to have in-depth knowledge of the particular woman, before being able to make a judgement about her chances of developing PND. Of all eatiology mentioned and all studies presented, none comprehensively produces certainty of onset of PND.

2.4.3 Symptoms

The overall prevalence of PND is no greater in women after delivery than during pregnancy or other times during her reproductive life (Stocky & Lynch, 2000), but still postpartum mood disorders represent the most frequent form of maternal morbidity following delivery (Letourneau et al., 2006). Included in these postpartum mood disorders, is PND a serious condition, characterized by the disabling symptoms of dysphoria, emotional lability, sleep disturbance, confusion, significant anxiety, guilt, and suicidal ideation. Frequently, further exacerbating these symptoms, women experience low self-esteem, an inability to cope, feelings of incompetence, a loss of self, and social isolation (Ritter et al., 2000). The symptoms mentioned in APA (1994, p.320) are also changes in appetite or weight, changes in sleep and psychomotor activities, decreased energy, feelings of worthlessness or guilt, difficulties in thinking, concentrating or making decisions. These symptoms are not very different from symptoms of general depression (Stowe & Nemeroff, 1995). According to Sit & Wisner (2009) the presentation of PND vary, but mothers with major depression typically describe a diminished pleasure in interacting with people or formerly enjoyable activities as well as feelings of low self-efficacy, rather than having depressed mood. A depressed mother is less positive, less contingent, and shows less vocal and plays interaction to her child. Resulting from the mother's depressive symptoms, the infant shows less positive affection, and less contingent behaviour (Righetti-Veltema et al., 2003). Somatic symptoms include hyperventilation, chest pains, palpitations, headache and tingling of the limbs (Brockington, 2000), flushes, nausea, dizziness (PaNDa, 2001) and panic attacks (Welburn, 1980). Bozoky & Corwin (2002) emphasise that although fatigue is often viewed as a normal consequence of the physical adaptation to motherhood, it is a symptom that can predict depression. Some of the symptoms experienced in depression, such as weight loss, menstrual change, appetite change and low libido, may be normal in the postnatal period. Sleep disturbance is also common among new mothers, but early-morning wakening and an inability to return to sleep even when the child is asleep, are indicators of depression (Yonkers et al., 2001). Thus it is often hard to distinguish normal changes of the postnatal period from the pathology of PND. Another problem is that women often report only somatic symptoms, but above all try hard to appear less affected than they really are (Buist, 1997).

2.4.4 Screening and diagnosing

Still many mothers have little knowledge about depression, and may not be aware that they are depressed (Freeman et al., 2005). So screening is very important in identifying women at risk of, or with PND, because it is the first step in the pathway to treatment (Hanusa et al., 2008). It is estimated that approximately 50% of cases of PND go undetected by health workers (Hearn et al., 1998; Beck C.T. & Gable, 2001) and women rarely seek help on their own. Screening can be performed with different instruments (interviews, clinical examinations, psychological, sociological and psychiatric tests) but the most efficient tools are estimating and self-estimating scales that have been designed specifically for detecting PND. Some researchers claim that scales developed originally for general depression can be used (Lee et al., 2001), while others disagree as such scales do not consider the special characteristics of PND or rate normal physiological changes of the postnatal period as pathology (Cox, 1986).

The screening tools can be divided into those that are:

- antenatal (identifying women at risk of developing PND on the basis of risk factors, usually within the last trimester);
- postnatal (identifying women with symptoms of PND) and
- appropriate for use both, before and after the birth.

In most cases, women who are at risk for PND can be identified already during pregnancy so that appropriate follow-up can be initiated after delivery. However, as yet no antenatal tool has been shown to accurately predict all women who will develop PND (CRAG, 1996). This is because the predictive value of aetiological factors is not strong enough and consequently the clinical utility of any antenatal tool is of limited value (Cooper & Murray, 1997). With a sensitive tool it would be possible to detect women who are at risk of PND in pregnancy, as the prevalence of PND in late pregnancy is similar to postnatal rates (Joseffson et al., 2001; Evans et al., 2001). Some of the commonly used antenatal screening tools are:

Predictive Index for postpartum depression: This is a 17-item self-report questionnaire designed to detect the presence of risk factors for PND. The items are concerned with emotional and physical wellbeing in pregnancy, history of mood disorders and the quality of close relationships. It is estimated that 40% of women who had a score of 35, and 35% of women who scored 27, developed PND (Cooper et al., 1996). The scoring system is based on the mean scores of the EPDS for each risk factor. The index has a 39.8% positive predictive value (Webster et al., 2003). The sensitivity of the tool is 79% for the index and specificity is 50% (Cooper et al., 1996). The low predictive value has led some authors to suggest that it is ineffective (Cooper & Murray, 1997).

Antepartum Questionnaire: This is a self-administered, 24-item antenatal questionnaire. Questions refer to education, social and marital status, the woman's relationship with her mother, father and partner, pregnancy and medical history, social support, early family life, self-esteem, history of PND, anxiety and sadness. The scores in each section range from zero to six. If a woman scores above 46, she needs psychiatric evaluation (Beck C.T., 2002b) because she is at risk of developing PND. The tool has a sensitivity of 82% and a specificity of 78% (Posner et al., 1997).

Beck Depression Inventory (BDI): This is a self-reported antenatal screening technique, developed by A.T. Beck et al. (1961), with 21 items (each with four options), developed from the DSM IV classification (APA, 1994). Higher total scores indicate more severe depression (Affonso, 1992). Scores for individual items range from zero to three; a total score of 10-15 points indicates mild depression, 16-19 moderate depression, 20-29 moderate to severe depression and 30-63 severe depression (Beck C.T. et al., 1992). It can be used for general and PND (Spinelli, 1997), but some researchers state that it lacks factors which are specific to the experience of motherhood and does not measure anxiety, which is an important symptom for PND (Littlewood & McHugh, 1997; Beck C.T. & Gable, 2000; Vieira, 2002). It may give false positive or false negative results as symptoms of depression may be confounded with normal conditions of the postnatal period (Cox et al., 1987; Ugarriza, 2000) and is therefore more appropriate to detect general depression. It is also not sensitive enough to pick up minor forms of depression (O'Hara et al., 1984).The BDI has acceptable reliability and validity (Beck C.T., 1995b) but does not detect mild depression, which is in fact quite common (Whiffen, 1988). It only identifies three of the 11 themes that have been found to be important in PND (contemplating death, loss of interest and guilt), but it does address some symptoms that were not found to be of major importance, such as self-hatred and hypochondriasis (Beck C.T., 1995b). The BDI has 100% sensitivity and 99% specificity for identifying general depression, with 0.72 positive predictive value and a negative predictive value (the percentage of all those tested as negative who are correctly identified as negative) of 1 (Lasa et al., 2000). In the case of depression in pregnancy, it has (at a cut-off score of 16 or above) 83% sensitivity and 89% specificity, a positive predictive value of 0.5 and a negative predictive value of 0.98. As a result, some authors have recommended higher cut-off values for screening in pregnancy to exclude false positives (Holcomb et al., 1996).

Postpartal Emotional Disorders Questionnaire: Braverman & Roux (1978) defined 19 dichotomous questions on the basis of clinical practice and current knowledge of psychopathology. After testing this questionnaire, six items (risk factors) were selected as being the most important (marital status, history of PND, unwanted pregnancy, unplanned pregnancy, marital problems and feeling of being unloved by partner) and these together had 94% sensitivity and 85% specificity. The tool is used for antenatal screening (Beck C.T., 2002).

Petrick's checklist: This is a 16-question tool based on risk factors for PND that were identified from past studies and a literature review (Petrick, 1984). It is of use in the antenatal period (Beck C.T., 1996). The questions have three possible answers (yes/no/not applicable) and the checklist can be self-administered (Petrick, 1984). Items covered include anxiety in pregnancy, support from other women, emotional support from the family or partner, unplanned pregnancy, readiness to assume the role of a mother, personal or family history of depression, recent life changes, difficulty in accepting and making changes, fear of illness, acceptance of body changes of pregnancy, relationship with mother, plan to breast-feed, and sadness or emotional swings either pre-menstrually, in connection with hormonal contraception or after previous pregnancies (Beck C.T., 2002). Sensitivity and specificity have not been reported (Beck C.T., 2002).

Boyer's checklist: Using a literature review and the research base, Boyer identified 16 dichotomous questions for antenatal screening of PND (Beck C.T., 1996). The checklist includes risk factors such as emotional swings before menstruation, antenatal anxiety, family and friends' support, feeling of being unloved by the partner, unwanted pregnancy, previous PND, other mental illness or emotional problems, lack of control, nervous or worrying character, antenatal depression, unhappy childhood, dissatisfaction with current life situation, feelings of guilt when something bad happens and financial, housing or personal problems (Beck C.T., 2002). Pregnant women with three to six affirmative answers are said to be at risk, while those with more than six positive answers are at high risk. The tool's sensitivity and specificity have not been published (Beck C.T., 1996; Beck C.T., 2002).

Antenatal Screening Questionnaire (ASQ): The ASQ is a 10-item, self-administered, research-based tool that is able to predict 44% of women who will develop PND (Appleby et al., 1994). Eight items are scored from zero to three and two dichotomous questions scored zero or one point. The items cover the woman's psychiatric history, antenatal worries, unwanted pregnancy, life stress and social support (Beck C.T., 2002). In a study by Appleby et al. (1994), the ASQ was able to predict 44% of women who went on to become depressed. Authors were unable to find any details about the sensitivity and specificity of the tool and this needs further investigation.

Modified Antenatal Screening Questionnaire (MASQ): The MASQ, by Stamp et al. (1996), is a self-administered, nine-item, Likert-type checklist (Beck C.T., 2002), appropriate for antenatal screening. It takes into account a woman's relationship with her partner, her problem solving style, support from relatives and friends and her psychiatric history (Beck C.T., 1998). Each item is scored from zero to two points (Beck C.T., 2002) and a score of two or more indicates a risk of PND. The instrument has 81% sensitivity and 48% specificity for mild depression, with a positive predictive value of 34% and a negative predictive value of 89%. For severe depression it has 73% sensitivity and 43% specificity, with a 17% positive predictive value and 91% negative value (Stamp et al., 1996). As such it is appropriate only for screening for mild forms of PND (Beck C.T., 1998). The Crown Crisp Experiential Index (CCEI), which assess anxiety, was added to the MASQ. This tool includes 25 items. If a woman achieves 10 points on the CCEI, this is equivalent to 2 points on the MASQ and is sufficient to predict the woman to be at risk for PND (Beck C.T., 2002). The scale does not include items on childcare stress that are thought to be important in screening for PND (Beck C.T., 1998).

Despite multiple contacts with healthcare providers during the postpartum period, many women with PND are undiagnosed and untreated (Driscoll, 2006). Holden (1996) claims that postnatal detection may not be performed during the post-delivery stay in hospital, as signs do not appear until the sixth week and can be quite subtle. Therefore screening can be influenced by:

- the limited time of the health professional who is performing the detection;
- his/her willingness and qualification to use the screening tool;
- the mother's cooperation and trust (if she is afraid to be labelled as mentally ill and worries that her child will be taken away from her, she may not answer truthfully);
- the mother's ability and willingness to distinguish depressive symptoms from normal processes in the postnatal period (Richards, 1990).

To date, research in the PND has focused primarily on health professional screening methods of the detection of PND. As with antenatal tools used to detect PND there are also a number of postpartum screening tools used to detect PND. The two most frequently used tools for detection of PND in puerperium are discussed very briefly below:

Postpartum Depression Checklist (PDC): The PDC consists of a checklist of 11 symptoms of PND that were found in two qualitative studies (Beck C.T., 1992; Beck C.T., 1993) to be of importance: lack of interests, control, self-awareness, concentration and positive feelings, loneliness, uncertainty, anxiety, guilt, obsessive and suicidal thoughts. They are arranged in order of severity, starting with the least threatening, lack of concentration (Beck C.T., 1995b). If a mother is contemplating her own or her child's death, an immediate psychiatric referral is needed (Vieira, 2002). The screening technique should include an active dialogue between the health care professional and the mother, so that mild forms of PND can also be detected (Vieira, 2002). It is recommended for use at the discharge check at the gynaecologist, six weeks after the birth. It does not make a diagnosis but identifies a population that needs further evaluation. The interventions are derived directly from the symptoms reported (Beck C.T., 1995b). Sensitivity and specificity have not been reported for the PDC. However, false negatives may arise because some women find it difficult to confide their ambivalent feelings due to the social stigma attached to mental illness and unhappy motherhood (Vieira, 2002).

Postpartum Depression Screening Scale (PDSS): The PDSS is a 35-item, Likert-type scale, with five questions for each of seven dimensions: sleeping and eating problems, emotional lability, cognitive impairment, anxiety, guilt/shame, contemplation of harming oneself and loss of sense of self (Beck C.T. & Gable, 2000). It is based on previous studies by Beck C.T. (1993, 1996). This self-report questionnaire has 94% sensitivity and 98% specificity for severe PND, and 91% sensitivity and 72% specificity for any type of PND (i.e. both severe and mild depression) (Beck C.T. & Gable, 2001).

There are also screening tools that can be used both antenatally and postnatally:

Postpartum Depression Predictors Inventory (PDPI): Beck C.T. (1998) developed the PDPI which can, when used antenatally, predict women who are vulnerable to PND, but can also be administered postnatally (Beck C.T. & Gable, 2000). The tool was based on the results of a meta-analysis (Beck C.T., 1998). More recently, a revised inventory has been published which consists of 13 updated items (Beck C.T., 2001): self esteem, marital status, socio-economic status and unplanned/unwanted pregnancy are new and added to previously identified items such as antenatal depression and past history of depression, antenatal anxiety, social support, marital satisfaction and life stress. For postnatal screening, additional items are included: childcare stress, infant temperament and maternity blues. The scale should be regularly updated with regard to risk factors that are included or possible new ones (Beck C.T., 2002). PDPI is not scored. It is a list that should be used by a health worker in conversation with the woman (Beck C.T., 1998), ideally every trimester during pregnancy and periodically until the end of the first postnatal year (Beck C.T., 2002).

Edinburgh Postpartum Depression Scale (EPDS): This is a 10-item questionnaire that can be used with mothers or to detect PND in fathers (Murray & Cox, 1990; Clement, 1995; Littlewood & McHugh, 1997; Clement & Elliott, 1999). It has been specifically designed to detect PND (Beck C.T., 1992). From 21 items identified, 13 were confirmed after the pilot study and these were subsequently abbreviated into ten statements describing symptoms of depression (Cox, 1986). The statements have four possible answers, recorded on Likert-type scales, which range from zero to three (total maximum score 30). People completing the scale need to choose a response that describes their feelings in the past week (Tully et al., 1998; Vieira, 2002). Since the EPDS is a self-administered questionnaire, the questions need to be acceptable to users who may not regard themselves as unwell or in need of help (Cox et al., 1987). The tool includes symptoms such as the inability to laugh or to look forward to things with enjoyment, blaming oneself unnecessarily, feeling anxious or worried, feeling that things are too difficult to handle, difficulty sleeping because of unhappiness, feeling sad or miserable, crying, and having thoughts of harming oneself (Beck C.T., 1995b). With repeated use of the EPDS, the severity of the symptoms can be traced over time (Cox et al.,

1987). Screening at six to eight weeks postnatally and seven to nine months is recommended (Tully et al., 1998). It is said that women who score above a threshold of 12 to 13 are likely to be suffering from depressive illness, but if action is taken at the cut-off score of 9 to 10, the negative predictive value can be reduced to less than 10% (Cox et al., 1987; Wisner et al., 2002). It is recommended that a woman with a score of 9 or higher, or those who answer the tenth question positively (suicide intention), should have a clinical interview. It has been suggested that a score of five to ten needs further evaluation after 2-4 weeks (Wisner et al., 2002). The EPDS has 95% sensitivity (studies vary from 65-100%) and 93% specificity (studies vary from 49-100%) (Eberhard-Gran et al., 2001) and positive predictive values of 35.1% (Cooper & Murray, 1997) for severe depression to 83% overall (Cox et al., 1987), which is satisfactory when tested against the clinical psychiatric interview (Cox, 1986). The EPDS is simple to complete and does not require special psychiatric knowledge from the health care professional who interprets it. The only obstacle could be its use with women who have limited literacy, writing abilities or problems with understanding language of the tool (Cox et al., 1987). Beck C.T. (1995b) added as an obstacle the lack of an opportunity for women to describe their feelings fully. Shakespeare et al. (2003) found in their research that women prefer speaking about their feelings rather than filling out an EPDS questionnaire. Critics of the EPDS state that it does not consider symptoms such as loss of control, obsessive thoughts, loss of self, loneliness, lack of concentration (Vieira, 2002) and cognitive impairment (Powell-Kennedy et al., 2002). Of those symptoms that Beck C.T. (1992) identified to be important in PND, the EPDS includes only thoughts of harming oneself, anxiety and fear. Austin & Lumley (2003) have criticised the EPDS because in their study only 8% of women identified as being at risk actually developed PND. This contradiction could occur because the scale excludes some risk factors that are of importance in the development of PND. Lee et al. (2000) found the EPDS to be less reliable for severe PND, but found that the results could be improved by using it in combination with the General Health Questionnaire (GHQ), which raises the positive predictive value to 78%.

The characteristics of above mentioned tools suggest that the EPDS is currently the most evaluated and therefore appropriate screening tool for PND. It is also the most frequently used tool in studies on PND. New instruments are being developed all the time in order to screen more successfully for PND (for example the Brisbane mother baby scale by Webster et al. (2003) and other similar tools are currently being developed and tested). However, in a systematic review, Austin & Lumley (2003) found that no screening instrument met the criteria for routine application in the antenatal period, because the instruments under-estimated the contribution of key domains (predicting risk factors) and therefore had poor sensitivity and positive predictive values. The lack of consensus regarding the risk factors of PND makes it almost impossible to develop an antenatal screening tool that would cover all the issues that emerged from studies on predicting PND (Skočir, 2004). It is easier to screen postnatally for the symptoms. The specificity of the symptoms in the postnatal period means that instruments developed for detecting general depression are not recommended. A screening programme may open up also other problematic areas. Women with false positive scores may become anxious or consequently develop PND. For this reason it is necessary to explain to them that the purpose of the screening instrument is not to diagnose but only to select individuals that might need additional clinical judgement and psychiatric evaluation, and that the diagnosis of PND can only be determined through a clinical examination. Screening only identifies those individuals that have certain risk characteristics, it does not determine the presence of the disease or provide a diagnosis (Tiran, 2002).

The other problem is that although the identified antenatal risk factors are very amenable to prevention, they are not defined clearly enough and cannot provide a good basis for any predictive index for PND (Cooper & Murray, 1997). As a result, some researchers are of the opinion that early recognition and treatment would be more effective than preventive measures (Bowe & Watson, 2001). When PND is treated, the duration is much shorter than when it is untreated (England et al., 1994; CRAG, 1996; Barnett & Morgan, 1996; Wisner et al., 2002). It is estimated that the duration of PND can be shortened from seven months to six weeks through treatment (CRAG, 1996; O'Hara et al., 2000).

2.4.5 Prevention and treatment options

The symptoms of depression or anxiety that occur amongst childbearing women are similar to those that occur at other times of life, however the choices for treatment may differ during pregnancy or when a woman is breastfeeding. The treatment should fit the individual woman's symptoms (DeAngelis, 1997; Beck C.T., 2002) and consider her preferences with regard to the severity of the illness (Wisner et al., 2002), her possible resources for support and breastfeeding (Buist, 1997). Treating PND is challenging and requires specific knowledge and expertise. Includes a variety of options.

2.4.5.1 Pharmacotherapy

There is much dabate about when to offer pharmacological treatment to women with PND. National Institute for Clinical Excellence [NICE] guidlines state that (2007) antidepressants may be considered for use in women with mild, moderate or severe PND, particularly when they are unresponsive or reluctant to participate in non-drug management programmes. Misri & Kostaras (2002) say if the woman is at risk of suicide or infanticide, or has severe depression that does not respond to non-pharmacological treatment, then pharmacological therapy is recommended. Pharmacological treatments include use of antidepressants (Misri & Kostaras, 2002) and hormone therapies (Lawrie et al., 2003).

There are two types of *antidepressants:* (1) Tricyclic Antidepressants (TCA), e.g. imipramine, nortriptyline and sertraline; and (2) Selective Serotonin Re-uptake Inhibitors (SSRI), e.g. fluoxetine, sertraline and citalopram. TCAs work by blocking neuronal uptake of noradrenaline and serotonin, and SSRIs block the reabsorption of serotonin by the presynaptic neurone (Wylie et al., 2011). When choosing an antidepressant treatment, the following should be considered:

- prior response or adverse effects of woman to this medication: concurrent medications and risk interventions;
- maternal monitoring of drug absorption – the aim is that the lowest possible dose that provides complete control of the depressive symptoms is used;
- medication that is known to result in lower infant exposure should be used;
- infant's clinical status and drug concentrations should be regularly assessed;
- the mother should be advised to time her breast-feeding to when the medication is at its lowest levels in the breast milk (Misri & Kostaras, 2002, p. 908).

The risks, benefits and side effects (for the mother and the infant) must be carefully reviewed when making the decision about medication (Powell-Kennedy et al., 2002) and a careful follow-up and evaluation of benefits must be undertaken (Brockington, 2000).

Special care must be paid in pregnancy, as infant exposure to medication is greater through placental passage than through breast milk postnatally. All psychotropic medications are found in breast milk and are passed to the nursing infant, which is why potential side effects of an infant's exposure must be weighed against the effects of untreated depression (Misri & Kostaras, 2002). In pregnancy there is a risk of teratogenesis, while postnatally neonatal toxicity or the impact on neurodevelopment can be a risk. Usually medications are used only when a clear diagnosis has been made and there are no alternative solutions. It has been recommended that the lowest effective dose of the most established drug must be used for the shortest period (Schottish Intercollegiate Group Network [SIGN], 2002). However, some experts claim that in order to prevent a recurrence of the illness, treatment should continue for 9-12 months after the remission of symptoms (Stowe & Nemeroff, 1995). Wisner et al. (2002) recommend starting with half the dose for the first four days after childbirth, as women are more sensitive at that time and side effects might be more likely. The dose should then be gradually increased until remission. If the woman responds to the initial dose, lasting six to eight weeks, treatment with the same dose should be continued for at least six months until full remission. If there is no improvement in the first six weeks, re-evaluation of the medication must be carried out. This kind of treatment is also indicated to prevent a recurrence of PND in women with a history of depression (Wisner et al., 2002). It is important to reduce or discontinue the dose 2-4 weeks before the expected date of delivery, with recommencement after the birth, to prevent perinatal toxic symptoms and withdrawal syndrome (SIGN, 2002).

Hormonal therapy Already in 1971, Dalton has suggested that PND could be caused by having difficulty adjusting to the marked changes in hormone levels form late pregnancy to the early puerperium. Low progesterone levels after the birth were often attributed responsibility for causing PND (Nott et al., 1976), with this being the main rationale for implementing hormone treatment (Dennis et al., 1999). Dennis et al. (2009) Cochrane review of oestrogens and progestins for inhibiting and remedying PND, includes one perinatal trial by Lawrie et al. (1998), which reported a significant negative effect of progesteron on PND scores. As a consequence, progestogens (e.g. norethisterone), used to prevent PND, are now contraindicated (Wylie et al., 2011). Synthetic progesterons have adverse effects, and their use is associated with an increased risk of PND (Lawrie et al., 2003; Karuppaswamy & Vlies, 2003). Transdermal 17-ßestradiol is useful in chronic PND (Cooper & Murray, 1998) and is of modest value in the treatment of the late stages of severe PND (Lawrie et al., 2003, NICE 2007). Significant improvements have been noticed in the first month of treatment (Gregorie et al., 1996). Hormonal medications have not been validated yet for the prevention of recurrent PND (Boath & Henshaw, 2001; Lawrie et al., 2003).

2.4.5.2 Psychotherapeutic methods

In depression with mild to moderate symptoms, non-pharmacological treatment is proposed (Misri & Kostaras, 2002), but to be effective the mother must live within a supportive environment (Pinkofsky, 2000). Family involvement, such as marital therapy, has beneficial effects for all family members (Buist, 1997). However, PND which occurs in the context of a »continuum of psychiatric disturbance may be less likely to respond to these interventions« (Holden et al., 1989, p. 225). Psychotherapy can also be offered by health care providers in the primary sector, but these health care providers need to be educated for this approach (Holden et al., 1989; WHO, 1991; Yonkers et al., 2001). Because many women decline pharmacological treatments, these interventions are often the first line treatment (Austin, 2003). A further advantage of counselling is that it focuses on the problems behind PND and not just on easing the present symptoms (Welburn, 1980). While some experts believe that these therapies are unhelpful in the long term, they admit that there is an improvement in maternal mood right after the application (Cooper et al., 2003).

Interpersonal therapy (IPT) attempts to modify women's relationships or expectations about them. The women's expectations about motherhood and partners' support are discussed during counselling sessions and the therapy focuses on an identified problem area (Boath & Henshaw, 2001). The mother gains a feeling of competence and control through different methods such as psychoeducation, which is a strategy to reduce risk factors associated with the development of PND (Benner-Carson, 2000), role-playing and communication analysis (Beck C.T., 2002). Relief is achieved through discussing feelings, which raises self-esteem in coping with the problems (Dietch & Bunney, 2002). IPT is the only psychotherapy that has been shown to be effective in treating PND both before and after birth (Spinelli, 1997; Zlotnick et al., 2001). It is a good alternative to medication in pregnancy (Spinelli, 1997), even in more severe cases (Boath & Henshaw, 2001), and at the same time it promotes more realistic expectations about motherhood (Dietch & Bunney, 2002). It has been recommended also as a preventive measure for women at risk of PND (Zlotnick et al., 2001).

Cognitive behavioural therapy (CBT) aims to change negative thoughts that are a product of biased perception and to lift depression (Littlewood & McHugh, 1997). First the negative thoughts must be identified and then the process of alteration begins (Cox, 1986). With successful accomplishment of given tasks, the woman achieves self-esteem and sharpens her problem solving skills (Dietch & Bunney, 2002). It is indicated for mild to moderate forms of PND (SIGN, 2002; Meager & Milgrom, 1996). NICE (2007) recommends 4-6 sessions of CBT or IPT for pregnant women with symptoms of depression and/or anxiety who do not meet diagnostic criteria, but symptoms interfere with social functioning. Talking therapies such as Non-directive counselling and CBT, without concurrent use of antidepressants, are both methods of treating PND (Wiley et al., 2011).

Non-directive counselling is a less intensive form of psychotherapy. The art of efficient counselling is not to give advice (Welburn, 1980). The environment should be empathic but non-interfering, close but not intimate (Cox, 1986). Therapeutic listening and extra support are said to be helpful in the treatment of depression or can be an adjunctive therapy to pharmacological treatment (Holden et al., 1989). Non-directive counselling has been shown to be as effective as antidepressant treatment for mild to moderate forms of PND, although the effects of treatment are quicker with medication (Chilvers et al., 2001). Some researchers have found non-directive counselling to be more efficient than CBT (Boath & Henshaw, 2001). The key elements of counselling are:

- support and reassurance;
- education and practical advice in relation to parenting;
- open discussion about the effects of depression;
- advice on extra rest and relaxation (McCarthy, 1998).

Debriefing can be a form of therapeutic counselling. During the process, the mother integrates her childbirth experience into her life (Beck, C.T., 2002). An emotional »catharsis« that can be achieved by encouraging the woman to recollect her birth, can alleviate birth trauma, which could be a possible reason for post-traumatic stress disorder (Littlewood & McHugh, 1997) and consequently for PND. It is believed to be most effective if conducted 24 to 48 hours after the traumatic experience (Littlewood & McHugh, 1997). According to some authors, listening, support, understanding and explanation are helpful also after the non-traumatic birth, and Lavender & Walkinshaw (1998) recommend it as a routine procedure in postnatal maternity units. However, other researchers have found it to be unsuccessful in preventing and treating PND (Small et al., 2000; Bick, 2003; Priest et al., 2003).

Listening visits are also a form of talk therapy and may prevent PND if regularly performed in pregnancy (Clement, 1995). They lessen the mother's feelings of isolation (Gutteridge, 2001).

Self help and support groups may include peer-based meetings or telephone support from women who are past sufferers of PND. They may be helpful in easing depressive symptoms, and their support has been found to be acceptable to new mothers (Dennis, 2003). Support group meetings can be beneficial, particularly if childcare is provided during the meeting, so that mothers can focus on their own needs. Open discussion can reduce the stigma associated with PND (Mauthner, 1999), as well as the isolation, and at the same time provide an opportunity for knowledge to be disseminated. Cox (1986) found that self-help groups were more effective in reducing PND symptoms when they were facilitated by a health professional. The fact that all participants in the group have a similar experience provides them with feelings of confidence, honesty, freedom and trust, which are the necessary ingredients of a psychotherapeutic relationship (Gutteridge, 2001). These groups may help to raise self-esteem, which is, as previously said, an important factor in the development and persistence of PND (SIGN, 2002). However, some researchers have reported finding no improvement in PND after group interventions (Morgan et al., 1997).

Other psychological aspects of the puerperium, such as stress, should also be assessed and treated through conversation. With support, interventions directed towards women's chronic stressors could raise their self-esteem, which in turn decreases their depressive symptoms (Hall et al., 1996).

2.4.5.3 Other methods of treatment

Complementary methods are nowadays more and more popular method of treatment of postnatal mood disorders. *Partner involvement* studied Misri et al. (2000), who determined the impact of a supportive partner on PND scores. This trial showed the value of including the partner in psychotherapy sessions. Women whose partners had participated in psychotherapy sessions had lower EPDS scores at the final assessment which occurred 10 weeks after the first session than those whose partners had not participated (8.6% vs 14.7%). *St. John's wort* is said to be an effective treatment for mild to moderate depressive symptoms and it is said to be as effective as TCAs but with fewer side effects (SIGN, 2002). However, there is no research evidence regarding its safety which is why some experts suggest it should not be used in the childbearing period (Goldman & Koren, 2003b). *Hypnosis* has been indicated in the treatment of insomnia, as one of the symptoms of PND (Pajntar, 1994). *Acupuncture* has also been used in the treatment of PND (Schnyer et al., 2003), however, there is little evidence to support its effectiveness. *Massage therapy* has been found to be effective in treating the anxiety symptoms of depression (Boath & Henshaw, 2001). *Imagery, art and music therapy* can be used as adjuncts to other treatments (Boath & Henshaw, 2001). *Bright light therapy* has been shown to be effective in treating PND in 49% of cases in a study by Oren et al. (2002). One hour of diffuse white-fluorescent light morning sessions eased symptoms of depression and had no adverse effects (Corral et al., 2000; Oren et al., 2002). *Kangaroo (skin to skin) therapy* was found to be an effective treatment for PND and also a prophylactic measure against PND (Dombrowski et al., 2001). *Physical activity* is said by some authors to have good effects on postnatal depressive symptoms (Powell-Kennedy et al., 2002; Armstrong & Edwards, 2003) because of the effect of raised endorphines. Some researchers advocate that women with PND should exercise regularly as part of their treatment (Sampselle et al., 1999).

Electro convulsive therapy (ECT), a method of passing limited electric shocks through the brain, is believed by some to be a safe treatment option for severe types of depression where there is a danger of suicide (Misri & Kostaras, 2002). However, it is generally reserved for women with persistent PND and women who do not respond to psychotherapy or to pharmacotherapeutics (Dietch & Bunney, 2002; Bott, 1999). The major side effect is memory loss (Welburn, 1980). ECT is performed in the hospital as the severe forms of postnatal mood disorders need hospitalisation.

A combination of treatment methods is said to be the most effective intervention (Boath & Henshaw, 2001). It is not only the immediate initiation of the intervention that is successful, but the intervention also needs to be specifically matched to the individual depressive symptomatology of a mother (Beck C.T., 2002).

2.4.5.4 Place of treatment

It is stated that in 75% of cases, the treatment of PND could be managed within primary care (Yonkers et al, 2001). When hospital admission is necessary, the mother and the child should not be separated, but the safety of the child must be assured. Where hospital treatment is required, it is preferable that women are treated in out-patient clinics or mother-baby units (Aiken, 2000), which were first suggested by Thomas Main in 1948 (Buist, 1997). The majority of women with severe depression and psychosis need hospitalisation and in such cases these units are useful because the mother and child stay together and can improve their relationship.

For some women, admission to hospital may cause problems (single parents for example), which is why day-clinics (Littlewood & McHugh, 1997) or units for admission of the whole family have begun to be established (Barnett & Morgan, 1996). The mother can care for the child with the support of health workers who provide her with knowledge and raise her self-esteem. Besides that, mothers are observed if there is a risk of infanticide (Cox, 1986). However, the responsibility for the child must be clearly defined (SIGN, 2002) so that the role of the mother is not taken away by the nurses and midwives.

Mother-baby units are said to be more successful in the treatment of PND than primary care as they are specifically designed for this group of women. Shorter duration of hospitalisation, lower relapse rate and enhanced mother's self-confidence were observed in the study by Boath & Henshaw (2001). Available treatment otherwise depends on the nature and capacity of the clinic and the consistent availability of the nurse is of great value (Barnett & Morgan, 1996).

Mother-baby units are a more costly option than primary care, because more staff and space are needed. Every mother needs a nurse to take care of her and her baby in a private room, otherwise the dangers of infection and violence arise. It is absolutely necessary for the staff to have appropriate knowledge of mental health and childcare issues (Barnett & Morgan, 1996).

2.4.6 The effects of mother's postnatal mood disorder on other family members

PND affects the quality of life of mothers, has negative consequences for the emotional, behavioral, and cognitive functioning of their infants, and threatens the healthy functioning of the entire family (Sobey, 2002). It is therefore vital to assess what thoughts mother has been having and if she is concerned for herself. Family and/or partners may not know how to respond to these feelings and thoughts.The areas of risk that need to be assessed should include:

- suicide - thoughts, details and plans to take her own life
- self harm - behaviour that puts herself at risk physically (self mutilation) or the use of alcohol or drugs as a form of self medication, eg binging;
- harm baby - this may be due to neglect and difficulty caring for the baby, physical abuse as a result of the mother's inability to cope or emotional harm due to limited bonding and attachment.; infanticide is a rare but devastating outcome of PND;
- mental health risks - the woman may not be acknowledging the degree of severity of her symptoms and therefore not seeking or accepting treatment, she may be deteriorating rapidly.; her well-being can be put at increased risk (PaNDa, 2011).

Also Hagen (1999) says mothers with PND commonly have thoughts of harming themselves or their children. He claims these women exhibit fewer positive emotions and more negative emotions towards their children, are less responsive and less sensitive to infant cues, have a less successful maternal role attainment, and have consequently infants, who are less securely attached (Hagen, 1999). Beck C.T. (2002) writes that evidences show that this mood disorder not only has adverse effects on maternal-infant interaction during the first year of age, but may also have long-term effects on children over the age of one year. The mother's ongoing depression can cause harmful effects also for siblings and can contribute to emotional, behavioral, cognitive, and interpersonal problems later in life (Miller, 2002).

PND also has an effect on the couple's relationship. Living with a woman experiencing PND is difficult. Partners need a lot of support too and are known to be at risk of developing depression themselves. They often feel confused, lost and helpless (PaNDa, 2011). They can be the target of their partner's distress and irritability as she attempts to make sense of what is happening to her. There is an increase in marital discord and conflict, and men may become depressed when their wives or partners have PND (Burke, 2003).

2.5 The role of midwives in the prevention, detection and treatment of postnatal mood disorders

The literature suggests that 75% of cases of PND could be managed in the primary level of health care (Mc Carthy, 1998) and that midwives and nurses are capable of preventing, detecting and treating PND. Many studies have described midwives and nurses as having a pivotal role in the prevention, education, early recognition and support or treatment of women that suffer from PND (Leonard, 1998; Holden, 1996; Taylor, 1997; Wood et al., 1997; Gutteridge, 2001; Stewart & Henshaw, 2002). Lepper et al. (1994) claim that health workers can play an important role in fostering maternal mental and emotional health in pregnancy and the postnatal period. It is the provision of psychological care where midwives or other primary care providers are said to be of the most assistance (Powell-Kennedy et al., 2002). Trustful relationship is crucial if we want woman to reveal ambivalent or even negative feelings about the pregnancy, her baby and own mental health. Such relationship can be built only through time and therefore the continuity of care is very important. New models of midwifery care that enable continuity and individual care during pregnancy, labour and postpartum are appropriate. With continuous care, PND is early detected and diagnosed, woman is referred to proper professionals and therefore treatment results are better (Marks et al., 2003).

2.5.1 Pregnancy

Depression in pregnancy is as common as depression in puerperium, therefore a lot of interventions can be used already antenatally (Lee & Chung, 2007; Watson et al., 1984; Dennis & Creedy, 2004; Evans et al., 2001). Researchers have been testing different preventive measures in order to estimate their efficacy. One successful prevention procedure was found to be general social support - wide social netwwork (Elliott, 1989), especially when targeted at women at risk of PND (Zlotnick et al., 2001; Elliott et al., 2002). Eden (1989) found that it was helpful for a woman to have someone to talk to and to have continuity of carer.

Because history of previous depression is an important risk factor for PND, it is important that midwives discuss also mental health issues with pregnant woman (Altschuler et al., 2001; Nonacs & Cohen, 2002) during their antenatal examinations. Therefore it is important for midwives to have knowledge about the risk factors and symptoms of depression and to have protocols and referral systems for depressed woman that enable succesful management, such as Protokol, presented in Figure 1 (SIGN, 2002; Multidisciplinary working group for Grampian, 2001).

As shown in Figure 1 midwives can detect for PND, using interactive conversation or consultations (for example listening visits) (Holden, 1996; Welburn, 1980; Elliot et al., 2007; Jardri et al., 2010). They can also perform routinous detection in third trimester of pregnancy with screening tools like EPDS (Clement, 1995). However midwives need knowledge and skills to perform such therapeutic interventions (Taylor, 1997).

Although some authors claim that discussion about PND in pregnancy might predispose women to PND (Combes & Sconveld, 1992; Bowe & Watson, 2001; Austin, 2003b; Bick, 2003), it could be argued that the first possible action taken in pregnancy should be education about PND, as first time mothers have no experience about how they are likely to feel after the birth (Brockington, 2000; Shu-Shya & Yu-Ying, 2003). Dietch & Bunney (2002, p.144) suggest that: »Health care providers need to be proactive in educating women about their risks for depression, making them aware of symptoms«. Leaflets about the symptoms of PND and the different forms of postnatal mood disorders can be developed and distributed by antenatal health care providers (Brockington, 2000). Beck C.T. (2002) indicates that this can dispel unrealistic myths about motherhood that set women up to label themselves as failures as mothers, since mothers often have unrealistic expectations about labour and life in postnatal period (Renkert & Beam, 2001; Mivšek, 2007; Hillan, 1992).

TIME	ACTIVITY
ANTENATAL PERIOD	- Screening about risk factors - Counseling regarding and management of risk factors - Informations in antenatal classes
POSTNATAL PERIOD	Screening (EPDS)* Interventions on the basis of EPDS scores
ADDITIONAL INTERVENTIONS	- EPDS screening in pregnancy - Listening visits in pregnancy - Debriefing postnataly - Midwifery-led practice - …
* - Third trimester of pregnancy - 6 - 8 weeks postpartum - 6 months postpartum	EPDS Results: - 9-13 scores (reassessment after 2 weeks, referral to support group) - 14-19 scores (home visits, support, referral to therapist, GP notification, reassessment after 2 weeks) - 20 or more scores (support, referral to psychiatrist, GP notification, regular reassessment with EPDS)

Fig. 1. Protocol of activities and referrals that midwives can perform regarding pre- and post- nataly according to results of EPDS screening (adapted from SIGN, 2002 and Multidisciplinary working group for Grampian, 2001).

Welburn (1980) has noted that there is often a lack of discussion in antenatal classes about the feelings and emotions of pregnancy and the postnatal period, and this could be improved by the midwives who run these lessons. In particular, midwives could be of great help in fostering effective coping skills for new mothers and facilitating their adjustment to motherhood, according to McQueen & Mander (2003); among others also with practical skills about newborn care that would raise parents' self-confidence. It is therefore very important that competent health professionals conduct antenatal classes (Nolan, 1997; Svensson et al., 2009). It is also useful that antenatal classes continue postpartally as postpartum support groups to offer woman continuous support with concrete problems (Pairman et al., 2006).

2.5.2 Labour

Maternal and neonatal morbidity and mortality are often considered very powerful indicators of the quality of intrapartum care. Women's satisfaction with the experience and her feelings are often neglected (Oakley, 1993). Labour is exhausting, painful and potentially a frightening experience for women. A bad birth experience can be directly connected with PND (Welburn, 1980; Webster et al., 2003). As a consequence, midwives are in a position to indirectly prevent PND by offering adequate support, such as emotional support, comfort measures, advocacy, supporting the partner and information or advice (Hodnett, 1996).

Nicolson (1998) has pointed out that it is dangerous when midwives, in their eagerness to deliver a healthy child, forget the mother. During the process of labour the mother is supposed to be an active participant in making decisions. The concepts of choice, control and active participation are closely linked with the satisfaction with the birth (Green & Murray, 1994; Slade et al., 1993; Waldenström et al., 2000; Berg et al., 1996). Factors that affect the feeling of control are: continuity of care, informations about the labouring process and feeling of cooperation (Green & Murray, 1994; Slade et al., 1993; Waldenström et al., 2000; Waldenström, 1999; Berg et al., 1996; Brown & Lumley, 1998; Brown et al., 1994; Martin, 1994). These factors affect also woman-midwife relationship (Berg et al., 1996; Fraser, 1999; Hallsdorsdottir & Karlsdottir, 1996).

However, in some cases the mother may be totally at the mercy of health professionals and therefore she feels vulnerable (Welburn, 1980). Loss of control (over what is happening or over her own body) is related to the concept of choice and is said to be one of the basic triggers for PND (Beck C.T., 1995a; Nicolson, 1998). The loss can provoke unfulfilled expectations about the birth process. A dehumanising process, technology and lack of information during labour can contribute to fear that provokes stress in mothers. Even with lots of support, the maternity ward is still unknown territory, which is why home births are sometimes desired by women (Welburn, 1980). The support of a companion in labour (usually the partner) is a bond with the "known world" that can positively affect the psychosocial state of the woman (Wolman et al., 1994).

2.5.3 Postpartum

The period immediately after the birth is a time of emotional and physical vulnerability. Women need support (informational, physical, psychological and social) (Raphael-Leff, 1991; Hobfoll et al., 1995; Magill-Cuerden, 2000)). One of the interventions of psychological support presents concept of debriefing, so that every woman is given an opportunity to discuss her labour, which is said to be helpful (Taylor, 1997; Lavender & Walkinshaw, 1998). As its success largely depends on the quality of the relationship that the woman has with the person who provides the debriefing, it would be ideal if the midwife who cared for the woman antenatally continued to provide care after the birth. This would enable the midwife to use listening visits in pregnancy and debriefing at mutually convenient times (Littlewood & McHugh, 1997).

During the first few days, it is important to be sensitive for signs of postpartum blues, since when this condition lasts for a longer period, can develop into PND. With appropriate interventions we can prevent more serious mental conditions can be prevented (Price, 2007). According to Bozoky & Corwin (2002), PND can be prevented also through the treatment of postnatal fatigue, which untreated can lead to a downward spiral of postnatal mental disorders.

Some researchers blame early discharge for the development of PND, as women may not be prepared yet to take over the role of mother (Hickey et al., 1997). Ball (1982) states that the decision regarding the length of hospital stay should be made by the midwife in agreement with the woman. However, this can only be done if the midwife knows the mother from the pregnancy onwards. On the other hand, some studies have found that there is no connection between early discharge and PND if sufficient professional and non-professional support is available to the woman (Beck C.T. et al., 1992; Thompson et al., 2000). Returning home may be a pleasant but also a frustrating situation, when the mother feels responsible but lacks confidence with a new experience (Nicolson, 1998). Health professionals need to provide support in addition to professional help. This has been stated to be crucial (Nicolson, 1998), however, the term »psychosocial support« is hard to define. For some women this means practical help, while for others it means listening and emotional support (Clement & Elliott, 1999). Women must be given the opportunity to talk freely about their concerns regarding parenting difficulties. Sometimes they just need a sympathetic ear or social support; mobilising friends and the community could help in such cases (Powell-Kennedy, 2002). The attachment with the child can be a reason for anxiety in women, therefore should be presented in a positive and relaxed manner (Welburn, 1980). Midwives should avoid giving conflicting advice as they may cause uncertainty for mothers and lower their self-esteem. They must help women to adapt to a new role (Bott, 1999). If their help is appropriate, they can efficiently cut-down the incidence of PND. Frequent visits during the first two postpartum months that adress women's needs, decrease incidence of PND for 40% (De Craene, 2002).

For all women, preventive screening should be performed by community midwives or nurses at six weeks, four and eight months postnatally. In case of any abnormalities, nurses and midwives could, on the basis of the screening test results, refer a woman to her general practitioner (SIGN, 2002; Holden, 1996). Mothers at risk can be given special support and observation by their caregivers (Holden et al., 1989; Clement, 1995; Brockington, 2000). A number of researchers have found that nurses and midwives can manage non-psychotic depression using counselling (Holden et al., 1989; Prendergast & Austin 2001; Stewart & Henshaw, 2002) and they can play a major role in the treatment of PND (Taylor, 1997). Littlewood & McHugh (1997) identified seven basic principles in the philosophy of care for women with PND: assessment, relief of symptoms, maternal autonomy and self-esteem, individuality, universality, realism/ socialisation and group identity. These can all be incorporated within support groups (also for partners) and peer self-help groups that are facilitated by nurses and midwives. Educating women about PND and giving relevant advice can alleviate the guilt and shame they might be feeling (Powell-Kennedy et al., 2002; Brockington, 2002) and, in a non-judgemental environment, self-esteem may be enhanced. This is also an opportunity to check for signs of PND or identify mothers at risk (Cox, 1986). Health workers can also establish a support phone-line that would connect past sufferers of PND with women who currently have PND (Holden, 1996) or organise support groups for people who live with depressed women, to provide them with support and information (Gutteridge, 2000). It is said that depressive illness is the hardest mental condition for the family supporters, as it is quite capable of sapping the carer too (Kuipers & Bebbington, 1997).

In more serious conditions, where medical treatment is needed, nurses and midwives can help women to make informed choices about their treatment. They can suggest them about available resources and treatment options (Beck C.T., 2002; Wood et al., 1997). And if the hospitalization is needed, women should be hospitalized with a baby (in mother-baby units); midwife can play an important role in fostering the relationship between mother and a child with help, support and possitive feedback (praise can raise mother's self esteem about being a mother) (WHO, 1991; Barnett &Morgan, 1996).

The ideal approach is inter-disciplinary. Healing can be achieved only with collaborative work (Tully et al., 1998; Powell-Kennedy et al., 2002) and with the help of mental health professionals (Taylor, 1997). However one key element in this process is continuity of the caregiver, which can break the process of depersonalised care (Clement & Eliott, 1999). Such care needs to be flexible and individually planned for each woman (MacArthur et al., 2002).

Despite very firm arguments that midwives can manage mild prenatal and postnatal mood disorders, different studies (Eden, 1989; Stewart & Henshaw, 2002; Mivšek & Hundley, 2006; Keng, 2005; Ross-Davie et al., 2006; Sullivan et al., 2003; Isik & Bilgili, 2010) show that midwives are not prepared for this role.

3. Methodology

The question was, whether midwives are adequately prepared to take over the role of prevention and treatment of postnatal mood disorders and whether they feel prepared and willing to take over the responsibility for the care of these women.

Two stage study approach was therefore adopted. Results from 2003 study, which explored knowledge and opinions about PND of slovenian midwives and nurses who worked in the maternity hospital or in the community with pregnant, labouring or postnatal women (Skočir, 2004; Skočir & Hundley, 2006; Mivšek et al., 2008) showed very clearly that they lack knowledge about postnatal mental health issues and therefore do not feel competent to take over more active role in the management of postnatal mental health disorders. 134 nurses and midwives participated in the study (76,6% response rate). Only 24% of participants knew when PND usually bursts out and only 35% correctly estimated the incidence of PND. They often confused symptoms of postnatal blues or psychosys with PND and rarely knew about screening tests for PND (21%). Most of them knew only for medical options for the treatment of PND and were expressing low confidence in their ability to prevent, detect or manage postnatal mood disorders. 99% of them expressed the need for additional training in this field of care. One of main reccommendations of the study was that it is crucial to provide health workers who work with women in reproductive age with more information about mental health issues relevant for this population (Skočir, 2004; Skočir & Hundley, 2006; Mivšek et al., 2008).

3.1 Research design

On the basis of the recommendations from the study in 2003 (Skočir, 2004; Skočir & Hundley, 2006; Mivšek et al., 2008), the undergraduate study programme for midwives was implemented in 2004 with the lectures about postnatal mood disorders. The topic was included in the content of the subject Midwifery (in the third year of the study). Aim of the study was to explore whether educational intervention raised midwives' knowledge and affected their abilities in the management of postnatal mood disorders. The hypothesis tested in the study was that midwives, who finished the implemented midwifery study programme, with additional lectures in postnatal mental health issues, show greater knowledge about postnatal mood disorders, know to differentiate among different forms of postnatal mood disorders and express higher ability for management of those conditions. As the goal of this research was to explore and describe perceptions and experiences, a qualitative approach was adopted as suggested in the literature (Rezabek, 2000); questionnaire survey was used to gather the data.

3.2 Research tool

The questionnaire used was the same as in the study in 2003 (Skočir, 2004; Skočir & Hundley, 2006), developed on the basis of previously conducted focus groups (Skočir, 2004; Mivšek et al., 2008) and from concepts identified in the professional literature.

It covered four key topics: employment data, knowledge about PND, experiences and opinions about PND and demographic data. Most of the questions were closed with a variety of possible answers. Some questions included a category »other« to allow participants to add alternative responses. The section of the questionnaire that dealt with opinions and experiences used more subjective approaches and gave participants the opportunity to justify their choices with open questions. A 5-point Likert-type scale was used to assess participants' opinion of their role in the management of women with PND. In the last part of the questionnaire personal data were gathered and the participants were given an opportunity to add any information or explain things that they found unclear or irrelevant when completing the questionnaire (Skočir & Hundley, 2006).

3.3 Reliability and validity

For the purpose of better reliability and validity of the study it is reccommended to use questionnaires which were tested already in previous surveys. Therefore the questionnaire from the 2003 study was used. Face and content validity of the questionnaire were assessed through review by colleagues, academic supervisors and experts in the fiel das suggested in the literature (Polit et al., 2001). The pilot study further assessed validity and asked participants whether they understood the questionnaire or had any queries; on the basis of pilot study some modifications were made (Skočir, 2004; Skočir & Hundley, 2006).

3.4 Sampling and study setting

Five generations of midwifery students (150) attended the new midwifery study programme from 2004 in which curriculum of Midwifery was implemented with the lectures about postnatal mood disorders. Study sample was composed of those who already graduated (61).

3.5 Data collection

Questionnaires were distributed via mail in May 2011. The stamped envelope was included to the letter to raise response rate. Paricipant were asked to return the questionnaire within two weeks. After this time, written reminder was sent to all the participants and the deadline was extended fort two more weeks.

3.6 Ethical considerations

Participants'conntact details were obtained from the Faculty's register. Considering data protection act (Državni zbor RS, 1999), declaration was signed to use the data only for the purpose of the study and to assure participants of the study confidentiality. It was made clear to participants in the cover letter of the questionnaire that they could refuse to participate in the study and that their participation is anonimous.

3.7 Analysis

The data were collated and entered into the SPSS database. Answers to the open-ended questions were grouped in the same cathegories as the data from the 2003 study. Descriptive statistics were produced for all variables. Data were compared with the results from the study performed in 2003 (midwives with higher education only) (Skočir, 2004) and statistical tests for significant differences were calculated.

4. Results and discussion

The study included 9 participants (29%) in the 2003 group and 22 participants (79%) in the 2011 group; alltogether 31 participants (100%). They were all female, aged from 20 to 29 years with 1 to 5 years of working experience. They all had higher education; finished undergraduate midwifery study. Participants in 2011 study all undergone implemented programme which included 5 hour of lectures about postnatal mood disorders.

There were evident differences in the answers to the questions of both groups. The participants in 2011 group more often gave the expected answer to the question about the onset of PND; there was significant statistical difference between groups in this answer (2Î = 11,064, df = 3, sig.= 0,011), however there were no improvements in estimation of the incidence of PND and they also underestimated the possibility of prenatal depression (Figure 2).

There were no significant differences in citing and ranking symptoms of PND. Both groups most often exposed weepiness as the main problem (100% of participants in 2003 group and 97% of respondents in 2011 group). The next most commonly quoted symptom was insomnia (93%) in 2011 group and hallucinations (100%) in 2003 group. The third most common symptom, as perceived by the participants, was tiredness (83%) in 2011 group and guilt (100%) in 2003 group. It can be said that there was a slight improvement in differentiating PND and puerperal psychosys in the 2011 group of participants, not exposing hallucinations as the main symptom of PND.

Obvious differences were found in testing the statements about detection of PND, as shown in Figure 3. Answers about the PND screening tests showed statistically significant differences between groups (2Î = 20,728, df = 4, sig.= 0,000). Participants in 2011 group were more aware of screening possibilities for PND.

They were also more aware of existing support groups where they can refer woman who suffers from PND. Almost half of the participants in 2011 group knew support groups, while none of the participants in 2003 group knew any. The difference in the answers to this question was statistically significant (2Î =7,970, df = 1, sig.= 0,005).

As the first line treatment, participants in 2003 group cited counselling (50%), as the second option they quoted medicaments (25%) and psychotherapy (25%). In the 2011 group, participants choose psychotherapy (50%) or counselling (25%) as the first line treatment, than support groups (20%) and medicaments (4%).

Many participants already had experiences in caring for women with diagnosed PND or for women they suspected suffering PND. Despite similar length of work in both groups, participants in 2011 group reported more women with PND they cared for (35,7%) than

		ONSET OF PND				
		3-6 days after birth and lasts few days	after first year and lasts three years	3-6 days and lasts for a year	in the first year and lasts few months	Total
2011	f	1	1	10	10	22
	f (%)	4,5	4,5	45,5	45,5	100,0
2003	f	0	0	9	0	9
	f (%)	,0	,0	100,0	,0	100,0
Total	f	1	1	19	10	31
	f (%)	3,2	3,2	61,3	32,3	100,0
		INCIDENCE OF PND				
		2%	10-15%	40%		Total
2011	f	3	18	1		22
	f (%)	13,6	81,8	4,5		100,0
2003	f	0	9	0		9
	f (%)	,0	100,0	,0		100,0
Total	f	3	27	1		31
	f (%)	9,7	87,1	3,2		100,0
		PRESENCE OF DEPRESSIVE SYMPTOMS ALREADY IN PREGNANCY				
		yes, often	no	I don't know		Total
2011	f	12	8	2		22
	f (%)	54,5	36,4	9,1		100,0
2003	f	6	3	0		9
	f (%)	66,7	33,3	,0		100,0
Total	f	18	11	2		31
	f (%)	58,1	35,5	6,5		100,0

Fig. 2. Comparison of the answers about the facts on PND between the 2003 and 2011 group of participants. Expected answers are in blue fields.

participants in 2003 group (25%). Even if the difference is not statistically significant it might be that with the educational intervention they become more sensitive for women's mental health issues.

They were also mora active in the management of those cases. The only action that participants in 2003 group mentioned was that they talked with women with PND and gave advice, however participants from 2011 group also notified woman's general practitioner, talked with woman's partner and referred woman to the therapists.

		EXISTENCE OF SCREENING TESTS FOR PND					
		there isn't any	yes, it makes diagnosys	yes, but it doesn't make diagnosys	I don't know		Total
2011	f	2	1	15	4		22
	f (%)	9,1	4,5	68,2	18,2		100,0
2003	f	1	0	2	6		9
	f (%)	11,1	,0	22,2	66,7		100,0
Total	f	3	1	17	10		31
	f (%)	9,7	3,2	54,8	32,3		100,0
		PND SCREENING TESTS THAT YOU KNOW					
		not applicable	I don't know any	Beck C.T.'s	EPDS	csale euro5	Total
2011	f	0	0	1	2	4	22
	f (%)	,0	,0	14,3	28,6	57,1	100,0
2003	f	7	1	0	0	0	9
	f (%)	87,5	12,5	,0	,0	,0	100,0
Total	f	7	1	1	2	4	31
	f (%)	46,7	6,7	6,7	13,3	26,7	100,0

Fig. 3. Comparison of the answers about the screening on PND between the 2003 and 2011 group of participants. Expected answers are in blue fields.

Participants in 2003 group consider partner as the most appropriate person for detection of PND (75%) and psyciatrists or psychologists as the most important for the treatment (100%), while 2011 participants comprehend the responsibility for detection and treatment of PND more as a team work; their answers are very dispersed. They consider community nurses and midwives (31%), partner (24%), nurses and midwives in maternity hospital (10%) and psyhologists (10%) as the most appropriate people for detection of PND. As for the treatment, they quote psychologists and psyhiatrists (66%) as the most appropriate, followed by community midwives and nurses and family (8%). However the reasons for such choice are the same in both groups. The most important person for detection is chosen because of the regular contact with women and the most appropriate person for treatment is chosen on the basis of the professional knowledge about PND.

According to this, participants in 2011 group feel more responsible to take care for women's mental health in antenatal (education in antenatal classes), intrapartum (support during birth, debriefing) and postpartum period (detection, referrals, consultations). There were statistically significant differences between the groups particularly in the awareness of their role in support during labour (t= -2,778, df = 7,135, sig.= 0,027), their role in detection of PND (t= -2,953, df = 9,511, sig.= 0,015) and their role in discussing PND with woman in antenatal classes (2Î =11,476, df = 1, sig.= 0,001).

Perception of more active involvement in the management of mental health issues changed also their perception of the best place for treatment of woman with PND. More participants in 2011 group think that women should be treated at primary level of health care and stay at home with the family (66%) than in 2003 group (56%). However if they would need hospitalization due to severerity of illness, the participants in 2011 group prefer an option of woman to be accepted in psychiatry unit located in the maternity hospital (31%) over the hospitalization in psychiatric hospital that 2003 group of participants chose most often (25%); the answers showed statistically significant differences (2Î =5,513, df = 1, sig.= 0,019). Participants in 2011 also more frequently disagreed with the need of separation from the baby (83% vs 78% in 2003 group) as they felt that baby belongs with mother and that a childcare could be beneficial for woman (58% in comparison with 18% in 2003 group).

Participants felt more prepared to manage postnatal mental health issues after the educational intervention, as shown in Figure 4, however the difference in answers was not significant.

		SELF-CONFIDENCE FOR ACIVE ROLE IN ADDRESSING POSTNATAL HEALTH ISSUES			Total
		not	**confident, so I might could help**	**very confident, I could help the woman**	
2011	f	11	8	3	22
	f (%)	50,0	36,4	13,6	100,0
2003	f	8	1	0	9
	f (%)	88,9	11,1	,0	100,0
Total	f	19	9	3	31
	f (%)	61,3	29,0	9,7	100,0

Fig. 4. Comparison of the answers about confidence in managing postnatal mental health problems between the 2003 and 2011 group of participants.

Despite raised awareness about their role in the management of postnatal mental health and greater subtlety for women's mental needs, there was not so much improvement in participants' knowledge as we expected and also they alone expressed desire for additional information and training. Unexpectedly participants in 2011 more often quoted the need for theoretical knowledge (57%) than participants in 2003 group (25%). It is possible that the educational intervention made them more critical about their knowing. They also wished opportunities for practical skills on counselling and leading postnatal support groups (40% of participants in 2011 group). This might indicate that only theoretical intervention does not provide enough knowledge for midwives to be more confident. Counselling skills should be also provided as a practical component of additional training.

5. Conclusion

Approach to management of perinatal mental health issues is affected by the perception of cause. Doctors usually consider postnatal mood disorders pathological and treat them with medicines (Beck C.T., 2002b), while midwives, according to their philosophy about physiology of pregnancy, birth and the puerperium, might consider mild postnatal mood disorders normal response to the changes that occur in the puerperium and the demands of transition to parenthood. This argument support also the facts from the literature review; depression in postpartum is not more commom than in other periods of woman's life (CRAG, 1996), aetiology is unknown and varies individually, risk factors are therefore inconsistent (Robertson et al., 2004; Skočir, 2004; Wylie et al., 2011) and symptoms similar to normal changes of the puerperium (Yonkers et al., 2001). Some experts even claim that depression in the transition to motherhood is necessity - a kind o »recognition of imperfection« (Raphael-Leff, 1991) and »healthy rersponse to all the loss of the puerperium« (Stewart & Henshaw, 2002). It could be stated that PND as such does not exists and is a product of medicalization of another part of woman's reproductive mental health (as Pre-Menstrual Syndrome - [PMS]). Life changes that demand alteration of lifestyle, social roles and relationships can be stressful for the individuals (Stowe & Nemeroff, 1995). Postpartum period is demanding because of all the situations that need adaptations (Albers & Williams, 2002). And the feelings of loss, anxiety and depression might be considered a normal response to these changes. However it is not clear at which point mother's physical exhaustion and sadness due to adaptation to parenthood convert into pathology. The most certain assumtion is that the causes and the development of postnatal mood disorders are very complex and individual. Therefore it is very important that there is certain continuity of carer who accompanies woman and her family through pregnancy, birth and postpartum and establish a relationship with the family that can be helpful in addressing also mental health issues.

Few interventions to reduce psychological morbidity of postnatal women have been successful, but obvious individualized continuous midwifery care (for example new models of midwifery-led care) have benefits on mothers' postnatal psychological well-being (MacArthur et al., 2002; Yaffe, 2002). This should be considered in development of policies for antenatal and postnatal care provision (Bick, 2003). Midwives, responsible for this role, however need educational interventions (Wheatley et al., 2003; Barnes, 2002) that include theoretical informations and practical counselling skills. Results of the study show similar results as others (Jardri et al., 2010); midwives benefit from educational interventions (Jardri et al., 2010). They perform their role more successfully and feel more confident.

6. References

[1] Affonso, D. D. (1992). Postpartum depression: A nursing perspective on women's health and behaviors. *J Nurs Scholarsh, 24,* pp. 215–221.

[2] Affonso, D. D., De, A. K., Horowitz, J. A. and Mayberry, L. J. (2000) An international study exploring levels of postpartum depressive symptomatology, *J psychosom Res,* 49(3), pp. 207-16.

[3] Aiken, C. (2000) "*Post-natal depression: At home no one hears you scream*". London: Jessica Kingsley publishers Ltd.

[4] Albers, L., Williams, D. (2002.) Lessons for US postpartum care. *The Lancet,* 359(2), pp.370-1.

[5] Altschuler, L., Cohen, L., Moline, L., Kahn, D., Carpenter, D., Docherty, J. (2001). *The expert consensus guideline series: Treatment of depression in women*. New York: McGraw Hill.

[6] APA (1994) *Diagnostic and statistical manual of mental disorders IV* 4th ed. Washington: APA.

[7] APA. (2000). *Diagnostic and statistical manual of mental disorders IV* 4th TR. Washington, DC: Task Force.

[8] Appleby, L., Gregoire, A., Platz, C., Prince, M. and Kumar, R. (1994) "Screening women for high risk of postnatal depression", *Journal of psychosomatic research,* 38(6), pp. 539-45.

[9] Armstrong, K. and Edwards, H. (2003) "The effect of exercise and social support on mothers reporting depressive symptoms: a pilot randomised controlled trial", *International journal of mental health nursing,* 12(2), pp. 130-8.

[10] Austin, M. P. and Lumley, J. (2003) "Antenatal screening for postnatal depression: a systematic review", *Acta Psychiatr Scand,* 107(1), pp. 10-7.

[11] Ball, J. (1982) "Stress and the postnatal care of women", *Nursing times,* 78(45), 1904-7.

[12] Barnes, S.L. (2002). What midwives need to know about postpartum depression. *MIDIRS Midwifery digest*, 61,pp. 18-9.

[13] Barnett, B., Morgan, M. (1996). Postpartum psychiatric disorders - who should be admitted and to which hospital? *Aust N Z J Psychiatry,* 30(6),pp. 709-14.

[14] Beck C.T., A. T., Ward, C. H., Mendelson, M., Mock, J. and Erbaugh, J. (1961). An inventory for measuring depression, *Arch gen psychiatry,* 4, pp.561-71.

[15] Beck C.T. (1992) The lived experience of postpartum depression: a phenomenological study, *Nurs Res,* 41(3), pp.166-70.

[16] Beck C.T.(1993). Teetering on the edge: a substantive theory of postpartum depression, *Nurs Res,* 42(1), pp.42-8.

[17] Beck C.T. (1995a) Screening methods for postpartum depression, *JOGNN,* 24(4), pp.308-12.

[18] Beck C.T. (1995b) Perceptions of nurses' caring by mothers experiencing PND, *JOGNN,* 24(9), pp.819-25.

[19] Beck C.T. (1996). A meta-analysis of predictors of postpartum depression, *Nurs Res,* 45(5), pp.297-303.

[20] Beck C.T. (1998). A check list to identify women at risk for developing postpartum depression, *JOGNN,* 27(1), pp.39-45.

[21] Beck C.T. (2002) Theoretical perspectives of postpartum depression and their treatment implications, MCN, 27(5), pp.282-7.

[22] Beck C.T., Gable, R. K. (2000). Postpartum depression screening scale: development and testing, *Nurs Res,* 49(5), pp.272-82.

[23] Beck C.T., Gable, R. K. (2001). Further validation of the postpartum depression screening scale, *Nurs Res,* 50(3), pp.155-64.

[24] Beck C.T., Reynolds, M. A. and Rutowsky, P. (1992). Maternity blues and postpartum depression, *JOGNN,* 21(4), pp.287-93.

[25] Benner-Carson, V. (2000) *Mental health nursing* 2nd ed. Philadelphia: W.B. Saunders company.

[26] Berg, M., Lundgren, I., Hermansson, E., Wahlberg, V. (1996). Women's Experience of The encounter with the Midwife during Childbirth. *Midwifery,* 12, pp.11-15.

[27] Bick, D. (2003). Strategies to reduce postnatal psychological morbidity: the role of midwifery services. *Dis manage health outcomes,* 11(1), pp.11-20.

[28] Boath, E. and Henshaw, C. (2001) "The treatment of PND: a comprehensive literature review, *Journal of reproductive and infant psychology*, 19(3), pp.215-48.
[29] Bott, J. (1999) Reflections on PND, *British journal of midwifery*, 7(1),pp. 45-8.
[30] Bozoky, I. and Corwin, E. J. (2002) Fatigue as a predictor of postpartum depression, *JOGNN*, 31(4), pp.436-43.
[31] Bowe, S. and Watson, A. (2001) "Perinatal depression: a randomised controlled trial of an antenatal education intervention for primiparas", *Obstetrical and gynaecological survey*, 56(10), pp. 597-9.
[32] Braverman, J. and Roux, J. (1978) "Screening for the patients at risk for postpartum depression", *Obstetrics and gynaecology*, 52(6), pp. 731-6.
[33] Brochington, J. F. (1992) Disorders specific to the puerperium, *Int J Ment Health*, 21(2),pp. 41-5.
[34] Brochington, J. F., Martin, C., Brown, G. W., Goldberg, D. and Margison, F. (1990) Stress and puerperal psychosis, *Br J psychiatry*, 157(september), pp.331-4.
[35] Brockington, J. (2000) Postnatal illness: the professional view, In: Aiken, C. *PND – at home no one hears you scream*. London: Jessica Kingsley publishers Ltd.
[36] Brown, S., Lumley, J. (1998). Changing Childbirth: Lessons from an Australian Survey of 1336 women. *Br J Obstet Gynecol*, 105, pp.143-155.
[37] Brown, S., Lumley, J., Small, R., Astbury, J. (1994). *Missing Voices: The Experience of Motherhood.* New York: Oxford University Press.
[38] Buist, A. (1997) Postpartum psychiatric disorders - guidelines for management, *CNS drugs*, 8(2), pp.113-23.
[39] Burke, L. (2003). The impact of maternal depression on familial relationships. *International Review of Psychiatry, 15*, pp. 243-255.
[40] Campbell, S.B., Cohn, J.F., Flanagan, C. ,Popper S., Meyers, T. (1992) Course and correlates of postpartum depression during the transition to parenthood. *Dev. Psychopathol.*4, pp. 29–47.
[41] Clarke-Akalanue, E. and Myles, P. (2002) Supporting new mothers with PND: an evaluation, *J Community Nurs*, 16(12), pp.18-20.
[42] Clement, S. (1995) Listening visits in pregnancy: a strategy for prevention PND, *Midwifery*, 11(2), pp.75-80.
[43] Clement, S. and Elliott, S. (1999) Psychological health before, during and after chilbirth, In: Marsh, G. and Renfrew, M. *Community based maternity care*. Oxford: Oxford university press.
[44] Chilvers, C., Dewley, M., Fielding, K., Gretton, V., Miller, P., Palmer, B., Weller, D., Churchill, R., Williams, I., Bedi, N., Duggan, C., Lee, A. and Harrison, G. (2001) "Antidepressant drugs and generic counselling for treatment of major depression in primary care: randomised trial with patient preference arms", *British medical journal*, 322(7289), pp. 722-75.
[45] Combes, G. and Schonveld, A. (1992) *Life will never be the same again: A review of antenatal and postnatal health education*. London: Health education authority.
[46] Cooper, P.J., Murray, L. (1998) Clinical review - fortnightly review: postnatal depression. BMJ,316(7148),pp.1884–6.
[47] Cooper, P.J., Murray, L., Hooper, R., West, A.(1996) The development of a predictive index for postpartum depression. Psychol Med 26,pp. 627– 634.

[48] Cooper, P.J., Murray, L., Wilson, A. and Romaniuk, H. (2003) "Controlled trial of the short and long term effect of psychological treatment of post-partum depression", *British journal of psychiatry,* 182 (May), pp. 412-9.

[49] Corral, M., Kuan, A. and Kostaras, D. (2000) "Bright light therapy's effect on postpartum depression", *American journal of psychiatry,* 157(2), pp. 303-4.

[50] Cox, J. L. (1986) *PND – a guide for health professionals.* Edinburgh: Churchil Livingstone.

[51] Cox, J.L. (2004) Postnatal mental disorder: towards ICD-11 *World Psychiatry.* 2004, 3(2),pp.96–97.

[52] Cox, J. L., Holden, J. M., Sagovsky, R. (1987) Detection of postnatal depression. Development of the 10-item Edinburgh Postnatal Depression Scale. *Br J Psychiatry,* 150,pp. 782 *-786.*

[53] CRAG working group on maternity services (1996) Report on detection and early intervention in PND. URL www.show.scot.nhs.uk/crag – last accessed December 2010

[54] Dalton, K. and Herxheimer, A. (1997) Treatment of PND. Additional information would enhance value of the study, *BMJ,* 315(7100), p.122.

[55] Dean, C. and Kendall, R. E. (1981) The symptomatology of puerperal illness, *Br J Psychiatry,* 139(2), pp.128-33.

[56] DeAngelis, T. (1997) There's a new hope for women with postpartum blues, *American psychological association monitor,* p.22.

[57] De Caene, M. (2002). Postnatal pampering: Midwives can prevent postpartum depression. Psychology today. May 2002.

[58] Dennis, C., Hodnett, E., Kenton, L. *et al.* (2009). Effect of peer support on prevention of postnatal depression among hig risk women: Multisite randomised controlled trial. *BMJ, 338* (a3064), pp. 1–9.

[59] Dennis, C.L., Creedy, D. (2004). Psychosocial and psychological interventions for preventing postpartum depression. *Cochrane database of systematic reviews,*p. 4.

[60] Dietch, K. V. and Bunney, B. (2002) The silent disease: diagnosing and treating depression in women, *Lifeliness,* 6(2), pp.140-5.

[61] Dombrowski, M. A. S., Anderson, G. C., Santori, C. and Burkhammer, M. (2001) "Kangaroo (skin to skin) care with a postpartum women who felt depressed", *American journal of maternal child nursing,* 26(4), pp. 214-6.

[62] Driscoll JW. (2006) Postpartum depression. How nurses can identify and care for women grappling with this disorder. *AWHONN Lifelines,*10(5), pp.400-409.

[63] Državni zbor RS (1999) Zakon o varstvu osebnih podatkov. Uradni list RS, 59/99.

[64] Eden, C. (1989) Midwives knowledge and management of PND, *Aust J Adv Nurs,* 7(1),pp. 35-42.

[65] Eberhard-Gran, M., Eskild, A., Tambs, K., Opjordsmoen, S. and Samuelsen, S. (2001) "Review of validation studies of the EPDS", *Acta Psychiatr Scand,* 104(4), pp. 243-9.

[66] Elliot, S., Ross-Davie, M., Sarkar, A., Green, C. (2007). Detection and intial assesment of mental disorder:the midwifes role. *The british journal of midwifery,* 15(12), pp.759-764.

[67] England, S. J., Ballard, C. and George, S. (1994) "Chronicity in postnatal depression", *European journal of psychiatry,* 8(2), pp. 93-6.

[68] Fraser, D.M. (1999). Women's Perceptions of Midwifery Care: A Longitudinal Study to Shape Curriculum Development. *Birth,* 26, pp.99-107.

[69] Freeman, M. P., Wright, R., Watchman, M., Wahl, R. A., Sisk, D. J., Fraleigh, L., Weibbrecht, J. M. (2005). Postpartum depression assessments at well-baby visits: Screening feasibility, prevalence, and risk factors. *Journal of Women's Health, 14*(10), pp. 929-935.

[70] Gatrell, C. (2005). *Hard labour: the sociology of parenthood.* New York: Open University Press.

[71] Glover, V., Liddle, P., Taylor, A., Adams, D., Landler, M. (1994) Mild hypomania (the highs) can be a feature of the first post-partum week: association with later depression. Br J psychiatry ,164(4),pp.517-521.

[72] Goldman, R. D. and Koren, G. (2003) "Mothersrisk update: Taking St. John's wort during pregnancy", *Canadian family physician,* 49(january), pp.29-30.

[73] Green, J.M., Murray, D. (1994). The use of EPDS in research to explore the relationship between antenatal and postnatal dysphoria. In: Cox, J., Holden, J, eds. *Perinatal psychiatry – use and misuse of the EPDS.* London: Royal College of psychiatrists.

[74] Gregoire, A. J., Kumar, R., Everitt, B., et al(1996)Transdermal oestrogen for treatment of severe postnatal depression. Lancet, 347,pp.930–933.

[75] Gutteridge, K. (2001) The Tamworth PND support group, *Midwifery digest,* 11(1), Suppl. 1, pp.17-9.

[76] Hagen, E. H. (1999). The functions of postpartum depression. *Evolution and Human Behavior, 20,*pp. 325-359.

[77] Hall, L. A., Kotch, J. B., Browne, D. and Rayens, M. K. (1996) "Self esteem as a mediator of the effects of stressors and social resources on depressive symptoms in postpartum mothers", *Nursing research,* 45(4), pp. 231-8.

[78] Hallsdorsdottir, S., Karlsdottir, A.I. (1996). Journeying through Labour and Delivery: Perceptions of Women who have Given Birth. *Midwifery,* 12, pp.48-61.

[79] Hanusa, B.H., Scholle, S.H., Haskett, R.F., Spadaro, K., Wisner, K.L.(2008) Screening for depression in the postpartum period: A comparison of three instruments. *J Women's Health,* 17,pp.585-596.

[80] Hearn, G., Iliff, A., Jones, I., Kirby, A., Ormiston, P., Parr, P., Rout, J. and Wardman, L. (1998) PND in the community, *Br J Gen Pract,* 48(428), pp.1064-6.

[81] Hewitt, C. E., Gilbody, S. M., Brealey, S., Paulden, M., Palmer, S., Mann, R. (2009) Methods to identify postnatal depression in primary care: an integrated evidence synthesis and value of information analysis. *Health Technol Assess,*13(36), pp.1-145, 147-230.

[82] Hickey, A. R., Boyce, P. M., Ellwood, D. and Morris-Yates, A. D. (1997) Early discharge and risk for PND, *Med J Aust,* 167(5), pp.244-7.

[83] Hirst, K.P., Moutier, C.Y.(2010) Postpartum major depression. *Am Fam Physician,*82(8),pp.926-33.

[84] Hillan, E. (1992). Issues of delivery in midwifery care. *JAN,* 17(3),pp.274-278.

[85] Hobfoll, S.E., Ritter, C., Lavin, J., Hulsizer, M.R., Cameron, R.P. (1995). Depression prevalence and incidence among inner-city pregnant and postpartum women. *J Consult Clin Psychol,* 63(3), pp.445-53.

[86] Hodnett, E. (1996) Nursing support of the labouring women, *JOGNN,* 25(3), pp.257-64.

[87] Holcomb, W.L.J., Stone, L.S., Lustman, P.J., et al., 1996. Screening for depression in pregnancy: characteristics of the Beck Depression Inventory. *Obstet Gynecol,* 88 (6), pp.1021- 1025.

[88] Holden, J. M., Sagovsky, R. and Cox, J. L. (1989) Counselling in a general practice setting: controlled study of health visitor intervention in treatment of PND, *BMJ*, 298(6668), pp.223-6.

[89] Holden, J. M. (1991). Postnatal depression: its nature, effects, and identification using the Edinburgh Postnatal Depression scale. *Birth, 18*(4),pp. 211-221.

[90] Holden, J. (1996) "The role of health visitors in postnatal depression", *International review of psychiatry*, 8(1), pp. 79-86.

[91] Huang, Y. and Mathers, N. (2001) PND - biological or cultural? A comparative study of postnatal women in United Kingdom and Taiwan, *JAN*, 33(3), pp.279-87.

[92] Isik, S.N., Bilgili, N. (2010). PND: midwives' and nurses' knowledge and practices. *Embase Erciyes Tip Dergisi*, 32(4), pp.265-74.

[93] Jardri, R., Maron, M., Pelta, J., Thomas, P., Codaccioni, X., Goudemand, M., Delion, P. (2010). Impact of midwives training on PND screening in the first week post delivery: a quality improvement report. *Midwifery*, 26, pp.622-9.

[94] Jones, I., Craddock, N., (2001). Familiality of the puerperal trigger in Bipolar Disorder: results of a family study. *American Journal of Psychiatry*,158,pp.913-917.

[95] Josefsson, A., Berg, G., Nordin, C. and Sydsjo, G. 2001. Prevalence of depressive symptoms in late pregnancy and postpartum. *Acta Obstetricia et Gynecologica Scandinavica*, 80(3), pp.251-255.

[96] Karuppaswamy, J. and Vlies, R. (2003) "The benefit of oestrogens and progestogens in postnatal depression", *Journal of obstetrics and gynaecology*, 23(4), pp. 341-6.

[97] Kitzinger, S. (1994). *Me, matere*. Ljubljana: Ganeš.

[98] Keng, S. L. (2005). Malaysian midwives' views on PND. *British journal of midwifery*, 13(2), pp.78-86.

[99] Kuipers, E. and Bebbington, P. (1997) *Living with mental illness – a book for relatives and friends*. 2nd rev ed. London: Souvenir press.

[100] Lavender, T. and Walkinshaw, A. (1998) "Can midwives reduce postpartum psychological morbidity?: a randomised trial", *Birth*, 25(4),pp. 215-9.

[101] Lawrie, T. A., Herxheimer, A. and Dalton, K. (2003) "Oestrogens and progestogens for preventing and treating postnatal depression", *The Cochrane library*, 4(2).

[102] Lee, D. T. S., Yip, A. S. K., Chiu, H. F. K. and Chung, T. K. H. (2000) "Screening for postnatal depression using the double –test strategy", *Psychosom Med*, 62(2), pp. 258-263.

[103] Letourneau NL, Fedick CB, Willms JD, Dennis CL, Hegadoren K, Stewart MJ. *Longitudinal study of postpartum depression, maternal-child relationships and children's behaviour to 8 years of age.* In: Devore D, editor. Parent-child relations: New research. New York: Nova Science; 2006. pp. 45–63.

[104] Lasa, L., Ayuso-Mateos, J. Vázquez-Barquero, J. L., et al *(2000)* The use of the Beck Depression Inventory to screen for depression in the general population: a preliminary analysis. *J Affect Disord*, 57,pp. 261 -265.

[105] Lavender, T. and Walkinshaw, A. (1998) Can midwives reduce postpartum psychological morbidity?: a randomised trial, *Birth*, 25(4), pp.215-9.

[106] Lazzare J. *The mother knot.* Durham: Duke University Press, 1997.

[107] Leathers, S. J., Kelley, M. A. and Richman, J. A. (1997) Postpartum depressive symptomatology in new mothers and fathers: parenting, work and support, *The journal of nervous and mental disease*, 185(3), pp.129-39.

[108] Lee, D. T. S., Chung, T. K. H. (2007). PND: an update. *Best Pract Res Clin Obstet Gynecol*, 21(2), pp.183-191.

[109] Leonard, L. G. (1998) Depression and anxiety disorders during multiple pregnancy and parenthood, *JOGNN*, 27(3), pp.329-37.

[110] Lepper, H. S., DiMatteo, M. R. and Tinsley, B. J. (1994) Postpartum depression: how much do obstetric nurses and obstetricians know?, *Birth*, 21(3), pp.149-54.

[111] Ličina, M. and Radovanovič, M. (1995) Poporodne duševne motnje s poudarkom na poporodni depresivni simptomatiki, *Medicinski razgledi*, 34(2), pp.245-63.

[112] Littlewood, J. and McHugh, N. (1997) *Maternal distress and PND: the mith of Madonna.* Houndsmills: MacMillan.

[113] LoCicero, A. K., Weiss, D. M. and Issokson, D. (1997) Postpartum depression: a proposal for prevention through an integrated care and support network, *Applied and preventive psychology*, 6(4), pp.169-78.

[114] MacArthur, C., Winter, H. R., Bick, D. E., Knowles, H., Lilford, R., Henderson, C., Lancashire, R. J., Braunholtz, D. A. and Gee, H. (2002) Effects of redesigned community postnatal care on women's health four months after birth: a cluster randomised controlled trial, *The Lancet*, 359(2), pp.378-85.

[115] Magill-Cuerden, J. (2000). Social support in pregnancy: midwives have a crucial role to play. *British journal of midwifery*, 14(3), p.156.

[116] Marks, M. N., Siddle, K., Warwick, C. (2003). Can we prevent PND? A randomised controlled trial to assess the effect of continuity of midwifery care on rates of PND in high risk women. *J Matern Fetal Neonatal Med*, 13 (2), pp.119-223.

[117] Martin, C. (1994). How do you Count Maternal Satisfaction? A User-commissioned Survey of Maternity Services. In: Roberts, H., ed. *Women's Health Counts*. London: Routledge, pp.6-20.

[118] Mauthner, N. S. (1999) Feeling low and feeling really bad about feeling low: women's experiences of motherhood and postpartum depression, *Canadian psychology*, 40(2), pp.143-61.

[119] McCarthy, A. (1998) PND: diagnosis and treatment. *Ir Med J*, 91 (2S),p. 44.

[120] McIntosh, J. (1993) Postpartum depression: women's help-seeking behaviour and perceptions of cause, *JAN*, 18(2), pp.178-84.

[121] McQueen A, Mander R. (2003). Tiredness and fatigue in the postnatal period. *JAN*, 42(5), pp.463-9.

[122] Meager, I. and Milgrom, J. (1996) "Group treatment for postpartum depression", *Australian and New Zeland journal of psychiatry*, 30(6), pp. 852-60.

[123] Miller, L. J. (2002). Postpartum depression. *Journal of the American Medical Association, 287*, pp. 762-765.

[124] Misri, S. , Kostaras, X. (*2002*). Benefits and risks to mother and infant of drug treatment for postnatal depression. *Drug Safety*, 25 (13),pp. 903-911.

[125] Mivšek AP, Hundley V, Kiger A. (2008). Slovenian midwives' and nurses' views on postnatal depression: an exploratory study. *Int Nurs Rev* 55(3), pp. 320-6.

[126] Mivšek, A.P. (2007). Slovenske ženske v pričakovnju: Sprijaznjene in pasivne ali aktivne in zadovoljne uporabnice obporodnega zdravstvenega varstva. In: Drglin Z, ed. *Rojstna mašinerija. Sodobne obporodne vednosti in prakse na Slovenskem.* Koper: Založba Annales, pp.73-103.

[127] Multidisciplinary working group for Grampian (2001). PND protocol for Grampian: redesign of PND pathway. Unpublished draft.

[128] Murray, D. (1996) Oestrogen and PND, *The Lancet*, 347(9006),pp. 918-9.
[129] Murray ,L., Cox, J.L. (1990) Screening for depression during pregnancy with the Edinburgh Depression Scale (EPDS). *J. Infant Reprod. Psychol*,8, pp. 99–107.
[130] Musek, J. (1995). *Ljubezen, družina, vrednote*. Ljubljana: EDUCY.
[131] Nastran Ule, M. (1993). *Psihologija vsakdanjega življenja*. Ljubljana: Znanstveno in publicistično središče.
[132] National Institute for Health and Clinical Excellence (2007) Intrapartum care. London: National Institute for Health and Clinical Excellence.
[133] Nicolson, P. (1998) *Post-natal depression; psychology, science and the transition to motherhood*. London: Routledge.
[134] Noh, P. (1987) Postpartum depression: extent, timing and persistence of emotional disorders following childbirth. *Br J Psychiatry*, 151, pp. 523-7.
[135] Nolan, M.L. (1997). Antenatal education- where next? *JAN*, 25, pp.1198-1204.
[136] Nonacs, R., Cohen, L.S. (2002). Depression during pregnancy: diagnosis and treatment options. *J Clin Psychiatry*, 63 Suppl 17, pp.24–30.
[137] Nott, P.N., Franklin, M. ,Armitage, C., Gelder, M.G. (1976) Hormonal changes and mood in the puerperium, *Br. J. Psychiatry*, 128, pp. 379–383.
[138] Oakley, A. (1993). *Becoming a Mother*. Oxford: Martin Robertson.
[139] O'Hara, M.W., Neunaber, D.J., Zekoski, E.M. (1984) Prospective study of postpartum depression: prevalence, course, predictive factors. *J Abnorn Psychol*, 93, pp. 158–181.
[140] O'Hara, M. W., Zekoski, E. M., Phillips, L. H. and Wright, E. J. (1990) Controlled prospective study of postpartum mood disorders: comparison of childbearing and non-childbearing women, *J Abnorm Psychol*, 99(1), pp.3-15.
[141] O'Hara, M. W., Stuart, S., Gorman, L. L. and Wenzel, A. (2000) "Efficacy of interpersonal psychotherapy for postpartum depression", *Archives of general Psychiatry*, 57(11), pp. 1039-45.
[142] Okano, T., Nagata, S., Hasegawa, M., Nomura, J. and Kumar, J. (1998) Effectiveness of antenatal education about PND: a comparison of two groups of Japanese mothers, *Journal of mental health*, 7(2),pp. 191-8.
[143] Oren, D. A., Wisner, K. L., Spinelli, M., Epperson, C. N., Peindl, K. S., Terman, S. U., J. and Terman, M. (2002) "An open trial of morning light therapy for treatment of antepartum depression", *The American journal of psychiatry*, 159(4), pp. 666-9.
[144] Pairman, S., Tracy, J., Pincombe, C., Thorogood, F., eds. (2006). *Midwifery: Preparation for Practice*. Sydney: Elsevier.
[145] Pajntar, M. (1994) Poporodne depresije. In: Pajntar, M. and Novak Antolič, Ž. (1994) *Nosečnost in vodenje poroda*. Ljubljana: Cankarjeva založba.
[146] Paykel, E. S. (2000) Classification of postpartum disorders in ICD 10 and DSM IV, The Marce society newsletter, p.9.
[147] Pitt, B. (1968) Atypical depression following childbirth, *Br J Psychiatry*, 114(516), pp.1325-35.
[148] Petrick, J. (1984) "Postpartum depression: identification of high risk mothers", *Journal of obstetrics, gynaecology and neonatal nursing*, 13(1), pp. 37-40.
[149] Pinkofsky, H. B. (2000) "Effects of antipsychotics on the unborn child: what is known and how should this influence prescribing", *Paediatric drugs*, 2(2), pp.83-90.
[150] Polit, D. F., Beck C.T., C.T. and Hungler, B. P. (2001). *Essentials of nursing research-methods, appraisal and utilization*. 5th ed. Philadelphia: Lippincot.

[151] Posner, N.A., Unterman, R.R., Williams, K.N., Williams, G.H. (1997), Screening for postpartum depression. An antepartum questionnaire. *J Reprod Med,* 42. pp. 207–215.

[152] Postnatal and Antenatal Depression Association (PaNDa). Information package. Melbourne: PaNDa, 2001.

[153] Powell-Kennedy, H., Beck C.T., C.T. and Driscoll, J. W. (2002) A light in the fog: caring for women with postpartum depression, *J Midwifery Women Health,* 47(5), pp.318-30.

[154] Prendergast, J. and Austin, M. P. (2001) Early childhood nurse delivered cognitive behavioural counselling for post-natal depression, *Australian psychiatry,* 9(3), pp.255-9.

[155] Price, S., ed. (2007). *Mental health in pregnancy and childbirth.* Edinburgh: Elsevier Churchill Livingstone.

[156] Priest, S. R., Henderson, J., Evans, S. F. and Hagan, R. (2003) "Stress debriefing after childbirth: a randomised controlled trial", *Medical journal of Australia,* 178(11), pp.542-5.

[157] Raphael-Leff, J. (1991) Postnatal psychological complications, In: *Psychological processes of childbearing.* London: Chapman and Hall.

[158] Renkert, S., Nutbeam, D. (2001) Opportunities to improve maternal health literacy through antenatal education: an exploratory study. *Health Promot Int,* 16(4), pp.381-388.

[159] Rezabek, R. (2000) Online focus groups: electronic discussions for research. Forum: Qualitative social research. URL http://www.qualitative-research.net/fgs - last accessed February 2011.

[160] Richards, J.P. (1990) Postnatal depression: A review of recent literature. *Br J Gen Pract,* 40: pp. 472-476

[161] Righetti-Veltema, M., Bousquet, A., & Manzano, J. (2003). Impact of postpartum depression symptoms on mother and her 18-month old infant. *European Child and Adolescent Psychiatry,* 12. pp. 75-83.

[162] Ritter, C., Hobfoll, S. E., Lavin, J., Cameron, R. P., & Hulsizer, M. R. (2000). Stress, psychosocial resources, and depressive symptomatology during pregnancy in low-income inner-city women. *Health Psychology, 19,* pp. 576-585.

[163] Robertson, E., Grace, S., Wallington, T., Stewart, D.E. (2004). Antenatal risk factors for postpartum depression: A synthesis of recent literature. *Gen Hosp Psychiatry,* 26, pp. 289-295.

[164] Robling, S. A., Paykel, E. S., Dunn, V. J., Abbott, R., Katona, C. (2000). Long-term outcome of severe puerperal psychiatric illness: a 23 year follow-up study. *Psychological Medicine, 30,* pp. 1263-1271.

[165] Ross-Davie, M., Elliot, S., Sarkar, A., Green, L. (2006). A public health role in perinatal mental health: are midwives ready? *British journal of midwifery,* 14(6), pp. 330-4.

[166] Sampselle, C., Seng, J., Yeo, S., Killion, C. and Oakley, D. (1999) "Physical activity and postpartum well-being", *Journal of obstetrics gynaecologic and neonatal nursing,* 28(1), pp.41-9.

[167] Sand Sobey, W. (2002) Barriers to postpartum depression prevention and treatment: a policy analysis, *J Midwifery Women Health,* 47(5),pp. 331-6.

[168] Scott, A.I.F. (1997) Treatment of postnatal depression. *BMJ,* 315(7100), p. 122.

[169] Schnyer, R. N., Manber, R. and Fitzcharles, A. J. (2003) "Acupuncture treatment for depression during pregnancy: conceptual framework and two case reports", *Alternative health practitioner,* 8(1), pp. 40-53.

[170] Shu-Shya, H., Yu-Ying, F. (2003) Effectiveness of informational support in reducing the severity of postnatal depression in Taiwan. Journal of Advanced Nursing, 42(1), pp.30-36.

[171] SIGN (2002). PND and puerperal psychosis: a national clinical guideline 60. Edinburgh: SIGN.

[172] Simkin, P. (1996) The experience of maternity in a woman's life, *JOGNN*, 25(3), pp.247-52.

[173] Sit, D.K., Wisner, K.L.(2009) Identification of postpartum depression. *Clin Obstet Gynecol* ,52,pp.456–468.

[174] Skočir AP. (2004) Midwives' and nurses' perception of their role in the prevention, detection and management of postnatal depression. MSc thesis. Aberdeen: University of Aberdeen

[175] Skočir AP, Hundley V. (2006) Are Slovenian midwives and nurses ready to take on a greater role in caring for women with postnatal depression. Midwifery 22(1): 40-55.

[176] Slade, P., MacPherson, S.A., Hume, A., Maresh, M. (1993). Expectations, Experiences and Satisfaction with Labour. *Br J Clin Psychol*, 32,pp. 469-483.

[177] Small, R., Lumley, J., Donohue, L. S., Potter, A. and Waldenstrom, U. (2000) "Randomized controlled trial of midwife-led debriefing to reduce maternal depression after operative childbirth", *British medical journal*, 321(7268), pp. 1043-7.

[178] Sobey, W. S. (2002). Barriers to postpartum depression prevention and treatment: A policy analysis. *Journal of Midwifery and Women's Health, 47*, pp. 331-336.

[179] Spinelli MG. (1997) Interpersonal psychotherapy for depressed antepartum women: a pilot study. *Am J Psychiatry*,154(7), pp.1028-30.

[180] Stamp, G., Williams, A. and Crowther, C. (1996) "Predicting postnatal depression among pregnant women", *Birth*, 23(4), pp. 218-23.

[181] Stewart, C. and Henshaw, C. (2002) Midwives and perinatal health, *British journal of midwifery*, 10(2), pp.117-21.

[182] Stocky A. ,Lynch J. (2000). Acute psychiatric disturbance in pregnancy and puerperium. *Balliere's Clin. Obstet. Gynaecol.* 14 ,pp. 73–87

[183] Stowe, Z. N. and Nemeroff, C. B. (1995) Women at risk for postpartum onset major depression, *Am J Obstet Gynecol*, 173(2), pp.639-45.

[184] Stowe, Z.N., Levy, S.T., Nemeroff, C.B. (1997)Comorbid postpartum depression and bereavement: a complicated case. *Am J Psychiatry*, 154(3), pp. 418-422.

[185] Stuchbery, M., Matthey, S. and Barnett, B. (1998) PND and social support in Vietnamese, Arabic and anglo-celtic mothers, *Soc Psychiatry Psychiatr Epidemiol*, 33(10), pp.483-90.

[186] Sullivan, A., Raynor, M., Oates, M. (2003). Why mothers die: perinatal mental health. *British journal of midwifery*, 11, pp.310-2.

[187] Svensson, J., Barclay, L., Cooke, M. (2009). Randomised-controlled trial of two antenatal education programmes. *Midwifery*, 25, pp.114-124.

[188] Taylor, A. (1997) Out of the blues, *Nursing times*, 93(45), p.18.

[189] Thompson, J. F., Roberts, C. L., Currie, M. J. and Ellwood, D. A. (2000) Early discharge and PND - a prospective cohort study, *Med J Aust*, 172(11), pp.532-5.

[190] Thoits PA. Self-labelling processes in mental illness: the role of emotional deviance. *Am J Socio*, 91,pp.221–49.

[191] Thurtle, V. (1995) Post-natal depression: the relevance of sociological approaches, *JAN*, 22(3), pp.416-24.

[192] Tiran, D. (2002) "Bailliere's midwives' dictionary". 9th ed. Edinburgh: Elsevier science limited.

[193] Tully, C., Watson, C. and Abrams, A. (1998) PND: training health visitors to use the EPDS, *Community practitioner*, 71(6), pp.213-5.

[194] Yaffe, M. (2002). Midwife-led postnatal care with individual needs assessment may reduce PND. *Evidence-based healthcare*, 6(3), pp.117-8.

[195] Yonkers, K. A., Ramin, S. M., Rush, A. J., Navarrete, C. A., Carmody, T., March, D., Hartwell, S. F., Leveno, K. J. (2001). Onset and persistence of postpartum depression in an inner-city maternal health clinic system. *Am J Psychiatry*, 158(11), pp.1856-63.

[196] Ugarriza D.D. (2000), Postpartum depressed women's explanation of depression, *J Nurs Scholarsh*, 34 (3), pp. 227–233.

[197] Vieira, T. (2002) When happiness becomes grief: screening tools for postpartum depression. *AWHONN*,6(6), pp. 506-513.

[198] Waldenström, U. (1999). Experience of Labour and Birth in 1111 Women. *J Psychosom Res*, 47, pp.471-482 .

[199] Waldenström, U., Brown, S., McLachlan, H., Forster, D., Brennecke, S. (2000). Does Team Midwife Care Increase Satisfaction with Antenatal, Intrapartum, and Postpartum Care? A Randomised Controlled Trial. *Birth*, 27(3), pp.156-167.

[200] Watson, J. P., Elliott, S. A., Rugg, A. J., Brough, D. I. (1984). Psychiatric disorder in pregnancy and the first postnatal year. *Br J Psychiatry*, 144, pp.453–462.

[201] Webster, J., Pritchard, M. A., Creedy, D. and East, C. (2003) A simplified predictive index for the detection of women with risk for PND, *Birth*, 30(2), pp.101-8.

[202] Welburn, V. (1980) *PND*. Manchester University press: Trowbridge and Esher.

[203] Wheatley, S.L., Brugha, T.S., Shapiro, D.A. (2003). PATA PATA: midwives' experiences of facilitating a psychological intervention to identify and treat mild to moderate antenatal and PND. *MIDIRS: Midwifery digest* 13(4), pp.523-30.

[204] Whiffen, V. and Gotlib, I. H. (1993) Comparison of postpartum and nonpostpartum depression: clinical presentation, psychiatric history and psychosocial functioning, *J Consult Clin Psychol*, 61(3), pp.485-94.

[205] Whiffen, V. E., Johnson, S. M. (1998). An attachment theory framework for the treatment of childbearing depression.

[206] Wisner, K.L., Parry, B.L., Piontek, C. M.(2002). Postpartum depression.*N Engl J Med*,347,pp.194-199.

[207] WHO. (1991). *Evaluation of methods for the treatment of mental disorders – report of WHO scientific group on the treatment of psychiatric disorders*. Geneva: WHO.

[208] WHO. (1992) *International Classification of diseases – ICD 10*. Geneva: WHO library cataloguing in publication data.

[209] Wisner, K. L., Parry, B, L. and Piontek, C. M. (2002) Clinical practice - postpartum depression, *N Engl J Med*, 347 (3), pp.194-9.

[210] Wolman, W. L., Chalmers, B., Hofmeyr, G. J. and Nikodem, V. C. (1994a) Postpartum depression and companionship in the clinical birth environment: a randomised controlled trial, *Am J Obstet Gynecol*, 168(5), pp.1388-93.

[211] Wood, A. F., Thomas, S, P., Drppleman, P. G. and Meighan, M. (1997) The downward spiral of postpartum depression, *MCN*, 22(6), pp.308-16.

[212] Wylie, L., Hollins Martin, C. J., Martin, C. R., Rankin, J. (2011). The enigma of PND: an update. *J Psychiatr Ment Health Nurs*, 18, pp.48-58.

[213] Zavrl, N. (1999). *Očetovanje in otroštvo*. Ljubljana: Znanstveno in publicistično središče.

[214] Zlotnick, C., Johnson, S. L., Miller, W., Pearlstein, T. and Howard, M. (2001) "Postpartum depression in women receiving public assistance: pilot study of an interpersonal therapy oriented group intervention", *American journal of psychiatry*, 158(4),pp. 638-40.

13

Psychoeducation for Bipolar Mood Disorder

Mohammad Reza Fayyazi Bordbar and Farhad Faridhosseini
Mashhad University of Medical Sciences
Psychiatry and Behavioral Sciences Research Center
Iran

1. Introduction

There has been significant improvement in available treatments for Bipolar Mood Disorder (BMD) during the past few years. However, this disorder still causes difficulties for the patients, their families, and the society (Kleinman et al., 2003). Overall studies have demonstrated that this disease affects patients' entire family and may weaken its strengths and adaptive abilities (Barry, 2001). Unavoidable poor outcome of this illness and their recurrent nature have put BMD among the most debilitating disorders, and has lead researchers to pay more attention to its treatment (Miklowitz et al., 2004). Based on the results of different studies, this disorder stands at the 6th or 7th place among other debilitating disorders worldwide (Calabrese, et al. 2003; Chisholm, et al. 2005; Simon et al., 2006). According to World Health Organization (WHO), mood disorders are one of the most important worldwide health issues of the 21st century (Boyd, 2007).

Chronic and recurrent nature of BMD impacts several aspects of patients' lives, from their interpersonal relationships to the quality of their work. As an example, divorce rate is reported to be higher in this patient population (Ghoreishizadeh et al., 2008). Additionally, this disorder is costly to the society and its average yearly cost is 45 billion dollars in the United States (Rouget & Aubry, 2007). This yearly cost is reported to be 2055 million pounds in the UK, 86% of which is indirect costs (Gupta & Guest, 2002). Pharmacological treatment alone has not been successful for complete improvement in the symptoms (Bassili, 2009). Complaince is an issue in the treatment of BMD due to patients' lack of understanding of the disease process and its consequences, as well as the disorder itself in some cases disrupting patients' judgment. This along with the nature of the disorder, may lead to multiple relapses (Keck et al., 1997; Parikh et al., 1997). The relapse rate has been reported to be 30%-40% in one year, 60% in two years and 70% in five years (Otto et al., 2003; Rouget & Aubry, 2007).

Furthermore, quality of life in bipolar patients is impaired even during remission, and the suicide rates are higher. Suicide rate is reported to be as high as 20%-30% in bipolar patients (Bellivier et al., 2011; Valtonen et al., 2005). Therefore, maintenance pharmacological treatmentis highly recommended for these patients, as well as other forms of treatments (Ghaemi et al., 2004). Adjunct treatments along with pharmacotherapy are found to be effective in improvement of prognosis and helping patients in better adjustment with the disorder (Beynon et al., 2008).

2. Nonpharmacological treatments for BMD

Various non-pharmacological therapeutic approaches have been suggested for treatment of patients with BMD. One of these methods is Psychoeducation (PE). There are various programs focusing on education which have been found to be helpful in reducing relapse rate, burden of illness as well as improving symptoms and patients' function. Some of these programs are long term and some are short in duration. They all cover topics such as patients' awareness of the illness (which is recurrent and chronic), knowing more about the triggers of a new episodes, necessity of treatment and different treatment options, confronting mood swings, attaining better control over symptoms, awareness about the side effects of medications, effects of coffee and nicotin intake and importance of regular habits and life style (Colom & Vieta, 2006; Dashtbozorgi et al., 2009; Fayyazi et al., 2009; Sadock et al., 2009; Wheeler, 2010a). Psychoeducation may be done in a variety of ways, such as individual treatment, group therapy, multi-family therapy, and patient - family dyadic therapy (Wheeler, 2010a). These methods will be discussed in details below.

Another reportedly successful treatment method is cognitive behavioral therapy (CBT). Originally CBT was used for treatment of unipolar depression, but it was later used for BMD patients and many therapeutic programs and techniques were introduced to treat these patients. These programs mainly focus on relaxation strategies, cognitive restructuring, psychoeducation, problem solving strategies and therapeutic interventions to treat patients with co-morbidities (Colom & Vieta, 2006; Miklowitz et al., 2007; Wheeler, 2010b). Some studies have demonstrated efficacy in relapse prevention. However well designed studies with longer follow-up are needed to assess effectiveness of CBT in improvement disease outcome. What seems to be the main core of therapeutic elements in CBT is psychoeducation (Scott et al., 2006).

Interpersonal therapy (IT) is another type of non-pharmacological treatment for BMD which focuses on interpersonal aspects of patients. One type of IT is called interpersonal and social rhythm therapy (IPSRT) which is designed to help BMD patients with their daily activities such as exercise, eating and sleeping, as well as their social relationships (Colom & Vieta, 2006; Miklowitz et al., 2007; Sadock et al., 2009). This type of treatment includes four steps. First, patient's history is taken. Then, general information regarding bipolar disorder and its effects on interpersonal aspects of patient's life is discussed. This is followed by therapist's focus on special techniques for mood stability and gaining control over symptoms. Once patient's symptoms are under control, they will be helped to recognize important factors affecting mood stability (Stresing, 2010).

These treatments can be used in group therapies. Also, researchers have found families' involvement to be important and effective, especially in psychoeducational therapies, since bipolar mood disorders affect both patients and their families. Research done by Dore and colleagues, focuses on the influence of relatives and family members on patient's occupation, financial and residential status, marital status, child care, social and even recreational activities. (Dore & Romans, 2001) It was found that burden of the disease on family members may affect outcome of the disease (Perlick et al., 2008). Several long term and short term programs have been designed for family members of patients. According to Fayyazi and colleagues, one session of educational intervention at the time of patients' discharge from the hospital can increase pharmacological compliance, as well as decreasing one year relapse rate of the disease (Fayyazi et al., 2009). Additionally, two-session educational therapy for families has shown to help patients (Miklowitz et al., 2000) although most therapeutic educational programs designed for family members take longer duration (Colom & Vieta, 2006).

3. Psychoeducation

In 1998 Goldman defined psychoeducation as "education or training of a person with a psychiatric disorder in subjects that serve the goals of treatment and rehabilitation". According to Goldman, the goal of psychological education is to enhance patients' acceptance of their disease, as well as their participation in therapy and improvement of coping mechanisms when facing problems caused by their disease (Goldman, 1998 as cited in Atri & Sharma, 2007). In 2006, National Institute for Health and Clinical Excellence (NICE) introduced the following definition for psychoeducation (PE):

> *"Any structured group or individual program that addresses an illness from a multi dimensional viewpoint including familial, social, biological and pharmacological perspectives, as well as providing service users and carers with information support and management strategies"* (The National Institute for Health and Clinical Excellence [NICE], 2006 as cited in Getachew et al., 2009).

Based on this definition, psychological education for more than one session seems to increase patients' knowledge of their disease. These sessions are structureed and may be done in groups. Many programs suggested by research groups are designed in group format (Colom & Vieta, 2006). The goal is for patients to learn about their diseases and increase awareness, by learning self management techniques and adaptation to psychosocial problems caused bytheir illness. In this method, the focus is on improving patients' strengths, resources and adaptation skills (Beynon et al., 2008).

Psychoeducation is not only applied to use for patients with bipolar and other psychiatric disorders, but also for chronic medical diseases such as diabetes and ischemic heart disease in which there is a significant necessity for changing lifestyle and patients' habits in order to control disease and attain better outcome (Colom & Vieta, 2009). Experimental evidence has demonstrated that psychoeducation could be a part of recovery for patients suffering from psychiatric disorders such as schizophrenia and bipolar disorder (Getachew et al., 2009). In 2009, Colom and colleagues found that six- month group psychoeducation for euthymic bipolar patients has long lasting prophylactic effects after 5 years of follow up (Colom et al., 2009). The effects of this treatment on recovery are independent of its influence on compliance with pharmacological treatment (Colom et al, 2003; Atri & Sharma, 2007).

The core elements leading to the therapeutic effects of these structured psychoeducation sessions for patients and their families are subject to further exploration. These factors may include the following:

1. Level of expressed emotions and hostility, as well as extreme exertion of control by the families may lead to relapse (Lefley, 2009). Therefore, educational sessions for family members can be effective in providing support of the patients by their family members (Miklowitz et al., 2003; Miklowitz, 2008a).
2. Participation in group psychoeducational sessions may help patients to experience less social isolation. Patients may have a perception that they are the only one suffering

from the disease and therefore, have no place in the society. Meeting others with similar concerns may lead to experiencing stigmatization and isolation in these patients. (Colom & Vieta 2006; Colom & Vieta 2009)

3. Patients' continuous participation provides them with more opportunity to stay in contact with their therapy team. Hence, they spend more in treatments which leads to better adherences to the medication treatment, and lowering of the relapse risk (Colom & Vieta 2006; Colom & Vieta 2009).
4. These sessions can also encourage patients to be more actively involved in their own treatment and experience better control over the situation (Colom & Vieta 2006; Colom & Vieta 2009).
5. It is important for patients to accept their therapists and approve the quality of care they received (Howard et al., 2003). Their participation in structured sessions may result in increased satisfaction about therapy and alliance with their therapists because of improvement in their perception of better quality of care (Colom & Vieta 2006; Colom & Vieta 2009).
6. Participating patients will interact with one another, which also helps them learn from each other and exchange their experiences, creating a sense of cohesion among them as well as acquiring new coping mechanisms. This may function as a new support network for both patients and their families (Colom & Vieta 2006; Colom & Vieta 2009).
7. Teaching stress management methods to patients and their families could help them in dealing with their illness and reducing interepisodic symptoms and improving individual's mental health (Miklowitz et al., 2003).
8. Changing the behaviors and life style may also help patients. Gaining healthier habits and quitting destructive ones, such as drug use, may lead to decrease in manic relapses by reducing precipitators of episodes. In 1980, Green and colleagues explained these sessions as, "Any combination of learning experiences designed to facilitate voluntary adaptation of behavior conductive to health." (Green et al., 1980 as cited in Atri & Sharma, 2007).
9. Participation of family members in these sessions may inform caregivers of the early warning signs of relapse, and therefore, encouraging patients to seek treatment in the right time (Miklowitz et al., 2003).

4. Psychoeducation for bipolar mood disorder

With presence of lithium in 1950s, hopes are increased for controlling and curing bipolar disorder or manic-depressive psychosis. In controlled clinical trials, the efficacy of this medicine as well as other mood stabilizers like valproate, carbamazepine, olanzepine, lamotrigine, etc. was proved. They are proved effective in controlling mood symptoms in acute episodes of the disorder plus in maintenance therapy to prevent further relapse of episodes (Vieta et al, 2011). Many of these medications have been medically proved to be effective based on evidences and they belong to level A of evidence which is to say there are many randomized clinical trials (RCT) studies proving their efficacy (Geddes & Briess, 2006). Yet, these studies are criticized regarding their external validity as to whether they are as effective in everyday clinical setting to help prevent relapse of episodes or the interfering factors reduce their efficacy in daily clinical practice (Simon, 2006; Ketter, 2008). It seems that the effectiveness of treatments is not as high as the efficacy the studies report (Amini et al., 2010). although medical treatments has efficacy on symptom relief, but not completely effective in reducing disease burden and patients are not able to regain their functions (Cakir & Ozerdem, 2010).In RCT studies, volunteering patients are of different characteristics from those who are encountered in daily clinical setting (McComb, 2011). It appears that next to RCT studies and clinical trials, naturalistic studies like cohort are needed to measure the medication's effectiveness in the daily clinical practice. One of the effective elements in everyday clinical trials is the extent of treatment compliance on behalf of the bipolar disorders. Treatment compliance is one of the prognostic predictors in bipolar disorder (Velligan, 2010). The influencing elements in low compliance are alcohol and drug abuse consumption, severity of symptoms, young age, negative attitude to treatment, insufficient information about the problem and inaccessibility of medical services (Sajatovic, 2011).

Sure enough, success in curing the disorder is bound to factors other than the efficacy of mood stabilizers. Non-pharmacological strategies in order to increase treatment compliance on the part of the patient and better psychosocial compatibility are also needed (Vieta, 2005; Velligan, 2010). These pressing needs were the reason why lithium clinics were established during the 1970s in Europe and America to discover the effect of lithium in maintenance therapy of bipolar disorder. Drug prescription and medical monitoring were not the only things done in these clinics as patients were also receiving psychoeducation, rehabilitation, cognitive therapy, and job counselling and systematic care was performed for the patients (Osher et al., 2010).

In the early 1990s, Peet and Harvey presented their empirical studies of the effect of psychoeducation regarding bipolar disorder. They proved that presenting the patients in the experimental group with a video tape and a handout can significantly increase their knowledge and their attitude toward the treatment compared to the patients in control group (Peet & Harvey, 1991). They demonstrated that increased knowledge and positive attitude toward treatment improved compliance (improvement was defined as the number of forgotten taking drugs and the level of lithium in erythrocytes) (Peet & Harvey, 1991). Prior to implementing psychoeducation as an independent realm in cognitive therapy and psychological interventions for patients diagnosed with bipolar disorder, the patients had received supportive group therapy in lithium clinics. In the 1980s, a number of studies showed the effect of group therapy on increasing the understanding of the patient about the problem, lowering the stigma and reducing the times getting hospitalized. In Van Gent's study, patients reported increased self confidence and problem solving ability (Van Gent et al., 1988). In Volkmar's research, 47 sessions of group therapy and a two-year follow up resulted in a reduction in getting hospitalized (Volkmar et al., 1981). Still, the number of studies done in the 1980s was limited and they were mostly methodologically flawed with no control groups or very small research populations.

Little by little, the format of the groups has been driven closer to groups with more specific targets and structures; for instance, in 1986, Kripke and Robinson held group therapies to improve problem solving and measured the level of lithium as an indicator of patients' compliance. The outcome factors included the relapse rate, and the level of social compatibility which were considered as signs of recovery (Kripke & Robinson, 1986 as cited in Colom & Vieta, 2006). Creating homogenous groups aiming at improving self management skills of the patients diagnosed with the disorder was the way that replaced the classic supportive group therapy in the early 90s. In 1993, Pollack suggested patients' group therapy getting started from the very first moment of getting hospitalized, the groups' being homogenous and the focus being on self-assistance and self management skills (Pollack, 1993; Pollack, 1995). It seems that, when paired with psychoedcation, group therapy approach can result patients to obtain benefits of both interventions to the point that they are able to manage their disorder, improve their interpersonal relations, alter their attitude toward the disorder and manage their sense of stigma (Colom & Vieta, 2006). In Garves's research on outpatients in 1993, He reported an increase in treatment adherence, a decrease in denying the problem, and a higher control over internal and external stressors (Garves, 1993).

4.1 Systematic care models

In the 1990s, planning of structured programs for patients' management which include group therapies based on psychoeducation was aimed by a number of research groups, some of which are introduced below.

4.1.1 Life goals program

Life Goals Program (LGP) is a manual based structured group therapy that is focused on systematic education and problem solving skills so that patients become able to manage the illness (Sajatovic et al., 2009). It is found that bipolar patients, who took part in this program, became more knowledgeable about the disorder and had better adherence (Bauer & McBride, 2003). This program, which was introduced by Bauer & MacBride in 1998, had two major goals in two phases: "to improve patient participation in medical model treatment (phase 1) and assist patients in meeting functional status goals (phase 2) "(Bauer et al., 1998). The first study in this field was carried out as an open trial on 29 patients (Bauer et al., 1998). In the following randomized clinical trials, the program proved to be effective on the disorder's outcome, patients' function and quality of life (Bauer et al., 2006a, 2006b). In these studies LGP was performed on the patients as one of the components of a systematic care model for patients (so called collaborative care model [CCM]). Although Sajatovic used LGP as stand-alone intervention in a randomized controlled trial on 164 outpatients and she found no effect on treatment attitude during a one year of follow up; moreover, the depressed patients were less benefited from this intervention (Sajatovic et al., 2009).

LGP consists of two phases. In the first phase, psychoeducation is held for the patients in 6 interactive weekly sessions. The groups are of 6-8 members. During the first phase, the importance of good compliance, the effect of non-adherence on the course of the disorder as well as the effect of maladaptive behaviours on the symptoms' relapse is discussed. Also, participants try to identify their own symptom profiles, early warning signs and triggers of previous relapses. In the second phase of treatment, which is optional, monthly group sessions are held in an unstructured format which aims on goal setting and problem solving (Bauer & McBride, 2006).

4.1.2 Systematic care program

It was designed by Greg Simon and his colleagues as a systematic disease management program which consists of a number of components: 1. Evaluating and programming care, 2. Structured monthly phone calls, 3. Providing feedback for the treatment team, 4.

Structured group psychoeducation on the format of LGP, discussed in the previous chapter 5. Providing supportive interventions, education and care coordination. Depending on needs, the program can last for 2 years and nurses and case managers in addition to psychiatrists are involved. In a randomized study, this program was effective in reducing relapses of manic episodes, but not for depressive episodes (Simon, 2005).

4.2 Psychoeducation without systematic care

In the late 1990s and the first decade of the 21st century controlled trials started to prove the efficacy of psychoeducation on course and prognosis of bipolar mood disorder. This section reviews those trials which have investigated efficacy of individual and group psychoeducation without a systematic care model.

4.2.1 Individual psychoeducation

In 1998, Perry designed the first (and up to now the only) randomized clinical trial to investigate the efficacy of individual psychoeducation for bipolar patients. 69 patients were randomly allocated to two drug-receiving groups with routine care and drug-treatment plus 7-12 psychoeducational sessions. The sessions were aimed to help patients diagnose the early warning signs and know how to manage these signs. During a 18 month follow up, a decrease of 30% in relapses of manic (but not depressive) episodes were reported (Perry et al., 1999).

4.2.2 Barcelona bipolar disorder project

Colom, Vieta and their colleagues organized the structured 90-minute weekly group psychoeducational sessions, held in a 6 month period (21 sessions). The sessions are interactive and are facilitated by a psychologist under the supervision of a psychiatrist. Colom & Vieta have suggested 9-12 patients for each group but it can be raised to 15, considering the dropout of participants (Colom & Vieta, 2006). Patients in the euthymic phase are eligible to take part in sessions. The first controlled trial by the research group got published in 2003 as a double blind randomized clinical trial in which 120 bipolar patients in euthymic phase were randomly allocated to two groups of 21 sessions of structural psychoeducation or unstructured supportive group therapy. The psychoeducation group experienced fewer relapses both regarding manic and depressive episodes at the end of 2-year follow up (in previous trials psychoeducation was more effective on preventing relapses of mania and mixed episodes rather than depressive episodes) (Colom et al, 2003). 99 patients out of the 120 who received group psychoeducation in previous study completed the whole 5 year-follow up. The number of relapses, time to the first relapse, days spent acutely ill, days of hospitalization and the number of being hospitalized all significantly decreased compared to the control group. These findings show continuation of prophylactic effect of the sessions on patients (Colom et al., 2009a). In post hoc subanalysis, prophylactic effect of psychoeducation sessions in subgroup of patients with Bipolar II disorders was also demonstrated (Colom et al., 2009b). The surprising fact of the first Barcelona study is that the number of drop-outs in the group treated with structural psychoeducation was more than the group receiving unstructured supportive group therapy (11.6% vs. 26.6%) while the level of lithium in patients during the 2-year follow up was higher among those being treated with psychoeducation (Miklowitz, 2008a).

4.2.3 The content of psychoeducation program for the patient

Based on the patients' needs and clinical experiences of skilful clinicians, most of patient-oriented psychoeducational packages are focusing on 5 important components in order to help patients cope effectively with their illness. The difference among various programs often lies in the number of sessions allocated to each component (Colom & Vieta, 2009):

1. **Illness awareness.** It aims to increase patient's insight, illness acceptance and decrease the stigma and humiliation felt by patients. In this phase, the patient is presented with a biological model of bipolar disorder as a brain disease (Colom & Vieta, 2006).
2. **Adherence enhancement.** Considering the fact that the most important cause of relapse in bipolar patients is non-adherence (Guscott & Taylor, 1994; Silverstone et al., 1998) and common reasons of non-adherence are incorrect information, fear of side effects and dependency, and lack of information about the illness and the necessity of treatment (Sajatovic et al., 2011), adherence enhancement is one of the main points of psychoeducation sessions.
   Many patients are not capable to follow the correct instruction of taking medicine and this has been pointed out as the main cause of non-adherence (Sajatovic et al., 2011); thus it is necessary that patient is kept informed about different kinds of drugs for bipolar disorder, the ways of monitoring their effects, and the side effects of each, so that they would know how to manage side effects and also gain insight to disadvantage of discontinuation of medication.
3. **Detection of early warning signs.** Recognizing the early warning signs can help patients to prevent full-blown episode and subsequent probable hospitalization.
4. **Substance misuse avoidance.** Considering that substance and drug misuse can worsen the course of bipolar disorder.
5. **Lifestyle regularity.** Regular habits and stress management constitute the core components of Interpersonal and Social Rhythm Therapy (IPSRT), in which the patient is educated how to regulate their sleep hours and daily functions, eating and taking medication along with stress management, so that relapse triggers would reduce (Colom & Vieta, 2009).

The content of Barcelona's psychoeducational program inspired research groups in differing countries to run the program in ways to meet their needs and cultural obligations. For instance in Familiarizing Bipolar Disorder program in Poland the content was summarized into 8 group sessions. The program was reported to be effective based on an observational study (Maczka et al., 2010). Tabatabaie and Mottaqhipour in Iran, also, designed 8 psychoeducational sessions with a systematic care program of family education sessions along with occupational therapy. The content of the sessions are listed in Table 1. The research on the program's efficacy is going on.

1-The group members' introduction and familiarity with the task
2-Knowing the illness's signs
3-Etiology, the disorder's onset and course
4-Treatment and follow up
5-Relapse and its prevention
6-Adaptation to the illness
7-Stress management and problem solving
8-Recovery

Table 1. Session of suggested psychoeducational program by Tabatabai – Mottaqhipour in Iran

It is worth mentioning that in all these packages and in Barcelona Model, learning process is based on adult learning model in which people talk about their experiences and learn from other, so they can receive peer support from the group members. The ultimate goal of psychoeducational group to change maladaptive behaviour, because providing information, by itself, cannot lead to any behavioural change (Colom & Vieta, 2006).

4.2.4 The systematic illness management skills enhancement programme-bipolar disorder (SIMSEP-BD)

This program is a dyadic psychoeducational group (companion-patient dyad) where the patient takes part with companions. It takes twelve 90-minute weekly sessions. This program was introduced by D'Souza & Rich (D'Souza & Rich, 2002) based on which a randomized controlled trial research was published in 2010 by D'Souza and his colleagues. In this trial 58 patients who had recently experienced remission after an acute episode were randomly assigned to two treatment as usual (TAU) and SIMSEP-BD groups and followed up for 60 weeks. During the follow up, SIMSEP-BD group showed fewer relapses, higher treatment adherence and less manic signs (D'Souza et al., 2010).

4.3 Who would benefit psychoeducational group therapy the most?

Such as pharmacotherapy and other methods of psychotherapy, there are factors which can predict the patient's response to psychoeducational program. In Even and his colleagues' trial in 2007 it was observed that younger age, shorter illness period, higher levels of education, more information on Lithium and less external locus of control are predictive factors of more cooperation in psychoeducational sessions. Especially with logistic regression it is viewed that external locus of control is an independent factor in determining the amount of patient's participation in the treatment (Even, 2007).

In 2010, in a post study by Even, it is showed up that patients' participation in psychoeducational sessions results in a change in locus of control and this change pertains for 24 months beyond the sessions being over which can be one of the main effective components of psychoeducational on bipolar patients (Event, 2010).

In a trial by Cakin and colleagues in 2009, the existence of mixed episodes, drug treatment adherence and blood levels suitable for mood stabilizers and limited number of mood episodes are the predictive factors of cooperation and higher motivation of patients and more participation on sessions (Cakir, 2009).

5. Psychoeducation with the family

Bipolar disorder is a relapsing illness and even using the appropriated pharmacotherapy relapse chances are still there. These relapses can cause occupational problems, patient's distorted interpersonal relationships, and conflicts with relatives (Coryell et al., 1993; Gitlin et al., 1995; Perlis et al., 2006). Among bipolar patients a severe role impairment related to depression is observed in 90% of patients in such a way that half of them are incapable of working and carrying out their activities (Suppes et al, 2001). The existence of bipolar mood signs leads to an increase in excitement and financial aspects of family. It is expected for this illness to be the main reason of lost years of healthy life, after cardiovascular disorders (Murry & Lopez, 1996).

On the other hand, stress factors within the family life can exacerbate bipolar disorder. Expressed Emotion (EE) is one of the best predictors of mood swing (Miklowitz et al, 1998, 2000). Expressed Emotion is an scale for the extent of the family and caregivers emotional attitudes to patient's psychiatric disorder. When talking of high EE, it is meant to be lots of critical comments, hostility-arising situations, or showing emotional over-involvement (like over protection, exaggerated emotional responses, or inordinate self-sacrifice) (Miklowitz, 2008b; Morris et al, 2007). Numerous trials have reported intense relationship between EE and poor outcome in schizophrenic patients, mood bipolar and other psychiatric disorders. In a trial in 2010 it was observed that family's levels of conflict in bipolar patients was higher than families of normal people and also impairment in cohesion and adaptability was higher in such families and things with families with high EE is even worse (Sullivan & Miklowitz, 2010). In a meta-analysis, EE had a high effect size in predicting psychosis in schizophrenic patients. But the mean effect size of EE in mood disorder patients was even higher (Butzlaff & Hooley, 1998). The risk of a relapse or re-hospitalization in patients with high EE is 5-9 times more than those with low EE (Miklowitz et al., 1998; Priebe et al, 1989).

Thus, it can be concluded that bipolar disorder affects family relationships and vice versa. When looking at bipolar disorder from biopsychosocial view point, drugs seem necessary to control biological aspects of the illness, but psychological interventions are quite mandatory specially when working on the family to control the psychological and family-social aspects of the illness. Mood disorder treatment has got 3 phases (Prien, 1993). Acute treatment phase in which the acute signs of illness are mostly controlled with pharmacotherapy and psychosocial interventions and these are mostly done in a hospital. In this phase, especially in the first and second episodes, the patient and their family experience a trauma, are shocked and should be understood so that they can cope with the situation. The next phase is stabilization or continuation phase in which drug therapy and psychological treatment simultaneously help the patient leave the critical phase behind so that the illness can be stabilized. And finally maintenance phase, the main goal of which is making the patient continue with the drug therapy and returning them to their family and their own activities in the long run. Although severe signs are rare in this phase, some depression signs do not leave the patient for a long time. In this phase many of families need help to enable them to maintain the patient's fine situation and help them return to their normal career and life. Differing types of family therapy programs have been designs for mood disorders and in most trials their role is defined as to regulate high adherence and relapse prevention. Some of these programs are briefly defined here:

5.1 Family focused therapy (FFT)

This treatment program was first designed for schizophrenic patients based in psychosocial interventions model. In this program patient's caregivers, who could include their father, mother, sister, brother, spouse, children or those people who leave close to them like friends, accompany the patient participate in treatment. The program has three main parts:

1. **Psychoeducation about bipolar disorder**. In this part the patient's family receive intensive training on three areas. a) The illness signs and course which include signs and symptoms of bipolar disorder, the way the last episode of illness has progressed in the patient, the recent events of patient's life, a description of patient's hospitalization experiences and different types of illness prognosis as well as its course. b) Etiology of

bipolar disorder including explanation on stress-diathesis model and the role of stress and life events, genetic preparation and biological factors, protective and risky factors. c) Intervention based on stress-diathesis model including drug and the mechanism of their influence, psychosocial interventions, how the family supports and helps with it, the pattern of patient's self-management and the relapse drills.

2. **Communication Enhancement training (CET).** In this phase the patient and their family are helped to make or rebuilt effective relationship patterns. In most major psychiatric disorders, the capacity of patient and their family in making relations faces deterioration and this problem is very intense when an acute episode of disorder blocks all normal relations and even in the recovery phase people do not know how much they can trust patient. In this phase, along with an interactive relation, it is taught how to make direct and clear positive or negative emotional connections. Verbal and nonverbal strategies of active listening are discussed and it is practice how to positively ask a family member to change a behavior of theirs.
3. **Problem solving Skills training.** During the last session of treatment, problem solving is practiced. The family members define a problem and brain storm different solutions to it, discuss the advantages and disadvantages, choose a way and operationalize the chosen solution and finally reassess different aspects of the problem. They start off with simple tiny examples which do not emotionally engage them and finally discuss the main issues (Miklowitz, 2008b; Morris et al, 2007).

Where these sessions are held depends on the patient and their situation, and can be the patient's house or a clinic. The program consists of 21 session with the first 7 sessions being psychoeducation, the following 7-10 session revolve around CET and the last 4-5 sessions are allocated to problem solving and finally the last session is planned to terminate the program. It goes without saying that booster sessions can be designed if needed. The sessions are held every week (for three months), then biweek (for three months) and finally they are held monthly (Miklowitz, 2008b; Morris et al, 2007). In these sessions six goals are programmed which are listed in the Table-2.

Assist the patient and relatives in:
1- Integrating the experiences associated with mood episodes in bipolar disorder
2- Accepting the notion of a vulnerability to future episodes
3- Accepting a dependency on mood-stabilizing medication for symptom control
4- Distinguishing between the patient's personality and his/her bipolar disorder
5- Recognizing and learning to cope with stressful life events that trigger recurrences of bipolar disorder
6- Reestablishing functional relationships after a mood episode

Table 2. The Six Objectives of Family-Focused Treatment

Following trials have sought to know if this treatment is effective or ineffective. The first study is Miklowitz and Goldstein's in 1990. He treated 9 adult patients for 9 months and compared the results with 23 other patients. His experimental group showed far fewer relapses than the control group (11% vs. 61%) (Miklowitz & Golsestein, 1990). Other controlled trials also aimed to control the effect of FFT on bipolar patients' outcome. In a study in 2000 which followed up BMD patients being treated with FFT for 2 years, fewer relapses and more periods of remission were traced compared to a group receiving crisis intervention (Miklowitz et al, 2000). Yet another study showed more compliance in patients treated by FFT compared to the control group (Miklowitz et al, 2003). Another study by Miklowitz in 2009 on adolescents with bipolar disorder showed that FFT intervened by reduction of EE leads to a significant reduction in depression and manic signs (Miklowitz et al, 2009). A study in Turkey was conducted to investigate the effect of FFT in a non-American society and showed that adding FFT to pharmacotherapy leads to more clinical recovery (Ozerdem et al, 2009).

Apart from the main illness signs, the amount of compliance, and relapses prevention, other goals have been aimed when practicing FFT. In a trial published in 2006 it was declared that FFT along with educating the family to recognize suicidal behaviors in bipolar patients and to control some of the environmental factors can reduce the suicidal risk. Further probation in this research area is suggested (Miklowitz & Taylor, 2006). As well, in a trial in September 2010, FFT interventions had led to a reduction in depression and healthy risk behavior in both caregivers and patients (Perlick et al, 2010).

5.2 Multifamily psychoeducational group therapy (MFPGS)

This type of intervention includes fewer sessions (6 sessions) in which semi-structural interventions are applied to a small group of patients and their families. These interventions are focused on educating people and improving their coping strategies. Patients and their families are encouraged to improve and increase their family interactions.

Not much information is in hand regarding the effect of this treatment. In a trial in 1995, four groups were intervened by MFPGS and compared to nine other groups. Although the amount of EE was less in MFPGS groups, this difference was not statistically significant (Honig et al, 1995). In another trial in 2004, 92 patients with BMD received the treatment in 3 groups: MFPGS plus pharmacotherapy, psycho-social interventions plus pharmacotherapy, and just pharmacotherapy. Time of recovery was set as the comparison scale and no significant difference between the groups was reported (Miller et al, 2004). In yet another trial on 165 adolescents with BMD and their families, who had been treated with MFPGS, the quality of utilized services, by modulation of the improvement of parents' beliefs about the treatment, showed a significant increase. Also, severity of mood signs in patients was improved which is due to utilize better services (Mnedenhall et al, 2009).

5.3 Psychoeducational family intervention

Other forms of psychoeducational interventions, specifically in families alone, have been presented. An example can be the case of a trial in Iran in which families received training on mood disorders, causes and signs, therapeutic approaches to prevent relapses, patterns of relation with the patient, managing the patient's crisis and expanding mental health in the family, all during 6 educational sessions. Family adaptation was significantly better in intervention group while the severity of symptoms of the patient's and global functioning showed no difference (Dashtbozorgi et al, 2009). Also, in another study, in which families were subject to twelve 90-minute educational sessions, recurrence of manic phase (and not depression) showed a significant decrease in a year (Reinares et al, 2008).

Some trials implemented even shorter forms of psychoeducational interventions. In a trial, the intervention of a 2-hour educational session for bipolar patients' families at the time of hospital discharge focusing on educational content of bipolar illness, treatment methods, risk factors, and open discussions about the illness showed that patients have a better compliance with the treatment in a year and experience fewer manic episodes (Fayyazi et al, 2009).

Despite the load of reports about the effectiveness of the intervention on families of bipolar patients, there is a major limitation that decreases its efficiency. Almost 40-60% of bipolar patients, depending on their differing cultures, are not living with their families and this deprives them of receiving this treatment.

6. Psychoeducation in special groups

6.1 Integrated group therapy for comorbidity of substance dependency

Weiss and colleagues arranged integrated group therapy program for patients with comorbidity of substance dependency and bipolar mood disorder which includes 20 weekly sessions in which cognitive behavioral model is integrated with recovery programs to help with prevention of substance use relapse for bipolar mood disorder. The reason behind this integration is the similarity of the two disorders in substance use relapse process and recovery from mood episodes. It seems that the behavioral and cognitive interventions for each disorder have an impact on the other one. During the sessions the shared subjects of the two disorders (like the effect of taking drugs on the process of bipolar disorder) are also discussed. In these sessions, patients' monitoring and checking in, including drug taking monitoring, treatment adherence, methods of coping with high risk situations and mood charts are also talked over (Weiss et al., 2007). In a randomized control trial on this group, fewer substance uses were reported in drug counseling group which was exposed to integrated group therapy, although other outcome factors had no significant differences (Weiss et al., 2007). In the next trial on the same group, the number of sessions were reduced to 12 which were conducted by a counselor specialized in substance use disorders. In this trial previous findings including lower substance uses and good clinical outcome were confirmed (Weiss et al., 2009).

6.2 Psychoeducation in children and adolescents

About 13-28% of patients with bipolar disorder experience their first episode before age of 13 and 50-66% have disease onset before age of 18 (Merikangas et al., 2007; Perlis et al., 2004). Early onset of disorder causes higher risk of suicide, poor function, and poor quality of life (Geller et al., 2000; Geller et al., 2004). Diagnosis and pharmacotherapy for bipolar disorder in children and adolescents have been the focus of attention in the past decade but few studies have investigated psychological interventions in this group. Two programs for family therapy, focused on psychoeducation, have been suggested for children and adolescents:

6.2.1 Family-focused treatment for adolescents (FFT-A)

Family-Focused Treatment for Adolescents (FFT-A), which has been designed based on adults' FFT and includes twenty one 50-minute sessions (12 weekly, 6 biweekly and 3 monthly sessions) in which patients, family and siblings attend. Just like adults' FFT, three phases are done during the sessions:

1. Encouraging the adolescents and their family to reach a better understanding of the disorder, including etiology and the course of disorder in children and adolescents.
2. Encouraging adherence to pharmacotherapy. 3.Guiding the patient and their family to prevent recurrence through detecting early signs of relapse and having quick contact with a psychiatrist.

In a randomized control trial in 2008 and through a 2-year follow up by Miklowitz and colleagues, 30 adolescents being treated with FFT-A were compared to 28 adolescents treated with Enhanced care. Although the groups revealed no statistical differences, the group receiving FFT-A experienced less depression during follow-up (Miklowitz et al., 2008).

6.2.2 Multifamily psychoeducational psychotherapy

In this intervention, 8 sessions are held with both the patient and their parents with three main goals being followed: social support, information and skill building. In a trial, 78 children got multifamily psyhoeducational psychotherapy, compared with 87 children being treated as usual. Results suggested that simultaneous use of brief psyhoeducational group therapy plus pharmacotherapy would lead to better outcome in children aged 8-12 with major mood disorder (Feristad et al., 2008).

7. Web-based programs

Education regarding the disorder and self-management methods is important and influential for patients but various psychoeducational programs are not accesible for everyone and some patients are not able to to take part in all weekly sessions. Due to such issues and along with widespread usage of internet by many people all around the world, the idea of online psychoeducation has been offered so that patients, caregivers, and mental health workers are provided with necessary information (Proudfoot et al., 2007). To actualize the goal, educational programs are being designed on the web. For instance, beating bipolar program has been offered by Barnes and Simpson, as an online psychological package (www.beatingbipolar.org) (Barnes et al., 2011). In 2009, this program was investigated under the title of Bipolar Interactive Psychoeducation Study. Almost 100 patients, diagnosed with bipolar disorder in euthymic phase were randomly allocated into two groups, received web based education or treatment as usual. The research was to compare the two groups on lifestyle along with present mood symptoms, the number of episodes, global function, functional impairment and insight. The results of the study has not yet been published to the best of our knowledge (Simpson et al., 2009).

Since 2008, Goudarzi, Mahdavi and colleagues have also entered educational information for patients on www.ismd.ir for three groups of caregivers, patients and mental health workers but no study has been conducted to determine its efficacy yet.

8. Conclusion

Experiencing acute mood episodes in bipolar mood disorder may lead to serious individual, interpersonal and social dysfunction. In the past decades development of pharmacotherapy has significantly improved control of depression, mania, and mixed episodes and great advancement has been achieved. However, patients with bipolar disorder still suffer from several instances of illness relapse, one important cause of them be stopping pharmacotherapy. Research has shown that psychological interventions especially psychoeducation can lead the to patients' better knowledge of the reality of this disorder and better grasp on the importance of treatment adherence. Through psychoeducation, they will also learn about the early signs of relapse and consequently will be more open to appropriate interventions. They will also learn to avoid triggers of relapse such as use of illicit drugs. Many controlled trials have assessed psyhoeducational interventions. Although this research line is still ongoing in terms of study of the details of these interventions and their efficacy, most studies have confirmed the role of such interventions in prevention of relapses and reduction of the negative effects of the illness on patients' function.

9. Acknowledgements

I sincerely endorse my appreciation to my colleague, Dr. Arash Javanbakht at the University of Michigan, Department of Psychiatry, who devotedly edited the final version of this chapter.

10. References

Amini, H.; Sharifi, V.; Nejatisafa, A.; Arbabi, M.; Tabatabaie, M.; Alimadadi, Z. et al. (summer 2009). One year follow-up of patients with bipolar disorder admitted to Roozbeh hospital. *Iranian Journal of Psychiatry and Clinical Psychology*. Vol. 15, No. 2, pp. 168-174, ISSN-1735-4315

Atri, A. & Sharma, M. (2007). Psychoeducation: Implications for the Profession of Health Education. *Californian Journal of Health Promotion*. Vol. 5, No. 4, pp. 32-39, ISSN 15458725

Barnes, E.; Simpson, S.; Griffiths, E.; Hood, K.; Craddock, N. & Smith, D. J. (2011). Developing an online psychoeducation package for bipolar disorder. *Journal of Mental Health*. Vol. 20, No. 1, pp. 21-31, ISSN 0963-8237

Barry, P. D. (2001). *Mental health and mental illness*. (7th), Lippincott Williams & Wilkins, ISBN 0781731380, NewYork

Bassili, R.; Stiller, A.; Radloff-Gabriel, D.; Milev, R.; & Jamieson, M. (2009). Pilot evaluation of a psychoeducational group for people with bipolar disorder. *Queen's Health Sciences Journal*, Vol. 9, No. 2, pp. 12-16

Bauer, M. S. & McBride, L. (2003). *The Life Goals Program: Structured Group Psychotherapy for Bipolar Disorder*, Springer, ISBN-13: 978-0826116949, New York

Bauer, M. S.; McBride, L.; Chase, C.; Sachs, G. & Shea, N. (1998). Manual-based group psychotherapy for bipolar disorder: a feasibility study. *Journal of Clinical Psychiatry. Vol.* 59, No. 9, pp. 449-455, ISSN 0160-6689

Bauer, M. S.; McBride, L.; Williford, W. O.; Glick, H.; Kinosian, B.; Altshuler, L. et al. (2006b). Collaborative Care for Bipolar Disorder: Part II. Impact on Clinical Outcome, Function, and Costs. *Psychiatric Services*. Vol. 57, No. 7, pp. 937-945, ISSN 1075-2730

Bauer, M. S.; McBride, L.; Williford, W. O.; Glick, H.; Kinosian, B.; Altshuler, L. et al. (2006a). Collaborative Care for Bipolar Disorder: Part I. Intervention and Implementation in a Randomized Effectiveness Trial. *Psychiatric Services*. Vol. 57, No. 7, pp. 927-936, ISSN 1075-2730

Bellivier, F.; Yon, L.; Luquiens, A.; Azorin, J. M.; Bertsch, J.; Gerard, S. (Jun 2011). Suicidal attempts in bipolar disorder: results from an observational study (EMBLEM). *Bipolar Disorders.* Vol. 13, No. 4, pp. 377-386, ISSN 1398-5647

Beynon, S.; Soares-Weiser, K.; Woolacott, N.; Duffy, S. & Geddes, J. R. (2008). Psychosocial interventions for the prevention of relapse in bipolar disorder: systematic review of controlled trials. *British Journal of Psychiatry.* Vol. 192, pp. 5–11, ISSN 0007-1250

Boyd, M. A. (2007). *Psychiatric nursing: Contemporary practice.* (4th), Lippincott Williams & Wilkins, ISBN 0781791693, Philadelphia

Butzlaff, R. L.; & Hooley, J. M. (1998). Expressed emotion and psychiatric relapse: A meta-analysis. *Archives of General Psychiatry,* Vol. 55, pp. 547–552 ISSN 0003990X

Çakir, S. & Özerdem, A. (2010). Psychotherapeuti c and Psychosocial Approaches in Bipolar Disorder: A Systemati c Literature Review. *Turkish journal of psychiatry,* Vol. 21, No. 2, pp. 143-54, ISSN 02136163

Cakir, S.; Bensusan, R.; Akca, Z. K. & Yazici, O. (Dec 2009). Does a psychoeducational approach reach targeted patients with bipolar disorder? *Journal of Affective Disorder,* Vol. 119, No.1-3, pp. 190-193, ISSN 0165-0327

Calabrese, J. R.; Hirschfeld, R. M.; Reed, M.; Davis, M.; Frye, M.; Keck, P. et al. (2003). Impact of bipolar disorder on a US community sample. *Journal of Clinical Psychiatry,* Vol. 64, pp. 425-432 ISSN 0160-6689

Chisholm, D.; Ommeren, M.; Ayuso-Mateos, J. L. & Saxena, S. (2005). Cost-effectiveness of clinical interventions for reducing the global burden of bipolar disorder. *British Journal of psychiatry,* Vol. 187: pp. 559-567, ISSN 0007-1250

Colom, F. & Vieta, E. (2006). *Psychoeducation Manual for Bipolar Disorder,* Cambridge University Press, ISBN-13 978-0-521-68368-5, New York.

Colom, F.; Vieta, E.; Martinez-Aran, A.; Reinares, M.; Goikolea, J. M.; Benabarre, A. et al. (2003). A randomized trial on the efficacy of group Psychoeducation in the prophylaxis of recurrences in bipolar patients whose disease is in remission. *Archives of General Psychiatry.* Vol. 60, pp. 402–407, ISSN 0003990X

Colom, F.; Vieta, E.; Sa´ nchez-Moreno, J.; Palomino-Otiniano, R.; Reinares, M.; Goikolea, J. M. et al. (2009a). Group psychoeducation for stabilised bipolar disorders: 5-year outcome of a randomized clinical trial. *British Journal of Psychiatry.* Vol. 194, pp. 260–265 ISSN 0007-1250

Colom, F.; Vieta, E.; Sanchez-Moreno, J.; Goikolea, J. M.; Popova, E.; Bonnin, C. M. et al. (2009b). Psychoeducation for bipolar II disorder: an explanatory 5-year outcome subanalysis. *Journal of Affective Disorders.* Vol. 112, pp. 30–35, ISSN 0165-0327

Colom, F. & Vieta, E. (2009). Psychoeducation for bipolar disorders, In: *Kaplan & Sadock's Comprehensive Textbook of Psychiatry, 9th Edition,* Sadock, B.J.; Sadock, V.A. & Ruiz, P. (Ed.), pp.1822-1838, Lippincott Williams & Wilkins, ISBN 0683301284, New York.

Coryell, W.; Scheftner, W.; Keller, M.; Endicott, J.; Maser, J. & Klerman, G. L. (1993). The enduring psychosocial consequences of mania and depression. *American Journal of Psychiatry,* Vol. 15, pp. 720–727, ISSN 0002-953X

Dashtbozorgi, B.; Ghadirian, F.; Khajeddin, N. & Karami, K. (Summer 2009). Effect of family psychoeducation on the level of adaptation and improvement of patients with mood disorders. *Iranian Journal of Psychiatry and Clinical Psychology,* Vol. 15, No. 2, pp. 193-200, ISSN 1359-1045

Dore, G. & Romans, S. E.: Impact of bipolar affective disorder on family and partners. *Journal of Affective Disorders* 2001; Vol. 67: pp. 147–158, ISSN 0165-0327

D'Souza, R. & Rich, D. (2002). A case-control study in the use of 'Illness Management Skills Enhancement Programme' for treatment adherence in patients with a bipolar disorder. *Bipolar Disorder*, Vol. 4, pp. 121-126 ISSN 1398-5647

D'Souza, R.; Piskulic, D. & Sundrama, S. (2010). A brief dyadic group based psychoeducation program improves relapse rates in recently remitted bipolar disorder: A pilot randomised controlled trial. *Journal of Affective Disorders*, Vol. 120, pp. 272–276, ISSN 0165-0327

Even, C.; Richard, H.; Thuile, J.; Friedman, S. & Rouillon, F. (March 2007). Characteristics of voluntary participants versus nonparticipants in a psychoeducation program for euthymic patients with bipolar disorder. *Journal of Nervous and Mental Disease*, Vol. 195, No. 3, pp. 262-265, ISSN 0022-3018

Even, C.; Thuile, J.; Kalck-Stern, M.; Criquillion-Doublet, S.; Gorwood, P.; Rouillon, F. (Jun 2010). Psychoeducation for patients with bipolar disorder receiving lithium: short and long term impact on locus of control and knowledge about lithium. *Journal of Affective Disorders*, Vol. 123, No.1-3, pp.299-302, ISSN 0165-0327

Fayyazi Bordbar, M. R.; Soltanifar, A. & Talaei, A. (Jun 2009). Short-Term Family-Focused Psycho-Educational Program for Bipolar Mood Disorder in Mashhad. *Iranian Journal of Medical Sciences*, Vol. 34, No. 2, pp. 104-109, ISSN 02530716

Fristad, M. A.; Verducci, J. S.; Walters, V. & Young, M. E. (2009). Impact of multifamily psychoeducational psychotherapy in treating children aged 8 to 12 years with mood disorders. *Arch Gen Psychiatry*, Vol. 66, No. 9, pp. 10131021 ISSN 0003990X

Geddes, J. & Briess, D. (2006). Bipolar disorder: Clinical evidence, In: *BMJ Clinical Evidence reviews*, 2007, Available from: clinicalevidence.bmj.com

Geller, B.; Bolhofner, K.; Craney, J. L.; Williams, M.; Delbello, M. P. & Gunderson, K. (2000). Psychosocial functioning in a prepubertal and early adolescent bipolar disorder phenotype. *Journal of the American Academy of Child and Adolescent Psychiatry*, Vol. 39, No. 12, pp. 1543-1548, ISSN 1527-5418

Geller, B.; Tillman, R.; Craney, J. L. & Bolhofner, K. (2004). Four-year prospective outcome and natural history of mania in children with a prepubertal and early adolescent bipolar disorder phenotype. *Archives of General Psychiatry*, Vol. 61, No. 5, pp. 459-467 ISSN 0003990X

Getachew, H.; Dimic, S. & Priebe, S. (2009). Is psychoeducation routinely provided in the UK? Survey of community mental health teams. *Psychiatric Bulletin*, Vol. 33, pp. 102-103, ISSN 0955-6036

Ghaemi, S. N.; Pardo, T. B. & Hsu, D. J. (2004). Strategies for preventing the recurrence of bipolar disorder. *Journal of Clinical Psychiatry*, Vol. 65, suppl. 10, pp. 16–23, ISSN 0160-6689

Ghoreishizadeh, M. A.; Deldoost, F. & Farnam, A. (2008). A psycho educational program for relapse prevention in bipolar disorder. *Research Journal of Biological Sciences*, Vol.3, No.7, pp. 786-789, ISSN 1815-8846

Gitlin, M. J.; Swendsen, J.; Heller, T. L. & Hammen, C. (1995). Relapse and impairment in bipolar disorder. *American Journal of Psychiatry*, Vol. 152, pp. 1635–1640, ISSN 0002-953X

Graves, J. S. (1993). Living with mania: A study of outpatient group psychotherapy for bipolar patients. *American Journal of Psychotherapy,* Vol. 47, No. 1, pp. 113-126, ISSN 0002-9564

Gupta, R. D. & Guest, J. F. (2002). Annual cost of bipolar disorder to UK society. *British Journal of Psychiatry,* 180: pp. 227-233, ISSN 0007-1250

Guscott, R. & Taylor, L. (June 1994). Lithium prophylaxis in recurrent affective illness. Efficacy, effectiveness and efficiency. *British Journal of Psychiatry,* Vol. 164, No.6, pp. 741-746, ISSN 0007-1250

Honig, A. S.; Hofman, A.; Hilwig, M.; Noorthoorn, E. & Ponds, R. (Apr 1995). Psychoeducation and expressed emotion in bipolar disorder: preliminary findings. *Psychiatry Research,* Vol. 56, No. 3, pp. 299-301, ISSN 0165-1781

Howard, P. B.; El-Mallakh, P.; Kay Rayens, M. & Clark, J. J. (Oct 2003). Consumer perspectives on quality of inpatient mental health services. *Archives of Psychiatric Nursing.* Vol. 17, No.5, pp. 205-217, ISSN 0883-9417

Keck, P. E. Jr.; McElroy, S. L.; Strakowski, S. M.; Bourne M. L. & West SA. (1997). Compliance with maintenance treatment in bipolar disorder. *Psychopharmacol Bull,* Vol. 33, No. 1, pp. 87-91, ISSN 0048-5764

Ketter, T.A. (2008). Incorporating trial data into bipolar disorder management. *Journal of Clinical Psychiatry,* Vol. 69, No.7, pp. 21-27, ISSN 0160-6689

Kleinman, L.; Lowin, A.; Flood, E.; Gandhi, G.; Edgell, E. & Revicki, D. (2003). Costs of bipolar disorder. *Pharmacoeconomics,* Vol. 21, No. 9, pp. 601-622 ISSN 1170-7690

Lefley, H. P. (2009). *Family psychoeducation for serious mental illnesses,* Oxford University Press, ISBN 978-0-19-534049-5, New York.

Maczka, G.; Grabski, B.; Gierowski, J. K. & Dudek, D. P. (2010). Group psychoeducation in the complex treatment of bipolar disorder--Cracow experiences. *Psychiatria Polska,* Vol. 44, No. 1, pp. 89-100, ISSN 0033-2674

McCombs, J. S.; Ganapathy, V. & Zolfaghari, S. (April 2011). Applying comparative effectiveness research methods in bipolar disorders. *Journal of Affective Disorder,* Vol. 130, No.1-2, pp. 145-54, ISSN 0165-0327

Mendenhall, A. N.; Fristad, M. A. & Early, T. J. (Jun 2009). Factors influencing service utilization and mood symptom severity in children with mood disorders: effects of multifamily psychoeducation groups (MFPGs). *Journal of Consulting and Clinical Psychology,* Vol. 77, No. 3, pp. 463-473, ISSN 0022-006X

Merikangas, K. R.; Akiskal, H. S.; Angst, J.; Greenberg, P. E.; Hirschfeld, R. M. A.; Petukhova, M. et al. (2007). Lifetime and 12-month prevalence of bipolar spectrum disorder in the National Comorbidity Survey replication. *Archives of General Psychiatry,* Vol. 64, No. 5, pp. 543-552 ISSN 0003990X

Miklowitz, D. J. (2008a). Adjunctive Psychotherapy for Bipolar Disorder: State of the Evidence. *American Journal of Psychiatry,* Vol. 165, No. 11, pp. 1408–1419, ISSN 0002-953X

Miklowitz, D. J. (2008b). *Bipolar disorder : A family-focused treatment approach.* (2th), Guilford Press, ISBN 978-1-59385-655-7, United States of America

Miklowitz, D. J.; Axelson, D. A.; Birmaher, B.; George, E. L.; Taylor, D. O.; Schneck, C. D. et al. (2008). Family-focused treatment for adolescents with bipolar disorder: Results of a 2-year randomized trial. *Archives of General Psychiatry,* Vol. 65, No. 9, pp. 1053-1061 ISSN 0003990X

Miklowitz, D. J.; Axelson, D. A.; George, E. L.; Taylor, D. O.; Schneck, C. D.; Sullivan, A.E. et al. (Jun 2009). Expressed emotion moderates the effects of family-focused treatment for bipolar adolescents. *Journal of the American Academy of Child and Adolescent Psychiatry,* Vol. 48, No. 6, pp. 643-51, ISSN 1527-5418

Miklowitz, D. J.; George, E. L.; Richards, J. A.; Simoneau, T. L. & Suddath, R. L. (Sep 2003). A randomized study of familyfocused psychoeducation and pharmacotherapy in the outpatient management of bipolar disorder. *Archives of General Psychiatry,* Vol.60, No. 9, pp. 904–912 ISSN 0003990X

Miklowitz, D. J. and Goldstein, M. J. (1990). Behavioral family treatment for patients with bipolar affective disorder. *Behavior Modification,* Vol. 14, pp. 457–489, ISSN 0145-4455

Miklowitz, D. J.; Otto, M. W.; Frank, f.; Reilly-Harrington, N. A.; Wisniewski, S. R.; Kogan, J. N. et al. (2007). Psychosocial Treatments for Bipolar Depression: A 1-Year Randomized Trial From the Systematic Treatment Enhancement Program. *Archives of General Psychiatry,* Vol. 64, pp. 419-427 ISSN 0003990X

Miklowitz, D. J.; Simoneau, T. L.; George, E. L.; Richards, J. A.; Kalbag, A. & Sachs-Ericsson, N. (Sep 2000) Family-focused treatment of bipolar disorder: 1- year effects of a psychoeducational program in conjunction with pharmacotherapy. *Biological Psychiatry,* Vol. 48, 582–592, ISSN 0006-3223

Miklowitz, D. J. & Taylor, D. O. (Oct 2006). Family-focused treatment of the suicidal bipolar patients. *Bipolar Disorders,* Vol. 8(5 Pt 2): pp. 640-651, ISSN 1398-5647

Miklowitz, D. J.; Wendel, J. S. & Simoneau, T. L. (1998). Targeting dysfunctional family interactions and high expressed emotion in the psychosocial treatment of bipolar disorder. *In Session: Psychotherapy in Practice,* Vol. 4, pp. 25–38, ISSN 1077-2413

Miller, I. W.; Solomon, D. A.; Ryan, C. E. & Keitner, G. I. (Nov 2004). Does adjunctive family therapy enhance recovery from bipolar I mood episodes? *Journal of Affective Disorders,* Vol. 82, No. 3, pp. 431-436, ISSN 0165-0327

Morris, C.; Miklowitz, D. & Waxmonsky, J. (2007). Family-focused treatment for bipolar disorder in adults and youth. *Journal of Clinical Psychology,* Vol. 63, No. 5, pp. 433-445, ISSN 0021-9762

Murray, C. J. L. & Lopez, A. D. (Eds.). (1996). *The global burden of disease: A comprehensive assessment of mortality and disability from diseases, injuries, and risk factors in 1990 and projected to 2020,* Harvard School of Public Health, ISBN 0674354486, Cambridge

Osher, Y.; Bersudsky, Y. & Belmaker R. H. (2010). The new lithium clinic. *Neuropsychobiology,* Vol. 62, No. 1, pp. 17-26, ISSN 0302-282X

Otto, M.; Reilly-Harrington, N. & Sachs, G. (2003). Psychoeducational and cognitive-behavioral strategies in the management of bipolar disorder. *Journal of Affective Disorders,* Vol. 73, No. 1-2, pp. 171-181, ISSN 0165-0327

Ozerdem, A.; Oguz, M.; Miklowitz, D. & Cimilli, C. (Sep 2009). Family focused treatment for patients with bipolar disorder in Turkey: A case series. *Family Process,* Vol. 48, No. 3, pp. 417-428, ISSN 0014-7370

Parikh, S. V.; Kusumakar, V.; Haslam, D. R.; Matte, R.; Sharma, V. & Yatham, L. N. (Aug 1997). Psychosocial interventions as an adjunct to pharmacotherapy in bipolar disorder. *Canadian Journal of Psychiatry,* Vol. 42 Suppl 2, pp. 74S-78S, ISSN 07067437

Peet, M. & Harvey, N.S. (Feb 1991). Lithium maintenance: A standard education program for patients. *British Journal of Psychiatry,* Vol. 158, pp.197-200, ISSN 0007-1250

Perlick, D. A.; Miklowitz, D. J.; Lopez, N.; Chou, J.; Kalvin, C.; Adzhiashvili, V.; et al. (Sep 2010). Family-focused treatment for caregivers of patients with bipolar disorder. *Bipolar Disorders,* Vol. 12, No. 6, pp. 627-637, ISSN 1398-5647

Perlick, D. A.; Rosenheck, R. A.; Miklowitz, D. J.; Kaczynski, R.; Link, B.; Ketter, T. et al. (Jun 2008). Caregiver burden and health in bipolar disorder: a cluster analytic approach. *Journal of Nervous and Mental Disease,* Vol. 196, No. 6, pp. 484-91, ISSN 0022-3018

Perlis, R. H.; Miyahara, S.; Marangell, L. B.; Wisniewski, S. R.; Ostacher, M.; DelBello, M. P. et al. (2004). Long-term implications of early onset in bipolar disorder: data from the first 1000 participants in the Systematic Treatment Enhancement Program for Bipolar Disorder (STEP-BD). *Biological Psychiatry,* Vol. 55, No. 9, pp. 875-881, ISSN 0006-3223

Perlis, R. H.; Ostacher, M. J.; Patel, J.; Marangell, L. B.; Zhang, H.; Wisniewski, S. R.; et al. (2006). Predictors of recurrence in bipolar disorder: Primary outcomes from the Systematic Treatment Enhancement Program for Bipolar Disorder (STEP-BD). *American Journal of Psychiatry,* Vol. 163, No. 2, pp. 217–224, ISSN 0002-953X

Perry, A.; Tarrier, N.; Morriss, R.; McCarthy, E. & Limb, K. (Jan 1999). Randomized controlled trial of efficacy of teaching patients with bipolar disorder to identify early symptoms of relapse and obtain treatment. *BMJ,* Vol. 318, pp. 149-153, ISSN 0959-8138

Pollack, L. E. (Jan 1995). Treatment of inpatients with bipolar disorders: a role for self-management groups. *Journal of Psychosocial Nursing and Mental Health Services,* Vol. 33, No. 1, pp.11-16, ISSN 0279-3695

Pollack, L. E. (Oct 1993). How do inpatients with bipolar disorder evaluate diagnostically homogeneous groups? *Journal of Psychosocial Nursing and Mental Health Services,* Vol. 31, No. 10, pp. 26-32, ISSN 0279-3695

Priebe, S.; Wildgrube, C. & Muller-Oerlinghausen, B. (1989). Lithium prophylaxis and expressed emotion. *British Journal of Psychiatry,* Vol. 154, pp. 396–399, ISSN 0007-1250

Prien, R. F. (1993). Maintenance treatment. In: *Handbook of affective disorders* Paykel, E. S. (Ed.), pp. 419–435, Guilford Press, ISBN 0898626749, New York

Proudfoot, J.; Parker, G.; Hyett, M.; Manicavasagar, V.; Smith, M.; Grdovic, S. et al. (2007). Next generation of self-management education: Web-based bipolar disorder program. *Australian and New Zealand Journal of Psychiatry,* Vol. 41, No.11, pp. 903-909, ISSN 0004-8674

Reinares, M. F.; Colom, J.; Sanchez-Moreno, C.; Torrent, C.; Martinez-Aran, A.; Comes, M. et al. (2008). Impact of caregiver group psychoeducation on the course and outcome of bipolar patients in remission: A randomized controlled trial. *Bipolar disorders,* 10(4): pp. 511-519, ISSN 1398-5647

Rouget, B. & Aubry, J. (2007). Efficacy of psychoeducational approaches on bipolar disorders: a review of the literature. *Journal of Affective Disorders,* Vol. 989, No. 1-2, pp. 11-27, ISSN 0165-0327

Sajatovic, M.; Levin, J.; Fuentes-Casiano, E.; Cassidy, K. A.; Tatsuoka, C. & Jenkins, J. H. (May-June 2011). Illness experience and reasons for nonadherence among individuals with bipolar disorder who are poorly adherent with medication. *Comprehensive Psychiatry,* Vol. 52, No. 3, pp. 280-287, ISSN 0010-440X

Sajatovic, M.; Biswas, K.; Kilbourne, A.; Fenn, H.; Williford, W. & Bauer, M.S. (2008). Factors associated with prospective long term treatment adherence among individuals with bipolar disorder. *Psychiatric Services*, Vol. 59, No.7, pp.753–759, ISSN 1075-2730

Sajatovic, M.; Davies, M. A.; Ganocy, S. J.; Bauer, M. S.; Cassidy, K. A. et al. (Sep 2009). A Comparison of the Life Goals Program and Treatment as Usual for Individuals with Bipolar Disorder. *Psychiatric services,* Vol. 60, No. 9, pp. 1182-1189, ISSN 1075-2730

Scott, J.; Paykel, E.; Morriss, R.; Bentall, R.; Kinderman, P.; Johnson, T. et al. (Apr 2006). Cognitive-behavioural therapy for severe and recurrent bipolar disorders: randomised controlled trial. *British Journal of Psychiatry*, Vol. 188, pp. 313-320, ISSN 0007-1250

Silverstone, T.; McPherson, H.; Hunt, N. & Romans, S. (Feb 1998). How effective is lithium in the prevention of relapse in bipolar disorder? A prospective naturalistic follow-up study. *Australian and New Zealand Journal of Psychiatry*, Vol.32, No.1, pp.61-66, ISSN 0004-8674

Simon, G. E.; Ludman, E. J.; Bauer, M. S.; Unützer, J. & Operskalski, B. (May 2006). Long-term effectiveness and cost of a systematic care management program for bipolar disorder. *Archives of General Psychiatry*, Vol. 63, No.5, pp. 500-508, ISSN 0003990X

Simon, G. E.; Ludman, E. J.; Unützer, J.; Bauer, M. S.; Operskalski, B. & Rutter, C. (Jan 2005). Randomized trial of a population-based care program for people with bipolar disorder. *Psychological Medicine,* Vol. 35, No. 1, pp. 13-24, ISSN 0033-2917

Simpson, S.; Barnes, E.; Griffiths, E.; Hood, K.; Cohen, D.; Craddock, N. et al. (2009). The Bipolar Interactive Psychoeducation (BIPED) study: trial design and protocol, In: *BMC Psychiatry* 2009, Available from:
http://www.biomedcentral.com/1471-244X/9/50.

Stresing, D (Jul 2010). Action-Oriented Therapy for Bipolar Disorder. In: *Everyday Health,* 22.08.2010, Available from
http://www.everydayhealth.com/bipolar-disorder/action-oriented-therapy-for-bipolar-disorder.aspx

Sullivan, A. E. & Miklowitz, D. J. (Feb 2010). Family functioning among adolescents with bipolar disorder. *Journal of Family Psychology*, Vol. 24, No. 1, pp. 60-7, ISSN 08933200

Suppes, T.; Leverich, G. S.; Keck, P. E.; Nolen, W. A.; Denicoff, K. D.; Altshuler, L. L. et al. (2001). The Stanley Foundation Bipolar Treatment Outcome Network: II. Demographics and illness characteristics of the first 261 patients. *Journal of Affective Disorders*, Vol. 67, pp. 45–59, ISSN 0165-0327

Valtonen, H.; Suominen, K.; Mantere, O.; Leppämäki, S.; Arvilommi, P. & Isometsä, E. T. (Nov 2005). Suicidal ideation and attempts in bipolar I and II disorders. *Journal of Clinical Psychology*, Vol. 66, No. 11, pp. 1456 -1462, ISSN 0021-9762

Van Gent, E. M.; Vida, S. L. & Zwart, F. M. (1988). Group therapy in addition to lithium therapy in patients with bipolar disorders. *Acta Psychiatrica Belgica*, Vol. 88, No. 5-6, pp. 405-418, ISSN 0300-9009

Velligan, D. I.; Weiden, P. J.; Sajatovic, M.; Scott, J.; Carpenter, D.; Ross, R. et al. (2010). Strategies for addressing adherence problems in patients with serious and persistent mental illness: recommendations from the expert consensus guidelines. *Journal of Psychiatric Practice,* Vol. 16, No. 5, pp. 306-324, ISSN 1527-4160

Vieta, E. (2005). Improving treatment adherence in bipolar disorder through psychoeducation. *Journal of Clinical Psychiatry*. Vol. 66, Supplement 1, pp. 24-29, ISSN 0160-6689

Vieta, E.; Günther, O.; Locklea,R. J.; Ekman, M.; Miltenburger, C.; Chatterton, M. L. et al. (Sep 2011). Effectiveness of psychotropic medications in the maintenance phase of bipolar disorder: a meta-analysis of randomized controlled trials. *International Journal of Neuropsychopharmacology*, Vol. 14, No. 8, pp. 1029-1049, ISSN 1461-1457

Volkmar, F. R.; Bacon, S.; Shakir, S. A. & Pfefferbaum, A. (1981). Group therapy in the management of manic-depressive illness. *American Journal of Psychotherapy*, Vol. 35, No. 2, pp. 226-234, ISSN 0002-9564

Weiss, D.; Griffin, M.L.; Jaffee, W.B.; Bender, R.; Graff, F.S.; Gallop, R.J. et al. (2009). A "community-friendly" version of Integrated Group Therapy for patients with bipolar disorder and substance dependence: A randomized controlled trial. *Drug and Alcohol Dependence*, Vol. 104, No.3, pp. 212–219, ISSN 0376-8716

Weiss, R. D.; Griffin, M. L.; Kolodziej, M. E.; Greenfield, S. F.; Najavits, L. M.; Daley, D. C. et al. (2007). A randomized trial of integrated group therapy versus group drug counseling for patients with bipolar disorder and substance dependence. *American Journal of Psychiatry*. Vol. 164, pp. 100–107, ISSN 0002-953X

Wheeler, R. B. (Jul 2010a). Psychoeducation Therapy for Bipolar Disorder. In: *Everyday Health*, 22.08.2010, Available from http://www.everydayhealth.com/bipolar-disorder/psychoeducation-therapy-for-bipolar-disorder.aspx

Wheeler, R. B. (Jul 2010b). Cognitive-Behavioral Therapy for Bipolar. In: *Everyday Health*, 22.08.2010, Available from http://www.everydayhealth.com/bipolar-disorder/cognitive-behavioral-therapy-for-bipolar-disorder.aspx

14

Recent Therapies in Depression

Sangita Saini[1],
Anil Shandil[2] and S. K. Singh[3]
[1]*P.D.M. College of Pharmacy, Bahadurgarh, Haryana*
[2]*L.L.R.M Medical College, Meerut, U.P*
[3]*Guru Jambheshwar University of Science and Technology, Hisar, Haryana*
India

1. Introduction

Depression is a significant health problem. It affects men and women of all ages and social backgrounds. Around one in five of the population of world will experience depression at some point in their lives. Prevalence is higher in women than men. It can range in severity from a mild disturbance to a severe illness with a high risk of suicide. The impact of the disorder will also be experienced by family, friends and colleagues. In 2006/07 there were around 500,000 general practitioner consultations with depression and other affective disorders. Over half of those with depression do not seek formal treatment. As well as the personal and social consequences of depression there are also negative economic effects. Depression is associated with sickness absence and prevents many people seeking, maintaining or returning to employment. In an economic analysis the total loss of output due to depression and chronic anxiety in England in 2002/3 was estimated at £12 billion.5. The most common intervention for depression is prescribed antidepressant medication. Managing depression in the primary care and general practice setting is challenging. The common reasons leading to this challenge are:

- The constraints of time
- Insufficient training at the undergraduate level
- Somatisation of symptoms

The use of pharmacological agents often enables the physician to circumvent time constraints. Difficulties arise when knowledge about the properties and use of these agents is lacking. The other factor that often hinders the management of the depressed patient is that psychological and social issues are inadequately addressed. Psychological and social interventions should accompany pharmacological therapy in order to bring about remission of symptoms in the depressed patient. Appropriate management of depression requires knowledge of the severity and type of depressive disorders. Depressive disorders can be categorised as mild, moderate or severe. Severe depression can result in psychosis. A chronic mild to moderate depression lasting more than two years is termed dysthymia. Depression can also be part of bipolar disorder. There is a variation in the management of each of the above.

Mild depression can be handled in primary care and does not need pharmacological therapy. The watchful waiting, guided self-help, exercise and brief psychological interventions. Moderate to severe depression can also be handled in the primary care setting with medication, psychological interventions and social support. Severe depression with psychosis and risk of suicide, recurrent atypical and treatment-resistant depression should be managed by mental health specialists.

Therapies used to treat depression has been divided into parts

i. Pharmacological
ii. Non-Pharmacological

Pharmacological

The drugs used in mental depression are classified as

1. First generation antidepressants
    a. Tricyclic antidepressant eg imipramine, despiramine, amitriptyline, nortryptyline, doxepine, clomipramine, trimipramine
    b. Monoamine oxidase inhibitors egPhanalzine, Pargyline
    c. Lithium carbonate (antimanic drug)
2. Second generation antidepressants
    a. Selective serotonin reuptake inhibitors
    b. Selective norepinephrine reuptake inhibitors
    c. Serotonin norepinephrine reuptake inhibitors
    d. Atypical antidepressants
    e. Reversible inhibitors of MAO-A

Non pharmacological

Non pharmacological can be sub divided into three therapies

1. Psychological therapy
2. Self help
3. Structured exercise

1. Psychological therapies

Behavioural activation
Cognitive behavioural therapy
Counselling
Couple-focused therapy
Family therapy
Hypnotherapy
Interpersonal therapy
Mindfulness based cognitive therapy
Music therapy
Problem solving therapy
Psychodynamic psychotherapy
Reminiscence therapy
Other psychological therapies
Common factors in psychological therapies

2. Self help

Self help support groups
Guided self help
Computerised self help

3. Exercise and lifestyle modification

Exercise
Lifestyle modification

4. Herbal remedies and nutritional supplements

Folate
*Hypericum*extract (St John's wort)
Inositol
Polyunsaturated fatty acids
S-adenosyl-L-methionine
Other nutritional supplements and herbal remedies

5. Complementary and alternative therapies

Acupuncture
Animal assisted therapy
Homeopathy
Light therapy
Massage therapy
Yoga

2. Psychological therapies

2.1 Psychological therapies

Although there are some studies comparing psychological therapies for depression, the majority of studies involve comparisons of psychological therapies with prescribed antidepressant medication treatment. There is very less evidence to support detailed recommendations on the number of therapy sessions required for efficacy, maintenance of effect or prevention of relapse. Practitioners delivering psychological therapies should be trained to approve levels of competency, participate in continuing professional development and be registered with the appropriate governing body. Several types of psychotherapy are used in individual and group settings and with families. Patients must be medically stable to be able to participate meaningfully in any type of psychological therapy. Thus, a patient who has required hospitalization for refeeding and to stabilize his/her medical condition will ordinarily not be able to participate in therapy until after he/she has recovered sufficiently to enable cognitive function to return to normal. A given psychologist or psychiatrist may use several different approaches tailored to the situation. Two types of psychotherapy used for the patient with stabilized medical condition; Cognitive behavior therapy (CBT) and behavior therapy (BT)

Cognitive therapy is the most extensively researched psychological treatment for nonpsychotic unipolar outpatient depressive disorders.This is a type of psychotherapeutic treatment that attempts to change a patient's feelings and behaviors by changing the way the patient thinks about or perceives his/her significant life experiences whereas behavior Therapy (BT) is a type of psychotherapy that uses principles of learning to increase the frequency of desired behaviors and/or decrease the frequency of problem behaviors. A meta-analysis of 16 studies found behavioural activation to be effective in reducing depressive symptoms in adults and older adults compared to treatment as usual and waiting list control, and as effective as cognitive therapy. This is consistent with the conclusions of a study incorporating behavioural activation therapy as part of a larger meta-analysis specifically in patients aged over 50 with depression.

Denmen concluded that cognitive behavior therapy was more effective than behavior therapy, other psychotherapies, and pharmacotherapy. An important advantage to cognitive behavior therapy might be in treating depression in patients with personality disorder, who are recognized as responding less well to all forms of therapy.

The cognitive and behavioral therapies have evolved as an alternative to more traditional nondirective and insight oriented modes of psychotherapy. The family of cognitive and behavioral therapies includes a diverse group of interventions. Nevertheless, the treatments shareseveral pragmatic and theoretical assumptions. First, these therapies emphasize psychoeducation: patients learn about the nature of their difficulties and are providedreasons for use of particular treatment strategies. Second, the cognitive and behavioral therapies typically employ homework and self-help assignments to provide patients theopportunity to practice therapeutic methods that enhance the generalization of newly acquired skills outside of the therapy hour. Third, objective assessment of psychiatricillness is an integral part of treatment, and the selection of therapeutic strategies derives logically from such assessments. Fourth, the therapeutic methods used are structured and directive, and as such require a high level of therapist activity (often they are described in treatment manuals). Fifth, for most disorders, the cognitive and behavioral therapies are time-limited interventions. Sixth, and perhaps most important, these therapies are built on empirical evidence that validates their theoretical orientation and guides the choice of therapeutic techniques. Specifically, learning theories (i.e., classical, operant, and observational models of learning) and the principles of cognitive psychology are relied on heavily in constructing cognitive-behavioral treatment models.

Counselling

Counselling is the skilled and principled use of relationships which develop self knowledge, emotional acceptance and growth, and personal resources. The overall aim is to live more fully and satisfyingly. Counselling may be concerned with addressing and resolving specific problems, making decisions, coping with crises, working through inner feelings and inner conflict, or improving relationships with others. The counsellor's role is to facilitate the patient's work in ways that respect the patient's values, personal resources, and capacity for self determination.

During a study of 12 months on the patient, counselling was as effective as antidepressants but antidepressants may result in more rapid recovery and are likely to be chosen by those who are more severely depressed.

Couple-focused therapy

Couple-focused therapy has the twofold aim of modifying negative interaction patterns and increasing mutually supportive aspects of couple relationships, thus changing the interpersonal context linked to depression. A review identified eight studies evaluating the effect of marital therapy on depression. A variety of treatment models were subsumed within the marital therapy approach, including CBT, emotion-focused, interpersonal and systemic therapy. A variety of control comparisons were used, including CBT, interpersonal therapy, drug therapy, combined individual and drug therapy. Duration of treatment ranged from 10-20 weeks and follow up ranged from post-test to two years. Studies were characterised by small sample size, lack of intention to treat analysis and high numbers lost to follow up. The review concluded there was no evidence to support marital therapy being any more or less effective than one to one therapies or drug therapy in the treatment of depression, even when associated with marital distress. In comparison to no/minimal treatment the outcome for depression was better in the marital therapy group, although this was based on only two small studies.A couple-focused approach should be considered where the current relationship is contributing to the depression, or where involvement of a partner is considered to be of potential therapeutic benefit.

Family therapy

Family therapy helps people in a close relationship help each other. It enables family members to express and explore difficult thoughts and emotions safely, to understand each other's experiences and views, appreciate each other's needs, build on family strengths and make useful changes in their relationships and their lives.

Hypnotherapy

Hypnosis is used in this therapy as main techniques. When Cognitive Behaviour Therapy supplemented by hypnotherapy produced a significantly larger reduction in depressive symptoms than CBT alone.

Interpersonal therapy

A time-limited intervention, which aims to reduce symptoms by working on improving the quality of the patient's interpersonal relationships. IPT focuses on specific interpersonal problem areas such as grief, role transition and interpersonal disputes. A positive therapeutic alliance is encouraged and a range of therapeutic strategies are employed to encourage the open expression of affect and problem resolution. Patient literacy is not required. A study examining achievement of complete remission in major depression found that interpersonal therapy (IPT), CBT and medication were equally effective.

Music therapy

A therapeutic approach where music-making forms the primary basis for communication. A review of five small, diverse and poor quality studies concluded that music therapy on its own or as an adjunct to psychological therapies, is acceptable to people with depression and is associated with improvements in mood. The small number and poor methodological quality of studies mean that it is not possible to be confident about its effectiveness.

Problem solving therapy

A brief focused psychological intervention that is delivered by an individual trained in problem solving approaches. These are often highly individualised and have a pragmatic focus, in which the professional and individual work through a series of defined steps to clarify the person's problems, desired goals, generate potential solutions and help to implement the chosen solution

Psychodynamic psychotherapy

Based on psychodynamic theories of development and of the mind and includes attention to unconscious as well as conscious mental processes. The approach places emphasis on the importance of the therapeutic relationship, including transference and counter transference, how difficulties from the past can be repeated in the therapeutic relationship as well as in current relationships and therefore understood and changed. The therapy involves both expressive and supportive elements.

Reminiscence therapy

Entails a progressive return to an awareness of past experiences, both successful and unsuccessful, so that salient life experiences may be re-examined and re-integrated. The life review process gives older people opportunities to place their accomplishments in perspective, to resolve lingering conflicts, and to find new significance and meaning in their lives, thereby relieving the despair and depression that often accompany ageing.

2.2 Self help

Self help support groups

There is no standard definition of support groups in the literature. No studies were identified on self help groups as a stand-alone intervention for patients with depression. Practitioners referring patients to self help groups should consider the following parameters of good practice:

Groups should be:

- linked to an organisation or well established group that can offer the necessary
- resources, support and promotion of the groups
- subject to regular review and evaluation
- held in accessible, non-stigmatising and welcoming venues
- recovery-focused and with clear confidentiality policies maintained by members and facilitators
- led by facilitators who are trained in listening, conflict management and facilitation skills; and who are supervised and supported themselves.

Self help

Computerised self help Online or computer based packages of self help material. Guided self help Self help interventions which incorporate some form of therapist support.Self help interventions Self help interventions cover a range of interactive packages,paper or web-based written self help materials.

2.3 Exercise and lifestyle modification

Exercise

Exercise is a subset of physical activity, which is any movement of the body that results in energy expenditure rising above resting level, and includes activities of daily living, domestic chores, gardening and walking whereas Structured exercise that is undertaken three or more times a week for 30-40 minutes at an intensity sufficient to provide an energy expenditure of 70-80% of heart rate reserve; this equates to the public health dose of accumulating 30 minutes of moderate

Lifestyle modification

1. reducing alcohol consumption
2. No good quality evidence was identified on the effect of reducing alcohol consumption ondepressive symptoms.
3. Examination of alcohol consumption as a causative factor in depression was outside the scopeof the guideline.
4. Primary care interventions for patients with alcohol dependence, hazardous or harmful drinkingare described in SIGN 74.62
5. reducing caffeine intakeNo good quality evidence was identified on the effects of reducing caffeine intake on depressivesymptoms.
6. return to work
7. good practice in lifestyle advice for patients with depression

2.4 Herbal remedies and nutritional supplements

This section considers herbal remedies and nutritional supplements which have been subjected to evaluate their efficacy in the treatment of depression. They are not licensed medications and have not been subjected to the rigorous regulatory approval process required for prescription medications. In addition to this there are issues around quality control and lack of standardisation of herbal remedies and nutritional supplements.

Folate

A well conducted systematic review of folate for depression was identified. There was only one study of folate as a stand-alone treatment for depression. This did not find significant benefit.

***Hypericum* extract (St John's Wort)**

A perennial herb of the genus Hypericum.Inositol An isomer of glucose. It is a naturally occuring compound which is widely available as a dietary supplement. Polyunsaturated fatty acids (PUFAs)

"Essential fatty acids" that humans cannot synthesise de novo; intake is dependent on dietary sources such as fish and seafood. The examples most studied are the omega-3 fatty acids eicosapentaenoic acid (EPA) and docosahexaenoic acid (DHA). S-adenosyl-Lmethionine

(SAMe) A coenzyme involved in methyl group transfers. It is available as a nutritional supplement. Selenium A non-metallic element which rarely occurs in its elemental state in nature.

Inositol

A good quality systematic review of small, short term RCTs reported that current evidence is unclear whether or not inositol is of benefit in the treatment of depression.

Polyunsaturated fatty acids

Five systematic reviews of the use of polyunsaturated fatty acids (PUFAs) in the treatment of patients with depression were identified. Most trials included in the reviews examined the use of PUFAs as supplements to antidepressant medication with only two small RCTs examining the use of PUFAs as a stand-alone treatment of depression.

s-adenosyl-L-methionine

One well conducted systematic review of 28 small and heterogeneous studies found a modest benefit of S-adenosyl-L-methionine (SAMe) over placebo in the treatment of depression. There were no significant differences in outcome when SAMe was compared with tricyclic antidepressants.

Chromium

A mineral that humans require in trace amounts.Folate Folic acid and folate (the anionic form) are forms of the water soluble vitamin B9. These occur naturally in food and can also be taken as supplements.

Ginseng

A perennial plant which grows in eastern Asia. The root extract is widely available as a herbal remedy. Ginkgo biloba Ginkgo biloba, also known as the Maidenhair tree, is a unique species of tree, the fruits and seeds of which are used in traditional Chinese medicine. Leaf extracts are available as supplements.

2.5 Complementary and alternative therapies

Acupuncture

A family of procedures involving the stimulation of anatomical locations on or in the skin by a variety of techniques. There are a number of different approaches to diagnosis and treatment in acupuncture that incorporate medical traditions from China, Japan, Korea, and other countries. Depression is widely experienced in our communities. In clinical depression, people report a lack of interest in life and activities which they otherwise normally enjoy. This can be accompanied by other symptoms including weight loss, over-eating, feelings of uselessness, sleep disturbance, self neglect and social withdrawal, insomnia or hypersomnia (sleeping too much), loss of energy, low self esteem and poor concentration. Acupuncture has a long history of use in China and Japan. Traditional Chinese medicine theory describes a state of health maintained by a balance of energy in the body. Acupuncture involves the insertion of fine needles into different parts of the body to correct the imbalance of energy in the body. There are a range of styles of acupuncture from traditional/classical acupuncture, auricular acupuncture, trigger point acupuncture, and single point acupuncture.

Animal assisted therapy

A therapy that uses dogs or other pets to improve the physical and mental health of patients with certain acute or chronic diseases. There is evidence from one systematic review that the introduction of animal assisted activities may have beneficial effects on the severity of depressive symptoms in older people resident in nursing homes and psychiatric institutions. The degree to which the benefits found are a result of animal contact or human contact with the animal facilitator is unclear and requires further investigation.

Homeopathy

Homeopathy A system of medicine which is based on treating the individual with highly diluted substances given mainly in tablet form, which trigger the body's natural system of healing.

Light therapy

Treatment that involves regular use of a certain spectrum of lights in a light panel or light screen that bathes the person in that light. Light therapy is also used to treat conditions such as seasonal affective disorder (seasonal depression).

Massage therapy

The manipulation of the soft tissues of the body - the muscles, tendons and ligaments. The therapeutic value of massage has been recorded in several studies. It has been shown to reduce stress and anxiety; relax muscles; aid in circulation, digestion, and excretion; and reduce pain perception. There are many different types of massage -- effleurage, deep tissue, and relaxation massage. Even the simplest massage may convey to the recipient a feeling of being cared for. Both maternal and infant massage have been evaluated in the treatment of postpartum depression.

Aromatherapy

A therapy based on the use of very concentrated "essential" oils from the flowers, leaves, bark, branches, rind or roots of plants with purported healing properties. Aromatherapy involves the administration of pure essential oils of fragrant plants either through breathing the aromatic vapours using an aroma diffuser, or absorbing diluted oils through the skin in a bath or during massage.

Yoga

An ancient system of breathing practices, physical exercises and postures, and meditation, intended to integrate the practitioner's body, mind, and spirit.

Reiki

A hands-on alternative healing technique that involves the exchange of energy between practitioner and patient to restore mental, physical, emotional, and spiritual balance.

Reflexology

Involves massage of reflex areas found in the feet and the hands.

T'ai Chi

A Chinese exercise system that uses slow, smooth, body movements to achieve a state of relaxation of both body and mind. Thought field therapy Involves tapping with the fingers at meridian points on the upper body and hands.

3. Conclusion

In the present review, a number of drug and non-drug strategies group setting may be considered as a treatment option to reduce relapse in patients with depression who have had three or more episodes. Whereas, problem solving therapy and short term psychodynamic psychotherapy may be considered as a treatment option in depressed patients. Structured exercise may be considered as a treatment option for patients with depression. No applicable evidences were identified on which to base a recommendation for complementary and alternative therapies.

4. References

[1] Bedi N, Chilvers C, Churchill R, Dewey M, Duggan C, Fielding K, Gretton V, Miller P, Harrison G, Lee A, Williams I (2000). Assessing effectiveness of treatment of depression in primary care: Partially randomised preference trial. Br. J. Psychiatry, 177: 312-318.

[2] Alisa W (2008). Neurolinguistic psychotherapy: A postmodern Alladin A, Alibhai A (2007). Cognitive hypnotherapy for depression: An empirical investigation. Int. J. Clin. Exp. Hypn., 55: 147-66.

[3] Dimidjian S, Davis KJ (2009). Newer variations of cognitive-behavioral therapy: behavioral activation and mindfulness-based cognitive therapy. Curr. Psychiatry Rep., 11: 453-458.

[4] Cuijpers P, van Straten A, Warmerdam L (2007). Behavioral activation treatments of depression: A meta-analysis. Clin. Psychol. Rev., 27: 318-326.

[5] Denman C (2001). Cognitive–analytic therapy. Adv. Psychiatr. Treat., 7: 243-252.

[6] Hensley PL, Nadiga D, Uhlenhuth EH (2004). Long-term effectiveness of cognitive therapy in major depressive disorder. Depress. Anxiety, 20: 1-7.

[7] Klausner EJ, Clarkin JF, Spielman L, Pupo C, Abrams R, Alexopoulos GS (1998). Late life depression and functional disability: The role of goal-focused group psychotherapy. Int. J. Geriatr. Psychiatry, 13: 707-716.

[8] Leichsenring F (2001). Comparative effects of short-term psychodynamic psychotherapy and cognitive-behavioral therapy in depression: a meta-analytic approach. Clin. Psychol. Rev., 21: 401- 419.

[9] Michael B, Gillian EH (2001). Counselling and interpersonal therapies for depression: towards securing an evidence-base. Br. Med. Bull., 57: 115-132.

[10] Luty SE, Carter JD, McKenzie JM, Rae AM, Frampton CMA, Mulder RT, Joyce PR (2007). Randomised controlled trial of interpersonal psychotherapy and cognitive behavioural therapy for depression. Br. J. Psychiatry, 190: 496-502.

[11] Barbato A, D'Avazo B (2007). Marital therapy for depression (Cochrane Review). In: The Cochrane Library, London: Wiley, p. 4.

[12] Allison S, Stacey K, Dadds V, Roeger L, Wood A, Martin G (2003). What the family brings: gathering evidence for strengths-based work. J. Fam. Ther., 25: 263-284.

[13] de Mello MF, de Jesus Mari J, Bacaltchuk J, Verdeli H, Neugebauer R (2005). A systematic review of research findings on the efficacy of interpersonal therapy for depressive disorders. Eur. Arch. Psychiatry Clin. Neurosci., 255: 75-82.

[14] Maratos AS, Gold C, Wang X, Crawford MJ (2008). Music therapy for depression (Cochrane Review). In: The Cochrane Library, London: Wiley, p. 1.

[15] Cuijpers P, van Straten A, Warmerdam L (2007). Problem solving therapies for depression: A meta-analysis. Eur. Psychiatry, 22: 9-15.

[16] Simpson S, Corney R, Beecham J (2003). A randomized controlled trial to evaluate the effectiveness and cost-effectiveness of psychodynamic counselling for general practice patients with chronic depression. Psychol. Med., 33: 229-239.
[17] Watt LM, Cappeliez P (2000). Integrative and instrumental reminiscence therapies for depression in older adults: Intervention strategies and treatment effectiveness.Aging Ment. Health, 4: 166-177.
[18] Singh NA, Clements KM, Singh MA (2001). The efficacy of exercise as a long-term antidepressant in elderly subjects: a randomized, controlled trial. J. Gerontol. A Biol. Sci. Med. Sci., 56: 497-504.
[19] Nabkasorn C, Miyai N, Sootmongkol A, Junprasert S, Yamamoto H, Arita M, Miyashita K (2006). Effects of physical exercise on depression, neuroendocrine stress hormones and physiological fitness in adolescent females with depressive symptoms. Eur. J. Public Health, 16: 179-184.
[20] Mather AS, Rodriguez C, Guthrie MF, McHarg AM, Reid IC, McMurdo ME (2002). Effects of exercise on depressive symptoms in older adults with poorly responsive depressive disorder: randomized controlled trial. Br. J. Psychiatry, 180: 411-415.
[21] Dunn AL, Trivedi MH, Kampert JB, Clark CG, Chambliss HO (2005). Exercise treatment for depression: efficacy and dose response. Am. J. Prev. Med., 28: 1-8.
[22] Waite LW, Holder MD (2003). Assessment of the Emotional Freedom Technique: An Alternative Treatment for Fear. Sci. Rev. Mental Health Pract., 2: 20-26.
[23] Leo RJ, Ligot Jr JSA (2007). A systematic review of randomized controlled trials of acupuncture in the treatment of depression. J. Affect. Disord., 97: 13-22.
[24] Smith CA (2005). Acupuncture for depression (Cochrane Review). In: Cochrane Library, Issue 2. London: Wiley.4
[25] Souter MA, Miller MD (2007). Do animal-assisted activities effectively treat depression?A meta-analysis. Anthrozoos, 20: 167-180.
[26] Mukaino Y, Park J, White A, Ernst E (2005). The effectiveness of acupuncture for depression - A systematic review of randomized controlled trials. Acupunct. Med., 23: 70-76.
[27] Pilkington K, Kirkwood G, Rampes H, Fisher P, Richardson J (2005).Homeopathy for depression: a systematic review of the research evidence. Homeopathy, 94: 153-163.
[28] Even C, Schroder CM, Friedman S, Rouillon F (2008). Efficacy of light therapy in nonseasonal depression: A systematic review. J. Affect. Disord., 108: 11-23.
[29] Hou WH, Chiang PT, Hsu TY, Chiu SY, Yen YC (2010). Treatment effects of massage therapy in depressed people: a meta-analysis. J. Clin. Psychiatry, 71: 894-901.
[30] van der Watt G, Janca A (2008). Aromatherapy in nursing and mental health care.Contemp. Nurse., 30: 69-75.
[31] Pilkington K, Kirkwood G, Rampes H, Richardson J (2005). Yoga for depression: The research evidence. J. Affect. Disord., 89: 13-24.
[32] Lee MS, Pittler MH, Ernst E (2008). Effects of Reiki in clinical practice: a systematic review of randomized clinical trials. Int. J. Clin. Pract., 62: 947-954.
[33] Ernst E (2009). Is reflexology an effective intervention? A systematic review of randomised controlled trials. Med. J. Aust., 191: 263-266.

[34] Wang C, Bannuru R, Ramel J, Kupelnick B, Scott T, Schmid CH (2010). Tai Chi on psychological well-being: systematic review and metaanalysis. BMC Complement. Altern. Med., 10: 23.

15

Deep Brain Stimulation for Treatment-Resistant Depression: A State-of-the-Art Review

Lucas Crociati Meguins
Section of Neurosurgery,
Department of Neurological Sciences,
Hospital de Base da Faculdade de Medicina
de São José do Rio Preto (FAMERP),
São José do Rio Preto, São Paulo
Brazil

1. Introduction

Depression is a common underdiagnosed, undertreated and costly mood disorder associated with substantial morbidity and mortality (1-3). It is related to the emotion of sadness or irritability and accompanied by several psycophysiological changes, such as disturbances in sleep, appetite or sexual desire; constipation; loss of the ability to experience pleasure in work or with friends; crying; suicidal thoughts; and slowing of speech and action (4-6). Approximately 5% of the population has major depression at any given time, with men experiencing a lifetime risk of 7%-12%; and women 20%-25% (3, 5).

Depressive mood is one of the most common - and potentially dangerous - psychiatric complications of every chronic disease because it often worsens the overall clinical condition. The prevalence of depression in patients with chronic conditions ranges from 25 to 33 percent and is frequent among people hospitalized with cancer, who have had a recent heart attack or stroke, or have diabetes or Parkinson's disease (7). Besides, even between non-hospitalized patients, depression has an important effect on life and may limit the social activities of individuals. People who are depressed tend to be more withdrawn from their community, friends and family, are less likely to work, present a decreased productivity and retire earlier. The relevant social-economic and clinical impacts of mood disorder, especially depression, on patient's life have drawn attention to possible therapeutic mechanisms that could improve the quality of life and outcomes of depressive patients.

The clinical approach of patients suffering of depressive mood has evolved considerably throughout years and includes mainly supportive care combined with pharmacotherapy and psychotherapy (8, 9). Globally, treatment planning requires coordination of short-term strategies to induce remission associated with longer term maintenance designed to prevent recurrence. The wide variety of drugs available to treat depressive patients has given the physician a large spectrum of therapeutic possibilities for a rationale approach.

Approximately 35 million adults in Unite States population, at least once in lifetime, experience an episode of major depression (3). The essential feature of a major depressive event is defined as a persistent period of at least two weeks in which there is either depressive mood or the loss of interest or pleasure in nearly all activities (DSM-IV). The available treatment currently used includes selective serotonin reuptake inhibitors (SSRI), tricyclic antidepressants (TCA), selective norepinephrine reuptake inhibitors (NRI) and monoamine oxidase inhibitors (MAOI). However, only 47% of patients respond well to antidepressant treatment and only 33% of patients achieve remission in the first line of treatment with a widely used selective serotonin reuptake inhibitors (10). Additionally, relapse within 6-12 months is verified in approximately 50% of those who remit (11). Therefore, while currently available treatments are amongst the most widely prescribed drugs, they fail to have an effect on many patients and have incomplete effects for many others. Treatment-resistant depression has an estimated prevalence of 2-5% and continued depressive symptoms are directly responsible to ongoing functional impairment, increased utilization of health care resources, a high risk of suicide and an overall increased mortality (12-14). An effective treatment is then necessary, but there is very limited evidence for the efficacy of any specific therapeutic intervention in chronic, treatment-refractory depression.

Electroconvulsive therapy (ECT) is the main procedure indicated to patients resistant to antidepressive medications and consists of the induction of convulsive crises by means of applying an electric current through the brain for therapeutic purpose (15). The remission rate is of depression with electroconvulsive therapy varies, generally, from 50% to 80% (15). However, ECT may have important side effects, especially cognitive impairment, that can limit its clinical use (16) and it is associated with a high relapse rate (15, 17).

Repetitive transcranial magnetic stimulation (rTMS) was recently approved by FDA (Food and Drug Administration) for the treatment of depression that has not responded to one antidepressant medication. This procedure is well-tolerated and has few side effects. Its mechanism of therapeutic action is linked to the principal of electromagnetic induction discovered by Faraday in 1831 by wich electrical energy and magnetic fields are related. According to Pascual-Leone, an increased neuronal excitability linked to the increase of magnetic field may produce electrical currents in certain areas of the brain, especially the left prefrontal cortex, and reduce the depressive symptoms (18). rTMS has been described as a generally safe technique with consistent statistically significant antidepressant results (19, 20), however is associated with relatively low response rate and the maintenance of benefits over time is relatively unknown (21).

Vagus nerve stimulation (VNS) was approved in 2005 for the treatment of patients failing to respond to, at least, four antidepressant drugs and refers to electrical stimulation of the cervical portion of the left vagus nerve through a surgically implanted electrode and pulse generator. Surgical risks are relatively minor and long-term treatment is usually well tolerated. However, it has been noted that only a substantial minority of patients with extremely difficult-to-treat depressive disorder benefit from VNS and 21%-50% of patients who responded to VNS failed to maintain at least a 40% decrease in baseline depression severity over 1-2 years (22-24).

Deep brain stimulation (DBS) has been described as a promising therapy to severe and treatment-resistant depression. Since 2005, when Mayberg and colleagues (25) have reported that the subgenual cingulated region (Brodman area 25) is a potential target to electrical stimulation with effective reverse of symptoms in patients suffering treatment-resistant depression, a large number of studies have been conduced to explore other potential sites in the brain that could be involved on the mechanisms of depressive mood disorders.

This procedure is characterized by a stereotactical implantation of an intracranial electrode that is connected to a pulse generator/battery pack (26). In 1947, at Temple University in Philadelphia, the neurologist Spiegel and the neurosurgeon Wycis first described a stereotactic apparatus and its use in humans to perform ablative procedures (27). The great knowledge achieved from the treatment of patients with movement disorders with DBS, such as Parkinson's disease, primary dystonia, and essential tremor, allowed its experimental use also on patients with treatment-resistant depression with encouraging results. Due to its reversible, revisable and adjustable potential, DBS represents an alternative and additional therapeutic approach to treatment-resistant depression.

The initial evidence that a focal disrupting pathological activity in limbic-cortical pathways was present in patients with negative mood was observed by Mayberg and colleagues, in 1999 (28). Using functional neuroimaging studies, they consistently demonstrated the involvement of the subgenual cingulate (Cg25) (Brodmann area 25) in both the acute sadness and antidepressant treatment effects. A markedly increased blood flow was noted to the Cg25 during induction of transient normal sadness and, in contrast, a decreased activity was reported with clinical response to different antidepressant treatments, which suggested a critical role for this region in modulating negative mood states. This scientific observation raised the hypothesis that a therapeutic intervention on specific sites of the limbic-cortical circuits could effectively reverse the symptoms of otherwise treatment-resistant depression. Mayberg and colleagues, in 2005 (29), on an open-label study, applied bilateral DBS on the subcallosal cingulate white matter in six patients with treatment-resistant depression and observed a clinically significant antidepressant response in four after 6 months, which is known as the first published report of DBS for treatment-resistant depression.

Based on the presence of neuronal dysregulation in limbic-cortical circuits and positive lesional effects, various target areas for depressive disorders have been discussed. Schlaepfer and colleagues, in 2008 (30), stimulating the nucleus accumbens of 3 patients, observed after one week, 42% reduction in symptoms. Malone and colleagues, in 2009 (31), stimulating the ventral internal capsule and ventral striatum of 15 patients, noted a distinct reduction in symptoms by 57% during 12 months. Lozano and colleagues, in 2008 (32), stimulating the subgenual area of cingulum of 20 patients, observed reduction in symptoms by an average 48% during 12 months, with 2 non-responders and 7 remission. Other targets for DBS in treatment resistant depression have been proposed, including globus pallidus internus, inferior thalamic peduncle, lateral habenula and the rostral cingulated cortex (33, 34) (Table 1).

The ideal patient that may benefit from DBS is still a matter of controversy. The failure to two antidepressant treatments in the current episode appears to have a predictive validity. In STAR*D, remission rates in successive stages of treatment were 36.8% in the first treatment, 30.6% in the second treatment, 13.7% in the third treatment, and 13.0% in

YEAR	REFERENCE	PLACE	PATIENTS	STIMULATION SITE	SYMPTOMS REDUCTION
2005	Mayberg *et al* (28)	Canada	6	Subgenual area of cingulum	71%
2008	Schlaepfer *et al* (29)	Germany	3	Nucleus accumbens	42%
2008	Lozano *et al* (31)	Canada	20	Subgenual area of cingulum	48%
2009	Malone *et al* (30)	USA	15	ventral internal capsule/ventral striatum	57%
2011	Puigdemont *et al* (35)	Spain	8	Subgenual area of cingulum	50%
2011	Kennedy *et al* (36)	Canada	20	Subcallosal cingulate gyrus	64.3%

Table 1. deep brain stimulation in patients presenting treatment-resistant depression.

the forth treatment (11). Additionally, the likelihood of remission is markedly lower for patients for whom two or more adequate antidepressant trials failed (11). Therefore, in some studies, to qualify for DBS, patients had to be in an episode of major depression for at least 1 year and had to have failed to respond to at least four treatments (32). Exclusion criteria usually includes substance abuse, severe personality disorder and psychotic depression (35).

The potential for adverse effects from DBS for treatment-resistant depression has not been completely studied. Lozano and colleagues, in 2008 (32), reported increased depression in 10% of patients that underwent DBS of the subgenual area of cingulum. Moreover, Malone and colleagues, in 2009 (31), describe hippomanic episodes, increased suicidility and syncope on patients submitted to DBS of internal capsule/ventral striatum. Future studies must carefully point the most commons adverse effects and complications, and the specific targets related to it as well.

DBS is a recently proposed therapeutic approach to treatment-resistant depression and, therefore, remains with many unsolved questions. Future investigations must correctly address which patient benefits most from the surgical procedure and which is the best therapeutic target or, if not possible, which target best correlates to specific symptoms of depression. Large, randomized placebo-controlled studies of DBS for treatment-resistant are necessary to describe the potential risks and complications from operation. Additionally, as the study of DBS progresses, complementary information must put into evidence which is the best adjunctive therapy, the potential interactions of pharmacotherapy and the relative safety and efficacy of the various DBS targets.

This review aimed to describe the context in which DBS has emerged as an experimental therapy for refractory depression, present a brief historical note, discuss the different targets in these pathological neuro-psychiatric disorder, review the preliminary results of some clinical studies, and discuss the pros and cons of the method, as well as the possible adverse effects and future perspectives.

2. Acknowledgment

The author appreciates the remarkable contribution made by Dr. Antônio Ronaldo Spotti, Dr. Carlos Eduardo Dall'aglio Rocha, and Dr. Fábio de Nazaré Oliveira, all from the Department of Neurological Sciences, Hospital de Base da Faculdade de Medicina de São José do Rio Preto (FAMERP), São José do Rio Preto, São Paulo, Brazil, reviewing the chapter.

3. References

[1] Sumanathissa M, De Silva VA, Hanwella R. Prevalence of major depressive episode among patients with pre-dialysis chronic kidney disease. Int J Psychiatry Med. 2011;41(1):47-56.

[2] Botega NJ, Mitsuushi GN, Azevedo RC, Lima DD, Fanger PC, Mauro ML, Gaspar KC, Silva VF. Depression, alcohol use disorders and nicotine dependence among patients at a general hospital. Rev Bras Psiquiatr. 2010 Sep;32(3):250-6.

[3] Kessler RC, Berglund P, Demler O, Jin R, Koretz D, Merikangas KR, Rush AJ, Walters EE, Wang PS; National Comorbidity Survey Replication. The epidemiology of major depressive disorder: results from the National Comorbidity Survey Replication (NCS-R). JAMA. 2003 Jun 18;289(23):3095-105.

[4] Castelo de Souza, D, Meguins LC, Meguins EMC. Major Depression in a Brazilian Amazon Woman with Down Syndrome: A Case Report . The Internet Journal of Mental Health. 2008; Volume 5: Number 2.

[5] Belmaker RH; Agam G. Major depressive disorder - mechanisms of disease. N Engl J Med. 2008; 358(1): 55-68.

[6] Sullivan PF, Neale MC, Kendler KS. Genetic epidemiology of major depression: review and meta-analysis. Am J Psychiatry. 2000 Oct;157(10):1552-62.

[7] Rifkin A. Depression in physically ill patients. Don't dismiss it as 'understandable'. Postgrad Med. 1992 Sep 1;92(3):147-9, 153-4.

[8] Smiga SM, Elliott GR. Psychopharmacology of depression in children and adolescents. Pediatr Clin North Am. 2011 Feb;58(1):155-71, xi.

[9] Dupuy JM, Ostacher MJ, Huffman J, Perlis RH, Nierenberg AA. A critical review of pharmacotherapy for major depressive disorder. Int J Neuropsychopharmacol. 2011 Feb 24:1-15.

[10] Trivedi MH, Rush AJ, Wisniewski SR, Nierenberg AA, Warden D, Ritz L, Norquist G, Howland RH, Lebowitz B, McGrath PJ, Shores-Wilson K, Biggs MM, Balasubramani GK, Fava M; STAR*D Study Team. Evaluation of outcomes with

citalopram for depression using measurement-based care in STAR*D: implications for clinical practice. Am J Psychiatry. 2006 Jan;163(1):28-40.

[11] Rush AJ, Trivedi MH, Wisniewski SR, Nierenberg AA, Stewart JW, Warden D, Niederehe G, Thase ME, Lavori PW, Lebowitz BD, McGrath PJ, Rosenbaum JF, Sackeim HA, Kupfer DJ, Luther J, Fava M. Acute and longer-term outcomes in depressed outpatients requiring one or several treatment steps: a STAR*D report. Am J Psychiatry. 2006 Nov;163(11):1905-17.

[12] Miller IW, Keitner GI, Schatzberg AF, Klein DN, Thase ME, Rush AJ, Markowitz JC, Schlager DS, Kornstein SG, Davis SM, Harrison WM, Keller MB. The treatment of chronic depression, part 3: psychosocial functioning before and after treatment with sertraline or imipramine. J Clin Psychiatry. 1998 Nov;59(11):608-19.

[13] Crown WH, Finkelstein S, Berndt ER, Ling D, Poret AW, Rush AJ, Russell JM. The impact of treatment-resistant depression on health care utilization and costs. J Clin Psychiatry. 2002 Nov;63(11):963-71.

[14] Murphy JM, Monson RR, Olivier DC, Sobol AM, Leighton AH. Affective disorders and mortality. A general population study. Arch Gen Psychiatry. 1987 May;44(5):473-80.

[15] Antunes PB, Rosa MA, Belmonte-de-Abreu PS, Lobato MI, Fleck MP. Electroconvulsive therapy in major depression: current aspects Rev Bras Psiquiatr. 2009 May;31 Suppl 1:S26-33.

[16] Little A. Treatment-resistant depression. Am Fam Physician. 2009 Jul 15;80(2):167-72.

[17] Sackeim HA, Prudic J, Devanand DP, Decina P, Kerr B, Malitz S. The impact of medication resistance and continuation pharmacotherapy on relapse following response to electroconvulsive therapy in major depression. J Clin Psychopharmacol. 1990 Apr;10(2):96-104.

[18] Pascual-Leone A, Tormos JM, Keenan J, Tarazona F, Cañete C, Catalá MD. Study and modulation of human cortical excitability with transcranial magnetic stimulation. J Clin Neurophysiol. 1998 Jul;15(4):333-43.

[19] Aliño JJ, Jiménez JL, Flores SC, Alcocer MI. Efficacy of transcranial magnetic stimulation (TMS) in depression: naturalistic study. Actas Esp Psiquiatr. 2010 Mar;38(2):87-93.

[20] Kozel FA, George MS. Meta-analysis of left prefrontal repetitive transcranial magnetic stimulation (rTMS) to treat depression. J Psychiatr Pract. 2002 Sep;8(5):270-5.

[21] O'Reardon JP, Solvason HB, Janicak PG, Sampson S, Isenberg KE, Nahas Z, McDonald WM, Avery D, Fitzgerald PB, Loo C, Demitrack MA, George MS, Sackeim HA. Efficacy and safety of transcranial magnetic stimulation in the acute treatment of major depression: a multisite randomized controlled trial. Biol Psychiatry. 2007 Dec 1;62(11):1208-16.

[22] Cristancho P, Cristancho MA, Baltuch GH, Thase ME, O'Reardon JP. Effectiveness and safety of vagus nerve stimulation for severe treatment-resistant major depression in clinical practice after FDA approval: outcomes at 1 year. J Clin Psychiatry. 2011 Jan 11.

[23] Nahas Z, Marangell LB, Husain MM, Rush AJ, Sackeim HA, Lisanby SH, Martinez JM, George MS. Two-year outcome of vagus nerve stimulation (VNS) for treatment of major depressive episodes. J Clin Psychiatry. 2005 Sep;66(9):1097-104.

[24] Sackeim HA, Brannan SK, Rush AJ, George MS, Marangell LB, Allen J. Durability of antidepressant response to vagus nerve stimulation (VNS). Int J Neuropsychopharmacol. 2007 Dec;10(6):817-26.

[25] Mayberg HS, Lozano AM, Voon V, McNeely HE, Seminowicz D, Hamani C, Schwalb JM, Kennedy SH. Deep brain stimulation for treatment-resistant depression. Neuron. 2005 Mar 3;45(5):651-60.

[26] Holtzheimer PE 3rd, Mayberg HS. Deep brain stimulation for treatment-resistant depression. Am J Psychiatry. 2010 Dec;167(12):1437-44.

[27] Hariz MI, Blomstedt P, Zrinzo L. Deep brain stimulation between 1947 and 1987: the untold story. Neurosurg Focus. 2010 Aug;29(2):E1.

[28] Mayberg HS, Liotti M, Brannan SK, McGinnis S, Mahurin RK, Jerabek PA, Silva JA, Tekell JL, Martin CC, Lancaster JL, Fox PT. Reciprocal limbic-cortical function and negative mood: converging PET findings in depression and normal sadness. Am J Psychiatry. 1999 May;156(5):675-82.

[29] Mayberg HS, Lozano AM, Voon V, McNeely HE, Seminowicz D, Hamani C, Schwalb JM, Kennedy SH. Deep brain stimulation for treatment-resistant depression. Neuron. 2005 Mar 3;45(5):651-60.

[30] Schlaepfer TE, Cohen MX, Frick C, Kosel M, Brodesser D, Axmacher N, Joe AY, Kreft M, Lenartz D, Sturm V. Deep brain stimulation to reward circuitry alleviates anhedonia in refractory major depression. Neuropsychopharmacology. 2008 Jan;33(2):368-77.

[31] Malone DA Jr, Dougherty DD, Rezai AR, Carpenter LL, Friehs GM, Eskandar EN, Rauch SL, Rasmussen SA, Machado AG, Kubu CS, Tyrka AR, Price LH, Stypulkowski PH, Giftakis JE, Rise MT, Malloy PF, Salloway SP, Greenberg BD. Deep brain stimulation of the ventral capsule/ventral striatum for treatment-resistant depression. Biol Psychiatry. 2009 Feb 15;65(4):267-75.

[32] Lozano AM, Mayberg HS, Giacobbe P, Hamani C, Craddock RC, Kennedy SH. Subcallosal cingulate gyrus deep brain stimulation for treatment-resistant depression. Biol Psychiatry. 2008 Sep 15;64(6):461-7.

[33] Kuhn J, Gründler TO, Lenartz D, Sturm V, Klosterkötter J, Huff W. Deep brain stimulation for psychiatric disorders. Dtsch Arztebl Int. 2010 Feb;107(7):105-13.

[34] Holtzheimer PE, Mayberg HS. Deep brain stimulation for psychiatric disorders. Annu Rev Neurosci. 2011;34:289-307.

[35] Malone DA Jr. Use of deep brain stimulation in treatment-resistant depression. Cleve Clin J Med. 2010 Jul;77 Suppl 3:S77-80.

[36] Puigdemont D, Pérez-Egea R, Portella MJ, Molet J, de Diego-Adeliño J, Gironell A, Radua J, Gómez-Anson B, Rodríguez R, Serra M, de Quintana C, Artigas F, Alvarez E, Pérez V. Deep brain stimulation of the subcallosal cingulate gyrus: further evidence in treatment-resistant major depression. Int J Neuropsychopharmacol. 2011 Jul 22:1-13.

[37] Kennedy SH, Giacobbe P, Rizvi SJ, Placenza FM, Nishikawa Y, Mayberg HS, Lozano AM. Deep brain stimulation for treatment-resistant depression: follow-up after 3 to 6 years. Am J Psychiatry. 2011 May;168(5):502-10.